Advances in Experimental Medicine and Biology
Volume 1008

More information about this series at http://www.springer.com/series/5584

M.R.S. Rao
Editor

Long Non Coding RNA Biology

Springer

Editor
M.R.S. Rao
Molecular Biology and Genetics Unit
Jawaharlal Nehru Centre for Advanced Scientific Research
Bangalore
India

ISSN 0065-2598 ISSN 2214-8019 (electronic)
Advances in Experimental Medicine and Biology
ISBN 978-981-13-5342-0 ISBN 978-981-10-5203-3 (eBook)
DOI 10.1007/978-981-10-5203-3

Printed on acid-free paper

This Springer imprint is published by Springer Nature
The registered company is Springer Nature Singapore Pte Ltd.
The registered company address is: 152 Beach Road, #21-01/04 Gateway East, Singapore 189721, Singapore

Foreword

In recent years, our functional understanding of the genome has changed dramatically. The discovery of pervasive transcription and the prevalence of noncoding RNAs (ncRNAs) have challenged the traditional definition of genes and what should be considered a functional region of the genome. Noncoding RNAs are defined as transcripts that do not encode proteins. Formally, this classification includes components of the translation machinery such as ribosomal RNA (rRNA) and transfer RNA (tRNA). However, the term ncRNA is also commonly used to describe other noncoding transcripts implicated in the regulation of gene expression. In particular, this category includes short regulatory ncRNAs (e.g., microRNAs, small interfering RNA) and longer transcripts about which much less is known. This book focuses on this latter class of long ncRNAs (lncRNAs).

The rise of modern genomics was fundamental for uncovering the diversity of lncRNAs, their omnipresence across the tree of life, and their varied mechanisms of action. lncRNAs regulate gene expression at all levels, from modifying the epigenetic status of chromatin to modulating the stability of protein-coding mRNAs in the cytoplasm. Some lncRNAs exert their effects locally (e.g., by regulating the expression of neighboring genes in *cis*), while others affect multiple loci across the genome. Sometimes the presence of an lncRNA is essential for its own functional impact. This is the case for lncRNAs acting as scaffolds for the assembly and recruitment of various functional elements to chromatin. In other situations, lncRNAs function through their own transcription. For example, antisense lncRNA transcripts (lncRNAs overlapping gene-coding regions on the opposite DNA strand) may inhibit or modulate the expression of the sense mRNA either by direct transcriptional interference or by recruiting chromatin modifiers. Thus, even if lncRNAs are constantly removed by the cellular machinery, they may still have a functional impact on the transcriptome through their generation.

lncRNAs also have roles beyond chromatin. Some act as sponges sequestering diverse regulatory elements (e.g., miRNAs) and thus decreasing their impact on target mRNAs. Others modulate how the cell recognizes mRNA molecules and thus regulate translation efficiency, cellular localization, or stability. These examples illustrate the diverse mechanisms of action that lncRNAs use to regulate gene expression. However, despite many examples describing functional roles of lncRNAs, for the vast majority, the mechanisms of action remain unknown.

lncRNAs have been found in nearly all organisms studied, suggesting that they are a fundamental component of gene expression. Interestingly, the ratio of lncRNAs (in comparison to canonical mRNAs) is higher in more complex organisms, suggesting that they contribute to the increased regulatory complexity in higher eukaryotes. In budding yeast, it has been shown that lncRNA expression can result in chains of overlapping sense and antisense transcripts that lead to the rewiring of regulatory gene expression networks. This process, in which lncRNAs spread regulatory signals across the genome, is thought to be even more complex in higher eukaryotes. Furthermore, the boundaries between coding and noncoding RNAs are often vague; in some cases, previously identified lncRNAs have been shown to encode short peptides with potential biological functions. This implies that lncRNAs may also serve as evolutionary intermediates for the generation of new genes.

The last decade of genomics has revealed a surprising diversity of lncRNAs with diverse mechanisms of action. However, we have so far only seen the tip of the iceberg in a whole new field of biology. In the coming years, the development of new tools such as high-throughput long-read sequencing promises to boost our ability to discover and characterize lncRNAs. The development of novel approaches to assay the functional impact of such lncRNAs will allow us to better understand their phenotypic consequences and mechanisms of action. In addition, the detailed biochemical and molecular characterization of lncRNAs will be essential for our understanding of fundamental cell biology, evolution, health, and disease.

Lars M. Steinmetz, PhD
Professor of Genetics, Stanford University School of Medicine,
Stanford, CA, USA
Co-Director, Stanford Genome Technology Center, Stanford, CA, USA
Principle Investigator and Senior Scientist, European Molecular Biology
Laboratory (EMBL), Heidelberg, Germany

Vicent Pelechano, PhD
Assistant Professor and SciLifeLab Fellow,
Department of Microbiology, Tumor and Cell Biology,
Science for Life Laboratory, Karolinska Institute,
Solnavägen, Solna, Sweden

Preface

RNA has been an interesting molecule in biology displaying a multitude of functions in all cellular systems. According to the RNA world hypothesis, it is proposed that RNA was the first macromolecule to arise in the universe even before DNA and proteins came into existence. However, modern biology has seen a great influence of DNA as the information storage house, while the proteins are really the final functional actors in various cellular functions. RNA was thought to have only a supporting role in maintaining the functional homeostasis. Messenger RNA carries the information encoded in DNA. Ribosomal RNA organizes the ribosome structure to facilitate the translation of the messenger RNAs. Transfer RNAs carry the amino acids to be incorporated into the proteins. Other small RNA molecules like 5S RNA and 5.8S RNA contribute to ribosome function. Small nuclear and nucleolar RNAs (snRNAs and snoRNAs) play significant roles in splicing machinery. For a long period of time, it was a common notion that the entire DNA in eukaryotic cells is not transcribed, and the junk DNA hypothesis prevailed for a few decades. However, subsequent to the completion of the Human Genome Project and the advent of NGS technology, we have seen that much of the DNA is pervasively transcribed to generate a large repertoire of RNA molecules which are now being classified as non--coding RNAs. This includes both short (22–33 nucleotides in length) and long RNA species (>200 nucleotides in length). Presently it is estimated that there are more genes coding for long noncoding RNA genes than the estimated ~25,000 protein--coding genes in humans and mice. It is becoming increasingly clear that long noncoding RNAs (lncRNAs) have ubiquitous biological function(s) in almost every aspect of cellular biology, regulating gene expression and contributing to the final functional output of protein-coding genes. lncRNAs are found in all eukaryotes from unicellular organisms to higher mammals. The present book is an attempt to briefly describe the most recent developments in the area of long noncoding RNA biology which is one of the emergent hot topics in the field of molecular and cellular biology today.

The first chapter by Jarroux, Morillon, and Pinskaya introduces the area of lncRNAs beginning with the history and their discovery. This is followed by a description of their characteristic features in comparison with messenger RNAs.

They also discuss the different classes of lncRNAs based on their genomic organization, cellular localization, and functions. This chapter lays an excellent foundation for the ensuing chapters in the book. The second chapter by Kanduri and associates discusses the role played by lncRNAs in genome organization and chromatin organization and regulation along with their varied mechanisms of action. They also discuss the role of lncRNAs in nucleolar and centromere functions. This chapter is followed by an exhaustive description of lncRNAs in an invertebrate model system, *Drosophila melanogaster*, by Subhash Lakhotia (Chap. 3). This model organism, owing to its richness in genetics and genetic approaches, has contributed significantly toward our understanding of the biology of lncRNAs. This chapter also includes a vast amount of information on the lncRNA *hsrw* that was discovered in the author's laboratory. Zapulla et al. have focused, in the fourth chapter, on the functions of lncRNAs in the widely studied unicellular eukaryote *Saccharomyces cerevisiae*. From transcription regulation to acting as a scaffold in the telomere region, the chapter sheds light upon how these single-celled organisms have evolved to add an extra layer of regulation in the form of lncRNAs. In the fifth chapter, Wang and Chekanova have categorized and detailed the lncRNAs involved in gene expression regulation in the plant kingdom. With an emphasis on the well-characterized plant *Arabidopsis thaliana*, the chapter deals with diverse roles that the lncRNAs play in plants. From acting as miRNA sponges and decoys to transcriptional silencing, they are involved in the coordination of crucial physiological phenomena such as flowering.

The sixth chapter by Mishra and associates discusses the role of lncRNAs in genomic imprinting, dosage compensation, body patterning, tissue development, and organogenesis and also the role of lncRNA in the etiology of human diseases and disorders. The following chapter (Chap. 7) by Felley-Bosco and Arun Renganathan summarizes the recent developments on the role of lncRNAs in cancer and their therapeutic potential. It is becoming apparent that every type of human cancer is associated with one or several of the lncRNAs and expression of many of them is perturbed in any given cancer. Current efforts worldwide are being directed toward developing these lncRNAs as biomarkers as well as potential targets for molecular intervention. Chapter 8 by Debosree Pal and M. R. S. Rao discusses the role of long noncoding RNAs in the biology of stem cells both with respect to their pluripotent properties and cell fate specification and differentiation. Stem cells are increasingly perceived as promising candidates for regenerative medicine, and the importance of the long noncoding RNAs in the biological features of stem cells provides additional opportunities for better understanding of the players involved in the maintenance of stemness as well as their differentiation properties. It is likely that lncRNAs may play a significant role in the application of stem cell technologies in regenerative medicine in the near future. In Chap. 9, Clark and Blackshaw describe the present scenario of the role of lncRNAs in brain function and more particularly in nervous system development. At present, neuroscience is one of the most challenging areas of human biology, and in this context, our understanding of the functional role of lncRNAs in nervous system development will be very valuable in the future prospects of neural function and neurodegenerative diseases.

The concluding chapter by Tripathi and associates provides a detailed yet comprehensive view about the technologies and methods that have been developed in the last decade or so to decode the functions of these lncRNAs. Experimental methods related to sequencing of these transcripts, understanding their localization, and solving their complex structures have been dealt with in this chapter, along with a listing of the bioinformatic and computational methods as well.

I am extremely thankful to all the contributors for taking their valuable time off and assembling each of the chapters very eloquently. The field of long noncoding RNA biology is expanding very fast and is one of the areas wherein a large number of papers are being published in the scientific literature today. In this context, all the chapters have been written very lucidly incorporating very up-to-date information. In such a rapidly expanding area covering almost every aspect of cellular biology, it is difficult to avoid any overlap of some of the information between the chapters. I would urge the readers to bear with this.

I should also express my deep sense of gratitude to Steinmetz and Pelechano who have been in the forefront of human genomics sciences for consolidating their thoughts in this emergent area of lncRNA biology in their foreword. I would like to personally thank Ms. Debosree Pal who has been involved with me from the beginning of the conceptualization of this book until the final execution of the project. Many of my former graduate students and postdoctoral fellows have contributed significantly toward the growth of long noncoding RNA biology in my laboratory to take the mrhl RNA biology to where it is today. I am extremely grateful for their efforts and the intellectual inputs in making mrhl RNA as one of the important molecules in epigenetically regulating gene expression during mammalian spermatogenesis. Finally, I would like to acknowledge all the help I received at every step of the execution of this project from Dr. Suvira Srivastava and her colleagues at Springer, New Delhi.

Bangalore, India M.R.S. Rao

Contents

About the Editor

M.R.S. Rao is one of the most prominent scientists in India, currently working as honorary professor at Jawaharlal Nehru Centre for Advanced Scientific Research, Bangalore, India. In 2010, he received the country's fourth highest civilian award "Padma Shri" from the government of India for his outstanding contribution to the field of biomedical sciences.

Prof. Rao obtained his Ph.D. from the Indian Institute of Science, Bangalore, in 1973 and subsequently worked in different capacities at various renowned international institutes/universities such as Baylor College of Medicine, USA; Harvard Medical School, USA; and Ludwig Institute for Cancer Research, University of California at San Diego, USA. His area of interest spans RNA biology, molecular and cellular biology, genomics and cancer biology, molecular genetics of type II diabetes and cancer genetics, and chromatin biology. He has made seminal contributions in delineation of the biology and role of long noncoding RNA molecules.

He has published over 150 articles in respected national and international journals and is a member of various prestigious national and international societies/associations. He has received numerous national and international awards during his distinguished scientific career.

Contributors

Vijay Suresh Akhade Department of Medical Biochemistry and Cell Biology, Institute of Biomedicine, Sahlgrenska Academy, University of Gothenburg, Gothenburg, Sweden

Seth Blackshaw, Ph.D. Johns Hopkins University School of Medicine, Baltimore, MD, USA

Julia A. Chekanova School of Biological Sciences, University of Missouri—Kansas City, Kansas City, MO, USA

Brian S. Clark, Ph.D. Johns Hopkins University School of Medicine, Baltimore, MD, USA

Emanuela Felley-Bosco Laboratory of Molecular Oncology, Division of Thoracic Surgery, University Hospital Zürich, Zürich, Switzerland

Evan P. Hass Department of Biology, Johns Hopkins University, Baltimore, MD, USA

Julien Jarroux ncRNA, epigenetic and genome fluidity, Institut Curie, Centre de Recherche, CNRS UMR 3244, PSL Research University and Université Pierre et Marie Curie, Paris, France

Chandrasekhar Kanduri Department of Medical Biochemistry and Cell Biology, Institute of Biomedicine, Sahlgrenska Academy, University of Gothenburg, Gothenburg, Sweden

Subhash C. Lakhotia Cytogenetics Laboratory, Department of Zoology, Banaras Hindu University, Varanasi, India

Rakesh K. Mishra CSIR-Centre for Cellular and Molecular Biology, Hyderabad, India

Antonin Morillon ncRNA, epigenetic and genome fluidity, Institut Curie, Centre de Recherche, CNRS UMR 3244, Université Pierre et Marie Curie, Paris, France

Rachel O. Niederer, Ph.D. Massachusetts Institute of Technology, Cambridge, MA, USA

Debosree Pal Molecular Biology and Genetics Unit, Jawaharlal Nehru Centre for Advanced Scientific Research, Bangalore, India

Rashmi U. Pathak CSIR-Centre for Cellular and Molecular Biology, Hyderabad, India

Marina Pinskaya ncRNA, epigenetic and genome fluidity, Institut Curie, Centre de Recherche, CNRS UMR 3244, Université Pierre et Marie Curie, Paris, France

M.R.S. Rao Molecular Biology and Genetics Unit, Jawaharlal Nehru Centre for Advanced Scientific Research, Bangalore, India

Arun Renganathan Molecular Biology and Genetics Unit, Jawaharlal Nehru Centre for Advanced Scientific Research, Bangalore, India

Parna Saha CSIR-Centre for Cellular and Molecular Biology, Hyderabad, India

Shreekant Verma CSIR-Centre for Cellular and Molecular Biology, Hyderabad, India

Hsiao-Lin V. Wang School of Biological Sciences, University of Missouri—Kansas City, Kansas City, MO, USA

David C. Zappulla, Ph.D. Department of Biology, Johns Hopkins University, Baltimore, MD, USA

Chapter 1
History, Discovery, and Classification of lncRNAs

Julien Jarroux, Antonin Morillon, and Marina Pinskaya

Abstract The RNA World Hypothesis suggests that prebiotic life revolved around RNA instead of DNA and proteins. Although modern cells have changed significantly in 4 billion years, RNA has maintained its central role in cell biology. Since the discovery of DNA at the end of the nineteenth century, RNA has been extensively studied. Many discoveries such as housekeeping RNAs (rRNA, tRNA, etc.) supported the messenger RNA model that is the pillar of the central dogma of molecular biology, which was first devised in the late 1950s. Thirty years later, the first regulatory non-coding RNAs (ncRNAs) were initially identified in bacteria and then in most eukaryotic organisms. A few long ncRNAs (lncRNAs) such as *H19* and *Xist* were characterized in the pre-genomic era but remained exceptions until the early 2000s. Indeed, when the sequence of the human genome was published in 2001, studies showed that only about 1.2% encodes proteins, the rest being deemed "non-coding." It was later shown that the genome is pervasively transcribed into many ncRNAs, but their functionality remained controversial. Since then, regulatory lncRNAs have been characterized in many species and were shown to be involved in processes such as development and pathologies, revealing a new layer of regulation in eukaryotic cells. This newly found focus on lncRNAs, together with the advent of high-throughput sequencing, was accompanied by the rapid discovery of many novel transcripts which were further characterized and classified according to specific transcript traits.

In this review, we will discuss the many discoveries that led to the study of lncRNAs, from Friedrich Miescher's "nuclein" in 1869 to the elucidation of the human genome and transcriptome in the early 2000s. We will then focus on the biological relevance during lncRNA evolution and describe their basic features as genes and transcripts. Finally, we will present a non-exhaustive catalogue of lncRNA classes, thus illustrating the vast complexity of eukaryotic transcriptomes.

Keywords Non-coding RNA • Classification • RNA World • Central dogma

---

J. Jarroux • A. Morillon (✉) • M. Pinskaya
ncRNA, epigenetic and genome fluidity, Institut Curie, Centre de Recherche, CNRS UMR 3244, PSL Research University and Université Pierre et Marie Curie, Paris, France
e-mail: julien.jarroux@curie.fr; antonin.morillon@curie.fr; marina.pinskaya@curie.fr

M.R.S. Rao (ed.), *Long Non Coding RNA Biology*, Advances in Experimental Medicine and Biology 1008, DOI 10.1007/978-981-10-5203-3_1

The deep complexity of eukaryotic transcriptomes and the rapid development of high-throughput sequencing technologies led to an explosion in the number of newly identified and uncharacterized lncRNAs. Many challenges in lncRNA biology remain, including accurate annotation, functional characterization, and clinical relevance. All these topics will be thoroughly discussed throughout the book. But to start with, we will detail the discovery of RNA as life's indispensable molecule. The long journey for the biological characterization of non-coding RNAs is summed up in Fig. 1.1, and this history will be described over the first half of this chapter, from the DNA to the first non-coding transcripts. Then, we will discuss how global genomic and transcriptomic studies changed our view on the role of RNA in regulatory circuits, biodiversity, and complexity. Finally, we will include a summary of the extensive classification of lncRNAs.

1.1 A Hundred-Years History of RNA Biology

Before the ever-expanding catalogues of lncRNAs that we have today, a long experimental and theoretical journey was required to prove the importance of RNA molecules in cell biology. It began in 1869 with the discovery of nucleic acids, and it took over a hundred years for researchers to finally identify non-coding transcripts and begin proposing regulatory roles for them.

1.1.1 From "Nuclein" to Nucleic Acids and to the Double Helix

At the end of the nineteenth century, a few pivotal discoveries foreshadowed the molecular biology era. In 1869, Friedrich Miescher isolated a material from nuclei that he called "nuclein" and which he described as highly acidic: in fact he had discovered DNA [1]. In contrast with proteins that were the main focus at the time, its content was low in sulfur and very high in phosphorus and could not be digested by protease treatment. Later, once the chemical composition of the "nuclein" isolated from different organisms had been discovered, it was realized that "thymus nucleic acid" consisted of DNA, while "yeast nucleic acid" was composed of RNA. In the early 1900s, several scientists proposed the chemical composition and the first structures for DNA and RNA, though the biological differences between these two molecules were still not apparent. Ironically, the discovery of "nuclein" by Miescher happened only a few years after Gregor Mendel published his work on the laws of heredity in 1866, but nevertheless many scientists thought proteins were the carriers of genetic information. Thus, the link between Mendel's model and

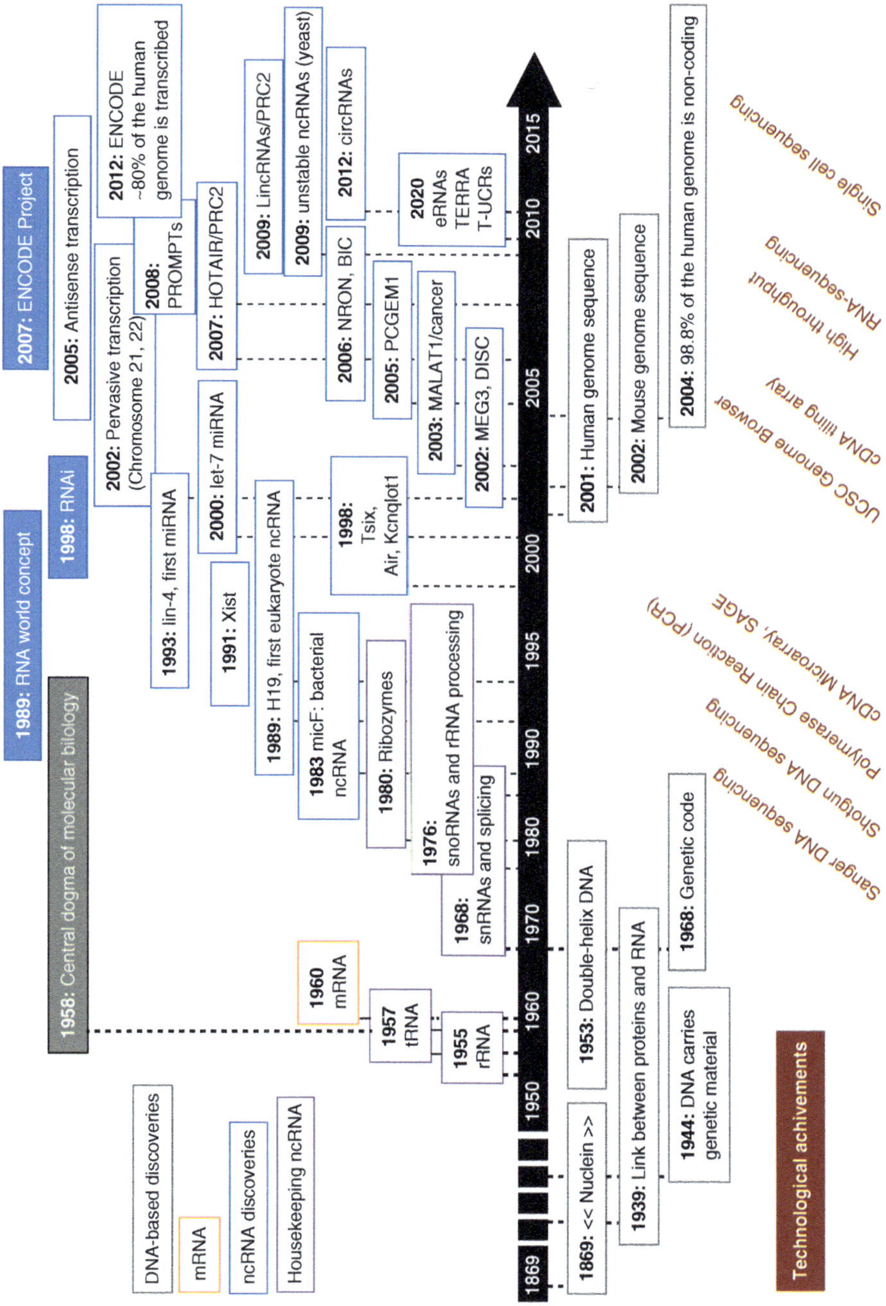

Fig. 1.1 The timeline of principle discoveries in nucleic acid biology and, in particular, eukaryotic ncRNAs

Miescher's "nuclein" remained missing until 1944 when Oswald Avery proposed DNA as a carrier of genetic information [2].

The link between DNA and RNA was established in the late 1950s as Elliot Volkin and Lawrence Astrachan thoroughly described RNA as a DNA-like molecule synthesized from DNA. This discovery was then further elaborated into a molecular concept of RNA and DNA synthesis [3, 4]. Indeed, following the X-ray crystallographic studies of Rosalind Franklin and the establishment of the double-helix structure of DNA by James Watson and Francis Crick in 1953, it was proposed in 1961 that RNA could be an intermediate molecule in the information flow from DNA to proteins [5]. First devised in 1958 by Francis Crick and then by François Jacob and Jacques Monod, the *Central Dogma of Molecular Biology* comprised transcription of a DNA gene into RNA in the nucleus followed by protein synthesis in the cytoplasm. It was also stated that the information flow can only proceed from DNA to RNA and then from RNA to protein, but never from protein to nucleic acids [5]. The mediating role of RNA became a new focus of research which has been pivotal for the development of modern molecular biology.

1.1.2 A Central Role for RNA in Cell Biology: The RNA World Concept

In 1939, Torbjörn Caspersson and Jean Brachet showed independently that the cytoplasm is very rich in RNA. They also showed that cells producing high amount of proteins seemed to have high amounts of RNA as well [5]. This was a first hint for the requirement of RNA during protein synthesis and its role as a link between DNA and proteins. In 1955, Georges Palade identified the very first ncRNA that makes part of the very abundant cytoplasmic ribonucleoprotein (RNP) complex: the ribosome. In his "Central Dogma" Crick also theorized that there was an "adapter" molecule for the translation of RNA to amino acids. This second class of ncRNAs was discovered in 1857 by Mahlon Hoagland and Paul Zamecnik: the transfer (t)RNA. In 1960, François Jacob and Jacques Monod first coined the term "messenger RNA" (mRNA) as part of their study of inducible enzymes in *Escherichia (E.) coli.* Indeed, they showed the existence of an intermediate molecule carrying the genetic information leading to protein synthesis. Shortly after, the work of Crick helped establish that the genetic code is a comma-less, non-overlapping triplet code in which three nucleotides code for one amino acid. It was later deciphered *in vitro* as well as *in vivo* and shown to be universal across all living organisms [6]. In the late 1960s, rather different from mRNAs, a new class of short-lived nuclear RNAs was found: heterogeneous nuclear (hn)RNAs. These long RNA molecules, which were in fact precursors for mature rRNAs and mRNAs, led to the study of rRNA processing and the discovery of splicing [7, 8]. During that period, small nuclear (sn)RNAs which are part of the spliceosome, the RNP machinery responsible for intron splicing from pre-mRNAs, were discovered [9]; as well as small nucleolar (sno) RNAs, which are involved in the processing and maturation of ribosomal RNAs in the nucleolus [10].

Although Jacob, Monod, and Crick had already mentioned independently that RNA was not just a messenger, many scientists considered it as a mere unstable intermediating molecule, overlooking the active roles of other classes of ncRNAs. However, this view partially changed in 1980 when Thomas Cech and Sidney Altman discovered that RNA molecules could act as catalysts for a chemical reaction. Initially, Cech's group found an intron from an mRNA in *Tetrahymena thermophila* that is able to perform its own splicing through an RNA-catalyzed cleavage [11]. Subsequently, Altman's group showed that the RNA component of the ribonucleoprotein RNase P is responsible for its activity in degrading RNA [12]. These RNA enzymes were called ribozymes and have been shown since then to be key actors of the genetic information flow, making part of both the ribosome and the spliceosome [13, 14].

The discovery of catalytic RNA also led scientists to develop the RNA World theory, which states that prebiotic life revolved around RNA, since it appeared before DNA and protein. Indeed, the extensive studies of its roles in cell biology revealed that RNA is necessary for DNA replication and that its ribonucleotides are precursors for DNA's deoxyribonucleotides. Moreover, as it was previously mentioned, RNA plays an important role in every step of protein synthesis, both as scripts (mRNAs) and actors (ncRNAs: rRNAs, tRNAs, etc.) (Fig. 1.2) [15]. Remarkably, the latter ones are constitutively expressed in the cell and are necessary for vital cellular functions, constituting a class of housekeeping ncRNAs. Being

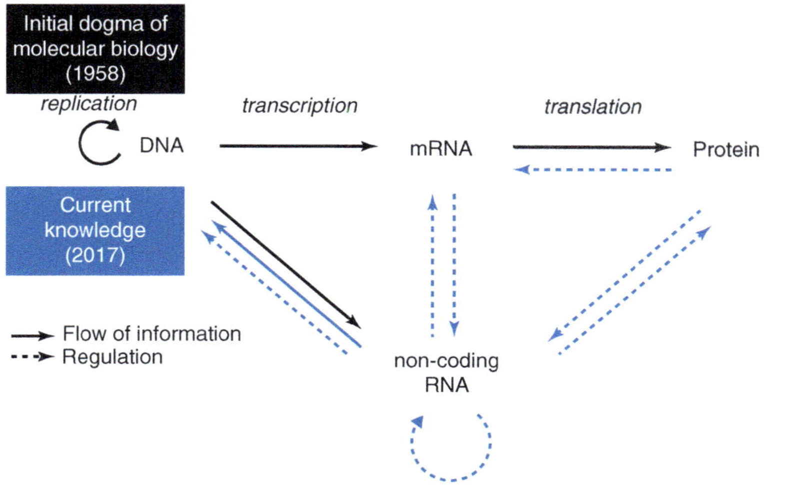

Fig. 1.2 Initial and current dogma of molecular biology

extensively studied, housekeeping ncRNAs are the subject of many specialized publications and will not be described here. Instead, other classes of regulatory ncRNAs that were discovered in the early 1990s will be discussed. These ncRNAs are characterized by very specific expression during certain developmental stages, in certain tissues or disease states, and play multiple roles in gene expression regulation.

1.1.3 Bacterial sRNAs: Pioneers of Regulatory ncRNAs

The very first regulatory ncRNA to be discovered and characterized was *micF* from the bacteria *E. coli.* It was described as the first RNA regulating gene expression through sense-antisense base pairing in 1984 by the team of Masayuki Inoue [16] and represents the major class of bacterial regulatory ncRNAs, small (s) RNAs. The *micF* ncRNA was shown to repress the translation of a target mRNA encoding a porin (outer membrane protein F, OmpF), involved in passive transport through the cell membrane. First discovered through multicopy plasmid experiments, the transcript was isolated 3 years later and shown to be an independent gene. When transcription of *micF* is activated, it inhibits the expression of the ompF gene at both mRNA and protein levels. Subsequently, following the characterization of the RNA duplex structure *in vitro*, *micF* was shown to bind to the ribosome-binding site (RBS) of the *ompF* mRNA, thus inhibiting ribosome binding and translation.

More recently, it was shown that the regulation of gene expression by *micF* through base pairing extends to other genes, among which is the *lrp* mRNA [17]. Lrp (leucine-responsive protein) is a transcription factor that vastly regulates gene expression in *E. coli* in response to osmotic changes and nutrient availability. Remarkably, Lrp regulates *micF* expression as well, thus creating a feedback and proving the important role of *micF* in global gene regulation and metabolism. The same mechanisms were also found in *Salmonella*, supporting the evolutionary conservation of this regulatory pathway [18]. Since then many other sRNAs ranging in length from 50 to 500 nucleotides (nt) have been discovered, including *trans-* or *cis*-encoded ncRNAs, RNA thermometers, and riboswitches. They all act by pairing, thus inhibiting translation of targeted mRNAs and inducing their degradation.

1.1.4 MicroRNAs and RNA Interference

In the early 1990s, several scientists observed independently and in different eukaryotic organisms, through experiments of transgene co-expression or viral infection, an intriguing phenomenon of RNA-mediated inhibition of protein synthesis. The regulatory effects of these RNA molecules reshaped the views of RNA as a mere messenger. The very first studies described the phenomenon as "co-suppression" in plants, as "posttranscriptional gene silencing" in nematodes, or as "quelling" in fungi, but none of them suspected RNA to be the key actor until the identification of the first micro (mi)RNA in the nematode *Caenorhabditis* (*C.*) *elegans* in 1993 by Victor Ambros and coworkers. Ambros discovered that the *lin-4* gene produces small RNAs of 22 and 61 nt from a longer non-protein-coding precursor. The longer RNA forms a stem-loop structure, which is cut to generate the shorter RNA with antisense complementarity to the 3′-untranslated region (UTR) of the *lin-14* transcript [19]. The *lin-4* RNA pairing to *lin-14* mRNA was proposed as a molecular mechanism of "posttranscriptional gene silencing", thus decreasing LIN-14 protein levels at first larval stages of nematode development [20]. Michael Wassenegger observed a similar phenomenon occurs in plants which he described as "homology-dependent gene silencing" or "transcriptional gene silencing"; this process is mediated by the incorporation of viroid RNA which induces the methylation of the viroid cDNA and gene silencing [21]. Ultimately the entire process of RNA-mediated gene silencing was elucidated in 1998 by Andrew Fire and Craig Mello in similar experiments with the *unc-22* gene of *C. elegans*.

In 2000, another essential miRNA was identified in *C. elegans*. This miRNA, *let-7*, was shown to have homologues in several other organisms, including humans [22, 23]. The biogenesis as well as the molecular mechanisms of miRNA-mediated gene silencing has been extensively characterized. In 2001, Thomas Tuschl showed that, in *C. elegans*, long double-stranded RNA is processed into shorter fragments of 21–25 nts. Since this discovery, it has been demonstrated that premature transcripts in the nucleus are processed into hairpin-structured RNA by the Drosha-containing microprocessor complex and then exported to the cytoplasm where they are cleaved into a double-stranded RNA by Dicer. One of the strands of this double-stranded RNA is loaded to the RISC complex and then targeted to an mRNA molecule by complementarity, thus inducing translational repression [23]. This simplified scheme constitutes the mechanistic basis of RNA interference (RNAi) and presently unites all gene silencing phenomena at transcriptional and posttranscriptional levels, mediated by small ncRNAs including miRNAs, small interfering (si)RNAs, and Piwi-interacting (pi)RNAs, all of which are processed from double-stranded RNA precursors [24, 25].

Although the focus on RNAi resulted in a breakthrough for modern biology and biotechnology, as well as provided a deeper understanding of gene regulation, development, and disease, the relevance of lncRNAs remained largely unexplored. Nevertheless, some lncRNAs were investigated in the late 1980s such as *H19* and *Xist,* the milestones of dosage compensation in mammals.

1.2 LncRNA Discovery in the Pre-genomic Era

In the 1980s, scientists were using differential hybridization screens of cDNA libraries to clone and study genes with tissue-specific and temporal patterns of expression. Initially, efforts were focused on genes producing known proteins;

subsequently, an *a posteriori* approach was adopted without regard to the coding potential of RNA. Through this approach, the first non-coding gene was discovered, *H19*, even though at that time it was first classified as an mRNA [26].

1.2.1 **H19**: *The Very First Eukaryote lncRNA Gene*

In the late 1980s, elegant genetic and molecular studies discovered a phenomenon of genomic imprinting or parent-of-origin-specific expression which constitutes part of the dosage compensation mechanisms. Independently, two imprinted genes were identified: the paternally expressed protein-coding *Igf2r* and the maternally expressed *H19*. Both genes were localized to mouse chromosome 7 in proximity to each other forming the *H19/IGF2* cluster [27, 28]. What made *H19* unusual was the absence of translation even though the gene contained small open reading frames. *H19* showed high sequence conservation across mammals, and the abundant transcript presented features of mRNAs: transcribed by RNA polymerase II, spliced, 3′ polyadenylated, and localized to the cytoplasm [29]. The expression of *H19* in transgenic mice revealed to be lethal in prenatal stages, suggesting not only that the dosage of this lncRNA is tightly controlled but that it has an important role in embryonic development. However, the function of *H19* as an RNA molecule in its own right remained a mystery until the functional characterization of another lncRNA involved in dosage compensation in mammals, *Xist*. Since that time, *H19* has been thoroughly investigated and represents the prototype of a multitasking lncRNA.

1.2.2 X Inactivation: Existence of **Xist**

In living organisms, sex can be determined by many ways; it is defined in mammals by the X and Y chromosomes, while males only have one X and Y chromosome, females have two X chromosomes in their karyotype. However, the X chromosome carries many genes, most of which have functions that are not involved in sex determination. Hence, there is a need for dosage compensation between males and females. Although the mechanism of choice in *Drosophila* is to double the transcription of the single X chromosome in males, it is the opposite in mammals: one of the female X chromosomes is inactivated. This phenomenon, called X-chromosome inactivation (XCI), was first discovered in mouse by Mary Lyon in 1961 [30] and further generalized to other mammals. XCI is established early in development and is initiated by a unique locus, the X-inactivation center (*Xic*).

In the early 1990s, this locus was found to produce a long non-coding RNA, *Xist* (X-inactive-specific transcript). It is expressed at very low levels in both, male and female, mouse undifferentiated embryonic stem (ES) cells. Upon differentiation *Xist* expression is activated in a monoallelic way in female cells, from the future inactive X (Xi) to initiate the onset of random XCI. Being retained in the nucleus, *Xist* triggers gene silencing in *cis* by physically localizing and spreading broadly on the future Xi [31–33]. In contrast to *H19* and other lncRNAs involved in dosage compensation, *Xist* is highly unusual since it triggers the silencing of the entire chromosome. The propagation of *Xist* along the Xi, called "coating," implicates the RNA wrapping around the X and the recruitment of multiple factors, including the polycomb repressive complexes 1 and 2 (PRC1 and PRC2). This triggers a cascade of chromatin changes and a global spatial reorganization of the Xi and, ultimately, the stable repression of nearly all Xi-linked genes throughout development and adult life [34]. While *Xist* expression is critical for the initiation of the XCI, in somatic cells, *Xist* and the whole *Xic* were shown to be dispensable for the maintenance of silencing in mouse [35]. In 1999 human *XIST*, ectopically expressed from the artificially inserted transgene on mouse autosomes, was demonstrated to function as *Xic* and to initiate XCI even in undifferentiated mouse ES cells, unlike the mouse counterpart. This result suggested differences in the developmental regulation of *Xist* and in the initiation of the XCI process between mouse and humans. In addition, the inactivation by ectopic human *XIST* was observed only in a portion of mouse male ES cells, thus confirming that *Xist* is not a unique actor of stable inactivation [36]. Indeed, the *Xic* was initially defined in mouse as the minimal region of the X chromosome that contains all sequences both necessary and sufficient for the initiation of XCI. Xic extends over 1 Mb, and transcriptomic studies revealed that this region contains several protein-coding and non-coding genes, including *Linx*, *Ftx*, and others. Remarkably, some non-coding genes within *Xic* and beyond show poor primary sequence conservation between human and mouse, and this includes the sequence of *Xist* itself [37]. In particular, the *Tsix* lncRNA is an antisense transcript which overlaps the whole *Xist* gene and its promoter in mouse. In humans, key regulatory elements are truncated, and the transcript overlaps *XIST* only at 3′ end. These differences abolish the *TSIX* function in transcriptional repression of *XIST* on the future active X in humans [38, 39]. Recently, another lncRNA, *XACT*, was discovered in human ES cells. This gene is located within the intergenic region, outside of *Xic*, and it is not conserved in mice. In female human ES cells, *XACT* is expressed from and coats both X chromosomes. This lncRNA seems to be specific to pluripotent cells and is proposed to ensure peculiar control of XCI in humans [40]. The biogenesis of *Xist*, its structure, and the molecular mechanism of XCI have been the focus of many studies in different mammals and extensively documented in other publications [34].

The pioneering studies of *H19* and *Xist* revolutionized our view of non-protein-coding gene functions and on the biological relevance of lncRNAs in general. These examples demonstrated the complexity and versatility of regulatory circuits orchestrated by a single lncRNA. They also stimulated the discovery and suggested potential mechanisms for other, yet uncharacterized, non-coding transcripts. A global effort toward lncRNA identification and characterization began in the 2000s, as a plethora of novel non-coding transcripts during the sequencing of the complete human genome.

1.3 From Non-coding Genome to Non-coding Transcriptome: The Genomic Era

Our modern view of eukaryotic transcriptomes was preceded by comprehensive investigations of genomic DNA and the discovery that, in addition to protein-coding (PC) sequences and regulatory elements essential for PC gene (PCG) transcription, the majority of the genome contains sequences that were considered to be useless evolutionary fossils. To differentiate these sequences from PC sequences, this DNA was named non-coding and referred to as selfish or junk DNA for almost 20 years [41].

1.3.1 The Human Genome Project: Genomic DNA Is Mostly Non-coding

In 1978, using the sequencing technique he had developed, Frederick Sanger generated the first ever full genomic sequence: the viral genome of the bacteriophage ϕX174 [42]. Since then, Sanger sequencing has been routinely used worldwide, and its discovery and development earned the Nobel Prize in Chemistry for Sanger, along with Walter Gilbert. During the following years, several viral genomes were sequenced and by the end of 1990, a worldwide sequencing effort, the Human Genome Project (HGP), was established by the National Institute of Health (NIH, USA) to completely sequence the human genome. In parallel, the American biochemist and entrepreneur Craig Venter founded his own company and sought private funding to achieve the same goal. This put pressure on the public groups involved in the HGP, and the race to unravel the human genome began. The first bacterial genome was published in 1995 [43]. It was followed in 1999 by the sequence of the euchromatic portion of human chromosome 22 [44], which covered approximately 65% of what is now known to be the full chromosome 22. This sequence was thought to contain 545 protein-coding genes (whether known or predicted), with PC exons spanning a mere 3% of the full sequence.

Finally, using clone-by-clone methodology, the first draft of the complete human genome was published in *Nature* in 2001 covering 96% of the euchromatin [45], followed the next day by Craig Venter's publication in Science of the whole-genome sequence obtained by the shotgun-cloning method [46]. Regular updates completed the human genome sequence in 2003. In the meantime, the genomes of several other organisms had already been released, notably yeast [47], pufferfish [48], worm [49], fruit fly [50], and mouse [51], thus allowing comparative studies to be performed.

The first surprise from this comprehensive genomic sequencing effort was the rather low number of PCGs compared to what was initially expected. Indeed, early studies that looked at the repartition of CpG islands predicted 70,000–80,000 genes in the human genome [52], a figure close to the well-admitted 100,000 genes from the mid-1980s. However, the HGP predicted around 31,000 PCGs in 2001 reduced to 22,287 PCGs in 2004 [45, 53]. In general, only 1.2% of the human genome represents PC exons, whereas 24 and 75% were attributed to intronic and intergenic non-coding DNA.

1.3.2 Pervasive Transcription and the Dark Matter of the Genome

The HGP also revealed that most of the genome is actually transcribed, whether it encodes proteins or not. Indeed, a tiling array with oligonucleotide probes spanning human chromosomes 21 and 22 revealed that 90% of detected cytosolic polyadenylated transcripts map to non-coding genomic regions and not to exons [54]. Similar results were found by the FANTOM and RIKEN consortia when analyzing the transcriptome in both human [55] and mouse [56]. They sequenced more than 60,000 full-length cDNAs from mouse in a standardized manner to generate accurate maps of the 5′ and 3′ boundaries of all transcripts, thus defining transcription start (TSS) and termination (TTS) sites. Remarkably, cap analysis gene expression (CAGE) sequencing, a technique that sequences 5′ ends of capped transcripts, revealed over 23,000 ncRNAs originating from both sense and antisense transcription representing approximately two thirds of the mouse genome [57]. For the first time, antisense transcription was proposed to contribute to the regulation of gene expression at transcriptional level in mammals.

These results were later confirmed by even larger-scale studies conducted in humans by the ENCODE (Encyclopedia of DNA Elements) consortium. This project compiled over 200 experiments in its pilot phase [58] and up to 1640 datasets from 147 different cell lines in its later release [59]. Through various sequencing techniques, landscapes of DNase I hypersensitive sites, histone modifications, transcription factor binding sites, and the whole transcriptome were defined. Conclusions from these studies estimated that 93% of the human genome is actively transcribed and associated with at least one primary transcript (i.e., coding and non-coding exons and introns); among these transcripts, approximately 39% of the genome represented PCGs (from promoter to poly(A) signal) and 1% protein-coding exons, while the other 54% mapped outside of PCGs (Fig. 1.3). However, many lncRNAs overlap with PCG annotations in both sense, coding and antisense strands. More recently, the mouse counterpart of the ENCODE Consortium confirmed previous reports by publishing a similar analysis which showed that 46% of the mouse genome produces mRNAs while at least 87% of its genome is transcribed [60, 61].

Many studies aiming to characterize non-coding transcription were also performed in other eukaryotes, including *Saccharomyces cerevisiae*. Even in this primitive unicellular eukaryote, about 85% of the genome is transcribed [62]. This phenomenon is often referred to as "pervasive transcription" and is widespread among eukaryotes. An expanding body of literature details its function [63, 64]. The identification and characterization of non-coding transcripts as unique ncRNAs extended the former definition of a "gene" beyond its coding function. Furthermore,

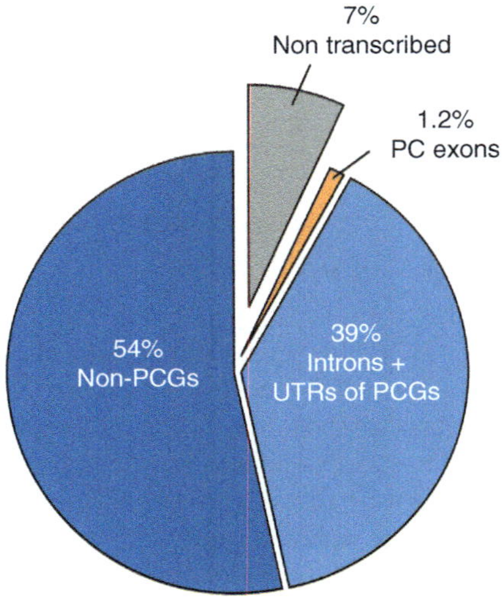

Fig. 1.3 Proportion of transcribed protein-coding and non-coding sequences (introns, UTRs, and others) in the human genome according to ENCODE [59]

the discovery of the non-coding genome and transcriptome gave rise to heated debates in the scientific community concerning the biological significance and functional relevance of these non-coding DNA and RNA, still perceived as a junk [63, 65, 66]. These debates challenged the Central Dogma of Watson and Crick, promoting ncRNAs to the epicenter of the cellular processes as a driver of biological complexity through evolution.

1.4 Non-coding RNAs: Junk or Functional

Polemics around the biological and functional relevance of lncRNAs were oriented toward understanding the origin, conservation, and diversification of lncRNA species across evolution.

1.4.1 Origin of lncRNA Genes

Non-coding genes were proposed to arise through various mechanisms including DNA-based or RNA-based duplications of existing genomic sequences, the metamorphosis of PCGs by loss of protein-coding potential, transposable element exaptation, or non-coding DNA exaptation [67]. Homologous non-coding genes arise from duplications of already existing lncRNA genes. Pseudogenes are an example of PCG metamorphosis during which a duplicated ancestral open reading frame had accumulated disruptions destroying its potential to be translated. Once transcribed, pseudogenes often produce lncRNAs, as in the case of *PTENP1*.

Pseudogenization of a PCG, due to mutations deleterious to translation, can also produce lncRNA genes that do not have an apparent protein-coding "homologue". An example is *Xist* which is derived from an ancestral *Lnx3* gene and which has acquired several frame-shifting mutations during early evolution of placental mammals [68]. Exaptation or co-option of RNA-derived transposable elements (TE) into non-coding genes is another frequent mechanism of lncRNA origination. In humans TEs constitute a large portion of the genome (40–45%) [45]. Most of them are genomic remnants that are currently defunct but are often embedded into non-coding transcripts. TEs are considered as major contributors to the origin and diversification of lncRNAs in vertebrates [69]. Together with local repeats, they provide lncRNA genes with TSS, splicing, polyadenylation, RNA editing, RNA binding sites, nuclear retention signals or particular secondary structures for protein binding [70–72].

Finally, pervasive transcription of the genome may generate cryptic RNAs that, if maintained through evolution, can give rise to lncRNA genes with novel functions. In particular, exaptation of non-coding sequences into lncRNAs can occur through the acquisition of regulatory elements within a silent region, thereby promoting transcription. However, the *de novo* origin of lncRNAs remains difficult to prove and is represented by few examples, such as the testis-specific lncRNA *Poldi* [73]. Interestingly in humans, the testis and cerebral cortex are the most enriched tissues for the expression of PCGs and non-coding genes of *de novo* origin. This particularity was suggested to contribute to phenotypic traits that are unique to humans, such as an improved cognitive ability [74, 75].

1.4.2 Evolutionary Conservation of lncRNAs

Genomic and transcriptomic studies across the eukaryotic kingdom allowed the analysis of the primary sequence conservation of protein-coding and non-coding loci. These studies revealed that the human genome is highly dynamic, and only 2.2% of its DNA sequence is subjected to conservation constraints [76]. Remarkably, non-coding genes are among the least conserved with more than 80% of lncRNA families being of primate origin [77]. This finding raised skepticism regarding the functionality and biological relevance of lncRNAs and initiated a search for other conservation constrains [78, 79]. If the criterion of primary sequence conservation is too restrictive in regard to lncRNA genes, other features such as structure, function, and expression from syntenic loci constitute multidimensional factors that are more applicable for evolutionary studies of lncRNAs [80]. Recently, a study looking at the non-coding transcriptome of 17 different species (16 vertebrates and the sea urchin) showed that although the body of non-coding genes tends not to be conserved, short patches of conserved sequences could be found at their 5′ ends. This confirmed a higher conservation of TSS and synteny, as well as expression patterns in different tissues, especially in those involved in development [81]. Indeed, the most conserved are developmentally regulated lncRNAs of the lincRNAs subfamily.

These lncRNAs have a remarkably strong conservation of spatiotemporal and syntenic loci expression, suggesting that it is selectively maintained and crucial for developmental processes [77, 82, 83].

1.4.3 Role of lncRNAs in Biological Diversity

The identification of new lncRNAs in the last decade continues to increase and, as anticipated in the past, largely exceeds that of protein-coding transcripts. The diversity of the non-coding transcriptome is considered as an argument to explain the remarkable phenotypic differences observed among species given a relatively similar numbers of protein-coding genes among fruit fly (13,985; BDGP release 4), nematode worm (21,009; Wormbase release 150), and human (23,341; NCBI release 36) [84]. In 2001, John Mattick and Michael Gagen proposed, for the very first time, that non-coding transcripts named "efference" RNA, together with introns, constitute an endogenous network enabling dynamic gene-gene communications and the multitasking of eukaryotic genomes. In contrast to core proteomic circuits, this higher-order regulatory system is based on RNA and operates through RNA-DNA, RNA-RNA, and RNA-protein interactions to promote the evolution of developmentally sophisticated multicellular organisms and the rapid expansion of phenotypic complexity. A direct correlation between the portion of non-coding sequences in the genome and organism complexity was hypothesized [85, 86]. Interestingly comparative genomics allowed the identification of a few regions in the human genome that have high divergence when compared to other species [87, 88]. These human accelerated regions (HAR) contain many lncRNA genes and have been suggested to be involved in the acquisition of human-specific traits during evolution. In 2006, a first lncRNA from these regions was shown to be expressed during cortical brain development [89]. Since then, many mutations involved in diseases were identified in these non-coding regions and shown to be associated with regulatory elements in the brain [90]. A more recent study showed that mutations of HAR enhancer elements could be involved in the development of autism, thus supporting the hypothesis that some HAR could be involved in human-specific behavioral traits and cognitive or social disorders when mutated [91]. However, the functionality of non-coding transcripts was and still remains hotly debated. Nevertheless, the conception of developmental and evolutionary significance has stimulated an exhaustive molecular characterization of lncRNA genes and transcripts.

1.5 The General Portrait of lncRNA Genes and Transcripts

lncRNAs have been identified in all species which have been studied at the genomic level, including animals, plants, fungi, prokaryotes, and even viruses. Genome-wide studies continue to enlarge the catalogue of lncRNAs continuously reshaping the specific features of lncRNAs as transcription units. Here, we will summarize the main features of lncRNAs that distinguish them from mRNAs (Table 1.1).

Table 1.1 Comparison of lncRNA and mRNA features

Feature	lncRNA	mRNA
Transcription		
	RNA polymerase II	RNA polymerase II
	RNA polymerase III (*B2-SINE*; *NDM29* [101, 102])	
	RNA polymerases IV and V (plants, [103])	
Chromatin modifications		
H3K4me3	Low (eRNA, PROMPTs)	High
	High (others)	
H3K4me1	High (eRNA, PROMPTs)	Low
	Low (others)	
H3K27ac	High	Low
H3K36me	Moderate/high	High
H3K79me2	Enriched (bidirectional lncRNAs)	Low
H3K27me3	Present at bivalent and repressed promoters	Present at bivalent and repressed promoters
5′-Cap	Present (7-methylguanosine, m7G)	Present (m7G)
Poly(A) tail	Present or not	Present
	Bimorphic	
Length	200—>100 kb (10 kb mean)	5 kb mean
Exon-intron composition	Yes	Yes
	Exons are longer	
Splicing	Yes or less efficient	Yes
	No (macro lncRNA, vlincRNAs)	
RNA stability	Variable, globally lower than mRNA	Variable
	Highly unstable (eRNA, XUTs, CUTs, PROMPTs)	
Evolutionary conservation	High (lincRNAs)	High
	Low or not conserved (others)	
Protein-coding potential	Non or very low (sORFs)	Yes
Structure	Versatile, multi-modular	Kozak hairpin at the 5′ end
Subcellular localization	Nucleus	Cytosol
	Cytosol	
	Mitochondria	
Expression specificity	high, including interindividual variability of expression	low to high
Transcript abundance	Very low or low	Moderate to high
	High (for few)	

1.5.1 Coding Potential of lncRNA Genes

As dictated by the acronym, lncRNA genes do not encode proteins. Cytosol-localized lncRNAs were found associated with mono- or polyribosomal complexes [92], but this association is not necessarily linked to translation but rather proposed to determine lncRNA decay [93, 94]. Some lncRNAs include short open reading frames (sORFs) and undergo translation, though only a minority of such translation events results in stable and functional peptides [95, 96]. This is the case of *DWORF*, a muscle-specific lncRNA that encodes a functional peptide of 34 amino acids [97–100]. Proteomic studies will undoubtedly introduce a new "coding" aspect to lncRNAs, expanding our conception of "coding" and leading to a possible concept of bifunctionality.

1.5.2 LncRNA Transcription and Transcript Organization

The majority of eukaryotic lncRNAs are produced by RNA polymerase II, with some exceptions, for example, the murine heat-shock induced B2-SINE RNAs [101] or the human neuroblastoma associated *NDM29* [102], which are synthesized by RNA polymerase III. However, the last two examples are not strictly considered as lncRNAs because the transcript length is below the arbitrary threshold of 200 nts. In plants, two specialized RNA polymerases, Pol IV and Pol V, transcribe some lncRNA genes [103]. Many lncRNAs are capped at the 5′ end, except those processed from longer precursors (intronic lncRNAs or circRNAs). However, some ambiguities exist concerning the presence of a cap, especially for highly unstable and low-abundant transcripts, since they can't be captured by the CAGE-seq technique. LncRNAs may or may not be 3′-end polyadenylated; in addition, they may also be present as both forms, such as bimorphic transcripts like *NEAT1* and *MALAT1* [104, 105]. LncRNAs with a polyadenylation signal have higher stability than those that are poorly or not polyadenylated, with the exception of lncRNAs bearing specific 3′-end structures as in case of *MALAT1* [106]. Of note, poly(A)+ transcriptomic studies exclude the possibility of discovery of non-polyadenylated transcripts and introduce a quantitative bias in the identification of such lncRNAs. This point should be taken into account in comparative studies or in selection of RNA-seq strategies, favoring the use of total RNAs instead of the more customary used poly(A)+ RNA fraction.

Similar to PCGs, transcription of many lncRNA genes requires canonical factors assisting the RNA polymerase machinery such as the pre-initiation complex (PIC), Mediator, transcription elongation complex, and also specific transcription factors that in turn could define the specificity of lncRNA expression in different biological contexts. However, some particularities in lncRNA promoters have been demonstrated. In humans lncRNA promoters are more enriched in A/T mono-, di-, and trinucleotide stretches and are characterized by reduced CG and almost depleted AT skews (CG and AT compositional strand biases); this is contrary to PCGs suggesting a distinct regulation of transcription for these two groups of genes [107]. Promoters of PROMPTs are devoid of transcription initiation factors such as TAFI, TAFII, p250, and E2F1 and are believed to initiate transcription without the use of conventional PIC [108]. eRNAs require the Integrator complex for the 3′-end cleavage of primary transcripts [109], and lncRNA precursors of small ncRNAs were shown to be processed by specific endonucleases [110, 111]. Some unstable lncRNAs such as yeast NUTs and CUTs are terminated by the Nrd1-dependent pathway, thus targeting them for rapid degradation by the exosome [112–114].

LncRNA genes can have a multi-exonic composition with similar splicing signals as PCGs and therefore could undergo splicing into several different isoforms with distinct functional outcomes and clinical relevance [115–117]. However, they usually comprise fewer and slightly longer exons than PCGs [118, 119].

1.5.3 Chromatin Signatures of lncRNAs Genes

As RNA polymerase II transcribes most of the lncRNA genes, their genomic regions present a chromatin organization resembling that of PCGs, with some differences. This could be due to the globally low expression of lncRNAs, which is a consequence of either low rate of transcription, lower stability, or both. Globally, lncRNA TSS reside within the DNase I hypersensitive sites suggesting nucleosome depletion from this region. LncRNA promoters have lower levels of histone H3K4 trimethylation (H3K4me3), which is in accordance with their low transcription rate. eRNAs and PROMPTs present high levels of histone H3K4 monomethylation (H3K4me1) and K27 acetylation (H3K27ac) at promoters, which is considered as a specific signature of enhancer- and promoter-associated unstable transcripts; these signatures exist in the following ratios: H3K4me3 over H3K4me1 as a mark of PROMPTS and H3K4me1 over H3K4me2 as a mark of eRNAs [120]. The body of most lncRNA genes with the exception of eRNAs and PROMPTs is marked by histone H3K36 trimethylation (H3K36me3). In yeast, sense-antisense transcription was reported to be associated with particular chromatin architecture: reduced histone H2B ubiquitination, H3K36me3, and histone H3K79 trimethylation, as well as increased levels of H3ac, chromatin remodeling enzymes, histone chaperones, and histone turnover [121]. In mouse, bidirectional transcription, which is often associated with developmental genes and genes involved in transcription regulation, was found to harbor high H3K79 dimethylation (H3K79me2) and elevated RNA polymerase II levels. This signature is characteristic of intensified rates of early transcriptional elongation within a region transcribed in both directions [122].

It is anticipated that single cell studies will resolve the problem of signal variability in a population of cells, allowing transcriptional events to be directly linked to specific chromatin modifications. Such efforts have already been initiated for transcriptome profiling [123–125] but remain challenging for epigenomic studies [126].

1.5.4 Expression Pattern of lncRNAs: Stability, Specificity, and Abundance

Several genome-wide studies addressed lncRNA stability and, depending on the employed experimental approach, revealed some discrepancy for different species of lncRNAs. In mouse, the measurements of the lncRNA half-life (t½) and decay rates were performed through transcription inhibition by actinomycin B treatment. In this case, lncRNAs showed a half-life range from 30 min to 48 h, which is similar to mRNAs; however, a mean *t*½ of 4.8 versus 7.7 h for mRNAs suggests that lncRNAs possess a lower stability. A high percentage of lncRNAs was classified as unstable (*t*½ < 2 h), e.g., *Neat1*, and a few as highly stable (*t*½ > 12 h) [127]. Comparison of the stability of different lncRNA species revealed that intronic or promoter-associated lncRNAs are less stable than either intergenic, antisense, or 3′ UTR-associated lncRNAs. Single-exon transcripts, a class of nuclear-localized lncRNAs, are overrepresented among unstable transcripts. In human HeLa cells, the same approach of transcriptional inhibition was used and revealed that antisense lncRNAs are more stable than mRNAs (median $t_{1/2} = 3.9$ versus 3.2 h, respectively), whereas intronic lncRNAs included both stable ($t_{1/2} > 3$ h) and unstable ($t_{1/2} < 1$ h) transcripts with the $t_{1/2}$ median of 2.1 h [128]. Recently discovered circular RNAs are examples of highly stable lncRNAs with the median $t_{1/2}$ of 18.8–23.7 h and which is at least 2.5 times longer than their linear counterparts [129].

Nuclear and cytoplasmic exosomes, cytoplasmic Xrn1, and nonsense-mediated decay (NMD), as well as RNAi pathways, are known to control lncRNA abundance in the cell. Circular RNAs are intrinsically protected from any exonucleolytic- or polyadenylation-dependent decay pathways. Of note, actinomycin D treatment has a large impact on cells, and this can particularly influence lncRNA decay because of the very high sensitivity of lncRNAs to stress. Indeed, the measurements of t½ for single lncRNAs could significantly vary from one experiment to another, pointing to the necessity of multiple approaches including *de novo* RNA labeling to achieve more accurate and confident conclusions.

Multiple transcriptome profiling globally highlighted a highly specific spatio-temporal, lineage, tissue- and cell-type expression patterns for lncRNAs compared to PCGs; only a minority are ubiquitously present across all tissues or cell types, such as *TUG1* or *MALAT1* [105, 130, 131]. Curiously, the brain and testis represent a very rich source of uniquely expressed lncRNAs supporting the hypothesis that such transcripts are important for the acquisition of specific phenotypic traits [82, 130]. The ubiquitously expressed lncRNAs are often highly abundant, whereas specific lncRNAs present in one tissue or cell type tend to be expressed at low levels [132]. Moreover, interindividual expression analysis in normal human primary granulocytes revealed increased variability in lncRNA abundance compared to mRNAs [133]. Some disease-associated single-nucleotide polymorphisms (SNPs) within lncRNA genes and their promoters were linked to altered lncRNA expression, thus supporting their functional relevance in pathologies [134].

The high specificity of lncRNA expression argues in favor of important regulatory roles that these molecules can play in different biological contexts, including normal and pathological development.

1.5.5 Subcellular Localization of lncRNAs

Globally, unlike mRNAs, many lncRNAs have nuclear residence with focal or dispersed localization pattern (*NEAT1*) [135]. However, others were also found both in the nucleus and in the cytosol (*TUG1*, *HOTAIR*) or in the cytosol exclusively (*DANCR*) [105]. Multiple determinants, such as a specific RNA motif (*BORG*) [136] or RNA-protein assemblies, may dictate the subcellular localization of lncRNAs and define their function [137]. Remarkably, environmental changes or infection can induce lncRNA delocalization (or active trafficking) from one cellular compartment to another, as in the case of stress-induced lncRNAs [138]. HuR and GRSF1 modulate nuclear export and mitochondrial localization of the nuclear-encoded *RMRP* lncRNA [139].

1.5.6 Structure of lncRNAs

RNA is a highly flexible and dynamic molecule that adopts complex secondary structures. The folding of lncRNAs defines their cellular decay and functional versatility, enabling their nuclear localization, stability, and interaction with proteins [140]. A growing number of examples demonstrate that the RNA secondary structure constitutes the primary functional unit and evolutionary constraint bypassing poor interspecies lncRNA sequence conservation [141]. One such example is the lncRNA *HOTAIR* which exists only in mammals, sharing 58% of homology between human and mouse [142, 143]. Covariance analysis across 33 mammalian sequences of *HOTAIR* revealed a significant number of covariant base pairs and half-flips, which maintained a similar structure regardless of the changed sequence; this was especially true in regions surrounding proposed protein-binding segments of the lncRNA [144]. On the other hand, low sequence conservation that induces changes in structure can drive acquisition of new functions and specialization of the lncRNA-mediated regulatory circuit. This is the case of human accelerated region 1 (HAR1)-derived lncRNAs expressed in developing neocortex in primates where the capacity to form a stable cloverleaf-like structure has arisen only in humans [89, 145]. However, we are still far from understanding the function of this lncRNA in human brain development.

Numerous structure prediction tools, such as Rfold, have been developed to give guidance for further functional studies. Structural analysis of RNA has increased our understanding of mechanistic aspects of lncRNA action; however, X-ray crystallography, nuclear magnetic resonance (NMR), and cryo-electron microscopy require purified and stable, nearly static, molecules and are not adapted to highly dynamic and flexible RNA. Very recently, new technologies based on high-throughput sequencing have evolved enabling both an *in vitro* and *in vivo* view of RNA conformation [140].

1.6 Classification of lncRNAs

Advances in deep sequencing technologies gave rise to a plethora of novel transcripts requiring a universal standardized system for lncRNA classification and functional annotation. The state of lncRNA annotations is still at its beginning, and different classifications based on their length, transcript properties, location in respect to known genomic annotations, regulatory elements, and function have been proposed. Here, we review a non-exhaustive cataloguing of eukaryotic lncRNAs summarized in Table 1.2.

Table 1.2 Classification of lncRNAs (adapted from [279])

lncRNA category	Abbreviation	Examples	Refs
Classification according to lncRNA length			
Long non-coding RNA	lncRNA		[105, 118]
Large non-coding RNA			
Very long intergenic ncRNA	vlincRNA		[280]
Macro lncRNA		*Xist, Airn, Kcnq1ot1*	[148]
Classification according to lncRNA location with respect to PCGs			
Intergenic lncRNAs			
Long intergenic ncRNA	lincRNA	*MALAT1, NEAT1, GAS5, CYRANO*	[83, 258, 281, 282]
Large intervening ncRNA		*Frigidair, lincRNA-COX2, XACT*	[40, 150]
Long intervening ncRNA			[283]
Antisense lncRNAs			
Natural antisense transcripts	NAT	*ZEB2NAT*	
	cis-NAT	*BACE1-AS*	
	trans-NAT		
Antisense lncRNA	asRNA		
	ancRNA		
SINE B2 containing RNAs	SINEUP	*AS-Uchl1*	[160]
Bidirectional lncRNAs: Long upstream antisense transcript	LUAT		[122]
Upstream antisense RNA	uaRNA		[170]

Intronic lncRNAs			
Totally intronic RNA	TIN		[178]
Circular intronic RNA	ciRNA	*ci-ankrd52*	[175]
Circular RNA	circRNA	*CDR1as/ciRS-7, cANRIL*	[182, 184, 285, 286]
Exonic circRNA	ecircRNA		
Exon-intron circRNA	EIcircRNA		
Stable intronic sequence RNA	sisRNA	*sisR-1*	[287–289]
Switch RNAs			[177]
Overlapping sense transcripts	Sense ncRNA	*HLXB9-lncRNA, SOX2-OT*	[153, 181]
Classification according to lncRNA residence within specific DNA regulatory elements and loci			
Pseudogenes		*PTENP1, Lethe*	[193, 265, 290]
Telomeres and subtelomeres		*TERRA*	[203]
		subTERRA	[206–209, 291]
Centromeres		*centromeric alpha-satellite RNA*	[210–212, 214]
Transcripts from ultraconserved regions	T-UCR	*Uc.283+A*	[195, 196, 200, 292]
		Evf2	
rDNA loci		*PAPAS*	[215]
Promoter-associated ncRNAs	pancRNA	*CCND1-lncRNA*	[218, 223]
	PALRs		
Promoter upstream transcripts	PROMPT		[108]
Upstream antisense RNA	uaRNA[a]		[170]
Enhancer lncRNA	eRNA	*IL1β-eRNA, FOXC1e*	[216, 293]
3′-UTR-associated RNAs	uaRNA[a]		[224]
Classification according to lncRNA biogenesis pathway			
Stable unannotated transcript	SUT		[163]
Cryptic unstable transcripts	CUT	*PHO84 CUT, PROMPT, eRNAs*	[163, 225, 228], [219]
Cytoplasmically degraded CUTs	CD-CUT		[229]
Meiotic unannotated transcripts	MUT		[230, 231]
Nrd1-unterminated transcripts	NUT		[113]
Xrn1-sensitive unannotated transcripts	XUT	*RTL, XUT1678 (ARG1-AS)*	[94, 226, 227]
Classification according to lncRNA subcellular localization or origin			
Nuclear lncRNAs		*NEAT1*	[128]
GAA repeat-containing RNAs	GRC-RNA		[240]
Chromatin-enriched RNAs	cheRNA		[237]
Chromatin-associated RNAs	CAR		[238]
Mitochondrial ncRNAs	mtncRNA	*ASncmtRNA-2*	[243, 294]

(continued)

Table 1.2 (continued)

Classification according to lncRNA function			
Scaffold lncRNA		*HOTAIR, LINP1, NORAD*	[251, 252, 295]
Architectural lncRNAs	arcRNAs	*NEAT1*	[254]
Guide lncRNA		*MEG3, Khps1*	[256, 257]
Ribo-activator	ncRNA-a	*SRA, Lnc-DC, NeST*	[259, 260], [262, 263]
	eRNAs	*FOXC1e*	[217]
Ribo-repressor or decoy		*GAS5, CCND1-lncRNA, PANDA, Lethe*	[168, 223, 258, 290]
Competing endogenous RNA	ceRNAs	*PTENP1, HULC, CDR1as/ciRS-7*	[182, 265, 267]
lncRNA precursors	endo-si-lncRNA	*H19* (*miR-675*)	[110, 268, 270, 271]
	pi-lncRNA	*MALAT1* (*mascRNA*)	
	mi-lncRNA	*P5CDH-SRO5*	
Classification according to lncRNA association with specific biological processes			
Hypoxia induced	HINCUT		[197, 275]
Senescence-associated lncRNAs	SAL	*vlincRNAs, SALNR*	[147, 274]
Stress-induced lncRNAs	si-lncRNA		[138]
Non-annotated stem transcripts	NAST		[276]
Prostate cancer-associated transcripts	PCATs	*PCA3, PCAT1*	[277]

[a]In the literature, the term "uaRNA" has been attributed to two distinct groups of transcripts: upstream antisense RNAs and UTR-associated RNAs

1.6.1 Classification According to lncRNA Length

By convention, a length of 200 nt constitutes a bottom line for discrimination of long or large ncRNAs from small or short ncRNAs. However, lncRNAs vary significantly in size, and those that exceed the length of 10 kb belong to the groups of **very long intergenic (vlinc)RNAs** and **macro lncRNAs**. These transcripts possess some particular features that distinguish them from other lncRNAs: they are poorly or not spliced, weakly polyadenylated at 3′ end, and are produced by particular genomic loci. The majority of vlincRNAs are localized in close proximity or within PCG promoters on the same or opposite strand and function in *cis* as positive regulators of nearby gene transcription. Interestingly, some vlincRNA promoters harbor LTR sequences that are highly regulated by three major pluripotency-associated transcription factors, suggesting a possible role in early embryonic development [146]. Others are specifically induced by senescence and are required for the maintenance of senescent features that in turn control the transcriptional response to environmental changes [147]. Macro lncRNAs are often antisense to PCGs and are produced from imprinted clusters in a parent-of-origin-specific manner. Macro lncRNAs silence nearby imprinted genes either through their lncRNA product triggering epigenetic chromatin modifications or by a transcriptional interference mechanism [148].

1.6.2 Classification According to lncRNA Location with Respect to PCGs

This attribute is commonly used by the GENCODE/Ensembl portal in transcript biotype annotations, but also employed on an individual scale by consortia and laboratories for newly assembled lncRNA transcripts. Initially transcripts are classified as either intergenic or intragenic (Fig. 1.4). **Long or large intergenic non-coding (linc)RNAs** do not intersect with any protein-coding and ncRNA gene annotations. This category also includes the adopted GENCODE and homonymous biotype of long or large intervening ncRNAs that were originally defined by specific histone H3K4-K36 chromatin signatures within evolutionary conserved genomic loci [149, 150]. LincRNAs are usually shorter than PCGs, transcribed by RNA polymerase II, 5′ capped, 3′ polyadenylated, and spliced. Although several highly conserved lincRNAs exist, the majority possess modest sequence conservation comprising short, 5′-biased patches of conserved sequence nested in exons [81]. Highly conserved lincRNAs are believed to contribute to biological processes that are common to many lineages, such as embryonic development [77], while others are proposed to assure phenotypic and functional variations at individual and interspecies levels. Many, if not most, lincRNAs are localized in the nucleus where they exercise their regulatory functions. One such example is *lincRNA-p21* which is induced by p53 upon DNA damage [151]. *lincRNA-p21* physically associates with and recruits the nuclear factor hnRNP-K to specific promoters mediating p53-dependent transcriptional responses.

Intragenic lncRNAs overlap with PCG annotations and can be further classified into antisense, bidirectional, intronic, and overlapping sense lncRNAs.

Antisense lncRNAs, asRNAs or ancRNAs, were first discovered in single gene studies, but the recent development of stranded tiling and RNA-seq technologies has identified them as a common genome-wide feature of eukaryotic transcriptomes [152–154]. This group encompasses so-called natural antisense transcripts, **NATs**, which are in turn subdivided into ***cis*-NATs**, which affect the expression of the corresponding sense transcripts and into ***trans*-NATs**, which regulate expression of non-paired genes from other genomic locations [155–157]. A very recent

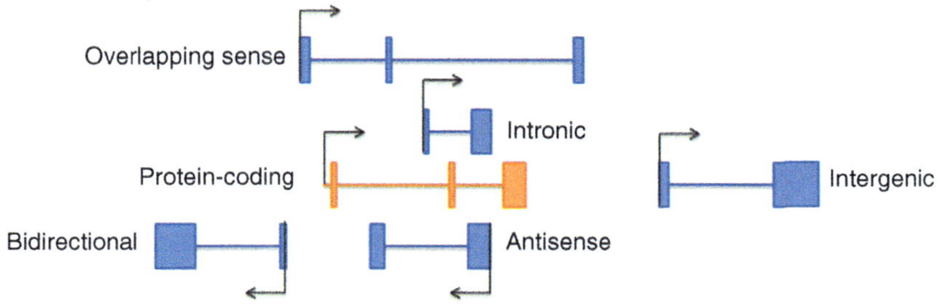

Fig. 1.4 Annotation of non-coding transcripts according to their genomic position relative to a protein-coding gene (*orange box*, protein-coding exon; *blue box*, non-coding exon)

study has pointed to a higher specificity of expression and an increased stability of asRNAs compared to lincRNAs and sense intragenic lncRNAs [128]. Due to sequence complementarity to sense-paired mRNAs or pre-mRNAs, asRNAs can act through RNA-RNA pairing, thereby ensuring specific targeting of the asRNA regulatory activity. This is the case of *BACE1-AS* that is highly expressed in Alzheimer's disease patients. It stabilizes the *BACE1* mRNA resulting in an increased expression of the *BACE1*-encoded beta-secretase and the accumulation of amyloid-beta peptides in the brain [158]. Antisense transcription across intron regions has been shown to regulate the local chromatin organization and environment, thus affecting co-transcriptional splicing of sense-paired pre-mRNAs [159]. Some NATs contain the inverted short interspersed nuclear element B2 (SINEB2), such as *AS-Uchl1* [160]. These NATs, called **SINEUPs**, are able to stimulate sense mRNA translation through lncRNA-mRNA pairing thanks to a complementary 5′ overlapping sequence to the paired-sense protein-coding gene. Recently, SINEUPs were proposed as a synthetic reagent for biotechnological applications and in therapy of haploinsufficiencies [161, 162]. In spite of the poor evolutionary conservation of sense-antisense transcription, some subgroups of lncRNAs, such as senescence-associated vlincRNAs and macro lncRNAs in mammals or XUTs in yeast, are mostly constituted of antisense transcripts, which suggests potential antisense-mediated regulatory pathways in control of cellular homeostasis, stress response, and disease [154].

The discovery of bidirectional transcription as an intrinsic feature of the eukaryotic transcriptional machinery has given rise to the identification of **bidirectional lncRNAs** [153, 163–166]. Originating from the opposite strand of a PCG strand, these transcripts do not overlap or only partially overlap with the 5′ region of paired PCGs, as is the case of promoter-associated (pa)ncRNAs, long upstream antisense transcripts (LUATs), and upstream antisense transcripts (uaRNA) [122, 167–170]. Presently, the number of bidirectional lncRNAs is largely underestimated not only because of the inaccurate annotation of transcriptional start sites (TSS) and promoters in the genome but also because of the highly unstable nature of these ncRNAs and the corresponding difficulty to detect them. Genomic studies have revealed that bidirectional promoters display distinct sequences and epigenetic features; moreover, they can be found near genes involved in specific biological processes such as developmental transcription factors or cell cycle regulation [122, 168, 169, 171, 172]. An imbalance in bidirectional transcription constitutes an endogenous fine-tuning mechanism that is particularly operative when facultative gene activation or repression is required [173, 174].

Intronic lncRNAs are restricted to PCG introns and could be either stand-alone unique transcripts or by-products of pre-mRNA processing. Examples of pre-mRNA-derived intronic transcripts are **circular intronic (ci)RNAs** produced from lariat introns which have escaped from debranching [175] and sno-lncRNAs produced from introns with two embedded snoRNA genes [176]. Such lncRNAs are proposed to positively regulate the transcription of the host PCG or its splicing by accumulating near the transcription locus. Another example of intronic lncRNAs of lariat origin, named **switch RNAs**, is produced by transcription through the immunoglobulin switch regions. They are folded into G-quadruplex structures to bind and recruit the activation-induced cytidine deaminase AID to DNA in a sequence-specific manner, thereby ensuring proper class switch recombination in the germ line [177]. Stand-alone intronic transcripts, expressed independently of the PCG hosts, are believed to be the most prevalent class of intronic lncRNAs, including so-called totally intronic ncRNAs, **TINs** [178, 179]. Expression of a certain TIN is activated during inflammation, but the exact function of these lncRNAs is still poorly understood [180].

Overlapping sense transcripts encompass exons or whole PCGs within their introns without any sense exon overlap and are transcribed in the same sense direction. This annotation includes the GENCODE-adopted homonymous biotype and has been attributed to a number of transcripts, denoted as "*GENENAME-OT*." One such example is *SOX2-OT* that harbors in its intron one of the major pluripotency regulators, the *SOX2* gene. *SOX2-OT* is dynamically expressed and is alternatively spliced not only during differentiation but also in cancer cells where it was proposed to regulate *SOX2* [181].

Intronic and overlapping sense lncRNAs could form **circular lncRNAs (circRNAs)** due to head-to-tail noncanonical splicing [182, 183]. Some sequence features such as the presence of repetitive elements within introns could be decisive for activation of noncanonical splicing and generation of a circular RNA molecule [184]. For example, Alu elements within introns are proposed to participate in RNA circularization via RNA-RNA pairing [185]. Remarkably, such events seem to be tissue or cell type specific, restricted to a certain developmental stage or pathological context [186, 187]. More generally, circRNAs function in the cytosol as miRNA sponges, as the case of CDR1as/ciRS-7 which is an RNA sponge of miR-7 [182, 183]. Some circRNAs, termed **exon-intron circRNAs (EIciRNAs)**, still contain unspliced introns and are retained in the nucleus, where they are able to interact with U1 snRNP and promote transcription of their parental genes [188]. The most remarkable property of circRNAs is their high stability which makes them eligible as potent diagnostic markers and therapeutic agents [189].

1.6.3 Classification According to lncRNA Residence Within Specific DNA Regulatory Elements and Loci

In addition to PCGs, mammalian genomes contain tens of thousands of pseudogenes, which are genomic remnants of ancient PCGs that have lost their coding potential throughout evolution. Importantly, many of them are transcribed in both sense and antisense directions into lncRNAs. Given high sequence similarity with parental genes, **pseudogene-derived lncRNAs** can regulate PCG expression *via* RNA-RNA pairing by acting as miRNA sponges, by producing endogenous siRNAs, or by interacting with mRNAs [190–192]. *PTENP1*, a lncRNA pseudogene derived from the tumor-suppressor gene PTEN, was among the first reported noncoding miRNA sponges with a function in cancer [193].

Ultra-conserved regions (UCRs) are genome segments that exhibit 100% DNA sequence conservation between human, mouse, and rat. The human genome contains 481 UCRs within intragenic (39%), intronic (43%), and exonic (15%) sequences [194]. These regions are extensively transcribed into ***T-UCR* lncRNAs** [195, 196]. Remarkably, expression of *T-UCRs* is induced by cancer-related stresses such as retinoid treatment or hypoxia. They are aberrantly expressed in different cancers and some are associated with poor prognosis [196–198]. Given high specificity of expression, *T-UCRs* were proposed as molecular markers for cancer diagnosis and prognosis [199]. The function of *T-UCRs* is still poorly understood. *Evf2* (or *Dlx6as*) is an example of *T-UCR* which acts as a decoy. It interacts with the transcription activator DLX1 increasing its association with key DNA enhancers but also with the SWI-/SNF-like chromatin remodeler brahma-related gene 1 (BRG1) inhibiting its ATPase activity. As a result, *Evf2* induces chromatin remodeling and Dlx5/Dlx6 enhancers decommissioning with a final repression of transcription [200, 201].

Telomeres, which are protective nucleoprotein structures at the ends of chromosomes, are transcribed into non-coding **telomeric repeat-containing RNAs, *TERRA***, in all eukaryotes. This family of transcripts is generated from both Watson and Crick strands in a cell cycle-dependent manner [202, 203]. Formation of RNA-DNA hybrids by *TERRA* at chromosome ends promotes recombination and, hence, delays senescence. However, in cells lacking telomerase- and homology-directed repair, *TERRA* expression induces telomere shortening and accelerates senescence [204, 205]. Subtelomeric regions are also actively transcribed [206–208]. In budding yeast, this heterogeneous population of lncRNAs, named *subTERRA*, is transiently accumulating in late G2/M and G1 phases of the cell cycle in wild-type cells or in asynchronous cells deleted for the Xrn1 exoribonuclease [209]. The exact function of *subTERRA* is not yet clear though it has been proposed to have a regulatory role in telomere homeostasis.

Recent findings in different eukaryotes including human revealed that centromeric repeats are actively transcribed into lncRNAs during the progression from late mitosis to early G1 [210–214]. These **centromeric lncRNAs** physically interact with different centromere-specific nucleoprotein components, such as CENP-A/CENP-C and HJURP, and are required for correct kinetochore assembly and the maintenance of centromere integrity.

Ribosomal (r)DNA loci were shown to be transcribed by RNA polymerase II, antisense to the rRNA genes, into a heterogeneous population of lncRNAs, called ***PAPAS*** (promoter and pre-rRNA antisense). Their expression is induced in quiescent cells and triggers the recruitment of histone H4K20 methyltransferase Suv4-20h2 to ribosomal RNA genes for histone modification and transcriptional silencing [215]. PAPAS also allow heterochromatin formation and gene silencing in growth-arrested cells.

Promoters and enhancers constitute fundamental *cis*-regulatory elements for the control of PCG expression, serving as platforms for the recruitment of transcription factors and transcription machinery and the establishment of particular chromatin organization. Remarkably, many, if not all, functional enhancers and promoters are pervasively transcribed, respectively, into **eRNAs** and PALRs, in both sense and antisense directions. Transcribed enhancer and promoter regions possess particular histone modification signatures that distinguish them from other transcription units. Such signatures include increased histone H3K27ac and H3K4me1 as compared with other lncRNA and PCGs. The termination of enhancer-derived lncRNAs, eRNAs, depends on the Integrator complex which ensures 3′-end transcript cleavage. The result is that eRNAs are poorly or not polyadenylated and highly unstable. Their expression is specific to cell type, tissue, or stages of development and can be activated by external or internal stimuli. Enhancer transcription was proposed to mark functional, active enhancer elements. However, eRNA function as stand-alone transcripts is still controversial, and the function of only few eRNAs, such as *FOXC1e* or *NRIP1e* [216], has been demonstrated. Specifically, it is proposed that these eRNAs control promoter chromatin environment, enhancer-promoter looping, RNA polymerase II loading and pausing, and "transcription factor trapping"; all these events contribute to a robust transcription activation of nearby and distant genes [217].

Promoter-associated lncRNAs or PALRs are transcribed in sense and antisense directions at promoter regions and can partially overlap the 5′ end of a gene [218]. This class of transcripts includes highly unstable **PROMPTs** (promoter upstream transcripts) and **upstream antisense RNAs** (uaRNAs) that are more easily detectable in a context where the nuclear exosome has been depleted [108, 170, 219]. Polyadenylation-dependent degradation of PROMPTs was proposed to ensure directional RNA production from otherwise bidirectional promoters [220]. The presence of a splicing competent intron within uaRNAs was shown to facilitate gene looping placing termination factors at the vicinity of a bidirectional promoter for termination and thereby ensuring RNA polymerase II directionality toward a PCG [221]. Some PALRs were shown to negatively regulate transcription of the nearby genes. One such example is a PALR from the *CCND1 gene* promoter which represses transcription by recruiting TLS and locally inhibiting CBP/p300 histone acetyltransferase activity on the downstream target gene, cyclin D1 [222, 223].

The 3′-untranslated regions (UTRs) of eukaryotic genes can be transcribed into independent transcription units or **UTR-associated (ua)RNAs** [224]. They are generated either by an independent transcriptional event from the upstream PCG or by posttranscriptional processing of a pre-mRNA. Expression of uaRNAs is regulated in a developmental stage- and tissue-specific fashion and is evolutionarily conserved; nevertheless, the functional relevance of such transcripts has not yet been explored.

1.6.4 Classification According to lncRNA Biogenesis Pathways

In budding yeast, since many lncRNAs are highly unstable or "cryptic," the commonly employed classification of lncRNAs is based on their decay or biogenesis features. However, some so-called stable unannotated transcripts (**SUTs**) were identified in a wild-type genetic background [163]. Others are only detectable under specific stress conditions or in RNA-decay mutant strains. These latter transcripts are roughly divided into three classes: **cryptic unstable transcripts (CUTs)**, which are sensitive to the nuclear RNA decay pathway [163, 225]; **Nrd1-unterminated lncRNAs (NUTs)** [113]; and **Xrn1-sensitive unstable transcripts (XUTs)**, which are degraded by the cytoplasmic 5′–3′ exoribonuclease, Xrn1 [226, 227]. The majority of XUTs are transcribed antisense to PCGs. CUTs are often bidirectional or overlapping sense transcripts, but can also be antisense, as is the case of the *PHO84* CUT [228]. Beyond each class definition, there is a considerable overlap between CUTs and NUTs but also XUTs and SUTs [94, 112]. Some CUTs have been reported to escape nuclear RNA decay and are exported to the cytoplasm where they are taken in charge by Xrn1 or by nonsense-mediated mRNA decay (NMD), as is the case of cytoplasmically degraded CUTs or CD-CUTs [229]. CD-CUTs bear a 5′ extension originating upstream from the *bona fide* promoter and which partially or completely overlaps PCGs. CD-CUT transcription is proposed to control the expression of a subset of genes from subtelomeric regions and, in particular, metal homeostasis genes. Another subclass of CUTs includes meiotically induced lncRNAs, **meiotic unannotated transcripts (MUTs)**, that are degraded by the nuclear exosome Rrp6 and the exosome targeting complex TRAMP [230, 231]. The key difference between CUTs, XUTs, and SUTs is determined by their distinct subcellular fates. CUTs are transcribed and degraded in the nucleus, while SUTs and XUTs are exported to the cytoplasm where many XUTs are degraded by Xrn1 unless they escape degradation by pairing to complementary mRNAs [94]. In this case, they could be protected from NMD-mediated degradation and eventually translated into peptides, giving rise to new putatively functional molecules [232]. Notably, CUTs and XUTs are conserved among yeast species [233], (Wery et al., unpublished).

In other eukaryotes, some highly unstable lncRNAs have been reported, for example, above mentioned PROMPTs and eRNAs which could be considered to be human analogues of CUTs, since they are highly stabilized upon RNA exosome depletion [108, 234]. The RNA exosome is proposed to play a role in resolving deleterious RNA/DNA hybrids (R-loops) arising from active enhancers to prevent recombination. So far, the existence of mammalian XUTs has not been reported; however, in humans, XRN1 was shown to be sequestrated by some RNA viruses [235, 236]. Their genomic RNA possesses a structured module in the 3′-UTR that traps and inhibits XRN1 catalytic activity. This action gives rise to the stabilization of the subgenomic flavivirus (sf)RNA which is important for the pathogenicity of the virus but could also result in a global stabilization of transcripts, including yet uncovered, highly unstable lncRNAs analogous to yeast XUTs.

1.6.5 Classification According to lncRNA Subcellular Localization or Origin

Knowing the subcellular localization of a particular lncRNA provides important insights into its biogenesis and function. LncRNAs could be exclusively cytosolic (*DANCR* and *OIP5-AS1*) or nuclear (*NEAT1*) or have a dual localization (*HOTAIR*)

[128]. Several subgroups of lncRNAs with a precise subcellular localization have been defined, such as **chromatin-enriched (che)RNAs** [237] and **chromatin-associated lncRNAs, CARs** [238]. Many nuclear and chromatin functions have been proposed for such lncRNAs, including the assembly of subnuclear domains or RNP complexes, the guiding of chromatin modifications, and the activation or repression of protein activity [239]. **GAA repeat-containing RNAs, GRC-RNAs**, represent a subclass of nuclear lncRNAs that show focal localization in the mammalian interphase nucleus, where they are a part of the nuclear matrix. They have been suggested to play a role in the organization of the nucleus by assembling various nuclear matrix-associated proteins [240].

The mitochondrial genome is also transcribed into **mitochondrial ncRNAs, ncmtRNAs** [241–243]. Their biogenesis is dependent on nuclear-encoded mitochondrial processing proteins. After synthesis, some ncmtRNAs are exported from the mitochondria to the nucleus [244]. Importantly, expression of ncmtRNAs is altered in cancers promoting them as potential targets for cancer therapy [245, 246].

1.6.6 Classification According to lncRNA Function

To highlight a regulatory role, lncRNAs are often classified based on their function. Several archetypal activities of lncRNAs are used for classification: scaffolds, guides, decoys or ribo-repressors, ribo-activators, sponges, and precursors of small ncRNAs. Here we present examples of functional lncRNA classifications that regroup several lncRNAs into subclasses with a common operating mode.

LncRNA scaffolds function in the assembly of RNP complexes. The structural plasticity of lncRNAs allows them to adopt complex and dynamic three-dimensional structures with high affinity for proteins [247]. LncRNA scaffolds are often actors of epigenetic and transcriptional control of gene expression regulation. In this case, a lncRNA can act in *trans* or in *cis* in respect to its transcription site [248]. They are known to associate with a multitude of histone- or DNA-modifying and nucleosome remodeling complexes [249, 250]. LncRNA-mediated assembly of these complexes reshapes the epigenetic landscape and the organization of chromatin domains, thus allowing the modulation of all DNA-based processes including transcription, recombination, DNA repair, as well as RNA processing [159, 177, 251, 252]. *HOTAIR* is one example of a scaffold lncRNA which recognizes numerous targets. *HOTAIR* adopts a four-module secondary structure [144] which interacts in the nucleus with the PRC2 and Lsd1/REST/coREST complexes through its 5′ and 3′ modules, respectively [253]; it then targets them to specific genomic locations to affect histone modifications and gene silencing. In the cytoplasm, *HOTAIR* associates with the E3 ubiquitin ligases, Dzip3 and Mex3b, facilitating ubiquitination and proteolysis of their respective substrates, Ataxin-1 and Snurportin-1, in senescent cells [251].

Architectural lncRNAs (arcRNAs) represent a subclass of lncRNA scaffolds that are essential for the assembly of particular nuclear substructures [254].

Presently, five lncRNAs are classified as arcRNAs, and among them is *NEAT1*, which assembles more than 60 different RNA-binding proteins and transcription factors in paraspeckles [255]. ArcRNAs are highly enriched in repetitive sequences indicative of complex RNA folding that is essential for their scaffold function. They could be temporarily regulated by stress, during development, or in disease. ArcRNAs often sequester regulatory proteins, thereby changing gene expression. A detailed molecular role of scaffold and arcRNAs will be discussed in the forthcoming chapter.

Guide lncRNAs can recruit RNP complexes to specific chromatin loci. Remarkably, a guide function of one and the same lncRNA depends on the biological context (cell-/tissue-type, developmental stage, pathology) and often cannot be explained by a simple RNA/DNA sequence complementarity. For some lncRNA guides the formation of a triple helix structure between DNA and the lncRNA was experimentally proven, as in the case of *Khps1* which anchors the CBP/p300 complex to the proto-oncogene *SPHK1* [256]. Another example is *MEG3* which guides the EZH2 subunit of PRC2 to TGFβ-regulated genes [257].

lncRNA decoys play the role of **ribo-repressors** for protein activities through the induction of allosteric modifications, the inhibition of catalytic activity, or by blocking the binding sites. One classical example of a ribo-repressor lncRNA is *GAS5* (growth arrest-specific 5), which acts as a decoy for a glucocorticoid receptor (GR) by mimicking its genomic DNA glucocorticoid response element (GRE). The interaction of *GAS5* with GR prevents it from binding to the GRE and ultimately represses GR-regulated genes, thus influencing many cellular functions including metabolism, cell survival, and response to apoptotic stimuli [258].

lncRNAs can also act as **ribo-activators** essential for or enhancing protein activities. One such example is the *lnc-DC* lncRNA which promotes the phosphorylation and activation of the STAT3 transcription factor [259]. Another subclass is the lncRNA transcriptional co-activators, also called **activating ncRNAs (ncRNA-a)**, which possess enhancer-like properties [260]. They were shown to interact with and regulate the kinase activity of Mediator, hence facilitating chromatin looping and transcription [261]. In addition to Mediator-interacting RNAs, other lncRNAs are able to upregulate transcription and could also be considered as ncRNA-a. Among them is the steroid receptor RNA activator *SRA* which interacts with and enhances the function of the insulator protein CTCF [262], and *NeST* which binds to and stimulates the activity of a subunit of the histone H3 Lysine 4 methyltransferase complex [263].

Competing endogenous RNAs (ceRNAs), also known as lncRNA sponges, are represented by lncRNAs and circRNAs that share partial sequence similarity to PCG transcripts; they function by competing for miRNA binding and posttranscriptional control [264]. Pseudogene-derived lncRNAs represent an important source of ceRNAs as they are particularly enriched in miRNA response elements, as is the case of the already mentioned *PTENP1* [265]. The subcellular balance between ceRNA, one or multiple miRNAs, and mRNA targets constitutes a complex network allowing a fine-tuning of the regulation of gene expression during adaptation, stress response, and development [266, 267].

Many lncRNAs host small RNA genes and serve as **precursor lncRNAs** for shorter regulatory RNAs, in particular, those involved in the RNAi pathway (mi/si/piRNAs). Many lncRNAs were identified and functionally studied before their precursor function was known. Such is the case for *H19*, one of the first discovered lncRNA genes and which contains two conserved microRNAs, miR-675-3p and miR-675-5p. In undifferentiated cells, *H19* acts as a ribo-activator interacting with and promoting the activity of the ssRNA-binding protein KSRP (K homology-type splicing regulatory protein) to prevent myogenic differentiation [268]. During development, and, in particular, during skeletal muscle differentiation, *H19* is processed into miRNAs ensuring the posttranscriptional control of the anti-differentiation transcription factors Smad [269]. Some piRNA clusters were found to map to lncRNA genes, mostly in exonic but also in non-exonic regions enriched in mobile elements thereby constituting putative *pi-lncRNA* precursors [270]. Putative endo-siRNAs can be produced from inverted repeats within lncRNA genes or from any double-stranded lncRNA-RNA precursors originated from sense-antisense convergent transcription [271, 272]. Endo-siRNAs have been documented in many eukaryotes, including fly, nematode, and mouse. Overlapping and bidirectional transcription is an abundant and conserved phenomenon among eukaryotes [154, 218]. However, in mammals, processing of sense-antisense paired transcripts into siRNA and their functional relevance is still controversial and requires experimental evidence, specifically at the single cell level. LncRNA processing into small RNA molecules could depend on different cellular machineries such as RNase P- and RNase Z-mediated cleavage of the small cytoplasmic mascRNA from *MALAT1* [110] or Drosha-DGCR8-driven termination and 3′ end formation for lnc-pri-miRNAs [111]. The possible coexistence of two operational modes combining a long, precursor lncRNA and a derived small RNA adds additional complexity in lncRNA-mediated regulatory circuits.

1.6.7 Classification According to lncRNA Association with Specific Biological Processes

Examination of the non-coding transcriptome in different biological contexts has resulted in the discovery of lncRNAs specifically associated with particular biological states or pathologies. LncRNAs differentially expressed during replicative senescence represent **senescence-associated lncRNAs, or SAL** [273]. One such example, *SALNR*, is able to delay oncogene-induced senescence by its interaction with and inhibition of the NF90 posttranscriptional repressor [274]. Hypoxia, one of the classic features of the tumor microenvironment, induces the expression of many lncRNAs, in particular those from UCRs, named **HINCUTs** [197, 275]. Oxidative stress induces the production of stress-induced lncRNAs, **si-lncRNAs**, that accumulate at polysomes in contrast to mRNAs, which are depleted [138]. Deep sequencing transcriptome analysis of mammalian stem cells identified **non-annotated stem transcripts, or NASTs**, that appear to be important for maintaining pluripotency

[276]. Finally, with the progression of clinical and diagnostic studies, a growing number of specific disease-associated lncRNAs have been detected. An example is the **prostate cancer-associated transcripts (PCATs)**, such as *PCAT1*, that were shown to have a role in cancer biology but also as potent prognostic markers [277].

1.6.8 Future Challenges in lncRNA Annotation and Classification

Presently, the discovery of a novel lncRNA is an everyday occurrence, and proper annotation and classification are a necessity. In addition to catchy nicknames, various classifications of lncRNAs that rely on certain properties of the transcript, its origin, or possible function are proposed in oral and written communications. However, and in the aim of universalization, a "gold standard" of annotation should be sought. Repositories such as RNAcentral and other consortia are working on the challenging task of integrating the unambiguous annotations of all transcripts and genes, including numerical identifiers in addition to unique transcript names such as "GENENAME". Recently, John Mattick and John Rinn have proposed some rules for lncRNA annotation. In particular, it has been recommended to refer to intergenic lncRNAs as "LINC-X," where X represents a number and to all intragenic lncRNAs as "GENENAME" corresponding to overlapping PCG annotations with a prefix "AS-" for antisense, "BI-" for bidirectional, "OT-" for overlapping sense, and "INT-" for intronic transcripts in order to provide them with a positional criterion [278]. Respecting this guideline, *OT-SOX2-(1)* would correspond to the first isoform of the *SOX2-OT1* lncRNA overlapping in sense orientation the *SOX2* gene, while *HOTAIR* should take the name of *AS-HOXC11-(1)* to designate the largest lncRNA antisense to the *HOXC11* gene. However, the descriptive nickname of experimentally assigned lncRNAs should be preserved on condition of its uniqueness as a gene name. To avoid confusion, the renaming of transcripts should be accurately marked in all lncRNA repositories.

Identification, annotation, and classification are the first steps toward unraveling lncRNA biology. This work is still in its early days and requires novel thinking and methodologies, in parallel with the development of new and more accurate technologies and improved tools for the discovery and assembly of transcripts. In particular, the twenty-first century has been marked by the emergence of new technologies in regard to genomics and integrative system biology. These new approaches will allow researchers to build a comprehensive framework of regulatory circuits embedding both coding and non-coding transcripts, thereby deciphering a bit more the puzzle of life biodiversity and complexity.

Acknowledgments We thank Edith Heard, Mike Schertzer, and members of the lab for attentively reading the manuscript and apologize to colleagues whose works are not discussed and cited due to space limitation.

References

1. Dahm R (2005) Friedrich Miescher and the discovery of DNA. Dev Biol 278:274–288. doi:10.1016/j.ydbio.2004.11.028
2. Avery OT, MacLeod CM, McCarty M (1944) Studies on the chemical nature of the substance inducing transformation of pneumococcal types induction of transformation by a desoxyribonucleic acid fraction isolated from Pneumococcus type III. J Exp Med 79:137–158. doi:10.1084/jem.79.2.137
3. Ochoa S (1980) A pursuit of a hobby. Annu Rev Biochem 49:1–31. doi:10.1146/annurev.bi.49.070180.000245
4. Griffiths AJF, Miller JH, Suzuki DT, et al. An Introduction to Genetic Analysis. 7th edition. New York: W. H. Freeman; 2000. Transcription and RNA polymerase. Available from: https://www.ncbi.nlm.nih.gov/books/NBK22085/
5. Cobb M (2015) Who discovered messenger RNA? Curr Biol 25:R526–R532. doi:10.1016/j.cub.2015.05.032
6. Crick FHC (1968) The origin of the genetic code. J Mol Biol 38:367–379. doi:10.1016/0022-2836(68)90392-6
7. Lewis JB, Atkins JF, Anderson CW et al (1975) Mapping of late adenovirus genes by cell-free translation of RNA selected by hybridization to specific DNA fragments. Proc Natl Acad Sci 72:1344–1348
8. Berk AJ (2016) Discovery of RNA splicing and genes in pieces. Proc Natl Acad Sci 113:801–805. doi:10.1073/pnas.1525084113
9. Weinberg RA, Penman S (1968) Small molecular weight monodisperse nuclear RNA. J Mol Biol 38:289–304. doi:10.1016/0022-2836(68)90387-2
10. Zieve G, Penman S (1976) Small RNA species of the HeLa cell: metabolism and subcellular localization. Cell 8:19–31. doi:10.1016/0092-8674(76)90181-1
11. Kruger K, Grabowski PJ, Zaug AJ et al (1982) Self-splicing RNA: autoexcision and autocyclization of the ribosomal RNA intervening sequence of tetrahymena. Cell 31:147–157. doi:10.1016/0092-8674(82)90414-7
12. Guerrier-Takada C, Gardiner K, Marsh T et al (1983) The RNA moiety of ribonuclease P is the catalytic subunit of the enzyme. Cell 35:849–857. doi:10.1016/0092-8674(83)90117-4
13. Cech TR (2000) structural biology: enhanced: the ribosome is a ribozyme. Science 289:878–879. doi:10.1126/science.289.5481.878
14. Butcher SE (2009) The spliceosome as ribozyme hypothesis takes a second step. Proc Natl Acad Sci U S A 106:12211–12212. doi:10.1073/pnas.0906762106
15. Bernhardt HS (2012) The RNA world hypothesis: the worst theory of the early evolution of life (except for all the others). Biol Direct 7:23. doi:10.1186/1745-6150-7-23
16. Inouye M, Delihast N (1988) Small RNAs in the prokaryotes: a growing list of diverse roles. Cell 53:5–7. doi:10.1016/0092-8674(88)90480-1
17. Corcoran CP, Podkaminski D, Papenfort K et al (2012) Superfolder GFP reporters validate diverse new mRNA targets of the classic porin regulator, MicF RNA: new MicF targets. Mol Microbiol 84:428–445. doi:10.1111/j.1365-2958.2012.08031.x
18. Delihas N (2015) Discovery and characterization of the first non-coding RNA that regulates gene expression, *micF* RNA: a historical perspective. World J Biol Chem 6:272. doi:10.4331/wjbc.v6.i4.272
19. Lee RC, Feinbaum RL, Ambros V (1993) The C. elegans heterochronic gene lin-4 encodes small RNAs with antisense complementarity to lin-14. Cell 75:843–854. doi:10.1016/0092-8674(93)90529-Y
20. Wightman B, Ha I, Ruvkun G (1993) Posttranscriptional regulation of the heterochronic gene lin-14 by lin-4 mediates temporal pattern formation in C. elegans. Cell 75:855–862. doi:10.1016/0092-8674(93)90530-4
21. Wassenegger M, Heimes S, Riedel L, Sänger HL (1994) RNA-directed de novo methylation of genomic sequences in plants. Cell 76:567–576. doi:10.1016/0092-8674(94)90119-8

22. Ameres SL, Zamore PD (2013) Diversifying microRNA sequence and function. Nat Rev Mol Cell Biol 14:475–488. doi:10.1038/nrm3611
23. He L, Hannon GJ (2004) MicroRNAs: small RNAs with a big role in gene regulation. Nat Rev Genet 5:522–531. doi:10.1038/nrg1379
24. Montgomery MK (2004) RNA interference. In: Gott JM (ed) RNA Interf Ed Modif. Humana, Totowa, pp 3–21
25. Castel SE, Martienssen RA (2013) RNA interference in the nucleus: roles for small RNAs in transcription, epigenetics and beyond. Nat Rev Genet 14:100–112. doi:10.1038/nrg3355
26. Pachnis V, Belayew A, Tilghman SM (1984) Locus unlinked to alpha-fetoprotein under the control of the murine raf and Rif genes. Proc Natl Acad Sci U S A 81:5523–5527
27. Bartolomei MS, Zemel S, Tilghman SM (1991) Parental imprinting of the mouse H19 gene. Nature 351:153–155. doi:10.1038/351153a0
28. Barlow DP, Stöger R, Herrmann BG et al (1991) The mouse insulin-like growth factor type-2 receptor is imprinted and closely linked to the Tme locus. Nature 349:84–87. doi:10.1038/349084a0
29. Brannan CI, Dees EC, Ingram RS, Tilghman SM (1990) The product of the H19 gene may function as an RNA. Mol Cell Biol 10:28–36. doi:10.1128/MCB.10.1.28
30. Lyon MF (1961) Gene action in the X-chromosome of the mouse (Mus musculus L.) Nature 190:372–373
31. Borsani G, Tonlorenzi R, Simmler MC et al (1991) Characterization of a murine gene expressed from the inactive X chromosome. Nature 351:325–329. doi:10.1038/351325a0
32. Brown CJ, Ballabio A, Rupert JL et al (1991) A gene from the region of the human X inactivation centre is expressed exclusively from the inactive X chromosome. Nature 349:38–44. doi:10.1038/349038a0
33. Brockdorff N, Ashworth A, Kay GF et al (1991) Conservation of position and exclusive expression of mouse Xist from the inactive X chromosome. Nature 351:329–331. doi:10.1038/351329a0
34. Gendrel A-V, Heard E (2014) Noncoding RNAs and epigenetic mechanisms during X-chromosome inactivation. Annu Rev Cell Dev Biol 30:561–580. doi:10.1146/annurev-cellbio-101512-122415
35. Brown CJ, Willard HF (1994) The human X-inactivation centre is not required for maintenance of X-chromosome inactivation. Nature 368:154–156. doi:10.1038/368154a0
36. Heard E, Mongelard F, Arnaud D et al (1999) Human XIST yeast artificial chromosome transgenes show partial X inactivation center function in mouse embryonic stem cells. Proc Natl Acad Sci 96:6841–6846. doi:10.1073/pnas.96.12.6841
37. Chureau C, Prissette M, Bourdet A et al (2002) Comparative sequence analysis of the X-inactivation center region in mouse, human, and bovine. Genome Res 12:894–908. doi:10.1101/gr.152902
38. Lee JT, Lu N (1999) Targeted mutagenesis of Tsix leads to nonrandom X inactivation. Cell 99:47–57. doi:10.1016/S0092-8674(00)80061-6
39. Migeon BR, Lee CH, Chowdhury AK, Carpenter H (2002) Species differences in TSIX/Tsix reveal the roles of these genes in X-chromosome inactivation. Am J Hum Genet 71:286–293. doi:10.1086/341605
40. Vallot C, Huret C, Lesecque Y et al (2013) XACT, a long noncoding transcript coating the active X chromosome in human pluripotent cells. Nat Genet 45:239–241. doi:10.1038/ng.2530
41. Orgel LE, Crick FHC (1980) Selfish DNA: the ultimate parasite. Nature 284:604–607. doi:10.1038/284604a0
42. Sanger F, Coulson AR, Friedmann T et al (1978) The nucleotide sequence of bacteriophage phiX174. J Mol Biol 125:225–246
43. Fleischmann RD, Adams MD, White O et al (1995) Whole-genome random sequencing and assembly of Haemophilus influenzae Rd. Science 269:496–512
44. Dunham I, Shimizu N, Roe BA et al (1999) The DNA sequence of human chromosome 22. Nature 402:489–495. doi:10.1038/990031
45. Lander ES, Linton LM, Birren B et al (2001) Initial sequencing and analysis of the human genome. Nature 409:860–921. doi:10.1038/35057062

46. Venter JC (2001) The sequence of the human genome. Science 291:1304–1351. doi:10.1126/science.1058040
47. Goffeau A, Barrell BG, Bussey H et al (1996) Life with 6000 genes. Science 274:546, 563–546, 567
48. Crollius HR (2000) Characterization and repeat analysis of the compact genome of the freshwater pufferfish Tetraodon nigroviridis. Genome Res 10:939–949. doi:10.1101/gr.10.7.939
49. Waterston R, Sulston J (1995) The genome of Caenorhabditis elegans. Proc Natl Acad Sci U S A 92:10836–10840
50. Adams MD, Celniker SE, Holt RA et al (2000) The genome sequence of Drosophila melanogaster. Science 287:2185–2195
51. Chinwalla AT, Cook LL, Delehaunty KD et al (2002) Initial sequencing and comparative analysis of the mouse genome. Nature 420:520–562. doi:10.1038/nature01262
52. Antequera F, Bird A (1993) Number of CpG islands and genes in human and mouse. Proc Natl Acad Sci U S A 90:11995–11999
53. International Human Genome Sequencing Consortium (2004) Finishing the euchromatic sequence of the human genome. Nature 431:931–945. doi:10.1038/nature03001
54. Kapranov P (2002) Large-scale transcriptional activity in chromosomes 21 and 22. Science 296:916–919. doi:10.1126/science.1068597
55. The FANTOM Consortium (2005) The transcriptional landscape of the mammalian genome. Science 309:1559–1563. doi:10.1126/science.1112014
56. Okazaki Y, Furuno M, Kasukawa T et al (2002) Analysis of the mouse transcriptome based on functional annotation of 60,770 full-length cDNAs. Nature 420:563–573. doi:10.1038/nature01266
57. Katayama S, Tomaru Y, Kasukawa T et al (2005) Antisense transcription in the mammalian transcriptome. Science 309:1564–1566. doi:10.1126/science.1112009
58. ENCODE Project Consortium, Birney E, Stamatoyannopoulos JA et al (2007) Identification and analysis of functional elements in 1% of the human genome by the ENCODE pilot project. Nature 447:799–816. doi:10.1038/nature05874
59. ENCODE Project Consortium (2012) An integrated encyclopedia of DNA elements in the human genome. Nature 489:57–74. doi:10.1038/nature11247
60. Mouse ENCODE Consortium, Stamatoyannopoulos JA, Snyder M et al (2012) An encyclopedia of mouse DNA elements (Mouse ENCODE). Genome Biol 13:418. doi:10.1186/gb-2012-13-8-418
61. Yue F, Cheng Y, Breschi A et al (2014) A comparative encyclopedia of DNA elements in the mouse genome. Nature 515:355–364. doi:10.1038/nature13992
62. David L, Huber W, Granovskaia M et al (2006) A high-resolution map of transcription in the yeast genome. Proc Natl Acad Sci 103:5320–5325. doi:10.1073/pnas.0601091103
63. Dinger ME, Amaral PP, Mercer TR, Mattick JS (2009) Pervasive transcription of the eukaryotic genome: functional indices and conceptual implications. Brief Funct Genomic Proteomic 8:407–423. doi:10.1093/bfgp/elp038
64. Berretta J, Morillon A (2009) Pervasive transcription constitutes a new level of eukaryotic genome regulation. EMBO Rep 10:973–982. doi:10.1038/embor.2009.181
65. Mattick JS (2003) Challenging the dogma: the hidden layer of non-protein-coding RNAs in complex organisms. Bioessays 25:930–939. doi:10.1002/bies.10332
66. Clark MB, Choudhary A, Smith MA et al (2013) The dark matter rises: the expanding world of regulatory RNAs. Essays Biochem 54:1–16. doi:10.1042/bse0540001
67. Marques AC, Ponting CP (2014) Intergenic lncRNAs and the evolution of gene expression. Curr Opin Genet Dev 27:48–53. doi:10.1016/j.gde.2014.03.009
68. Duret L (2006) The Xist RNA gene evolved in Eutherians by pseudogenization of a protein-coding gene. Science 312:1653–1655. doi:10.1126/science.1126316
69. Ganesh S, Svoboda P (2016) Retrotransposon-associated long non-coding RNAs in mice and men. Pflüg Arch - Eur J Physiol 468:1049–1060. doi:10.1007/s00424-016-1818-5
70. Kapusta A, Kronenberg Z, Lynch VJ et al (2013) Transposable elements are major contributors to the origin, diversification, and regulation of vertebrate long noncoding RNAs. PLoS Genet 9:e1003470. doi:10.1371/journal.pgen.1003470

71. Johnson R, Guigo R (2014) The RIDL hypothesis: transposable elements as functional domains of long noncoding RNAs. RNA 20:959–976. doi:10.1261/rna.044560.114
72. Hacisuleyman E, Shukla CJ, Weiner CL, Rinn JL (2016) Function and evolution of local repeats in the Fire locus. Nat Commun 7:11021. doi:10.1038/ncomms11021
73. Heinen TJAJ, Staubach F, Häming D, Tautz D (2009) Emergence of a new gene from an intergenic region. Curr Biol 19:1527–1531. doi:10.1016/j.cub.2009.07.049
74. D-D W, Irwin DM, Zhang Y-P (2011) De novo origin of human protein-coding genes. PLoS Genet 7:e1002379. doi:10.1371/journal.pgen.1002379
75. Durruthy-Durruthy J, Sebastiano V, Wossidlo M et al (2015) The primate-specific noncoding RNA HPAT5 regulates pluripotency during human preimplantation development and nuclear reprogramming. Nat Genet 48:44–52. doi:10.1038/ng.3449
76. Rands CM, Meader S, Ponting CP, Lunter G (2014) 8.2% of the Human genome is constrained: variation in rates of turnover across functional element classes in the human lineage. PLoS Genet 10:e1004525. doi:10.1371/journal.pgen.1004525
77. Necsulea A, Soumillon M, Warnefors M et al (2014) The evolution of lncRNA repertoires and expression patterns in tetrapods. Nature 505:635–640. doi:10.1038/nature12943
78. Young RS, Ponting CP (2013) Identification and function of long non-coding RNAs. Essays Biochem 54:113–126. doi:10.1042/bse0540113
79. Ponting CP, Oliver PL, Reik W (2009) Evolution and functions of long noncoding RNAs. Cell 136:629–641. doi:10.1016/j.cell.2009.02.006
80. Diederichs S (2014) The four dimensions of noncoding RNA conservation. Trends Genet 30:121–123. doi:10.1016/j.tig.2014.01.004
81. Hezroni H, Koppstein D, Schwartz MG et al (2015) Principles of long noncoding RNA evolution derived from direct comparison of transcriptomes in 17 species. Cell Rep 11:1110–1122. doi:10.1016/j.celrep.2015.04.023
82. Washietl S, Kellis M, Garber M (2014) Evolutionary dynamics and tissue specificity of human long noncoding RNAs in six mammals. Genome Res 24:616–628. doi:10.1101/gr.165035.113
83. Ulitsky I, Shkumatava A, Jan CH et al (2011) Conserved function of lincRNAs in vertebrate embryonic development despite rapid sequence evolution. Cell 147:1537–1550. doi:10.1016/j.cell.2011.11.055
84. Willingham AT, Gingeras TR (2006) TUF Love for "Junk" DNA. Cell 125:1215–1220. doi:10.1016/j.cell.2006.06.009
85. Mattick JS, Gagen MJ (2001) The evolution of controlled multitasked gene networks: the role of introns and other noncoding RNAs in the development of complex organisms. Mol Biol Evol 18:1611–1630
86. Mattick JS (2001) Non-coding RNAs: the architects of eukaryotic complexity. EMBO Rep 2:986–991. doi:10.1093/embo-reports/kve230
87. Pollard KS, Salama SR, King B et al (2006) Forces shaping the fastest evolving regions in the human genome. PLoS Genet 2:e168. doi:10.1371/journal.pgen.0020168
88. Bird CP, Stranger BE, Liu M et al (2007) Fast-evolving noncoding sequences in the human genome. Genome Biol 8:R118. doi:10.1186/gb-2007-8-6-r118
89. Pollard KS, Salama SR, Lambert N et al (2006) An RNA gene expressed during cortical development evolved rapidly in humans. Nature 443:167–172. doi:10.1038/nature05113
90. Bae B-I, Tietjen I, Atabay KD et al (2014) Evolutionarily dynamic alternative splicing of GPR56 regulates regional cerebral cortical patterning. Science 343:764–768. doi:10.1126/science.1244392
91. Doan RN, Bae B-I, Cubelos B et al (2016) Mutations in human accelerated regions disrupt cognition and social behavior. Cell 167:341–354.e12. doi:10.1016/j.cell.2016.08.071
92. van Heesch S, van Iterson M, Jacobi J et al (2014) Extensive localization of long noncoding RNAs to the cytosol and mono- and polyribosomal complexes. Genome Biol 15:R6. doi:10.1186/gb-2014-15-1-r6
93. Juna Carlevaro-Fita, Anisa Rahim, Roderic Guigo, Leah Vardy, Rory Johnson (2015) Widespread localisation of long noncoding RNAs to ribosomes: Distinguishing features and evidence for regulatory roles. bioRxiv 013508; doi: https://doi.org/10.1101/013508

94. Wery M, Descrimes M, Vogt N et al (2016) Nonsense-mediated decay restricts LncRNA levels in yeast unless blocked by double-stranded RNA structure. Mol Cell 61:379–392. doi:10.1016/j.molcel.2015.12.020
95. Housman G, Ulitsky I (2016) Methods for distinguishing between protein-coding and long noncoding RNAs and the elusive biological purpose of translation of long noncoding RNAs. Biochim Biophys Acta 1859:31–40. doi:10.1016/j.bbagrm.2015.07.017
96. Andrews SJ, Rothnagel JA (2014) Emerging evidence for functional peptides encoded by short open reading frames. Nat Rev Genet 15:193–204. doi:10.1038/nrg3520
97. Banfai B, Jia H, Khatun J et al (2012) Long noncoding RNAs are rarely translated in two human cell lines. Genome Res 22:1646–1657. doi:10.1101/gr.134767.111
98. Ruiz-Orera J, Messeguer X, Subirana JA, Alba MM (2014) Long non-coding RNAs as a source of new peptides. Elife. doi:10.7554/eLife.03523
99. Ji Z, Song R, Regev A, Struhl K (2015) Many lncRNAs, 5′UTRs, and pseudogenes are translated and some are likely to express functional proteins. Elife. doi:10.7554/eLife.08890
100. Nelson BR, Makarewich CA, Anderson DM et al (2016) A peptide encoded by a transcript annotated as long noncoding RNA enhances SERCA activity in muscle. Science 351:271–275. doi:10.1126/science.aad4076
101. Espinoza CA, Goodrich JA, Kugel JF (2007) Characterization of the structure, function, and mechanism of B2 RNA, an ncRNA repressor of RNA polymerase II transcription. RNA 13:583–596. doi:10.1261/rna.310307
102. Massone S, Ciarlo E, Vella S et al (2012) NDM29, a RNA polymerase III-dependent non coding RNA, promotes amyloidogenic processing of APP and amyloid β secretion. Biochim Biophys Acta Res 1823:1170–1177. doi:10.1016/j.bbamcr.2012.05.001
103. Ariel F, Romero-Barrios N, Jégu T et al (2015) Battles and hijacks: noncoding transcription in plants. Trends Plant Sci 20:362–371. doi:10.1016/j.tplants.2015.03.003
104. Yang L, Duff MO, Graveley BR et al (2011) Genomewide characterization of non-polyadenylated RNAs. Genome Biol 12:R16. doi:10.1186/gb-2011-12-2-r16
105. Djebali S, Davis CA, Merkel A et al (2012) Landscape of transcription in human cells. Nature 489:101–108. doi:10.1038/nature11233
106. Wilusz JE, JnBaptiste CK, LY L et al (2012) A triple helix stabilizes the 3′ ends of long noncoding RNAs that lack poly(A) tails. Genes Dev 26:2392–2407. doi:10.1101/gad.204438.112
107. Alam T, Medvedeva YA, Jia H et al (2014) Promoter analysis reveals globally differential regulation of human long non-coding RNA and protein-coding genes. PLoS One 9:e109443. doi:10.1371/journal.pone.0109443
108. Preker P, Almvig K, Christensen MS et al (2011) PROMoter uPstream transcripts share characteristics with mRNAs and are produced upstream of all three major types of mammalian promoters. Nucleic Acids Res 39:7179–7193. doi:10.1093/nar/gkr370
109. Lai F, Gardini A, Zhang A, Shiekhattar R (2015) Integrator mediates the biogenesis of enhancer RNAs. Nature 525:399–403. doi:10.1038/nature14906
110. Wilusz JE, Freier SM, Spector DL (2008) 3′ End processing of a long nuclear-retained noncoding RNA yields a tRNA-like cytoplasmic RNA. Cell 135:919–932. doi:10.1016/j.cell.2008.10.012
111. Dhir A, Dhir S, Proudfoot NJ, Jopling CL (2015) Microprocessor mediates transcriptional termination of long noncoding RNA transcripts hosting microRNAs. Nat Struct Mol Biol 22:319–327. doi:10.1038/nsmb.2982
112. Fox MJ, Gao H, Smith-Kinnaman WR et al (2015) The Exosome component Rrp6 is required for RNA Polymerase II termination at specific targets of the Nrd1-Nab3 pathway. PLoS Genet 11:e1004999. doi:10.1371/journal.pgen.1004999
113. Schulz D, Schwalb B, Kiesel A et al (2013) Transcriptome surveillance by selective termination of noncoding RNA synthesis. Cell 155:1075–1087. doi:10.1016/j.cell.2013.10.024
114. Porrua O, Libri D (2015) Transcription termination and the control of the transcriptome: why, where and how to stop. Nat Rev Mol Cell Biol. doi:10.1038/nrm3943
115. Spurlock CF, Tossberg JT, Guo Y et al (2015) Expression and functions of long noncoding RNAs during human T helper cell differentiation. Nat Commun 6:6932. doi:10.1038/ncomms7932

116. Hoffmann M, Dehn J, Droop J et al (2015) Truncated isoforms of lncRNA ANRIL are overexpressed in bladder cancer, but do not contribute to repression of INK4 tumor suppressors. Non-Coding RNA 1:266–284. doi:10.3390/ncrna1030266
117. Meseure D, Vacher S, Lallemand F et al (2016) Prognostic value of a newly identified MALAT1 alternatively spliced transcript in breast cancer. Br J Cancer 114:1395–1404. doi:10.1038/bjc.2016.123
118. Derrien T, Johnson R, Bussotti G et al (2012) The GENCODE v7 catalog of human long noncoding RNAs: analysis of their gene structure, evolution, and expression. Genome Res 22:1775–1789. doi:10.1101/gr.132159.111
119. Bogu GK, Vizán P, Stanton LW et al (2016) Chromatin and RNA maps reveal regulatory long noncoding RNAs in mouse. Mol Cell Biol 36:809–819. doi:10.1128/MCB.00955-15
120. Marques AC, Hughes J, Graham B et al (2013) Chromatin signatures at transcriptional start sites separate two equally populated yet distinct classes of intergenic long noncoding RNAs. Genome Biol 14:R131. doi:10.1186/gb-2013-14-11-r131
121. Murray SC, Haenni S, Howe FS et al (2015) Sense and antisense transcription are associated with distinct chromatin architectures across genes. Nucleic Acids Res 43:7823–7837. doi:10.1093/nar/gkv666
122. Lepoivre C, Belhocine M, Bergon A et al (2013) Divergent transcription is associated with promoters of transcriptional regulators. BMC Genomics 14:914. doi:10.1186/1471-2164-14-914
123. Kim DH, Marinov GK, Pepke S et al (2015) Single-cell transcriptome analysis reveals dynamic changes in lncRNA expression during reprogramming. Cell Stem Cell 16:88–101. doi:10.1016/j.stem.2014.11.005
124. Liu SJ, Nowakowski TJ, Pollen AA et al (2016) Single-cell analysis of long non-coding RNAs in the developing human neocortex. Genome Biol. doi:10.1186/s13059-016-0932-1
125. Ma Q, Chang HY (2016) Single-cell profiling of lncRNAs in the developing human brain. Genome Biol. doi:10.1186/s13059-016-0933-0
126. Rotem A, Ram O, Shoresh N et al (2015) Single-cell ChIP-seq reveals cell subpopulations defined by chromatin state. Nat Biotechnol 33:1165–1172. doi:10.1038/nbt.3383
127. Clark MB, Johnston RL, Inostroza-Ponta M et al (2012) Genome-wide analysis of long noncoding RNA stability. Genome Res 22:885–898. doi:10.1101/gr.131037.111
128. Ayupe AC, Tahira AC, Camargo L et al (2015) Global analysis of biogenesis, stability and sub-cellular localization of lncRNAs mapping to intragenic regions of the human genome. RNA Biol 12:877–892. doi:10.1080/15476286.2015.1062960
129. Enuka Y, Lauriola M, Feldman ME et al (2016) Circular RNAs are long-lived and display only minimal early alterations in response to a growth factor. Nucleic Acids Res 44:1370–1383. doi:10.1093/nar/gkv1367
130. Ward M, McEwan C, Mills JD, Janitz M (2015) Conservation and tissue-specific transcription patterns of long noncoding RNAs. J Hum Transcr 1:2–9. doi:10.3109/23324015.2015.1077591
131. Li F, Xiao Y, Huang F et al (2015) Spatiotemporal-specific lncRNAs in the brain, colon, liver and lung of macaque during development. Mol Biosyst 11:3253–3263. doi:10.1039/C5MB00474H
132. Jiang L, Zhao L (2016) Identifying and functionally characterizing tissue-specific and ubiquitously expressed human lncRNAs. Oncotarget. doi:10.18632/oncotarget.6859
133. Kornienko AE, Dotter CP, Guenzl PM et al (2016) Long non-coding RNAs display higher natural expression variation than protein-coding genes in healthy humans. Genome Biol. doi:10.1186/s13059-016-0873-8
134. Kumar V, Westra H-J, Karjalainen J et al (2013) Human disease-associated genetic variation impacts large intergenic non-coding RNA expression. PLoS Genet 9:e1003201. doi:10.1371/journal.pgen.1003201
135. Cabili MN, Dunagin MC, McClanahan PD et al (2015) Localization and abundance analysis of human lncRNAs at single-cell and single-molecule resolution. Genome Biol 16:20. doi:10.1186/s13059-015-0586-4

136. Zhang B, Gunawardane L, Niazi F et al (2014) A novel RNA motif mediates the strict nuclear localization of a long noncoding RNA. Mol Cell Biol 34:2318–2329. doi:10.1128/MCB.01673-13
137. Chen L-L (2016) Linking long noncoding RNA localization and function. Trends Biochem Sci 41:761–772. doi:10.1016/j.tibs.2016.07.003
138. Giannakakis A, Zhang J, Jenjaroenpun P et al (2015) Contrasting expression patterns of coding and noncoding parts of the human genome upon oxidative stress. Sci Rep 5:9737. doi:10.1038/srep09737
139. Noh JH, Kim KM, Abdelmohsen K et al (2016) HuR and GRSF1 modulate the nuclear export and mitochondrial localization of the lncRNA *RMRP*. Genes Dev. doi:10.1101/gad.276022.115
140. Lu Z, Chang HY (2016) Decoding the RNA structurome. Curr Opin Struct Biol 36:142–148. doi:10.1016/j.sbi.2016.01.007
141. Johnsson P, Lipovich L, Grandér D, Morris KV (2014) Evolutionary conservation of long non-coding RNAs; sequence, structure, function. Biochim Biophys Acta 1840:1063–1071. doi:10.1016/j.bbagen.2013.10.035
142. He S, Liu S, Zhu H (2011) The sequence, structure and evolutionary features of HOTAIR in mammals. BMC Evol Biol 11:102. doi:10.1186/1471-2148-11-102
143. Bhan A, Mandal SS (2015) LncRNA HOTAIR: a master regulator of chromatin dynamics and cancer. Biochim Biophys Acta 1856:151–164. doi:10.1016/j.bbcan.2015.07.001
144. Somarowthu S, Legiewicz M, Chillón I et al (2015) HOTAIR forms an intricate and modular secondary structure. Mol Cell 58:353–361. doi:10.1016/j.molcel.2015.03.006
145. Beniaminov A, Westhof E, Krol A (2008) Distinctive structures between chimpanzee and humanin a brain noncoding RNA. RNA 14:1270–1275. doi:10.1261/rna.1054608
146. St Laurent G, Shtokalo D, Dong B et al (2013) VlincRNAs controlled by retroviral elements are a hallmark of pluripotency and cancer. Genome Biol 14:R73. doi:10.1186/gb-2013-14-7-r73
147. Lazorthes S, Vallot C, Briois S et al (2015) A vlincRNA participates in senescence maintenance by relieving H2AZ-mediated repression at the INK4 locus. Nat Commun 6:5971. doi:10.1038/ncomms6971
148. Guenzl PM, Barlow DP (2012) Macro lncRNAs: a new layer of *cis* -regulatory information in the mammalian genome. RNA Biol 9:731–741. doi:10.4161/rna.19985
149. Khalil AM, Guttman M, Huarte M et al (2009) Many human large intergenic noncoding RNAs associate with chromatin-modifying complexes and affect gene expression. Proc Natl Acad Sci 106:11667–11672. doi:10.1073/pnas.0904715106
150. Guttman M, Amit I, Garber M et al (2009) Chromatin signature reveals over a thousand highly conserved large non-coding RNAs in mammals. Nature 458:223–227. doi:10.1038/nature07672
151. Huarte M, Guttman M, Feldser D et al (2010) A large intergenic noncoding RNA induced by p53 mediates global gene repression in the p53 response. Cell 142:409–419. doi:10.1016/j.cell.2010.06.040
152. Goodman AJ, Daugharthy ER, Kim J (2013) Pervasive antisense transcription is evolutionarily conserved in budding yeast. Mol Biol Evol 30:409–421. doi:10.1093/molbev/mss240
153. Kapranov P (2005) Examples of the complex architecture of the human transcriptome revealed by RACE and high-density tiling arrays. Genome Res 15:987–997. doi:10.1101/gr.3455305
154. Wood EJ, Chin-Inmanu K, Jia H, Lipovich L (2013) Sense-antisense gene pairs: sequence, transcription, and structure are not conserved between human and mouse. Front Genet. doi:10.3389/fgene.2013.00183
155. Magistri M, Faghihi MA, St Laurent G, Wahlestedt C (2012) Regulation of chromatin structure by long noncoding RNAs: focus on natural antisense transcripts. Trends Genet 28:389–396. doi:10.1016/j.tig.2012.03.013

156. W-Y S, Xiong H, Fang J-Y (2010) Natural antisense transcripts regulate gene expression in an epigenetic manner. Biochem Biophys Res Commun 396:177–181. doi:10.1016/j.bbrc.2010.04.147
157. Yuan C, Wang J, Harrison AP et al (2015) Genome-wide view of natural antisense transcripts in Arabidopsis thaliana. DNA Res 22:233–243. doi:10.1093/dnares/dsv008
158. Faghihi MA, Modarresi F, Khalil AM et al (2008) Expression of a noncoding RNA is elevated in Alzheimer's disease and drives rapid feed-forward regulation of β-secretase. Nat Med 14:723–730. doi:10.1038/nm1784
159. Gonzalez I, Munita R, Agirre E et al (2015) A lncRNA regulates alternative splicing via establishment of a splicing-specific chromatin signature. Nat Struct Mol Biol. doi:10.1038/nsmb.3005
160. Carrieri C, Cimatti L, Biagioli M et al (2012) Long non-coding antisense RNA controls Uchl1 translation through an embedded SINEB2 repeat. Nature 491:454–457. doi:10.1038/nature11508
161. Zucchelli S, Fasolo F, Russo R et al (2015) SINEUPs are modular antisense long non-coding RNAs that increase synthesis of target proteins in cells. Front Cell Neurosci. doi:10.3389/fncel.2015.00174
162. Indrieri A, Grimaldi C, Zucchelli S et al (2016) Synthetic long non-coding RNAs [SINEUPs] rescue defective gene expression in vivo. Sci Rep 6:27315. doi:10.1038/srep27315
163. Xu Z, Wei W, Gagneur J et al (2009) Bidirectional promoters generate pervasive transcription in yeast. Nature 457:1033–1037. doi:10.1038/nature07728
164. Scruggs BS, Gilchrist DA, Nechaev S et al (2015) Bidirectional transcription arises from two distinct hubs of transcription factor binding and active chromatin. Mol Cell 58:1101–1112. doi:10.1016/j.molcel.2015.04.006
165. Wei W, Pelechano V, Järvelin AI, Steinmetz LM (2011) Functional consequences of bidirectional promoters. Trends Genet 27:267–276. doi:10.1016/j.tig.2011.04.002
166. Seila AC, Calabrese JM, Levine SS et al (2008) Divergent transcription from active promoters. Science 322:1849–1851. doi:10.1126/science.1162253
167. Hamazaki N, Uesaka M, Nakashima K et al (2015) Gene activation-associated long noncoding RNAs function in mouse preimplantation development. Development 142:910–920. doi:10.1242/dev.116996
168. Hung T, Wang Y, Lin MF et al (2011) Extensive and coordinated transcription of noncoding RNAs within cell-cycle promoters. Nat Genet 43:621–629. doi:10.1038/ng.848
169. Uesaka M, Nishimura O, Go Y et al (2014) Bidirectional promoters are the major source of gene activation-associated non-coding RNAs in mammals. BMC Genomics 15:35. doi:10.1186/1471-2164-15-35
170. Flynn RA, Almada AE, Zamudio JR, Sharp PA (2011) Antisense RNA polymerase II divergent transcripts are P-TEFb dependent and substrates for the RNA exosome. Proc Natl Acad Sci 108:10460–10465. doi:10.1073/pnas.1106630108
171. Hu H, He L, Khaitovich P (2014) Deep sequencing reveals a novel class of bidirectional promoters associated with neuronal genes. BMC Genomics 15:457. doi:10.1186/1471-2164-15-457
172. Sigova AA, Mullen AC, Molinie B et al (2013) Divergent transcription of long noncoding RNA/mRNA gene pairs in embryonic stem cells. Proc Natl Acad Sci 110:2876–2881. doi:10.1073/pnas.1221904110
173. Morris KV, Santoso S, Turner A-M et al (2008) Bidirectional transcription directs both transcriptional gene activation and suppression in human cells. PLoS Genet 4:e1000258. doi:10.1371/journal.pgen.1000258
174. Kambara H, Gunawardane L, Zebrowski E et al (2015) Regulation of interferon-stimulated gene BST2 by a lncRNA transcribed from a shared bidirectional promoter. Front Immunol. doi:10.3389/fimmu.2014.00676
175. Zhang Y, Zhang X-O, Chen T et al (2013) Circular intronic long noncoding RNAs. Mol Cell 51:792–806. doi:10.1016/j.molcel.2013.08.017
176. Yin Q-F, Yang L, Zhang Y et al (2012) Long noncoding RNAs with snoRNA ends. Mol Cell 48:219–230. doi:10.1016/j.molcel.2012.07.033

177. Zheng S, Vuong BQ, Vaidyanathan B et al (2015) Non-coding RNA generated following Lariat Debranching mediates targeting of AID to DNA. Cell 161:762–773. doi:10.1016/j.cell.2015.03.020
178. Nakaya HI, Amaral PP, Louro R et al (2007) Genome mapping and expression analyses of human intronic noncoding RNAs reveal tissue-specific patterns and enrichment in genes related to regulation of transcription. Genome Biol 8:R43. doi:10.1186/gb-2007-8-3-r43
179. Louro R, El-Jundi T, Nakaya HI et al (2008) Conserved tissue expression signatures of intronic noncoding RNAs transcribed from human and mouse loci. Genomics 92:18–25. doi:10.1016/j.ygeno.2008.03.013
180. St Laurent G, Shtokalo D, Tackett MR et al (2012) Intronic RNAs constitute the major fraction of the non-coding RNA in mammalian cells. BMC Genomics 13:504. doi:10.1186/1471-2164-13-504
181. Shahryari A, Jazi MS, Samaei NM, Mowla SJ (2015) Long non-coding RNA SOX2OT: expression signature, splicing patterns, and emerging roles in pluripotency and tumorigenesis. Front Genet. doi:10.3389/fgene.2015.00196
182. Memczak S, Jens M, Elefsinioti A et al (2013) Circular RNAs are a large class of animal RNAs with regulatory potency. Nature 495:333–338. doi:10.1038/nature11928
183. Hansen TB, Jensen TI, Clausen BH et al (2013) Natural RNA circles function as efficient microRNA sponges. Nature 495:384–388. doi:10.1038/nature11993
184. Kramer MC, Liang D, Tatomer DC et al (2015) Combinatorial control of *Drosophila* circular RNA expression by intronic repeats, hnRNPs, and SR proteins. Genes Dev 29:2168–2182. doi:10.1101/gad.270421.115
185. Hadjiargyrou M, Delihas N (2013) The intertwining of transposable elements and non-coding RNAs. Int J Mol Sci 14:13307–13328. doi:10.3390/ijms140713307
186. Rybak-Wolf A, Stottmeister C, Glažar P et al (2015) Circular RNAs in the mammalian brain are highly abundant, conserved, and dynamically expressed. Mol Cell 58:870–885. doi:10.1016/j.molcel.2015.03.027
187. Peng L, Yuan X, Li G (2015) The emerging landscape of circular RNA ciRS-7 in cancer (Review). Oncol Rep. doi:10.3892/or.2015.3904
188. Li Z, Huang C, Bao C et al (2015) Exon-intron circular RNAs regulate transcription in the nucleus. Nat Struct Mol Biol 22:256–264. doi:10.1038/nsmb.2959
189. Li J, Yang J, Zhou P et al (2015) Circular RNAs in cancer: novel insights into origins, properties, functions and implications. Am J Cancer Res 5:472–480
190. Milligan MJ, Lipovich L (2015) Pseudogene-derived lncRNAs: emerging regulators of gene expression. Front Genet. doi:10.3389/fgene.2014.00476
191. Zheng D, Frankish A, Baertsch R et al (2007) Pseudogenes in the ENCODE regions: consensus annotation, analysis of transcription, and evolution. Genome Res 17:839–851. doi:10.1101/gr.5586307
192. Grandér D, Johnsson P (2015) Pseudogene-expressed RNAs: emerging roles in gene regulation and disease. In: Morris KV (ed) Long non-coding RNAs human disease. Springer, Cham, pp 111–126
193. Poliseno L, Salmena L, Zhang J et al (2010) A coding-independent function of gene and pseudogene mRNAs regulates tumour biology. Nature 465:1033–1038. doi:10.1038/nature09144
194. Bejerano G (2004) Ultraconserved elements in the human genome. Science 304:1321–1325. doi:10.1126/science.1098119
195. Mestdagh P, Fredlund E, Pattyn F et al (2010) An integrative genomics screen uncovers ncRNA T-UCR functions in neuroblastoma tumours. Oncogene 29:3583–3592. doi:10.1038/onc.2010.106
196. Watters KM, Bryan K, Foley NH et al (2013) Expressional alterations in functional ultra-conserved non-coding rnas in response to all-transretinoic acid – induced differentiation in neuroblastoma cells. BMC Cancer. doi:10.1186/1471-2407-13-184
197. Ferdin J, Nishida N, Wu X et al (2013) HINCUTs in cancer: hypoxia-induced noncoding ultraconserved transcripts. Cell Death Differ 20:1675–1687. doi:10.1038/cdd.2013.119

198. Fassan M, Dall'Olmo L, Galasso M et al (2014) Transcribed ultraconserved noncoding RNAs (T-UCR) are involved in Barrett's esophagus carcinogenesis. Oncotarget 5:7162–7171. doi:10.18632/oncotarget.2249
199. Scaruffi P, Stigliani S, Moretti S et al (2009) Transcribed-ultra conserved region expression is associated with outcome in high-risk neuroblastoma. BMC Cancer. doi:10.1186/1471-2407-9-441
200. Feng J (2006) The Evf-2 noncoding RNA is transcribed from the Dlx-5/6 ultraconserved region and functions as a Dlx-2 transcriptional coactivator. Genes Dev 20:1470–1484. doi:10.1101/gad.1416106
201. Cajigas I, Leib DE, Cochrane J et al (2015) *Evf2* lncRNA/BRG1/DLX1 interactions reveal RNA-dependent inhibition of chromatin remodeling. Development 142:2641–2652. doi:10.1242/dev.126318
202. Feuerhahn S, Iglesias N, Panza A et al (2010) TERRA biogenesis, turnover and implications for function. FEBS Lett 584:3812–3818. doi:10.1016/j.febslet.2010.07.032
203. Porro A, Feuerhahn S, Reichenbach P, Lingner J (2010) Molecular dissection of telomeric repeat-containing RNA biogenesis unveils the presence of distinct and multiple regulatory pathways. Mol Cell Biol 30:4808–4817. doi:10.1128/MCB.00460-10
204. Balk B, Maicher A, Dees M et al (2013) Telomeric RNA-DNA hybrids affect telomere-length dynamics and senescence. Nat Struct Mol Biol 20:1199–1205. doi:10.1038/nsmb.2662
205. Balk B, Dees M, Bender K, Luke B (2014) The differential processing of telomeres in response to increased telomeric transcription and RNA–DNA hybrid accumulation. RNA Biol 11:95–100. doi:10.4161/rna.27798
206. Greenwood J, Cooper JP (2012) Non-coding telomeric and subtelomeric transcripts are differentially regulated by telomeric and heterochromatin assembly factors in fission yeast. Nucleic Acids Res 40:2956–2963. doi:10.1093/nar/gkr1155
207. Trofimova I, Chervyakova D, Krasikova A (2015) Transcription of subtelomere tandemly repetitive DNA in chicken embryogenesis. Chromosome Res 23:495–503. doi:10.1007/s10577-015-9487-3
208. Broadbent KM, Broadbent JC, Ribacke U et al (2015) Strand-specific RNA sequencing in Plasmodium falciparum malaria identifies developmentally regulated long non-coding RNA and circular RNA. BMC Genomics. doi:10.1186/s12864-015-1603-4
209. Kwapisz M, Ruault M, van Dijk E et al (2015) Expression of subtelomeric lncRNAs links telomeres dynamics to RNA decay in S. cerevisiae. Non-Coding RNA 1:94–126. doi:10.3390/ncrna1020094
210. Wong LH, Brettingham-Moore KH, Chan L et al (2007) Centromere RNA is a key component for the assembly of nucleoproteins at the nucleolus and centromere. Genome Res 17:1146–1160. doi:10.1101/gr.6022807
211. Quénet D, Dalal Y (2014) A long non-coding RNA is required for targeting centromeric protein A to the human centromere. Elife. doi:10.7554/eLife.03254
212. Blower MD (2016) Centromeric transcription regulates Aurora-B localization and activation. Cell Rep 15:1624–1633. doi:10.1016/j.celrep.2016.04.054
213. Chan FL, Marshall OJ, Saffery R et al (2012) Active transcription and essential role of RNA polymerase II at the centromere during mitosis. Proc Natl Acad Sci 109:1979–1984. doi:10.1073/pnas.1108705109
214. Rošić S, Köhler F, Erhardt S (2014) Repetitive centromeric satellite RNA is essential for kinetochore formation and cell division. J Cell Biol 207:335–349. doi:10.1083/jcb.201404097
215. Bierhoff H, Dammert MA, Brocks D et al (2014) Quiescence-induced LncRNAs trigger H4K20 trimethylation and transcriptional silencing. Mol Cell 54:675–682. doi:10.1016/j.molcel.2014.03.032
216. Li W, Notani D, Ma Q et al (2013) Functional roles of enhancer RNAs for oestrogen-dependent transcriptional activation. Nature 498:516–520. doi:10.1038/nature12210
217. Li W, Notani D, Rosenfeld MG (2016) Enhancers as non-coding RNA transcription units: recent insights and future perspectives. Nat Rev Genet 17:207–223. doi:10.1038/nrg.2016.4

218. Kapranov P, Willingham AT, Gingeras TR (2007) Genome-wide transcription and the implications for genomic organization. Nat Rev Genet 8:413–423. doi:10.1038/nrg2083
219. Preker P, Nielsen J, Kammler S et al (2008) RNA exosome depletion reveals transcription upstream of active human promoters. Science 322:1851–1854. doi:10.1126/science.1164096
220. Ntini E, Järvelin AI, Bornholdt J et al (2013) Polyadenylation site–induced decay of upstream transcripts enforces promoter directionality. Nat Struct Mol Biol 20:923–928. doi:10.1038/nsmb.2640
221. Agarwal N, Ansari A (2016) Enhancement of transcription by a splicing-competent intron is dependent on promoter directionality. PLoS Genet 12:e1006047. doi:10.1371/journal.pgen.1006047
222. Wang X, Arai S, Song X et al (2008) Induced ncRNAs allosterically modify RNA-binding proteins in cis to inhibit transcription. Nature 454:126–130. doi:10.1038/nature06992
223. Song X, Wang X, Arai S, Kurokawa R (2012) Promoter-associated noncoding RNA from the CCND1 promoter. In: Vancura A (ed) Transcription regulation. Springer, New York, pp 609–622
224. Mercer TR, Wilhelm D, Dinger ME et al (2011) Expression of distinct RNAs from 3′ untranslated regions. Nucleic Acids Res 39:2393–2403. doi:10.1093/nar/gkq1158
225. Neil H, Malabat C, d'Aubenton-Carafa Y et al (2009) Widespread bidirectional promoters are the major source of cryptic transcripts in yeast. Nature 457:1038–1042. doi:10.1038/nature07747
226. van Dijk EL, Chen CL, d'Aubenton-Carafa Y et al (2011) XUTs are a class of Xrn1-sensitive antisense regulatory non-coding RNA in yeast. Nature 475:114–117. doi:10.1038/nature10118
227. Berretta J, Pinskaya M, Morillon A (2008) A cryptic unstable transcript mediates transcriptional trans-silencing of the Ty1 retrotransposon in S. cerevisiae. Genes Dev 22:615–626. doi:10.1101/gad.458008
228. Camblong J, Beyrouthy N, Guffanti E et al (2009) Trans-acting antisense RNAs mediate transcriptional gene cosuppression in S. cerevisiae. Genes Dev 23:1534–1545. doi:10.1101/gad.522509
229. Toesca I, Nery CR, Fernandez CF et al (2011) Cryptic transcription mediates repression of subtelomeric metal homeostasis genes. PLoS Genet 7:e1002163. doi:10.1371/journal.pgen.1002163
230. Lardenois A, Liu Y, Walther T et al (2011) Execution of the meiotic noncoding RNA expression program and the onset of gametogenesis in yeast require the conserved exosome subunit Rrp6. Proc Natl Acad Sci 108:1058–1063. doi:10.1073/pnas.1016459108
231. Frenk S, Oxley D, Houseley J (2014) The nuclear exosome is active and important during budding yeast meiosis. PLoS One 9:e107648. doi:10.1371/journal.pone.0107648
232. de Andres-Pablo A, Morillon A, Wery M (2016) LncRNAs, lost in translation or licence to regulate? Curr Genet. doi:10.1007/s00294-016-0615-1
233. Vera JM, Dowell RD (2016) Survey of cryptic unstable transcripts in yeast. BMC Genomics. doi:10.1186/s12864-016-2622-5
234. Pefanis E, Wang J, Rothschild G et al (2015) RNA exosome-regulated long non-coding RNA transcription controls super-enhancer activity. Cell 161:774–789. doi:10.1016/j.cell.2015.04.034
235. Moon SL, Blackinton JG, Anderson JR et al (2015) XRN1 stalling in the 5′ UTR of hepatitis C virus and bovine viral diarrhea virus is associated with dysregulated host mRNA stability. PLoS Pathog 11:e1004708. doi:10.1371/journal.ppat.1004708
236. Chapman EG, Moon SL, Wilusz J, Kieft JS (2014) RNA structures that resist degradation by Xrn1 produce a pathogenic Dengue virus RNA. Elife. doi:10.7554/eLife.01892
237. Werner MS, Ruthenburg AJ (2015) Nuclear fractionation reveals thousands of chromatin-tethered noncoding RNAs adjacent to active genes. Cell Rep 12:1089–1098. doi:10.1016/j.celrep.2015.07.033
238. Mondal T, Rasmussen M, Pandey GK et al (2010) Characterization of the RNA content of chromatin. Genome Res 20:899–907. doi:10.1101/gr.103473.109

239. Singh DK, Prasanth KV (2013) Functional insights into the role of nuclear-retained long non-coding RNAs in gene expression control in mammalian cells. Chromosome Res 21:695–711. doi:10.1007/s10577-013-9391-7
240. Zheng R, Shen Z, Tripathi V et al (2010) Polypurine-repeat-containing RNAs: a novel class of long non-coding RNA in mammalian cells. J Cell Sci 123:3734–3744. doi:10.1242/jcs.070466
241. Rackham O, Shearwood A-MJ, Mercer TR et al (2011) Long noncoding RNAs are generated from the mitochondrial genome and regulated by nuclear-encoded proteins. RNA 17:2085–2093. doi:10.1261/rna.029405.111
242. Burzio VA, Villota C, Villegas J et al (2009) Expression of a family of noncoding mitochondrial RNAs distinguishes normal from cancer cells. Proc Natl Acad Sci 106:9430–9434. doi:10.1073/pnas.0903086106
243. Anandakumar S, Vijayakumar S, Centre for Advanced Study in Crystallography and Biophysics, University of Madras et al (2015) Mammalian mitochondrial ncRNA database. Bioinformation 11:512–514. doi: 10.6026/97320630011512
244. Landerer E, Villegas J, Burzio VA et al (2011) Nuclear localization of the mitochondrial ncRNAs in normal and cancer cells. Cell Oncol 34:297–305. doi:10.1007/s13402-011-0018-8
245. Vidaurre S, Fitzpatrick C, Burzio VA et al (2014) Down-regulation of the antisense mitochondrial non-coding RNAs (ncRNAs) is a unique vulnerability of cancer cells and a potential target for cancer therapy. J Biol Chem 289:27182–27198. doi:10.1074/jbc.M114.558841
246. Lobos-González L, Silva V, Araya M et al (2016) Targeting antisense mitochondrial ncRNAs inhibits murine melanoma tumor growth and metastasis through reduction in survival and invasion factors. Oncotarget. doi:10.18632/oncotarget.11110
247. Guo X, Gao L, Wang Y et al (2015) Advances in long noncoding RNAs: identification, structure prediction and function annotation. Brief Funct Genomics. doi:10.1093/bfgp/elv022
248. Quinn JJ, Chang HY (2015) Unique features of long non-coding RNA biogenesis and function. Nat Rev Genet 17:47–62. doi:10.1038/nrg.2015.10
249. Han P, Chang C-P (2015) Long non-coding RNA and chromatin remodeling. RNA Biol 12:1094–1098. doi:10.1080/15476286.2015.1063770
250. Davidovich C, Cech TR (2015) The recruitment of chromatin modifiers by long noncoding RNAs: lessons from PRC2. RNA 21:2007–2022. doi:10.1261/rna.053918.115
251. Yoon J-H, Abdelmohsen K, Kim J et al (2013) Scaffold function of long non-coding RNA HOTAIR in protein ubiquitination. Nat Commun. doi:10.1038/ncomms3939
252. Lee S, Kopp F, Chang T-C et al (2016) Noncoding RNA NORAD regulates genomic stability by sequestering PUMILIO proteins. Cell 164:69–80. doi:10.1016/j.cell.2015.12.017
253. Tsai M-C, Manor O, Wan Y et al (2010) Long noncoding RNA as modular scaffold of histone modification complexes. Science 329:689–693. doi:10.1126/science.1192002
254. Chujo T, Yamazaki T, Hirose T (2016) Architectural RNAs (arcRNAs): a class of long noncoding RNAs that function as the scaffold of nuclear bodies. Biochim Biophys Acta 1859:139–146. doi:10.1016/j.bbagrm.2015.05.007
255. Yamazaki T, Hirose T (2015) The building process of the functional paraspeckle with long non-coding RNAs. Front Biosci 7:1–47. doi:10.2741/715
256. Postepska-Igielska A, Giwojna A, Gasri-Plotnitsky L et al (2015) LncRNA Khps1 regulates expression of the proto-oncogene SPHK1 via triplex-mediated changes in chromatin structure. Mol Cell 60:626–636. doi:10.1016/j.molcel.2015.10.001
257. Mondal T, Subhash S, Vaid R et al (2015) MEG3 long noncoding RNA regulates the TGF-β pathway genes through formation of RNA–DNA triplex structures. Nat Commun 6:7743. doi:10.1038/ncomms8743
258. Kino T, Hurt DE, Ichijo T et al (2010) Noncoding RNA Gas5 is a growth arrest- and starvation-associated repressor of the glucocorticoid receptor. Sci Signal 3:ra8. doi:10.1126/scisignal.2000568
259. Wang P, Xue Y, Han Y et al (2014) The STAT3-binding long noncoding RNA lnc-DC controls human dendritic cell differentiation. Science 344:310–313. doi:10.1126/science.1251456

260. Ørom UA, Derrien T, Beringer M et al (2010) Long noncoding RNAs with enhancer-like function in human cells. Cell 143:46–58. doi:10.1016/j.cell.2010.09.001
261. Lai F, Orom UA, Cesaroni M et al (2013) Activating RNAs associate with mediator to enhance chromatin architecture and transcription. Nature 494:497–501. doi:10.1038/nature11884
262. Yao H, Brick K, Evrard Y et al (2010) Mediation of CTCF transcriptional insulation by DEAD-box RNA-binding protein p68 and steroid receptor RNA activator SRA. Genes Dev 24:2543–2555. doi:10.1101/gad.1967810
263. Gomez JA, Wapinski OL, Yang YW et al (2013) The NeST long ncRNA controls microbial susceptibility and epigenetic activation of the interferon-γ locus. Cell 152:743–754. doi:10.1016/j.cell.2013.01.015
264. Szcześniak MW, Makałowska I (2016) lncRNA-RNA interactions across the human transcriptome. PLoS One 11:e0150353. doi:10.1371/journal.pone.0150353
265. An Y, Furber KL, Ji S (2016) Pseudogenes regulate parental gene expression *via* ceRNA network. J Cell Mol Med. doi:10.1111/jcmm.12952
266. Thomson DW, Dinger ME (2016) Endogenous microRNA sponges: evidence and controversy. Nat Rev Genet 17:272–283. doi:10.1038/nrg.2016.20
267. Tay Y, Rinn J, Pandolfi PP (2014) The multilayered complexity of ceRNA crosstalk and competition. Nature 505:344–352. doi:10.1038/nature12986
268. Giovarelli M, Bucci G, Ramos A et al (2014) H19 long noncoding RNA controls the mRNA decay promoting function of KSRP. Proc Natl Acad Sci 111:E5023–E5028. doi:10.1073/pnas.1415098111
269. Dey BK, Pfeifer K, Dutta A (2014) The H19 long noncoding RNA gives rise to microRNAs miR-675-3p and miR-675-5p to promote skeletal muscle differentiation and regeneration. Genes Dev 28:491–501. doi:10.1101/gad.234419.113
270. Ha H, Song J, Wang S et al (2014) A comprehensive analysis of piRNAs from adult human testis and their relationship with genes and mobile elements. BMC Genomics 15:545. doi:10.1186/1471-2164-15-545
271. Carlile M, Swan D, Jackson K et al (2009) Strand selective generation of endo-siRNAs from the Na/phosphate transporter gene Slc34a1 in murine tissues. Nucleic Acids Res 37:2274–2282. doi:10.1093/nar/gkp088
272. Werner A (2013) Biological functions of natural antisense transcripts. BMC Biol 11:31. doi:10.1186/1741-7007-11-31
273. Abdelmohsen K, Panda A, Kang M-J et al (2013) Senescence-associated lncRNAs: senescence-associated long noncoding RNAs. Aging Cell 12:890–900. doi:10.1111/acel.12115
274. C-L W, Wang Y, Jin B et al (2015) Senescence-associated long non-coding RNA (*SALNR*) delays oncogene-induced senescence through NF90 regulation. J Biol Chem 290:30175–30192. doi:10.1074/jbc.M115.661785
275. Choudhry H, Harris AL, McIntyre A (2016) The tumour hypoxia induced non-coding transcriptome. Mol Aspects Med 47–48:35–53. doi:10.1016/j.mam.2016.01.003
276. Fort A, Hashimoto K, Yamada D et al (2014) Deep transcriptome profiling of mammalian stem cells supports a regulatory role for retrotransposons in pluripotency maintenance. Nat Genet 46:558–566. doi:10.1038/ng.2965
277. Prensner JR, Iyer MK, Balbin OA et al (2011) Transcriptome sequencing across a prostate cancer cohort identifies PCAT-1, an unannotated lincRNA implicated in disease progression. Nat Biotechnol 29:742–749. doi:10.1038/nbt.1914
278. Mattick JS, Rinn JL (2015) Discovery and annotation of long noncoding RNAs. Nat Struct Mol Biol 22:5–7. doi:10.1038/nsmb.2942
279. St. Laurent G, Wahlestedt C, Kapranov P (2015) The landscape of long noncoding RNA classification. Trends Genet 31:239–251. doi:10.1016/j.tig.2015.03.007
280. Laurent GS, Vyatkin Y, Antonets D et al (2016) Functional annotation of the vlinc class of non-coding RNAs using systems biology approach. Nucleic Acids Res 44:3233–3252. doi:10.1093/nar/gkw162

281. Lin R, Maeda S, Liu C et al (2007) A large noncoding RNA is a marker for murine hepatocellular carcinomas and a spectrum of human carcinomas. Oncogene 26:851–858. doi:10.1038/sj.onc.1209846
282. Clemson CM, Hutchinson JN, Sara SA et al (2009) An architectural role for a nuclear noncoding RNA: NEAT1 RNA is essential for the structure of paraspeckles. Mol Cell 33:717–726. doi:10.1016/j.molcel.2009.01.026
283. Nam J-W, Bartel DP (2012) Long noncoding RNAs in C. elegans. Genome Res 22:2529–2540. doi:10.1101/gr.140475.112
284. Beltran M, Puig I, Pena C et al (2008) A natural antisense transcript regulates Zeb2/Sip1 gene expression during Snail1-induced epithelial-mesenchymal transition. Genes Dev 22:756–769. doi:10.1101/gad.455708
285. Salzman J, Gawad C, Wang PL et al (2012) Circular RNAs are the predominant transcript isoform from hundreds of human genes in diverse cell types. PLoS One 7:e30733. doi:10.1371/journal.pone.0030733
286. Jeck WR, Sorrentino JA, Wang K et al (2013) Circular RNAs are abundant, conserved, and associated with ALU repeats. RNA 19:141–157. doi:10.1261/rna.035667.112
287. Gardner EJ, Nizami ZF, Talbot CC, Gall JG (2012) Stable intronic sequence RNA (sisRNA), a new class of noncoding RNA from the oocyte nucleus of Xenopus tropicalis. Genes Dev 26:2550–2559. doi:10.1101/gad.202184.112
288. Talhouarne GJS, Gall JG (2014) Lariat intronic RNAs in the cytoplasm of *Xenopus tropicalis* oocytes. RNA 20:1476–1487. doi:10.1261/rna.045781.114
289. Pek JW, Osman I, Tay ML-I, Zheng RT (2015) Stable intronic sequence RNAs have possible regulatory roles in *Drosophila melanogaster*. J Cell Biol 211:243–251. doi:10.1083/jcb.201507065
290. Rapicavoli NA, Qu K, Zhang J et al (2013) A mammalian pseudogene lncRNA at the interface of inflammation and anti-inflammatory therapeutics. Elife. doi:10.7554/eLife.00762
291. Vembar SS, Scherf A, Siegel TN (2014) Noncoding RNAs as emerging regulators of Plasmodium falciparum virulence gene expression. Curr Opin Microbiol 20:153–161. doi:10.1016/j.mib.2014.06.013
292. Liz J, Portela A, Soler M et al (2014) Regulation of pri-miRNA processing by a long noncoding RNA transcribed from an ultraconserved region. Mol Cell 55:138–147. doi:10.1016/j.molcel.2014.05.005
293. Ilott NE, Heward JA, Roux B et al (2014) Long non-coding RNAs and enhancer RNAs regulate the lipopolysaccharide-induced inflammatory response in human monocytes. Nat Commun. doi:10.1038/ncomms4979
294. Bianchessi V, Badi I, Bertolotti M et al (2015) The mitochondrial lncRNA ASncmtRNA-2 is induced in aging and replicative senescence in Endothelial Cells. J Mol Cell Cardiol 81:62–70. doi:10.1016/j.yjmcc.2015.01.012
295. Zhang Y, He Q, Hu Z et al (2016) Long noncoding RNA LINP1 regulates repair of DNA double-strand breaks in triple-negative breast cancer. Nat Struct Mol Biol 23:522–530. doi:10.1038/nsmb.3211

Chapter 2
Long Noncoding RNA: Genome Organization and Mechanism of Action

Vijay Suresh Akhade, Debosree Pal, and Chandrasekhar Kanduri

Abstract For the last four decades, we have known that noncoding RNAs maintain critical housekeeping functions such as transcription, RNA processing, and translation. However, in the late 1990s and early 2000s, the advent of high-throughput sequencing technologies and computational tools to analyze these large sequencing datasets facilitated the discovery of thousands of small and long noncoding RNAs (lncRNAs) and their functional role in diverse biological functions. For example, lncRNAs have been shown to regulate dosage compensation, genomic imprinting, pluripotency, cell differentiation and development, immune response, etc. Here we review how lncRNAs bring about such copious functions by employing diverse mechanisms such as translational inhibition, mRNA degradation, RNA decoys, facilitating recruitment of chromatin modifiers, regulation of protein activity, regulating the availability of miRNAs by sponging mechanism, etc. In addition, we provide a detailed account of different mechanisms as well as general principles by which lncRNAs organize functionally different nuclear sub-compartments and their impact on nuclear architecture.

Keywords Long noncoding RNA • Chromatin • Gene regulation • Genome organization • Epigenetics

V.S. Akhade • C. Kanduri (✉)
Department of Medical Biochemistry and Cell Biology, Institute of Biomedicine, Sahlgrenska Academy, University of Gothenburg, Gothenburg 40530, Sweden
e-mail: kanduri.chandrasekhar@gu.se

D. Pal
Molecular Biology and Genetics Unit, Jawaharlal Nehru Centre for Advanced Scientific Research, Jakkur P.O, Bangalore 560064, India

M.R.S. Rao (ed.), *Long Non Coding RNA Biology*, Advances in Experimental Medicine and Biology 1008, DOI 10.1007/978-981-10-5203-3_2

2.1 Introduction

Noncoding RNA (ncRNA), once thought as a part of transcriptional noise, now represents a novel regulatory layer in the transcriptional and posttranscriptional gene regulation. Their rise from transcriptional noise to novel regulators of gene expression is well supported by their documented functional roles in various aspects of gene regulation, including epigenetic regulation, X chromosome inactivation, genomic imprinting, nuclear and cytoplasmic trafficking, transcription, and mRNA splicing [1, 2]. Latest genome-wide annotation studies based on high-throughput transcriptomics from single cell embryo to differentiated tissue cell types reveal that more than two-thirds of the mammalian genome is transcribed encoding tens of thousands of different classes of small and long noncoding RNA (lncRNAs). lncRNAs represent the largest class among the noncoding RNA subtypes, and according to the latest estimates, there are about more than 58,084 transcripts, which outnumber the protein-coding RNAs. lncRNAs have also emerged as key regulators in a wide range of biological processes such as cell proliferation, cell cycle, metabolism, apoptosis, differentiation and maintenance of pluripotency, etc. [3, 4]. They have not only shown to typically regulate gene regulation, but lncRNAs employ diverse mechanisms for their function, and these would be described below in detail.

2.2 lncRNA and Chromatin Regulation

lncRNAs have emerged as crucial players in the regulation of transcription via modulation of chromatin [5–9]. lncRNAs regulate chromatin structure at different functional steps which include histone modifications, DNA methylation, and chromatin remodeling. Execution of this function by lncRNAs across a broad evolutionary spectrum is suggestive of their conserved role in chromatin regulation despite lack of primary sequence conservation.

2.3 lncRNAs Regulate Chromatin Structure via Histone Modifications

Interaction of lncRNAs with histone-modifying complexes is very prominent with respect to two polycomb repressive complexes, PRC1 and, in particular, PRC2, which mediates methylation of lysine 27 on histone 3 (H3K27me), a histone mark associated with poised or repressed transcriptional status. A general mechanistic representation of regulation of histone modifications by lncRNAs is depicted in Fig. 2.1a. The first documentation of lncRNA and PRC2 interaction was from the studies on mammalian X chromosome inactivation, where X-inactive specific

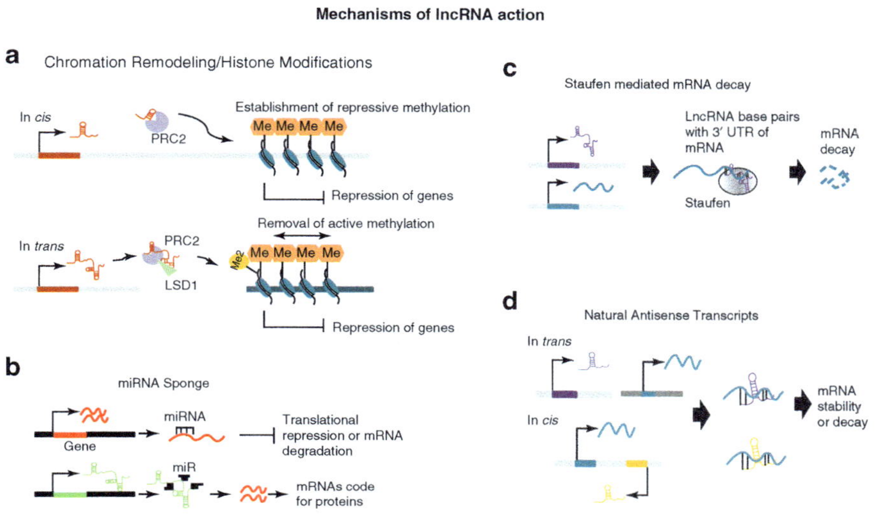

Fig. 2.1 Schematic representation of few mechanisms employed by lncRNAs in regulation of gene expression which include regulation of active and repressive histone marks (**a**), mRNA translation by acting as a miRNA sponge (**b**), Staufen-mediated decay of mRNA (**c**), and mRNA stability or decay by functioning as natural antisense transcripts (**d**)

transcript (*Xist*), a lncRNA that is highly expressed from the inactive X chromosomes in females (Xi), mediates the recruitment of PRC2 to the Xi to silence gene expression [10–12]. Another prominent example is HOTAIR lncRNA, which is transcribed from the HoxC gene cluster but mediates transcriptional gene silencing in the HoxD locus in trans via targeting PRC2 complex and the histone H3K4me1/2 demethylase LSD1 [13]. Recently, a murine lincRNA Pint (p53 induced noncoding transcript) was demonstrated to connect p53 activation with epigenetic silencing by PRC2 [14]. Pint controls cell survival and proliferation through regulating the TGF-β, MAPK, and p53 pathways. Pint mediates H3K27 trimethylation of target gene promoters via recruitment of PRC2. Also, the expression of PRC2 is required for the functional activity of Pint. Braveheart (Bvht), a heart associated lncRNA in mouse, is activated during early cardiac differentiation [15]. Bvht mediates epigenetic regulation of cardiac commitment through its interaction with SUZ12, a component of PRC2 complex during cardiomyocyte differentiation. Another similar example is the Fendrr lncRNA (Foxf1adjacent noncoding developmental regulatory RNA) which is required for heart and body wall development in mouse. Fendrr acts by modifying the chromatin signatures of its target genes through binding to both the PRC2 and TrxG/MLL complexes. Fendrr interacts with WDR5 which is a component of the MLL complexes that mediates H3K4 methylation associated with loci that are actively transcribed or primed for activation [16, 17]. lncRNA SRA (steroid receptor RNA activator) also participates in transcriptional regulation through complex formation with trithorax group (TrxG) and polycomb repressive complex 2 (PRC2) complexes [18]. Very recently, *HoxB* lincRNA is shown to be required for

HoxB gene activation and mesoderm specification (HoxB lincRNA cell reports reference). Mechanistically, *HoxB* lincRNA controls chromatin dynamics through the recruitment of Set1/MLL complexes. Interaction of WDR5 has been reported with more than 200 lncRNAs in mouse embryonic stem cells (mESCs) [19]. The importance of these interactions is reflected by the requirement of WDR5-lncRNA association for binding of WDR5 to chromatin. Taking into consideration more specific studies, two lncRNAs, HOTTIP and NeST, have been described to recruit WDR5 to their neighboring genes and thus enhance their transcription [20, 21].

Example of an lncRNA interacting with the PRC1 complex is FAL1 RNA (focally amplified lncRNA on chromosome 1) [22]. FAL1 interacts with BMI1, an essential subunit of PRC1. FAL1 not only regulates the protein stability of BMI1 but also regulates the association between BMI1 and the target promoter regions thereby modulating target gene expression. ANRIL, an lncRNA transcribed from the INK4B-ARF-INK4A tumor suppressor locus, was discovered in a family with inherited melanoma-neural system tumors [23]. ANRIL recruits CBX7 (a PRC1 component) via a POL II-dependent mechanism to its locus in order to repress the neighboring INK4B-ARF-INK4A genes [23, 24]. CBX7 binding to ANRIL contributes to CBX7 function, and disruption of CBX7-ANRIL interaction impacts the ability of CBX7 to repress the INK4b/ARF/INK4a locus and control senescence [24].

2.4 lncRNAs Regulate Chromatin Remodeling

In addition to interaction with histone-modifying enzymes and covalently modifying chromatin, lncRNAs also associate with chromatin remodeling complexes to regulate gene expression. The association of lncRNA *SChLAP1* (second chromosome locus associated with prostate-1) with the chromatin remodeling complex SWI/SNF in human prostate cancer cells is one such example [25]. Knockdown of lncRNA *SChLAP1* leads to impaired cell invasion and proliferation, while knockdown of SMARCB1 (component of SWI/SNF) promoted cancer progression, thus revealing their opposite functions in human prostate cancer cell lines. Also, *SChlAP1* interacts with the SNF5 subunit of the SWI/SNF complex and antagonizes the genome-wide localization and regulatory functions of the SWI/SNF complex, thereby resulting in genome-wide derepression of gene activity. Similarly, the lncRNA *Mhrt* (myosin heavy chain-associated RNA transcripts or Myheart) binds to BRG1, the ATPase subunit of the SWI/SNF complex, and prevents Brg1 from recognizing its genomic DNA targets [26]. Recently, a very interesting study comprehensively identified the *Xist* lncRNA interactome which comprised of cohesins, condensins, topoisomerases, RNA helicases, chromatin remodelers, and modifiers [27]. Stable knockdown of the components of SWI/SNF complex demonstrated that the Xist-SWI/SNF interaction is required for proper maintenance of PRC2 function on the inactive X chromosome (Xi). The Evf2 lncRNA also interacts with SWI/SNF-related chromatin remodelers Brg1 and Baf170 [28, 29]. In contrast to

SChlAP1, Mhrt, and Xist, Evf2 lncRNA increases the association of Brg1 with important regulatory enhancer regions but directly inhibits the ATPase and chromatin remodeling activities of Brg1.

lncRNAs and the SWI/SNF complex are not only shown in the context of gene repression but also reported to mediate gene activation. lncTCF7, which promotes the self-renewal of human liver cancer stem cells by activating Wnt signaling, interacts with and recruits the SWI/SNF complex to the TCF7promoter thereby activating TCF7 expression and subsequently Wnt signaling [30]. Linc-Cox2 is an important RNA for regulation of immune response genes, and its expression is highly induced in macrophages upon stimulation by LPS through the Toll-like receptors [31]. Recent studies have shed more light on the mechanism of transcriptional regulation by linc-Cox2. Upon LPS stimulation in macrophages, linc-Cox2 is assembled into the SWI/SNF complex and promotes the assembly of NF-KB subunits into the SWI/SNF complex [32]. This ultimately leads to increased recruitment of SWI/SNF complex and transactivation of the late primary inflammatory response genes. Linc-Cox2, also stimulated in response to TNF-α stimulation in murine intestinal epithelial cells, represses the transcription from the I12b promoter, a secondary late-responsive gene induced by TNF-α, through targeting Mi-2/NuRD repressor complex [33].

2.5 eRNAs and Higher-Order Chromatin Organization

Enhancer-derived lncRNAs (eRNAs) are also documented in the control of regulatory contacts between enhancers and the cognate promoter through chromosome looping [34, 35]. Until recently, eRNAs were considered primarily as by-products of transcription from enhancer-promoter interactions. However, with several eRNAs functionally implicated in diverse biological contexts, they are being considered as independent transcriptional units [34, 36, 37]. The classical enhancer RNAs (eRNAs) are short, unspliced transcripts that are a product of bidirectional transcription [34, 38]. The locus control region (LCR) of the b-globin cluster is the very first identified example of enhancer-linked transcripts [39, 40]. The eRNAs control gene expression, possibly by affecting looping between enhancers and promoters [41, 42]. For example, LUNAR1 is a T cell acute lymphoblastic leukemia (T-ALL)-specific lncRNA that is transcribed from the insulin-like growth factor 1 receptor (*IGF1R*) locus and that exhibits pro-oncogenic characteristics, such as stimulating T-ALL cell growth [43]. LUNAR1 itself activates the *IGF1R* locus in *cis* via chromosome looping, thus leading to sustained IGF1 signaling in T-ALL cells. Likewise, the lncRNAs PRNCR1 and PCGEM1 regulate androgen receptor-dependent gene activation programs by promoting enhancer-promoter looping in prostate cancer cells [44]. Also, there are lncRNAs with enhancer-like functions and are referred to as activating ncRNAs (ncRNA-a). These lncRNAs mediate DNA looping and chromatin enhancement via Mediator, a large transcriptional co-activating complex [45]. Two such eRNAs, ncRNA-a3 and ncRNA-a7, bring about the recruitment and activation of Mediator complex. Activity of these eRNAs is compromised by the knockdown of Mediator components and conversely loss of the eRNAs led to the decreased recruitment of Mediator and Pol II to the target genes. The contacting of eRNAs by the Mediator subunits is essential for the chromosomal looping between the enhancer and the target gene, as demonstrated by chromosome conformation capture method. Recently, using mouse cortical neurons, it was shown that some eRNAs facilitate the release of the negative elongation factor (NELF), thus competitively derepressing paused Pol II and enabling productive elongation of the target RNA [46]. Similarly, eRNAs may also "trap" certain RNA-binding transcription factors at enhancers, thereby sustaining transcription factor-mediated regulation [47].

lncRNAs transcribed from enhancer regions can also have inhibitory effects on their target genes. The promoter deletion of the lncRNA Haunt (also known as linc-Hoxa1) leads to upregulation of several genes of the neighboring HoxA gene cluster [48]. A similar mechanism was also found for Playrr (D030025E07Rik), an lncRNA-encoded upstream of the homeodomain transcription factor Pitx2 [49]. A CRISPR/Cas9-generated mutation resulting in Playrr RNA decay caused upregulation of Pitx2 expression. An enhancer involved in Pitx2 regulation in the gut overlaps the TSS of Playrr. The expression of Playrr interferes with the looping of the Pitx2 promoter to its enhancer at the Playrr locus and thus affecting the Pitx2 gene activation.

2.6 Long Noncoding RNAs in Genomic Imprinting

Genomic imprinting is a phenomenon in which only of one of the alleles of an inherited parental pair is active, while the other one is maintained in an inactive state. This differential expression of the inherited parental alleles depends on the parent of origin; in some cases, an allele of a gene might be paternally imprinted, whereas in other cases it would be maternally imprinted. Imprinting is generally achieved by histone and/or DNA modification of the particular locus, and lately lncRNAs have been known to play a role in this phenomenon [50].

Airn lncRNA is 108 kb long, nuclear localized transcript, transcribed in an antisense direction from 3.7 kb imprinting control element (ICE) in intron 2 of the insulin-like growth factor type-2 receptor (Igf2r) gene [51]. It has been shown to control the parent of origin-specific expression of three genes: *Slc22a2*, *Slc22a3*, and *Igf2r*. Paternal but not the maternal inheritance of 3.7 kb ICE deletion leads to biallelic expression of all three genes, including *Igf2r*, and reduction in birth weight of mice [52]. A similar phenotype was also observed when the *Airn* RNA was prematurely truncated by the insertion of a polyadenylation signal 3 kb downstream of its transcription start site. By performing *Airn* RNA-TRAP experiments in mouse liver and placental cell types, it was observed that *Airn* interacts physically with the *Scl22a3*, which is situated >230 kb away from the *Airn* locus, and recruits G9a methyltransferase to the *Scl22a3* promoter on the paternal chromosome in cis

[53, 54]. Based on these observations, the authors have hypothesized that *Airn* might act in a manner similar to *XIST* wherein direct interaction of the RNA with its target DNA loci in *cis* on the paternal chromosome would lead to the silencing of *Scl22a3* through the enrichment of a repressive histone mark H3K9me3.

Kcnq1ot1 is another lncRNA whose role in genomic imprinting is well established. It is a 91 kb RNA encoded from 1 Mb *Kcnq1/Cdkn1c* locus on chromosome 7 that harbors several protein-coding genes [55, 56]. *Kcnq1ot1* is expressed from the paternal chromosome, whereas it is repressed on the maternal chromosome by CpG methylation. Its expression on the paternal chromosome correlates with repression of eight to ten neighboring protein-coding genes which span over megabase region (reviewed in [57]). Like *Xist* and *Airn*, *Kcnq1ot1* acts in cis on the paternal chromosome to silence *Kcnq1*, *Cdkn1c*, *Phlda2*, and *Slc22a18* genes in all tissues, i.e., embryonic and extraembryonic (ubiquitously imprinted loci) and *Osbpl5*, *Tssc4*, and *Ascl2* genes only in the extraembryonic tissues (placental-specific imprinted loci). *Kcnq1ot1* also acts at the epigenetic level to impart silencing of the imprinted genes. *Kcnq1ot1* mediates silencing of neighboring genes on the paternal chromosome through establishing paternal allele-specific repressive histone modifications (H3K9me3 and H3K27me3) and DNA methylation at the target imprinted gene promoters by interacting with chromatin (PRC2 and G9a) and DNA (DNMT1) modification enzymes and guide them specifically to the target imprinted gene promoters in cis by acting as a scaffold. Previously, using episomal-based system coupled with cell culture experiments, 890 nucleotide functional RNA sequence was identified at the 5′ end of the *Kcnq1ot1* RNA. Using transgenic mouse model (Δ890 mice) [57], it has been shown that the 890 RNA sequence at the 5′ end of *Kcnq1ot1* RNA is required for establishing the repressive histone modifications and DNA methylation at the promoters of ubiquitously imprinted genes *Kcnq1*, *Cdkn1c*, *Slc22a18*, and *Phlda2*. Interestingly, the Δ890 mice showed a phenotype similar to Dnmt−/− mice. On further analysis, it was seen that DNA methylation at the paternal *Cdkn1c* and *Slc22a18* loci was significantly reduced in the Δ890 mice. Surprisingly, *Kcnq1* and *Phlda2* did not show any changes in DNA methylation patterns in the wild-type or the Δ890 mice implying that their methylation status could be established at a very early stage of development and then transmitted as a memory. Thus *Kcnq1ot1* lncRNA acts through two independent mechanisms: either by recruiting heterochromatic chromatin remodelers or by DNA methyltransferases specifically to imprinted genes in a locus- and tissue-specific manner [58].

In mouse chromosome 7, there exists a whole cluster of genes covering around 600 kb on the genome which undergo the phenomenon of imprinting. *H19* and *Igf2* are two genes belonging to this cluster; corresponding to the human locus 11p15.5, *H19* is expressed maternally, whereas *Igf2* is expressed paternally [59, 60]. This imprinted locus, also known as imprinted gene network (IGN), is regulated in expression by the *H19* gene, and the network is itself involved in the development of the embryo. *H19* encodes for a 2.3 kb lncRNA, and it has been shown that targeted deletion of the gene induces an overgrowth phenotype and relieves imprinting on *Igf2* and several other genes in the IGN [59, 61]. Monnier et al. addressed the mechanism of action of H19 and performed RNA immunoprecipitation experiments wherein they found that H19 interacts with methyl-CpG-binding domain protein 1 (MBD1) [60]. MBD1 binds to methylated DNA to recruit various histone deacetylases and lysine methyl transferases that impart histone modifications leading to gene silencing. Indeed it was observed that in MBD1 silenced cells, five of the IGN genes, i.e., *Igf2*, *Slc38a4*, *Dcn*, *Dlk1*, and *Peg1* underwent derepression, whereas *Gtl2*, *Cdkn1c*, and *Igf2r* did not thereby implying that only for these five genes interplay of H19 and MBD1 is important for the imprinting to occur. Further ChIP experiments revealed that MBD1 binds directly to the DNA methylated regions (DMRs) of *Igf2*, *Slc38a4*, and *Peg1*and this interaction is hindered in the absence of H19. It was also observed that when H19 is expressed in trans in cells that harbor a deletion of the H19 transcription unit, Igf2 undergoes a loss of its imprinting status. Together these studies stated that H19 interacts with MBD1 to maintain transcriptional repression and imprinting of the above genes through imparting repressive histone modifications such as H3K9me3 at these loci.

2.7 Posttranscriptional Regulation by lncRNAs

lncRNAs as sponges: An interesting mechanism of action of lncRNAs that has come into focus is their activity as microRNA (miRNA) sponges (Fig. 2.1b). lncRNAs that are known to exert such an action harbor sequences complementary to miRNA sequences thereby sequestering them and preventing them from binding to their targets. Such lncRNAs can arise from pseudogenes or be in the form of circular RNAs or be common intergenic lncRNAs possessing miRNA binding sites. Cesana et al. in 2011 have studied for the first time one such lncRNA *linc-MD1*, a muscle-specific lncRNA implicated in muscle differentiation [62]. Bioinformatic analysis showed the presence of two binding sites for miR-135 and one binding site for miR-133 on *linc-MD1*. Interestingly, both miRNAs target important transcription factors regulating myogenic differentiation: miR-135 targets MEF2C and miR-133 targets MAML1. Luciferase assay experiments, with the wild-type *linc-MD1* sequence or its mutant derivatives containing mutations on the respective miR binding sites, cloned downstream of the luciferase gene and revealed that upon miR overexpression, luciferase activity for the wild-type MD1 sequence is depleted but not for the mutant derivatives. This suggests that it is sequestration of miR-133 and miR-135 by linc-MD1 that acts to maintain the expression of MAML1 and MEF2C during myogenic differentiation.

PTENP1 is a pseudogene that shares a high similarity with *PTEN*, its 3′-UTR being 1 kb shorter than *PTEN* itself [63]. *PTENP1* possesses a binding site for miR-499-5p which can target both PTEN and *PTENP1*. In a high-fat-diet-fed mice, it was observed that PTENP1 levels were upregulated with a reduction in Akt/GSK signaling pathway and decreased glycogen contents. When high-fat-diet-fed mice were injected with shRNA against *PTENP1*, miR-499-5p levels did not change although Akt/GSK phosphorylation increased as did glycogen synthesis indicating a shift from the insulin resistance. However, the levels of PTEN, a target of miR-499-5p, were found to be significantly reduced that led to the activation of the Akt/GSK signaling pathway. These studies proposed a mechanism whereby *PTENP1* acts as a decoy for miR-499-5p to regulate the activity of *PTEN* in insulin resistance.

Studies by Liang et al. directed toward understanding the role of lncRNAs in osteogenesis revealed the involvement of *H19* in promoting osteoblast differentiation from human mesenchymal stem cells (hMSCs) in vitro [64]. Bioinformatic analyses showed that the *H19* sequence harbor binding sites for miR-141 and miR-22; however, their ectopic expression in cells did not affect the *H19* levels. Subsequent luciferase reporter assays however showed a reduction in luciferase expression, and ablation of this effect when the miR binding sites were mutated thereby establishing that these miRs act through translational repression rather than degradation of the target. With respect to osteoblast differentiation, it was observed that miR-141 and miR-22 were both downregulated during in vitro differentiation, and ectopic expression of miRNA mimics of miR-141 and miR-22 prevented proper osteogenic differentiation of hMSCs. Furthermore, β-catenin was validated as a common target of both these miRNAs. Interestingly, Wnt signaling plays a significant role during osteogenesis. When a luciferase vector containing the miR-141 and miR-22 binding sites were coexpressed with a *H19* overexpression vector, the luciferase activity was found to be upregulated, suggesting that *H19* is probably involved in decoying the miRNAs. *H19* was also shown to activate Wnt signaling and increase β-catenin levels in cells. These experiments indicated that *H19* may act as a sponge for miR-141 and miR-22 which otherwise act as negative regulators of Wnt signaling, thereby causing Wnt-mediated osteogenic differentiation of hMSCs.

In a recent investigation, an interplay between *BC032469*, *hTERT*, and miR-1207-5p and miR-1266 was investigated in the proliferation of gastric cancer (GC) cells [65]. It was found to be among the most differentially expressed panel of lncRNAs perturbed in hTERT positive versus negative gastric cancer tissues. hTERT is the rate-limiting subunit of telomerase, and its elevated expression is associated with several malignancies. Knockdown of *BC032469* in gastric cancer cell lines results in significant downregulation of *hTERT*, and it also had a negative effect on the cell proliferation. It was shown that miR-1207-5p and miR-1266 can target both *hTERT* and *BC032469*. *BC032469* knockdown resulted in *hTERT* downregulation, and this could be partially reversed by providing anti miR-1207-5p. Overexpression of miR-1207-5p attenuated the *BC032469*-mediated hTERT activation. These results indicate that lncRNA *BC032469* acts a miRNA sponge to sequester miR-1207-5p, which otherwise binds to and degrades its target hTERT, leading to upregulation of hTERT activity in GC that contributes to growth and progression of gastric cancer.

2.8 lncRNAs in Staufen-Mediated mRNA Decay

The half-life of an mRNA is vital to cell survival and disease manifestation in eukaryotes. In the cytoplasm, mRNA stability by nonsense-mediated mRNA decay (NMD), nonstop mRNA decay (NSD), and no-go mRNA decay (NGD) [66]. Cis-regulatory elements in the 3′UTR of mRNAs also regulate their stability, one such mechanism being through Staufen-mediated decay (SMD). STAU1 is an established protein effector of SMD, which binds to double-stranded RNA formed either by inter- or intramolecular base pairing. In the scenario of intermolecular base pairing, lncRNAs are known to play a function. Such lncRNA contains Alu elements within their sequences that base pair via partial complementarity to the 3′UTRs of target mRNAs, thereby activating STAU-mediated decay of mRNAs (Fig. 2.1c).

Studies by Gong and Maquat were focused on the identification of double-stranded RNA structures in the 3′UTR of SMD targets that were similar to the STAU1 binding site (SBS) of ARF1 mRNA [67]. Two well-established SMD targets, plasminogen activator inhibitor 1 (SERPINE1) and FLJ21870 (or, ANKRD57), contain only a single Alu element in their 3′UTRs. In parallel, they also screened for lncRNAs that have a single Alu element in their sequences. They concentrated their studies on the lncRNA AF08799 that is derived from chromosome 11 of humans and potentially binds to the 3′UTRs of SERPINE1 and FLJ21870. This lncRNA was referred to as *1/2-sbsRNA1* and was found to be unaffected in expression in the STAU1-depleted conditions and was also found to be not processed by the Dicer or Drosha machineries. Interestingly, not only STAU1 depletion but knockdown of *1/2-sbsRNA1* caused an upregulation in the levels of both SERPINE1 and FLJ21870 mRNAs. Luciferase reporter assays in which the luciferase gene harbored the 3′UTRs of either of the mRNAs showed an increase in activity when *1/2-sbs RNA1* was knocked down in cells in comparison to no UTR luciferase vector control. Co-immunoprecipitation experiments in lysates of HeLa cells revealed that *1/2-sbs RNA1* interacts directly with STAU1 and SERPINE1 or FLJ21870 as well as UPF1, a protein involved in SMD. Furthermore, depletion of STAU1 reduced the interaction of *1/2-sbs RNA1* with SERPINE1 or FLJ21870 mRNAs thereby proving that binding of *1/2-sbs RNA1* to these mRNAs generates a STAU1 binding site and the binding of STUA1 stabilizes the duplex for SMD.

Further characterization of lncRNAs involved in SMD by the same group led to the identification of three other lncRNAs such as *lncRNA_BC058830* (*1/2-sbsRNA2*), *lncRNA_AF075069* (*1/2-sbsRNA3*), and *lncRNA_BC009800* (*1/2-sbsRNA4*). *1/2-sbs RNA2* targeted CDCP1 mRNA, while *1/2-sbs RNA3* and *1/2-sbs RNA4* were found to target the 3′UTR of MTAP mRNA. Knockdown of these lncRNAs individually resulted in an expected upregulation of their target mRNAs as did STAU1 or UPF1 downregulation as well. Interestingly, none of these three lncRNAs targeted SERPINE1 mRNA, revealing their specificity of action. This study was a novel one aimed at understanding the role of lncRNAs in SMD.

2.9 Biology of Natural Antisense Transcripts

Natural antisense transcripts (NATs), as the name suggests, are endogenously occurring transcripts that are coded from the opposite or antisense strand to the host gene locus. The host gene can itself be a protein-coding or a noncoding one. NATs were first discovered in prokaryotes, are classified under lncRNAs because of their length, and are associated with ~7–30% of all genes in eukaryotes (reviewed in [68]). NATs can be either cis-NATs wherein they are encoded from the opposite strand of a gene and are complementary to the sense transcript of the same gene or trans-NATs wherein they are transcribed from a locus and show partial or full complementary to the transcript from a different locus on the same chromosome or different chromosome. Schematic representation of lncRNAs acting as NATs is depicted in Fig. 2.1d.

TSIX, the antisense transcript of X-inactive specific transcript (*XIST*), is a well-investigated NAT. To give effect to dosage compensation between males and females (females have an extra X chromosome as compared to males), *XIST* acts to coat the future inactive X chromosome in females, the inactivation in all embryonic tissues being random [69]. In order to prevent *XIST* from inactivating both the chromosomes, the future active X chromosome transcribes *TSIX*, the antisense transcript of *XIST*, thereby precluding *XIST* from acting on this chromosome. Studies performed by Zhao et al. revealed that within *Xist*, another short repeat RNA termed as *RepA*, is generated. *RepA* targets the polycomb repressive complex (PRC2) to the future inactive X chromosome and causes the establishment of repressive histone marks thereby leading to its inactivation [12]. *TSIX* acts to inhibit this interaction on the future active chromosome thus providing effect to dosage compensation. In the studies by Ohhata et al., transgenic mice containing a multiple polyadenylation signal at exon 4 of *Tsix* gene were developed; these mice prematurely terminate the transcription of *Tsix* [70]. The premature transcription termination of *Tsix* led to inappropriate activation of *Xist* on the allele which would otherwise have undergone silencing, and this activation was achieved by the establishment of active histone marks and loss of DNA methylation at the *Xist* promoter. The *Xist-Tsix* sense-antisense pair thus presents forth a wonderful paradigm for understanding the mode of action cis-NATs.

Faghihi et al. identified the antisense counterpart of *β-secretase 1* (*BACE1*), an enzyme central to Alzheimer's disease pathophysiology, and termed it as *BACE1-AS* [71]. A 2 kb-long transcript, it is encoded from the strand opposite to that of BACE1, and it acts to regulate the levels of BACE1 in a concordant manner. This implies that *BACE1-AS* positively co-regulates the activity of BACE1 which was evident from the observation that siBACE1-AS cells showed a downregulation of both *BACE1-AS* (as expected) along with BACE1 itself. The effect on BACE1 was indeed observed at both the mRNA and the protein levels, and it depended on the concentration of si*BACE1-AS*, implying that higher siRNA concentrations led to higher downregulation of the BACE1 transcript. Interestingly, a similar effect was observed in vivo when mouse brain was subjected to a continuous infusion of *siBACE1-AS*. Furthermore, in vitro experiments, when cells were exposed to various types of cell stressors like high temperature, serum starvation, hydrogen peroxide, etc. (which are known to induce AD pathology), it was observed that *BACE1-AS* levels were upregulated anywhere between 30 and 130% with a concomitant increase in BACE1 levels as well. In samples from patients undergoing AD, *BACE1-AS* was found to have significantly higher expression as was BACE1, supporting the fact that *BACE1-AS* forms RNA duplex structure with BACE1 and acts to stabilize the mRNA. Such action of a NAT to stabilize its host mRNA in AD has important implications because not only does it enhance the activity of BACE1 in AD patients thereby playing a role in AD pathology, but also it can serve as a biomarker to detect AD in patients.

Nkx2.2AS is yet another example of a NAT that exerts a positive effect on the levels of its corresponding sense transcript *Nkx2.2* [72]. Nkx2.2 drives the differentiation of neural stem cells toward the oligodendrocyte lineage (oligodendrocytes are a type of glial cells of the brain). Interestingly, when neural stem cell cultures were induced to differentiate by removal of Fgf-b, overexpression of *Nkx2.2 AS* led to the formation of a larger population of oligodendrocytes. It was also observed that *Nkx2.2 AS* overexpression in neural stem cells led to a significant increase in Nkx2.2 levels thereby proving that *Nkx2.2 AS* acts to stabilize and positively regulate Nkx2.2.

Brain-derived neurotrophic factor *(BDNF)* is a neurotrophin required for neuronal growth, maturation, plasticity, axonal, and dendritic differentiation processes as well as in learning and memory. The antisense transcript of BDNF is referred to as *BDNF-AS* and is encoded 200 kb downstream of *Bdnf* locus [73]. Knockdown of *BDNF-AS* by siRNA treatment increased the mRNA and protein levels of BDNF although it did not affect the mRNA stability per se. When neurospheres (in vitro cultures of neural stem cells) were subjected to si*BDNF-AS*, it was observed that endogenous BDNF levels increased along with higher neuronal differentiation and neurite outgrowth. These results were also corroborated with in vivo studies on mouse brain wherein antagoNAT (antagonist to NAT, in this case *BDNF-AS*) indeed caused increased cellular proliferation and upregulation of BDNF levels. Mechanistically, *BDNF-AS* acts to induce the establishment of repressive chromatin marks on the BDNF locus through the recruitment of EZH2 (a component of PRC2), exemplifying the action of regulation by a NAT through epigenetic modifications.

Genome-wide analysis by tiling arrays in *Arabidopsis* revealed the presence of around 37,000 NATs, and it was observed that almost 70% of the protein-coding genes are associated with potential NATs [74]. A custom synthesized array was designed to profile the expression of NATs in various organs of *Arabidopsis* like roots, leaves, and inflorescence which led to the identification of ~15,000 NAT pairs (sense-antisense transcript pairs) in all of the organs combined with some showing tissue-specific expression as well. In order to further understand the tissue-specific regulation of NATs in *Arabidopsis*, etiolated seedlings (seedlings grown in lack of light conditions) and de-etiolated seedlings (seedlings grown in continuous white light for 1 and 6 h) were subjected to expression analysis on arrays. Interestingly, many of the candidate NATs were regulated in expression by light. SPA1 which encodes for a light signaling repressor protein was upregulated after 1 h of light exposure to seedlings along with its concordant antisense transcript, *AT2TU076050*. Intriguingly, the upregulation was seen only in the cotyledons of the seedlings, establishing the organ-specific regulation of NAT expression. Again, the mRNA coding for the protein AT3G49970, a phototropic-responsive protein, was downregulated in hypocotyls, but its associated NAT, *AT3TU075200*, was seen to be upregulated. Such sense-antisense pairs in *Arabidopsis* were observed to be epigenetically regulated in response to light. NATs upregulated under light conditions displayed enrichment for active histone marks; the same was true for NATs that were induced under dark conditions. Based on that, it was proposed that light-induced stimuli in *Arabidopsis* are brought about either by changes in histone modifications on the loci of NATs themselves or by NAT-guided epigenetic changes on their corresponding sense loci.

2.10 lncRNA-Mediated Regulation of Protein Activity

lncRNAs are also involved in regulation of activity of proteins involved in processes other than transcription. Hellwig and Bass (2008) reported the role of *C. elegans* lncRNA *rncs-1* in the processing of small RNAs through binding and subsequent inhibition of the Dicer enzyme [75]. lncRNA *rncs-1 is* 800 nt long with its expression restricted to the hypodermis and intestine. It is transcriptionally regulated in response to food supply. *rncs-1* per se is not a substrate for Dicer due to the presence of highly branched structures at its termini, but rather *rncs-1* RNA competitively inhibits Dicer-mediated cleavage of dsRNA, as mRNA levels of several Dicer-regulated genes vary with the changes in *rncs-1* expression. An lncRNA termed as sfRNA (subgenomic flavivirus RNA) is produced by flaviviruses, such as West Nile virus, and is essential for the pathogenicity of the virus [76]. This lncRNA is a highly structured RNA of 0.3–0.5 kb, derived from the 3′ untranslated region of the viral genome as a product of incomplete degradation of viral genomic RNA by cellular ribonucleases. The presence of highly conserved RNA structures at the start of the 3′ untranslated region render this RNA resistant to nucleases. Mechanism of sfRNA function involves inhibition of the host XRN1 enzyme (5′ to 3′ exoribonuclease) during the viral RNA genome degradation and is essential for virus-induced cytopathicity and pathogenicity. Very recently, two lncRNAs have been shown to execute their biological function by the regulation of protein activity. Marchese FP et al. demonstrate the role of lncRNA *CONCR* (cohesion regulator noncoding RNA) in the regulation of sister chromatid cohesion [77]. *CONCR* is an MYC-activated, cell cycle-regulated lncRNA required for DNA replication and cell cycle progression. CONCR interacts with and regulates the activity of DDX11, a DNA-dependent ATPase and helicase, thereby affecting DNA replication and sister chromatid cohesion. Liu X et al. report the regulation of kinase signaling by lncRNA in the context of metabolic stress response. lncRNA neighbor of BRCA1 gene 2 (NBR2) is induced by the LKB1-AMPK pathway in the conditions of energy stress and also shown to interact with AMPK [78]. Interaction of lncRNA NBR2 with AMPK promotes its kinase activity during energy stress. Loss of NBR2 attenuates AMPK activation, leading to defective cell cycling, altered apoptosis/autophagy response, and increased tumor development. The hypoxia-regulated lncRNA linc-p21was shown to bind HIF-1α and VHL and disrupt the VHL-HIF-1α interaction which leads to increased accumulation of HIF-1α [79]. This positive feedback loop between HIF-1α and lincRNA-p21 promotes glycolysis under hypoxia, indicating the importance of lincRNA-p21 in the regulation of the Warburg effect in tumor cells.

In addition to the modulation of gene expression through effects on mRNAs, some of the lncRNAs also regulate the activity of transcription factors. For example, lncRNA Evf2 which is transcribed from the Dlx-5/6 ultraconserved region as an alternatively spliced form of Evf-1 RNA. lncRNA Evf2 forms a stable complex with the Dlx2 transcription factor and thereby enhances the transcriptional activation of the Dlx-5/6 enhancer mediated by Dlx2 [80]. A different dimension to the regulation of protein activity by lncRNA emerges from the sequestration of transcription factors by lncRNA Gas5 which is induced under the conditions of nutrient deprivation and cellular stress [81]. Gas5 binds to the DNA binding domain of glucocorticoid receptor (GR) and acts as a decoy by competing with GR-responsive elements (GREs) in gene promoters for binding to the DNA binding domain of the GR. Therefore, Gas5 lncRNA modulates the transcriptional activity of the GR and functions as a riborepressor of the GR. Another example is the Lethe lncRNA which is regulated by TNF-α. Lethe RNA interacts with NFkB effector subunit RelA in an inducible fashion and inhibits RelA from binding to the target gene promoter [82]. The NRON (non-coding repressor of NFAT) lncRNA binds the transport receptor importin-β, and knockdown of NRON leads to nuclear accumulation of the transcription factor NFAT (nuclear factor of activated T cells) [83]. Thus NRON lncRNA competes for importin-β binding with NFAT and thus indirectly represses transcription by inhibiting the nucleocytoplasmic shuttling of NFAT.

2.11 Nuclear Architecture and Long Noncoding RNAs

The eukaryotic nucleus is a highly compartmentalized organelle with a complex dynamic organization. In order to retain the genetic material within very small nuclear volume, the cell has to package the genomic DNA into chromatin. The extensive packaging has to be performed without compromising the functional activities like transcription, DNA replication, and repair. In other words a high degree of genomic plasticity has to be maintained for efficient readout, processing, maintenance, and transfer of the genetic information which is essential for the cell to adopt different functional states. The highly condensed genomic DNA in the chromatin is associated with histone as well as nonhistone proteins. Other than regulating the function of proteins, the role of RNA in the organization of chromatin is also being investigated. In 1965, Bonner and Huang were the first to report the association of RNA with chromatin [84]. Using pea bud chromatin, they demonstrated the association of small RNA approximately 40 nt in length with the native nucleohistone. In further studies spanning three decades, various architectural functions were proposed for nuclear RNAs. These included chromatin-associated RNAs as a structural component of heterochromatin [85], role in eukaryotic chromosome structure [86], and RNA as a component of nuclear matrix and a putative role of RNA for the structural integrity of the nuclear matrix [87–89].

Scheme of nuclear architecture: The nuclear organization reflects different domains or sub-compartments which share regulatory functions. These domains comprise of nuclear bodies and chromatin domains (reviewed in [90]). Nuclear bodies are harbored in the interchromatin space. Many of the nuclear bodies were characterized till date, including paraspeckles, cajal bodies, nucleoli, and polycomb bodies. These nuclear bodies are favorable sites for efficient biological functions. The organization of distinct nuclear bodies is similar to the organization of different organelles within the cytoplasm, and of note, the nuclear bodies lack a well-defined membrane separating them from their surroundings. Irrespective of the absence of a membrane, the nuclear bodies are structurally intact. Various studies are now shedding light on the assembly, maintenance, and regulation of nuclear bodies and also implicated a crucial role for lncRNAs in these aspects which will be discussed in detail in this review.

The chromatin domains are subdivided into transcriptionally active and inactive chromatin regions based on their gene expression status. With recent technological advancements particularly the chromosome conformation capture (3C) and other 3C-related methods such as 4C and HiC, it is clear that chromosomes occupy distinct areas in the nucleus termed as chromosome territories [91–93]. Gene-rich chromosomal regions tend to localize to the center of the nucleus and gene-poor regions near the periphery [94]. Chromatin domains with co-regulated genes often co-localize in vivo [95]. Genes located near the interphase chromosomal attachment regions to the nuclear matrix tend to be poorly transcribed [96]. Based on these observations, it can be inferred that an intimate relationship exist between the organization of genome in the nuclear space and gene activity.

2.12 lncRNAs and Nucleolar Function

The nucleolus is the nuclear domain for the rRNA synthesis, processing, and ribosomal assembly. At the onset of mitosis, i.e., during early prophase nuclear envelope breaks down and reassembles during telophase, the final stage of mitosis. This process is dependent on the RNA polymerase I transcription and the recruitment and activation of the pre-rRNA processing machinery [97]. The nucleolus contains a scaffold of tandem ribosomal DNA (rDNA) repeats. Each repeat comprises of an rDNA enhancer/promoter which is upstream of its ribosomal RNA (rRNA) coding sequence and separated by a large intergenic spacer (IGS) region. The IGS region contains a highly repetitive DNA. Characterization of different proteins in the nucleolus has shown that only 30% of the protein components have functions related to rRNA maturation. In addition to the nucleolar changes during cell cycle, the nucleolus also undergoes dramatic reorganization in response to different types of cellular stress [98, 99] (Fig. 2.2a). During different kinds of cellular stress such as acidosis, DNA damage, ribosomal stress, and serum starvation, there is preferential retention of different proteins (VHL, MDM2, DNMT1, HSP70) to the nucleolus. Therefore, in addition to being the site of rRNA maturation, the nucleolus is vital in the cellular response to stress. The nucleolar detained proteins, interact with different IGS regions downstream of the rRNA transcriptional start site. Interestingly,

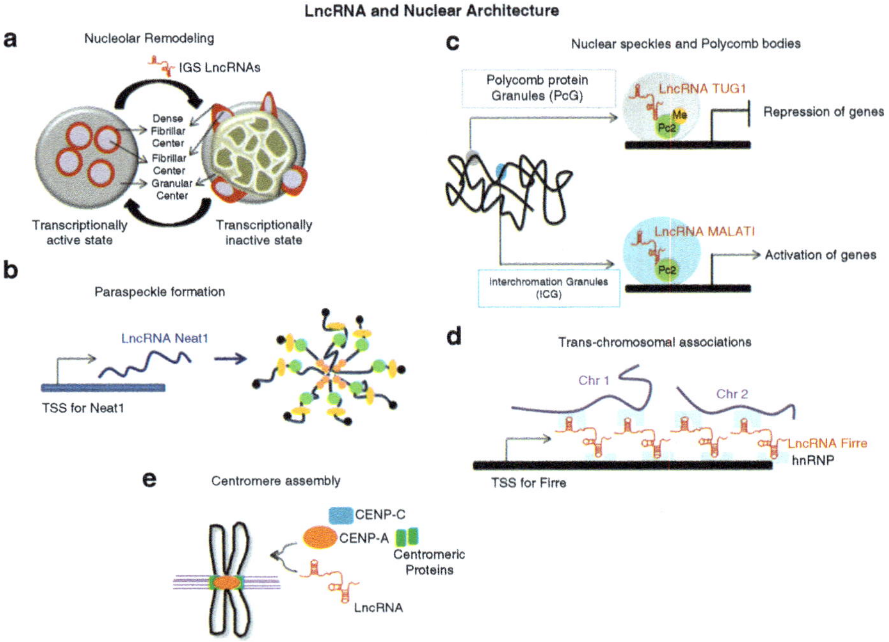

Fig. 2.2 Long noncoding RNAs and nuclear architecture. lncRNAs regulate diverse processes contributing to nuclear organization which include nucleolar remodeling (**a**), formation of paraspeckles (**b**), regulation of gene expression by association with nuclear speckles and polycomb bodies (**c**), topological organization of multichromosomal regions (**d**), and regulation of centromere assembly (**e**)

noncoding RNAs are expressed from IGS regions in response to the cellular stress that leads to the retention of the sequestered proteins at IGS loci [100]. Different IGS lncRNAs were reported to be transcribed, and each RNA responds to specific stress stimulus. Heat shock induces the transcription of IGS RNA located 16 kb ($IGS_{16}RNA$) and 22 kb ($IGS_{22}RNA$) downstream of the rDNA transcription start site. Similarly, acidosis or reduced extracellular pH induces $IGS_{28}RNA$ located 28 kb downstream of the rDNA transcription start site. These RNAs are transcribed by the RNA Pol I machinery and processed into 300–400 nucleotide products. Expression of each IGS transcript is independent from each other with respect to the different stress stimulus. They appear to have distinct functions as the ability of one IGS RNA to sequester its target proteins was not affected by knockdown of the other IGS RNA. Strikingly, the ectopic mislocalization of these IGS RNAs in other regions of the cell also results in protein retention which highlights their function in regulating protein dynamics or mobility. Furthermore, the functional role of these IGS lncRNAs is not only limited to protein sequestration but also in the remodeling of the nucleolus (Fig. 2.2a). Using different mammalian cell lines, Jacob MD et al. describe the involvement of IGS lncRNAs in bringing about the structural and functional adaptation of the nucleolus upon heat shock or acidosis [101]. During the latter stress conditions, the nucleolus undergoes reorganization involving de novo formation of the nucleolar detention center (DC). The DC is spatially, dynamically, and biochemically distinct sub-nucleolar structure with its architecture sustained by hydrophobic interactions among the IGS lncRNA-dependent sequestered proteins. Formation of DC leads to a reversible change in the distribution of nucleolar factors and also an arrest in the ribosome biogenesis. Formation of DC is correlated with the induction of IGS lncRNAs upon signal activation or exposure to environmental cues, and IGS RNA knockdown cells failed to form DC, and also they were unable to fully repress the nucleolar transcription upon heat shock. Therefore, the induction of IGS lncRNA by environmental cues acts as a molecular switch to regulate structure and function of the nucleolus.

Although RNA Pol I transcribes nucleolar rRNA genes, the inhibition of RNA Pol II also leads to downregulation of rRNA synthesis as well as disintegration of nucleoli [102–104]. This suggests a novel regulatory function for unknown RNA Pol II-dependent transcripts in the nucleolar organization and the Pol I transcriptional activity. In this regard, recent studies have characterized the functional role of RNA Pol II transcribed lncRNAs such as lncRNA PAPAS. The lncRNA PAPAS has been shown to be transcribed in the antisense orientation to the pre-rRNA coding region by RNA Pol II, which upon quiescence, directs the histone methyltransferase SUV4-20H2 to rRNA genes to induce histone H4 lysine 20 trimethylation (H4K20me3) and chromatin compaction [105]. The H4K20me3-mediated chromatin compaction was not just restricted to rRNA genes but was also present at other genomic elements like the IAP retrotransposons. In this case, the IAP retroelements were silenced by the IAP specific lncRNAs through the recruitment of Suv4-20 h2. In addition to quiescence, the PAPAS lncRNA also reinforces transcriptional repression of rRNA genes during hypotonic stress but by a different mechanism. During hypotonic stress, PAPAS lncRNA recruits the chromatin remodeling complex NuRD to the rDNA, leading to rDNA silencing. In another study, Herger MC et al. described the function of RNA Pol II transcripts originating from intronic Alu elements (aluRNAs) in nucleolar assembly and function. Earlier, Alu transcripts have been implicated in the regulation of gene expression and translation but not linked to the nucleolar organization [106, 107]. Through series of experiments, Herger MC et al. demonstrated that aluRNA interacts with the multifunctional nucleolar proteins nucleolin (NCL) and nucleophosmin (NPM) and tethered to chromatin and this is sufficient to target large genomic regions to the nucleolus [108]. This strongly suggests that the interaction of NCL and NPM with aluRNA is important to build up a functional nucleolus.

2.13 lncRNAs Are Critical Components of Paraspeckles

Paraspeckle is a prototypic example of mobile nuclear interchromatin subcompartments or nuclear bodies. Paraspeckles are observed in most mammalian cultured cell lines as subnuclear granules averaging about 360 nm in diameter [109]. The paraspeckle was first identified as a novel nuclear domain in a study to characterize the nucleolar proteome of different cultured human cells [110]. Paraspeckles are often located adjacent to splicing speckles and are marked by different proteins like PSPC1 (paraspeckle component protein 1), non-POU domain-containing octamer-binding protein (NONO, p54nrb), and SFPQ [111]. The requirement of RNA or RNA Pol II transcription in the maintenance of paraspeckle integrity was demonstrated by studies using the inhibitors like actinomycin D or D-ribofuranosylbenzimidazole (DRB) [111–113]. Several independent investigations have identified *NEAT1* lncRNA as the architectural RNA (arcRNA) of paraspeckles (Fig. 2.2b) [109–112]. Interestingly, ectopic nuclear accumulation of *NEAT1* lncRNA by tethering NEAT1 lncRNA in multiple copies to DNA leads to the de novo formation of paraspeckle [113]. In addition the role of NEAT1 lncRNA in the paraspeckle assembly was demonstrated by the direct visualization of the recruitment of the paraspeckle proteins [114]. NEAT1 has two isoforms: *NEAT1–1* (3.7 kb) and *NEAT1–2* (23 kb). In Neat1 knockout mouse embryonic fibroblasts, transient expression of Neat1–2, but not of Neat1–1, restores paraspeckle formation, demonstrating the architectural role of Neat1–2 in paraspeckle formation [115], which demonstrates a critical role for the NEAT-2 lncRNA in paraspeckle formation. However, overexpression of NEAT1–1 in cells expressing NEAT1–2 increases the number of paraspeckles, suggesting a supplementary role of NEAT1–1 in paraspeckle formation [115, 116]. Electron microscopy studies showed that the central region of NEAT1–2 is in the interior of the paraspeckle while the 5′ and 3′ terminus of NEAT1–2 at the paraspeckle periphery [109]. In contrast, NEAT-1 seems to localize primarily at the paraspeckle periphery. These observations collectively highlight an architectural role of NEAT1 transcripts in constraining the geometry of the paraspeckles.

2.14 lncRNAs in Nuclear Speckles and Polycomb Bodies

Nuclear speckles are nuclear domains enriched in pre-mRNA splicing factors, located in the interchromatin regions of the nucleoplasm of mammalian cells [117]. The constituents of nuclear speckles are dynamic on account of their continuous exchange with the nucleoplasm and active transcription sites. *MALAT1* lncRNA is found to be enriched in the nuclear speckles. The MALAT1 lncRNA interacts with serine/arginine (SR) splicing factors and modulates their phosphorylation and distribution to nuclear speckles, and knockdown of MALAT1 alters the splicing pattern of a subset of endogenous pre-mRNAs [118]. More importantly, the MALAT1 lncRNA has been shown to play an important role in large-scale nuclear architecture of the genome and consequently affect gene expression. This in part is achieved through lncRNA TUG1 which is localized to another distinct subnuclear body termed as polycomb group protein bodies (PcG bodies) (Fig. 2.2c). It has been proposed that activation of growth control genes occurs via inter-exchange between nuclear speckles and nuclear polycomb bodies via protein polycomb 2 (PC2) [119]. TUG1 and MALAT1 selectively interact with methylated and unmethylated Pc2 protein, respectively. Pc2 is present on growth control gene promoters. Under conditions of reduced cell growth, TUG1 lncRNA specifically interacts with methylated Pc2. This interaction leads to the repression of growth control genes via their recruitment to PcG bodies (Fig. 2.2c). However, in the presence of growth signals, MALAT1 lncRNA interacts with unmethylated Pc2, leading to relocation of the growth control genes to the "active" environment of nuclear speckles. Hence, these lncRNAs act as key factors in the spatial regulation of specific chromatin loci.

2.15 lncRNA and Topological Organization of Multichromosomal Regions

Multiple genes which are distributed on different chromosomes can often localize within shared regions of the nucleus. These regions located in the interchromosomal nuclear domains contain genes with shared functional roles or regulated by common distant regulatory elements [120–122]. A lncRNA termed linc-RAP-1 was first identified during a loss of function screen as being important for proper adipogenesis in the mouse adipocyte precursors [123]. This lncRNA was later termed Firre (functional intergenic repeating RNA element) and shown to localize within a single nuclear domain containing many genes previously implicated in energy metabolism [124]. Firre RNA localizes across a 5 Mb nuclear domain around its site of transcription on the X chromosome (Fig. 2.2d). This single nuclear domain not only includes the Firre transcription unit but also five other chromosomal loci located on different chromosomes in trans, including chromosomes 2, 9, 15, and 17. These trans chromosomal contacts require Firre RNA since deletion of the Firre locus results in the loss of spatial proximity between Firre and its trans chromosomal binding sites. Firre is localized in a punctate fashion in the nucleus. This localization is dependent on a unique 156-bp repeating RNA domain (RRD) in the Firre sequence which is also required for a physical interaction of Firre with the nuclear matrix factor hnRNPU (Fig. 2.2d). Knockdown of hnRNPU also leads to loss of spatial proximity between Firre locus and it trans chromosomal binding sites, an effect similar to the deletion of the Firre locus itself. Random integration of Firre into different chromosomal regions leads to the emergence of new nuclear foci, suggesting that Firre may be sufficient to create a nuclear domain at the integration sites [124]. Together, these observations suggest Firre lncRNA as a nuclear organization factor as it may serve to interface with and to modulate the topological organization of multiple chromosomes.

2.16 lncRNAs and Centromere Function

Centromeres are the chromosomal regions which are the platform for the formation of kinetochore and chromosome attachment to the mitotic spindle during cell division. Irrespective of their evolutionary conserved function, the centromere identity is established epigenetically rather than being defined by the underlying DNA sequence. CENP-A (also known as CID in *Drosophila melanogaster*) is a key epigenetic determinant of centromere identity [125]. Loss of CENP-A from cells results in the impairment of chromosome segregation, while CENP-A overexpression leads to the formation of ectopic centromeres and mislocalization of kinetochore proteins [126–128]. Deposition of CENP-A onto chromatin is carried out in a replication-independent manner unlike the canonical histones. Several studies from fission yeast to mammals; a consensus theme is emerging that lncRNA appear to play an important role in CENP-A regulation and deposition (Fig. 2.2e) [129–131]. The centromeric DNA in most organisms contains repetitive sequences or satellite sequences and form basis for characteristic pericentromeric heterochromatin [127]. Upon transcription, these repetitive and satellite sequences generate long noncoding RNAs, and these long noncoding transcripts have been shown to be involved in the initiation and maintenance of pericentromeric heterochromatin in *Drosophila*, Maize, mouse, and human cells (reviewed in [132]).

The centromeres of *D. melanogaster* contain satellite sequences which are mostly simple 5–12-bp-long repeats except for the centromere of X chromosome that contains a complex satellite repeat called as satellite III (SAT III) or 359-bp satellite and covers several megabase pairs of the acrocentric X chromosome with a 359-bp-long repeating unit [133, 134]. A recent study investigated the functional role of *D. melanogaster* SAT III region in the centromere regulation [135]. The SAT III region produces a long noncoding RNA (referred as SAT III RNA) that localizes to centromeric chromatin of the X chromosome as well as of autosomes during mitosis. SAT III RNA (~1.3 kb long) interacts with the inner kinetochore protein CENP-C and depends on CENP-C for its centromeric localization. Loss of SAT III RNA leads to mitotic defects and chromosome missegregation which is mostly attributed to the reduction of centromeric and kinetochore proteins during mitosis as well as the reduced levels of newly deposited CENP-A and CENP-C. Additionally, some SATIII RNA is also present at pericentromeres of mitotic chromosomes but not associated with chromatin. This pericentromeric SAT III RNA might also contribute to overall kinetochore structure. Hence, SAT III RNA is an integral part of centromere identity in *D. melanogaster* that influences centromere regulation epigenetically.

Centromeric repeats in maize are called CentC which are transcribed to produce transcripts that are about 900 bp long. These transcripts upon transcription associate with the maize CENP-A orthologue CENH3 and remain bound to the kinetochore and are thought to participate in the stabilization of centromeric chromatin [136]. In another recent study, Du et al. performed detailed genetic and biochemical characterization of maize inner kinetochore protein CENPC [137]. The DNA binding ability of CENPC is dependent on long single-stranded centromeric transcripts, which help in the recruitment of CENPC to the inner kinetochore.

The role of centromeric noncoding transcripts in mammalian models has been highlighted by many studies. Minor satellite repeats located on the mouse centromeres produce 4 kb-long transcripts and were implicated in regulating the centromeric function during stress response [138]. Knockdown of the transcribed murine centromeric minor satellite leads to defects in chromosome segregation [139]. These RNAs associate with CENP-A, as well as with components of the chromosomal passenger complex (CPC): Aurora B, Survivin, and INCENP [140]. The CPC is crucial for the regulation of chromosome-microtubule attachment and the activation of the spindle assembly checkpoint. Importantly, an RNA component is necessary for the observed interaction of Aurora B with CENP-A and for the Aurora B kinase activity. Interestingly, a very recent study by Michael Blower using Xenopus egg extracts showed that centromeric long noncoding RNAs bind to the CPC in vitro and in vivo [141]. His work demonstrates that the centromeric lncRNAs promote normal kinetochore function by regulating the localization and activation of the CPC.

In human cells, alpha-satellite repeats have been connected to centromere function [142]. Transcript derived from the alpha-satellite repeat is 1.3 kb long and mediates the localization of CENP-C and INCENP into the nucleolus in interphase cells, and subsequently during mitosis, there is relocation of CENP-C and INCENP to centromeres (Fig. 2.2e). In a recent report, Quenet and Dalal addressed the role of a centromeric long noncoding RNA in CENP-A loading [143]. This lncRNA co-immunoprecipitates with CENP-A and its loading factor HJURP in the chromatin fraction. Downregulation of the lncRNA leads to the loss of CENP-A and HJURP at the centromeres and thus causing severe mitotic defects.

2.17 Conclusion and Outlook

From the outlined literature, it is evident that lncRNAs influence gene expression at the transcription level in cis or trans either by acting as a molecular scaffold or a decoy. lncRNA, as a molecular scaffold, has been shown to influence gene expression by targeting chromatin remodelers to specific regions across the genome. Although this phenomenon is well characterized in several biological contexts, the reasons underlying the lncRNA-mediated targeting of chromatin remodelers to specific genomic regions across the genome are very poorly understood. Likewise, lncRNAs function as decoy, involves sequestering of chromatin remodelers and transcription factors from their site of action to regulate gene expression. Although the final outcome of lncRNA actions as molecular scaffold or a decoy on gene expression is similar, it is not clear about what molecular features of lncRNA that distinguish these contrasting functions. Besides, the scaffolding and decoy functions of lncRNAs also seem to control architecture of nuclear bodies such as paraspeckles, thus influencing the global transcriptional regulation. In our view, this would be one of the most interesting aspects, needing greater insights. Similarly, lncRNAs are also considered as regulatory framework to regulate the catalytic activity of several chromatin remodelers and transcription factors through interfering with their posttranslational modifications. This raises an important question of what features of lncRNAs that impedes or promotes the catalytic activity of transcription factors and chromatin remodelers. We propose that primary sequence together with its RNA secondary structures control various aspects of protein activity, probably by blocking the catalytically active sites. Though the technologies that address the secondary structures of RNAs have begun to be optimized, their contribution toward uncovering lncRNAs structures and their link to functions are still in their infancy. Hence, there is a greater emphasis required in optimizing technologies that can read physiological relevant secondary structures to understand their functions in various biological contexts such as molecular scaffold or decoys. Nonetheless, a decade long research on lncRNAs enabled us to know the power of dark matter of the genome in diverse biological functions. Further work should pave the way for understanding of hitherto unknown functions of lncRNAs in development and disease.

References

1. Mercer T, Mattick J (2013) Structure and function of long noncoding RNAs in epigenetic regulation. Nat Struct Mol Biol 20:300–307
2. Mercer T, Dinger M, Mattick J (2009) Long non-coding RNAs: insights into functions. Nat Rev Genet 10:155–159
3. Geisler S, Coller J (2013) RNA in unexpected places: long non-coding RNA functions in diverse cellular contexts. Nat Rev Mol Cell Biol 14:699–712
4. Wang X, Song X, Glass C, Rosenfeld M (2010) The long arm of long noncoding RNAs: roles as sensors regulating gene transcriptional programs. Cold Spring Harb Perspect Biol 3:a003756–a003756
5. Rinn J, Chang H (2012) Genome regulation by long noncoding RNAs. Annu Rev Biochem 81:145–166
6. Nagano T, Fraser P (2011) No-nonsense functions for long noncoding RNAs. Cell 145:178–181
7. Kugel J, Goodrich J (2012) Non-coding RNAs: key regulators of mammalian transcription. Trends Biochem Sci 37:144–151
8. Mondal T, Rasmussen M, Pandey G, Isaksson A, Kanduri C (2010) Characterization of the RNA content of chromatin. Genome Res 20:899–907
9. Mondal T, Subhash S, Vaid R, Enroth S, Uday S, Reinius B, Mitra S, Mohammed A, James A, Hoberg E et al (2015) MEG3 long noncoding RNA regulates the TGF-β pathway genes through formation of RNA–DNA triplex structures. Nat Commun 6:7743
10. Plath K (2003) Role of histone H3 lysine 27 methylation in X inactivation. Science 300:131–135
11. da Rocha S, Boeva V, Escamilla-Del-Arenal M, Ancelin K, Granier C, Matias N, Sanulli S, Chow J, Schulz E, Picard C et al (2014) Jarid2 is implicated in the initial Xist-induced targeting of PRC2 to the inactive X chromosome. Mol Cell 53:301–316
12. Zhao J, Sun B, Erwin J, Song J, Lee J (2008) Polycomb proteins targeted by a short repeat RNA to the mouse X chromosome. Science 322:750–756
13. Rinn J, Kertesz M, Wang J, Squazzo S, Xu X, Brugmann S, Goodnough L, Helms J, Farnham P, Segal E et al (2007) Functional demarcation of active and silent chromatin domains in human HOX loci by noncoding RNAs. Cell 129:1311–1323
14. Marín-Béjar O, Marchese F, Athie A, Sánchez Y, González J, Segura V, Huang L, Moreno I, Navarro A, Monzó M et al (2013) Pint lincRNA connects the p53 pathway with epigenetic silencing by the Polycomb repressive complex 2. Genome Biol 14:R104
15. Klattenhoff C, Scheuermann J, Surface L, Bradley R, Fields P, Steinhauser M, Ding H, Butty V, Torrey L, Haas S et al (2013) Braveheart, a long noncoding RNA required for cardiovascular lineage commitment. Cell 152:570–583

16. Grote P, Herrmann B (2013) The long non-coding RNA Fendrr links epigenetic control mechanisms to gene regulatory networks in mammalian embryogenesis. RNA Biol 10:1579–1585
17. Grote P, Wittler L, Hendrix D, Koch F, Währisch S, Beisaw A, Macura K, Bläss G, Kellis M, Werber M et al (2013) The tissue-specific lncRNA Fendrr is an essential regulator of heart and body wall development in the mouse. Dev Cell 24:206–214
18. Wongtrakoongate P, Riddick G, Fucharoen S, Felsenfeld G (2015) Association of the long non-coding RNA steroid receptor RNA activator (SRA) with TrxG and PRC2 complexes. PLoS Genet 11:e1005615
19. Yang Y, Flynn R, Chen Y, Qu K, Wan B, Wang K, Lei M, Chang H (2014) Essential role of lncRNA binding for WDR5 maintenance of active chromatin and embryonic stem cell pluripotency. Elife 3:e02046
20. Wang K, Yang Y, Liu B, Sanyal A, Corces-Zimmerman R, Chen Y, Lajoie B, Protacio A, Flynn R, Gupta R et al (2011) A long noncoding RNA maintains active chromatin to coordinate homeotic gene expression. Nature 472:120–124
21. Gomez J, Wapinski O, Yang Y, Bureau J, Gopinath S, Monack D, Chang H, Brahic M, Kirkegaard K (2013) The NeST long ncRNA controls microbial susceptibility and epigenetic activation of the interferon-γ locus. Cell 152:743–754
22. Hu X, Feng Y, Zhang D, Zhao S, Hu Z, Greshock J, Zhang Y, Yang L, Zhong X, Wang L et al (2014) A functional genomic approach identifies FAL1 as an oncogenic long noncoding RNA that associates with BMI1 and represses p21 expression in cancer. Cancer Cell 26:344–357
23. Pasmant E, Laurendeau I, Heron D, Vidaud M, Vidaud D, Bieche I (2007) Characterization of a germ-line deletion, including the entire INK4/ARF locus, in a melanoma-neural system tumor family: identification of ANRIL, an antisense noncoding RNA whose expression coclusters with ARF. Cancer Res 67:3963–3969
24. Yap K, Li S, Muñoz-Cabello A, Raguz S, Zeng L, Mujtaba S, Gil J, Walsh M, Zhou M (2010) Molecular interplay of the noncoding RNA ANRIL and methylated histone H3 lysine 27 by polycomb CBX7 in transcriptional silencing of INK4a. Mol Cell 38:662–674
25. Prensner J, Iyer M, Sahu A, Asangani I, Cao Q, Patel L, Vergara I, Davicioni E, Erho N, Ghadessi M et al (2013) The long noncoding RNA SChLAP1 promotes aggressive prostate cancer and antagonizes the SWI/SNF complex. Nat Genet 45:1392–1398
26. Han P, Li W, Lin C, Yang J, Shang C, Nurnberg S, Jin K, Xu W, Lin C, Lin C et al (2014) A long noncoding RNA protects the heart from pathological hypertrophy. Nature 514:102–106
27. Minajigi A, Froberg J, Wei C, Sunwoo H, Kesner B, Colognori D, Lessing D, Payer B, Boukhali M, Haas W et al (2015) A comprehensive Xist interactome reveals cohesin repulsion and an RNA-directed chromosome conformation. Science 349:aab2276
28. Bond A, VanGompel M, Sametsky E, Clark M, Savage J, Disterhoft J, Kohtz J (2009) Balanced gene regulation by an embryonic brain ncRNA is critical for adult hippocampal GABA circuitry. Nat Neurosci 12:1020–1027
29. Cajigas I, Leib D, Cochrane J, Luo H, Swyter K, Chen S, Clark B, Thompson J, Yates J, Kingston R et al (2015) Evf2lncRNA/BRG1/DLX1 interactions reveal RNA-dependent inhibition of chromatin remodeling. Development 142:2641–2652
30. Wang Y, He L, Du Y, Zhu P, Huang G, Luo J, Yan X, Ye B, Li C, Xia P et al (2015) The long noncoding RNA lncTCF7 promotes self-renewal of human liver cancer stem cells through activation of Wnt signaling. Cell Stem Cell 16:413–425
31. Carpenter S, Aiello D, Atianand M, Ricci E, Gandhi P, Hall L, Byron M, Monks B, Henry-Bezy M, Lawrence J et al (2013) A long noncoding RNA mediates both activation and repression of immune response genes. Science 341:789–792
32. Hu G, Gong A, Wang Y, Ma S, Chen X, Chen J, Su C, Shibata A, Strauss-Soukup J, Drescher K et al (2016) LincRNA-Cox2 promotes late inflammatory gene transcription in macrophages through modulating SWI/SNF-mediated chromatin remodeling. J Immunol 196:2799–2808
33. Tong Q, Gong A, Zhang X, Lin C, Ma S, Chen J, Hu G, Chen X (2015) LincRNA-Cox2 modulates TNF− -induced transcription of Il12b gene in intestinal epithelial cells through regulation of Mi-2/NuRD-mediated epigenetic histone modifications. FASEB J 30:1187–1197

34. Kim T, Hemberg M, Gray J, Costa A, Bear D, Wu J, Harmin D, Laptewicz M, Barbara-Haley K, Kuersten S et al (2010) Widespread transcription at neuronal activity-regulated enhancers. Nature 465:182–187
35. Ørom U, Derrien T, Beringer M, Gumireddy K, Gardini A, Bussotti G, Lai F, Zytnicki M, Notredame C, Huang Q et al (2010) Long noncoding RNAs with enhancer-like function in human cells. Cell 143:46–58
36. Koch F, Fenouil R, Gut M, Cauchy P, Albert T, Zacarias-Cabeza J, Spicuglia S, de la Chapelle A, Heidemann M, Hintermair C et al (2011) Transcription initiation platforms and GTF recruitment at tissue-specific enhancers and promoters. Nat Struct Mol Biol 18:956–963
37. Natoli G, Andrau J (2012) Noncoding transcription at enhancers: general principles and functional models. Annu Rev Genet 46:1–19
38. Andersson R, Gebhard C, Miguel-Escalada I, Hoof I, Bornholdt J, Boyd M, Chen Y, Zhao X, Schmidl C, Suzuki T et al (2014) An atlas of active enhancers across human cell types and tissues. Nature 507:455–461
39. Antoniou M, Grosveld F (1990) Beta-globin dominant control region interacts differently with distal and proximal promoter elements. Genes Dev 4:1007–1013
40. Ashe H, Monks J, Wijgerde M, Fraser P, Proudfoot N (1997) Intergenic transcription and transinduction of the human beta -globin locus. Genes Dev 11:2494–2509
41. Melo C, Drost J, Wijchers P, van de Werken H, de Wit E, Vrielink J, Elkon R, Melo S, Léveillé N, Kalluri R et al (2013) eRNAs are required for p53-dependent enhancer activity and gene transcription. Mol Cell 49:524–535
42. Zhang Y, Wong C, Birnbaum R, Li G, Favaro R, Ngan C, Lim J, Tai E, Poh H, Wong E et al (2013) Chromatin connectivity maps reveal dynamic promoter–enhancer long-range associations. Nature 504:306–310
43. Trimarchi T, Bilal E, Ntziachristos P, Fabbri G, Dalla-Favera R, Tsirigos A, Aifantis I (2014) Genome-wide mapping and characterization of notch-regulated long noncoding RNAs in acute leukemia. Cell 158:593–606
44. Yang L, Lin C, Jin C, Yang J, Tanasa B, Li W, Merkurjev D, Ohgi K, Meng D, Zhang J et al (2013) lncRNA-dependent mechanisms of androgen-receptor-regulated gene activation programs. Nature 500:598–602
45. Lai F, Orom U, Cesaroni M, Beringer M, Taatjes D, Blobel G, Shiekhattar R (2013) Activating RNAs associate with mediator to enhance chromatin architecture and transcription. Nature 494:497–501
46. Schaukowitch K, Joo J, Liu X, Watts J, Martinez C, Kim T (2014) Enhancer RNA facilitates NELF release from immediate early genes. Mol Cell 56:29–42
47. Sigova A, Abraham B, Ji X, Molinie B, Hannett N, Guo Y, Jangi M, Giallourakis C, Sharp P, Young R (2015) Transcription factor trapping by RNA in gene regulatory elements. Science 350:978–981
48. Yin Y, Yan P, Lu J, Song G, Zhu Y, Li Z, Zhao Y, Shen B, Huang X, Zhu H et al (2015) Opposing roles for the lncRNA haunt and its genomic locus in regulating HOXA gene activation during embryonic stem cell differentiation. Cell Stem Cell 16:504–516
49. Welsh I, Kwak H, Chen F, Werner M, Shopland L, Danko C, Lis J, Zhang M, Martin J, Kurpios N (2015) Chromatin architecture of the Pitx2 locus requires CTCF- and Pitx2-dependent asymmetry that mirrors embryonic gut laterality. Cell Rep 13:337–349
50. Kanduri C (2016) Long noncoding RNAs: lessons from genomic imprinting. Biochim Biophys Acta 1859:102–111
51. Wutz A, Barlow DP (1998) Imprinting of the mouse Igf2r gene depends on an intronic CpG island. Mol Cell Endocrinol 140:9–14
52. Sleutels F, Zwart R, Barlow D (2002) The non-coding Air RNA is required for silencing autosomal imprinted genes. Nature 415:810–813
53. Mohammad F, Mondal T, Kanduri C (2009) Epigenetics of imprinted long non-coding RNAs. Epigenetics 4:277–286
54. Nagano T, Mitchell J, Sanz L, Pauler F, Ferguson-Smith A, Feil R, Fraser P (2008) The air noncoding RNA epigenetically silences transcription by targeting G9a to chromatin. Science 322:1717–1720

55. Mohammad F, Mondal T, Guseva N, Pandey G, Kanduri C (2010) Kcnq1ot1 noncoding RNA mediates transcriptional gene silencing by interacting with Dnmt1. Development 137:2493–2499
56. Pandey R, Mondal T, Mohammad F, Enroth S, Redrup L, Komorowski J, Nagano T, Mancini-DiNardo D, Kanduri C (2008) Kcnq1ot1 antisense noncoding RNA mediates lineage-specific transcriptional silencing through chromatin-level regulation. Mol Cell 32:232–246
57. Kanduri C (2011) Kcnq1ot1: a chromatin regulatory RNA. Semin Cell Dev Biol 22:343–350
58. Mohammad F, Pandey G, Mondal T, Enroth S, Redrup L, Gyllensten U, Kanduri C (2012) Long noncoding RNA-mediated maintenance of DNA methylation and transcriptional gene silencing. Development 139:2792–2803
59. Ripoche M, Kress C, Poirier F, Dandolo L (1997) Deletion of the H19 transcription unit reveals the existence of a putative imprinting control element. Genes Dev 11:1596–1604
60. Monnier P, Martinet C, Pontis J, Stancheva I, Ait-Si-Ali S, Dandolo L (2013) H19 lncRNA controls gene expression of the Imprinted Gene Network by recruiting MBD1. Proc Natl Acad Sci 110:20693–20698
61. Brannan C, Dees E, Ingram R, Tilghman S (1990) The product of the H19 gene may function as an RNA. Mol Cell Biol 10:28–36
62. Cesana M, Cacchiarelli D, Legnini I, Santini T, Sthandier O, Chinappi M, Tramontano A, Bozzoni I (2011) A long noncoding RNA controls muscle differentiation by functioning as a competing endogenous RNA. Cell 147:358–369
63. Wang L, Zhang N, Wang Z, Ai D, Cao Z, Pan H (2016) Pseudogene PTENP1 functions as a competing endogenous RNA (ceRNA) to regulate PTEN expression by sponging miR-499-5p. Biochemistry 81:739–747
64. Liang W, Fu W, Wang Y, Sun Y, Xu L, Wong C, Chan K, Li G, Waye M, Zhang J (2016) H19 activates Wnt signaling and promotes osteoblast differentiation by functioning as a competing endogenous RNA. Sci Rep 6:20121
65. Lü M, Tang B, Zeng S, Hu C, Xie R, Wu Y, Wang S, He F, Yang S (2015) Long noncoding RNA BC032469, a novel competing endogenous RNA, upregulates hTERT expression by sponging miR-1207-5p and promotes proliferation in gastric cancer. Oncogene 35:3524–3534
66. Park E, Maquat L (2013) Staufen-mediated mRNA decay. Wiley Interdiscip Rev 4:423–435
67. Gong C, Maquat L (2011) lncRNAs transactivate STAU1-mediated mRNA decay by duplexing with 3′ UTRs via Alu elements. Nature 470:284–288
68. Su W, Xiong H, Fang J (2010) Natural antisense transcripts regulate gene expression in an epigenetic manner. Biochem Biophys Res Commun 396:177–181
69. Brown C, Hendrich B, Rupert J, Lafrenière R, Xing Y, Lawrence J, Willard H (1992) The human XIST gene: analysis of a 17 kb inactive X-specific RNA that contains conserved repeats and is highly localized within the nucleus. Cell 71:527–542
70. Ohhata T, Hoki Y, Sasaki H, Sado T (2007) Crucial role of antisense transcription across the Xist promoter in Tsix-mediated Xist chromatin modification. Development 135:227–235
71. Faghihi M, Modarresi F, Khalil A, Wood D, Sahagan B, Morgan T, Finch C, St. Laurent G III, Kenny P, Wahlestedt C (2008) Expression of a noncoding RNA is elevated in Alzheimer's disease and drives rapid feed-forward regulation of β-secretase. Nat Med 14:723–730
72. Tochitani S, Hayashizaki Y (2008) Nkx2.2 antisense RNA overexpression enhanced oligodendrocytic differentiation. Biochem Biophys Res Commun 372:691–696
73. Modarresi F, Faghihi M, Lopez-Toledano M, Fatemi R, Magistri M, Brothers S, van der Brug M, Wahlestedt C (2012) Inhibition of natural antisense transcripts in vivo results in gene-specific transcriptional upregulation. Nat Biotechnol 30:453–459
74. Wang H, Chung P, Liu J, Jang I, Kean M, Xu J, Chua N (2014) Genome-wide identification of long noncoding natural antisense transcripts and their responses to light in Arabidopsis. Genome Res 24:444–453
75. Hellwig S, Bass B (2008) A starvation-induced noncoding RNA modulates expression of Dicer-regulated genes. Proc Natl Acad Sci 105:12897–12902
76. Pijlman G, Funk A, Kondratieva N, Leung J, Torres S, van der Aa L, Liu W, Palmenberg A, Shi P, Hall R et al (2008) A highly structured, nuclease-resistant, noncoding RNA produced by flaviviruses is required for pathogenicity. Cell Host Microbe 4:579–591

77. Marchese F, Grossi E, Marín-Béjar O, Bharti S, Raimondi I, González J, Martínez-Herrera D, Athie A, Amadoz A, Brosh R et al (2016) A long noncoding RNA regulates sister chromatid cohesion. Mol Cell 63:397–407
78. Liu X, Xiao Z, Han L, Zhang J, Lee S, Wang W, Lee H, Zhuang L, Chen J, Lin H et al (2016) LncRNA NBR2 engages a metabolic checkpoint by regulating AMPK under energy stress. Nat Cell Biol 18:431–442
79. Yang F, Zhang H, Mei Y, Wu M (2014) Reciprocal regulation of HIF-1α and LincRNA-p21 modulates the Warburg effect. Mol Cell 53:88–100
80. Feng J (2006) The Evf-2 noncoding RNA is transcribed from the Dlx-5/6 ultraconserved region and functions as a Dlx-2 transcriptional coactivator. Genes Dev 20:1470–1484
81. Kino T, Hurt D, Ichijo T, Nader N, Chrousos G (2010) Noncoding RNA Gas5 is a growth arrest- and starvation-associated repressor of the glucocorticoid receptor. Sci Signal 3:ra8
82. Rapicavoli N, Qu K, Zhang J, Mikhail M, Laberge R, Chang H (2013) A mammalian pseudogene lncRNA at the interface of inflammation and anti-inflammatory therapeutics. Elife 2:e00762
83. Willingham A (2005) A strategy for probing the function of noncoding RNAs finds a repressor of NFAT. Science 309:1570–1573
84. Huang R, Bonner J (1965) Histone-bound RNA, a component of native nucleohistone. Proc Natl Acad Sci U S A 54:960–967
85. Paul I, Duerksen J (1975) Chromatin-associated RNA content of heterochromatin and euchromatin. Mol Cell Biochem 9:9–16
86. Pederson T, Bhorjee J (1979) Evidence for a role of RNA in eukaryotic chromosome structure. Metabolically stable, small nuclear RNA species are covalently linked to chromosomal DNA in HeLa cells. J Mol Biol 128:451–480
87. Nickerson J, Krochmalnic G, Wan K, Penman S (1989) Chromatin architecture and nuclear RNA. Proc Natl Acad Sci 86:177–181
88. Berezney R (1991) The nuclear matrix: a heuristic model for investigating genomic organization and function in the cell nucleus. J Cell Biochem 47:109–123
89. Ma H, Siegel A, Berezney R (1999) Association of chromosome territories with the nuclear matrix. J Cell Biol 146:531–542
90. Caudron-Herger M, Rippe K (2012) Nuclear architecture by RNA. Curr Opin Genet Dev 22:179–187
91. Dekker J (2002) Capturing chromosome conformation. Science 295:1306–1311
92. Dostie J, Richmond T, Arnaout R, Selzer R, Lee W, Honan T, Rubio E, Krumm A, Lamb J, Nusbaum C et al (2006) Chromosome conformation capture carbon copy (5C): a massively parallel solution for mapping interactions between genomic elements. Genome Res 16:1299–1309
93. Zhao Z, Tavoosidana G, Sjölinder M, Göndör A, Mariano P, Wang S, Kanduri C, Lezcano M, Singh Sandhu K, Singh U et al (2006) Circular chromosome conformation capture (4C) uncovers extensive networks of epigenetically regulated intra- and interchromosomal interactions. Nat Genet 38:1341–1347
94. Cremer T, Cremer M (2010) Chromosome territories. Cold Spring Harb Perspect Biol 2:a003889–a003889
95. Pombo A, Branco M (2007) Functional organisation of the genome during interphase. Curr Opin Genet Dev 17:451–455
96. Towbin B, Meister P, Gasser S (2009) The nuclear envelope—a scaffold for silencing? Curr Opin Genet Dev 19:180–186
97. Hernandez-Verdun D (2011) Assembly and disassembly of the nucleolus during the cell cycle. Nucleus 2:189–194
98. Boisvert F, Lam Y, Lamont D, Lamond A (2009) A quantitative proteomics analysis of subcellular proteome localization and changes induced by DNA damage. Mol Cell Proteomics 9:457–470
99. Boulon S, Westman B, Hutten S, Boisvert F, Lamond A (2010) The nucleolus under stress. Mol Cell 40:216–227

100. Audas T, Jacob M, Lee S (2012) Immobilization of proteins in the nucleolus by ribosomal intergenic spacer noncoding RNA. Mol Cell 45:147–157
101. Jacob M, Audas T, Uniacke J, Trinkle-Mulcahy L, Lee S (2013) Environmental cues induce a long noncoding RNA-dependent remodeling of the nucleolus. Mol Biol Cell 24:2943–2953
102. Haaf T, Ward D (1996) Inhibition of RNA polymerase II transcription causes chromatin decondensation, loss of nucleolar structure, and dispersion of chromosomal domains. Exp Cell Res 224:163–173
103. Sirri V, Hernandez-Verdun D, Roussel P (2002) Cyclin-dependent kinases govern formation and maintenance of the nucleolus. J Cell Biol 156:969–981
104. Burger K, Muhl B, Harasim T, Rohrmoser M, Malamoussi A, Orban M, Kellner M, Gruber-Eber A, Kremmer E, Holzel M et al (2010) Chemotherapeutic drugs inhibit ribosome biogenesis at various levels. J Biol Chem 285:12416–12425
105. Bierhoff H, Dammert M, Brocks D, Dambacher S, Schotta G, Grummt I (2014) Quiescence-induced LncRNAs trigger H4K20 trimethylation and transcriptional silencing. Mol Cell 54:675–682
106. Capshew C, Dusenbury K, Hundley H (2012) Inverted Alu dsRNA structures do not affect localization but can alter translation efficiency of human mRNAs independent of RNA editing. Nucleic Acids Res 40:8637–8645
107. Singer S, Männel D, Hehlgans T, Brosius J, Schmitz J (2004) From "junk" to gene: curriculum vitae of a primate receptor isoform gene. J Mol Biol 341:883–886
108. Caudron-Herger M, Pankert T, Seiler J, Nemeth A, Voit R, Grummt I, Rippe K (2015) Alu element-containing RNAs maintain nucleolar structure and function. EMBO J 34:2758–2774
109. Souquere S, Beauclair G, Harper F, Fox A, Pierron G (2010) Highly ordered spatial organization of the structural long noncoding NEAT1 RNAs within Paraspeckle nuclear bodies. Mol Biol Cell 21:4020–4027
110. Fox A, Lam Y, Leung A, Lyon C, Andersen J, Mann M, Lamond A (2002) Paraspeckles. Curr Biol 12:13–25
111. Fox A (2005) P54nrb forms a heterodimer with PSP1 that localizes to Paraspeckles in an RNA-dependent manner. Mol Biol Cell 16:5304–5315
112. Sasaki Y, Ideue T, Sano M, Mituyama T, Hirose T (2009) MEN/noncoding RNAs are essential for structural integrity of nuclear paraspeckles. Proc Natl Acad Sci U S A 106:2525–2530
113. Sunwoo H, Dinger M, Wilusz J, Amaral P, Mattick J, Spector D (2008) MEN/nuclear-retained non-coding RNAs are up-regulated upon muscle differentiation and are essential components of paraspeckles. Genome Res 19:347–359
114. Mao Y, Sunwoo H, Zhang B, Spector D (2010) Direct visualization of the co-transcriptional assembly of a nuclear body by noncoding RNAs. Nat Cell Biol 13:95–101
115. Naganuma T, Nakagawa S, Tanigawa A, Sasaki Y, Goshima N, Hirose T (2012) Alternative 3′-end processing of long noncoding RNA initiates construction of nuclear paraspeckles. EMBO J 31:4020–4034
116. Clemson C, Hutchinson J, Sara S, Ensminger A, Fox A, Chess A, Lawrence J (2009) An architectural role for a nuclear noncoding RNA: NEAT1 RNA is essential for the structure of Paraspeckles. Mol Cell 33:717–726
117. Spector D, Lamond A (2010) Nuclear speckles. Cold Spring Harb Perspect Biol 3:a000646–a000646
118. Tripathi V, Ellis J, Shen Z, Song D, Pan Q, Watt A, Freier S, Bennett C, Sharma A, Bubulya P et al (2010) The nuclear-retained noncoding RNA MALAT1 regulates alternative splicing by modulating SR splicing factor phosphorylation. Mol Cell 39:925–938
119. Yang L, Lin C, Liu W, Zhang J, Ohgi K, Grinstein J, Dorrestein P, Rosenfeld M (2011) ncRNA- and Pc2 Methylation-dependent gene relocation between nuclear structures mediates gene activation programs. Cell 147:773–788
120. Williams A, Spilianakis C, Flavell R (2010) Interchromosomal association and gene regulation in trans. Trends Genet 26:188–197
121. Lanctôt C, Cheutin T, Cremer M, Cavalli G, Cremer T (2007) Dynamic genome architecture in the nuclear space: regulation of gene expression in three dimensions. Nat Rev Genet 8:104–115

122. Spilianakis C, Lalioti M, Town T, Lee G, Flavell R (2005) Interchromosomal associations between alternatively expressed loci. Nature 435:637–645
123. Sun L, Goff L, Trapnell C, Alexander R, Lo K, Hacisuleyman E, Sauvageau M, Tazon-Vega B, Kelley D, Hendrickson D et al (2013) Long noncoding RNAs regulate adipogenesis. Proc Natl Acad Sci U S A 110:3387–3392
124. Hacisuleyman E, Goff L, Trapnell C, Williams A, Henao-Mejia J, Sun L, McClanahan P, Hendrickson D, Sauvageau M, Kelley D et al (2014) Topological organization of multichromosomal regions by the long intergenic noncoding RNA Firre. Nat Struct Mol Biol 21:198–206
125. Allshire R, Karpen G (2008) Epigenetic regulation of centromeric chromatin: old dogs, new tricks? Nat Rev Genet 9:923–937
126. Ahmad K, Henikoff S (2002) Histone H3 variants specify modes of chromatin assembly. Proc Natl Acad Sci U S A 99:16477–16484
127. Carroll C, Straight A (2006) Centromere formation: from epigenetics to self-assembly. Trends Cell Biol 16:70–78
128. Mendiburo M, Padeken J, Fulop S, Schepers A, Heun P (2011) Drosophila CENH3 is sufficient for centromere formation. Science 334:686–690
129. Bouzinba-Segard H, Guais A, Francastel C (2006) Accumulation of small murine minor satellite transcripts leads to impaired centromeric architecture and function. Proc Natl Acad Sci 103:8709–8714
130. Chen E, Saitoh S, Yanagida M, Takahashi K (2003) A cell cycle-regulated GATA factor promotes centromeric localization of CENP-A in fission yeast. Mol Cell 11:175–187
131. Ohkuni K, Kitagawa K (2011) Endogenous transcription at the centromere facilitates centromere activity in budding yeast. Curr Biol 21:1695–1703
132. Rošić S, Erhardt S (2016) No longer a nuisance: long non-coding RNAs join CENP-A in epigenetic centromere regulation. Cell Mol Life Sci 73:1387–1398
133. Lohe A, Hilliker A, Roberts P (1993) Mapping simple repeated DNA sequences in heterochromatin of Drosophila melanogaster. Trends Genet 9:379
134. Sun X (2003) Sequence analysis of a functional Drosophila centromere. Genome Res 13:182–194
135. Rošić S, Köhler F, Erhardt S (2014) Repetitive centromeric satellite RNA is essential for kinetochore formation and cell division. J Cell Biol 207:335–349
136. Topp C, Zhong C, Dawe R (2004) Centromere-encoded RNAs are integral components of the maize kinetochore. Proc Natl Acad Sci 101:15986–15991
137. Du Y, Topp C, Dawe R (2010) DNA binding of centromere protein C (CENPC) is stabilized by single-stranded RNA. PLoS Genet 6:e1000835
138. Bouzinba-Segard H, Guais A, Francastel C (2006) Accumulation of small murine minor satellite transcripts leads to impaired centromeric architecture and function. Proc Natl Acad Sci U S A 103:8709–8714
139. Ideue T, Cho Y, Nishimura K, Tani T (2014) Involvement of satellite I noncoding RNA in regulation of chromosome segregation. Genes Cells 19:528–538
140. Ferri F, Bouzinba-Segard H, Velasco G, Hube F, Francastel C (2009) Non-coding murine centromeric transcripts associate with and potentiate Aurora B kinase. Nucleic Acids Res 37:5071–5080
141. Blower M (2016) Centromeric transcription regulates Aurora-B localization and activation. Cell Rep 15:1624–1633
142. Wong L, Brettingham-Moore K, Chan L, Quach J, Anderson M, Northrop E, Hannan R, Saffery R, Shaw M, Williams E et al (2007) Centromere RNA is a key component for the assembly of nucleoproteins at the nucleolus and centromere. Genome Res 17:1146–1160
143. Quénet D, Dalal Y (2014) A long non-coding RNA is required for targeting centromeric protein A to the human centromere. Elife 3:e03254

Chapter 3
From Heterochromatin to Long Noncoding RNAs in *Drosophila*: Expanding the Arena of Gene Function and Regulation

Subhash C. Lakhotia

Abstract Recent years have witnessed a remarkable interest in exploring the significance of pervasive noncoding transcripts in diverse eukaryotes. Classical cytogenetic studies using the *Drosophila* model system unraveled the perplexing attributes and "functions" of the "gene"-poor heterochromatin. Recent molecular studies in the fly model are likewise revealing the very diverse and significant roles played by long noncoding RNAs (lncRNAs) in development, gene regulation, chromatin organization, cell and nuclear architecture, etc. There has been a rapid increase in the number of identified lncRNAs, although a much larger number still remains unknown. The diversity of modes of actions and functions of the limited number of *Drosophila* lncRNAs, which have been examined, already reflects the profound roles of such RNAs in generating and sustaining the biological complexities of eukaryotes. Several of the known *Drosophila* lncRNAs originate as independent sense or antisense transcripts from promoter or intergenic, intronic, or 5′/3′-UTR regions, while many of them are independent genes that produce only lncRNAs or coding as well as noncoding RNAs. The different lncRNAs affect chromatin organization (local or large-scale pan-chromosomal), transcription, RNA processing/stability, or translation either directly through interaction with their target DNA sequences or indirectly by acting as intermediary molecules for specific regulatory proteins or may act as decoys/sinks, or storage sites for specific proteins or groups of proteins, or may provide a structural framework for the assembly of substructures in nucleus/cytoplasm. It is interesting that many of the "functions" alluded to heterochromatin in earlier cytogenetic studies appear to find correlates with the known subtle as well as far-reaching actions of the different small and long noncoding RNAs. Further studies exploiting the very rich and powerful genetic and molecular resources available for the *Drosophila* model are expected to unravel the mystery underlying the long reach of ncRNAs.

ORCID ID: 0000-0003-1842-8411

S.C. Lakhotia
Cytogenetics Laboratory, Department of Zoology, Banaras Hindu University, Varanasi 221005, India
e-mail: lakhotia@bhu.ac.in

M.R.S. Rao (ed.), *Long Non Coding RNA Biology*, Advances in Experimental Medicine and Biology 1008, DOI 10.1007/978-981-10-5203-3_3

Keywords lncRNA • ncRNA • Hsromega • Omega speckles • Y chromosome • Promoter RNA • Antisense RNA

3.1 Introduction

"Studies of heterochromatic elements have suggested to one or another that 'heterochromatin' possesses a remarkable and even inconsistent galaxy of attributes (reviews in: SCHULTZ 1941, 1944, 1948, 1952; PONTECORVO 1944; RESENDE 1945; DARLINGTON 1947; VANDERLYN 1949; GOLDSCHMIDT 1949, 1955; BARIGOZZI 1950; HANNAH 1951; SHARMA and SHARMA 1958). For example, it has been suggested that *heterochromatin,* even though largely inert genetically, nevertheless acts (1) *on genes* to control mutation, to modify specific gene action, penetrance, and specificity, and to quantitatively affect the action of some or all genes; (2) *within chromosomes* to stabilize kinetochores and chromosomal ends, to affect breakability and rejoinability, to bring about or prevent pairing of homologs at meiosis, to regulate crossing over and chiasma localization, to produce specific, nonspecific, and reversed pairing and conjunctive properties, and to cause variegation and chromosomal 'stickiness'; (3) *transchromosomally* by controlling the dimensions of all chromosomes of a nucleus, by regulating crossing over, pairing, and disjunction of other chromosomes, and by controlling variegation brought about by genes located in other chromosomes; (4) *metabolically* by performing or mediating special syntheses of nucleic acids, proteins, nucleolar material,cytoplasm' , and energy-rich substances, and by controlling the transfer of substances across the nuclear membrane; (5) *on the cell* by playing a significant role in the control of cell size and by governing mitosis; (6) *on development* by regulating rates of growth and differentiation, by regulating the very nature of differentiation itself, and perhaps also by playing a role in sex determination; (7) *in speciation* by providing neutral (or genetically inert) anchorage and supplementary chromosomal parts for rearrangements, translocations, and increases in chromosomal arm numbers, no less also duplicate genes that may acquire new functions; and, finally, (8) *in theory* as the especialseat of the unorthodox' in genetic systems. And so on." K. W. Cooper [1]

The above attributes of heterochromatin listed by K. W. Cooper [1] in his classical paper on "general theory of heterochromatin" and heterochromatic elements in the genome of *Drosophila melanogaster* were suggested at a time (1959) when molecular biology was just incipient. These empirical suggestions about "functions" of heterochromatin clearly implicated its wide roles in eukaryotic genome organization and function. Interestingly, all the "inconsistent galaxy of attributes" listed above [1] seem to be generally applicable today to the diversity of noncoding RNAs that are receiving increasing attention and appreciation in recent years. Fortunately, compared to Cooper's times, we now have a much better understanding of the mechanistic underpinnings for "the especial 'seat of the unorthodox' in genetic systems."

The "noncoding" part of the genome has an interesting history to which the *Drosophila* model has contributed substantially and significantly. Very soon after the realization that the "Mendelian genes" are located at specific "loci" in a linear order on the fruit fly chromosomes [2], it was discovered that its large Y chromosome, although not involved in determination of sex, was essential for male fertility [3], and yet it was a "gene desert" with only the *bobbed* (now known to be the locus for rRNA genes) and the enigmatic *k1* and *k2* fertility factors mapping to this chromosome [1, 4–7]. Independent cytological studies by Heitz [8, 9] led to the discovery of heterochromatin as a distinct cytological entity, which remained condensed all through the cell's division cycle. Within a short time, studies carried out mostly on the heterochromatic Y chromosome and the pericentric heterochromatin of other chromosomes of *Drosophila* established a broad correlation between the "genetically inert" regions and heterochromatin. In parallel, observations on puffing in polytene chromosomes of *Drosophila* and *Chironomus* provided some of the first evidences for a general necessity of chromatin fiber to open up for the underlying gene to become active [10–12]. The near permanency of condensed state of heterochromatin, its genetic "inertness" and yet its remarkable, albeit "inconsistent galaxy of attributes" (see the quote from K. W. Cooper at the beginning), and its persistence in genomes of nearly all eukaryotes made heterochromatin an intriguing component of the genome to geneticists, cytologists, and evolutionary biologists [1, 13, 14].

Application of molecular biological methods to study gene activity in eukaryotes, especially on the lyonized inactive X chromosome in somatic cells of female mammals [15, 16], the dipteran polytene chromosome puffing [17], and the lecanoid chromosome system in coccids [18], further strengthened the belief that condensed chromatin was transcriptionally silent. Since heterochromatin was generally characterized by the absence of "Mendelian genes," it was suggested by some investigators that heterochromatic regions were dispensable [19]. Persistence of heterochromatin in almost all eukaryotes, in spite of its perceived genetic inactivity, lack of correlation between the c-value and biological complexity, and the superabundance of DNA in any genome or even at any gene locus became increasingly perplexing and led to the formulation of C-value paradox [20, 21]. Findings in the early days of molecular biology that most of the RNA synthesized in the nucleus was destroyed without being released to the cytoplasm for protein synthesis [22–26] added to the puzzle in the context of mRNA-dependent synthesis of proteins in cytoplasm.

With regard to the pervasive RNA synthesized in nucleus but not released to cytoplasm, Brown [13] suggested "Such observations would make sense if the genes in higher organisms were required to build complex machinery for their own control." Similarly, evolutionary biologists like Mayr, who were worried about the general and widespread belief implied in the "central dogma of molecular biology" [27, 28] that only the structural or protein-coding genes are of significance in organisms' lives and evolution, observed "day will come when much of population genetics will have to be rewritten in terms of the interaction between regulator and structural genes" [29]. While reviewing nature of heterochromatin, Shah et al. [14] commented "Genetic activity of DNA is usually studied in terms of transcription of specific RNA molecules, viz. mRNA, rRNA, tRNA. Is it necessary that DNA exerts its influence only through these RNA molecules? Probably not. Products of regulatory genes may not be translated into proteins, but the regulatory action is still achieved." They further stated "These observations indicate that some DNA in the genome may function in ways other than the classically established pathways of transcription and translation. Constitutive heterochromatin DNA may be an example in this category" [14].

Notwithstanding these early significant and prophetic, but mechanistically unclear, views about the importance of components of genome that did not code for proteins, concepts of "selfish" and "junk" DNA to explain the substantial presence of noncoding DNA in eukaryotic genomes [30–32] found almost immediate global acceptance because of the rather dogmatic belief in the "central dogma of molecular biology" [27, 28]. Consequently, significance of heterochromatin and the pervasive nuclear-restricted RNAs remained elusive. This obviously delayed the understanding of real significance of noncoding components of genome.

It is interesting that while early studies in *Drosophila* contributed significantly to the evolution of the general notion that genes function via their encoded proteins, some of the early challenges to the general belief that the "genetically inert" heterochromatic regions are transcriptionally silent also came from studies in *Drosophila*. The most remarkable was the discovery of intense transcriptional activity of heterochromatic and "gene-poor" Y chromosome of *Drosophila* when a series of studies in the 1960s by O. Hess, G. F. Meyer, and colleagues (reviewed in [33–35]) revealed the presence of transcriptionally active special structures derived from the Y chromosome in primary spermatocytes (*see* Sect. 3.3.3). A few years later, Lakhotia and Jacob [36] reported that the classical beta-heterochromatin in polytene nuclei [9] was transcriptionally as active as the typical euchromatic regions. A repetitive sequence, *Dm142*, located adjacent to the 1.688 satellite DNA, was also found to be transcribed [37]. One of the first identified long noncoding RNA (lncRNA) producing "euchromatic" genes was the heat shock inducible *93D* or *hsrω* gene [38–41]. The other early identified and better characterized lncRNAs in *Drosophila* were the roX1 and roX2 transcripts, which were found to be essential for the dosage compensation related hyperactivity of the single X chromosome in male flies [42, 43].

As the genomic studies progressed, the noncoding RNAs graduated from being "selfish" and trivial to real, and now they are widely recognized to be of great significance for generating the diversity and complexity in eukaryotes through diverse regulatory mechanisms [44–70]. It is significant that like many protein-coding mRNAs, a large number of *Drosophila* lncRNAs have been found to show specific patterns of subcellular localization [71].

With increasing diversity of known ncRNAs, a variety of names and classifications have been proposed [51, 60, 69]. A common empirical grouping is based on length of the transcripts such that those less than 100–200 nucleotides (nt) are grouped as small ncRNAs, while the longer ones are generally called long ncRNA (lncRNA). The lncRNAs have also been variously classified in relation to their genomic locations. Some lncRNAs overlap with protein-coding genes on the same or opposite strand and thus have been named sense or antisense lncRNA, respectively; those derived from introns of protein-coding genes have been grouped as intronic RNAs. Transcripts from the promoter regions of some genes have been named promoter RNA or pRNA. Bidirectional lncRNAs are oriented head to head with a protein-coding gene within ~1 kb, while intergenic lncRNAs (also called lincRNA) are transcribed from a region located in the interval between two protein-coding genes. Some lncRNAs which carry introns and/or polyA tails like in mRNAs but do not have recognized ORF have been grouped as mRNA-like ncRNAs or mlncRNAs [72, 73], while some RNAs that function as mRNAs as well as lncRNAs have been named coding and noncoding RNAs or cncRNAs [74]. A more recent suggestion [69] is that the different lncRNAs be named by their major mode of action at architectural, epigenetic, or translational levels in cells.

Widespread usages of bioinformatics and NGS technologies for RNA-sequencing [72, 75–84] have led to a continuous increase in the numbers of known or putatively identified *Drosophila* lncRNA genes and of the lncRNAs that are listed at the Flybase (www.flybase.org). Thus while the releases FB2014_03 (9 May 2014) and FB2015_05 (20 Nov 2015) listed 1778 and 2130 lncRNA genes, respectively, the recent Flybase release FB2016_03 (May 24, 2016) lists 2470 lncRNA genes and 2910 lncRNAs in *Drosophila melanogaster*. While describing an earlier release of the version R6.03Flybase, it was pointed out [82] that some of those annotated as lncRNA may have short ORFs and, therefore, their annotations include the comment "probable lncRNA gene; may encode small polypeptide(s)," while some protein-coding genes may, with more data, turn out to be also producing noncoding RNAs. There is also ambiguity in some cases about extended 3′UTR versus new lncRNA gene, and, therefore, several of the earlier proposed lncRNAs [72, 75, 76, 78] may not have been annotated in the Flybase as lncRNA [82]. A recent computational study [85] combined 462 novel lncRNAs with 4137 previously published lncRNAs in *Drosophila* to provide a curated dataset. Another recent study [83] reported 1077 lncRNAs in late embryonic and larval transcriptomes, with 646 being novel. It is obvious that the identities of lncRNAs in *Drosophila* (and other organisms) are in flux and many more lncRNAs are still awaiting discovery and adequate annotation.

Some of the better known lncRNA genes in *Drosophila* are reviewed here (see Table 3.1) to illustrate the diversity and complexity of their transcripts and the varied roles that they play in cellular networks.

Table 3.1 Diversity of actions of the experimentally studied lncRNAs in *Drosophila*

Transcript	Function/s	Section in this review
iab-2 through iab-8 in the *Ubx*-complex	Infra-abdominal domain (between *AbdA* and *AbdB* in the *Ubx* complex)specific expression in early embryos	3.2.1.1
bxd in the *Ubx*-complex	Regulation of transcription of *Ultrabithorax*	3.2.1.1
lincX (overlapping the cis-regulatory elements of *Scr*)	Activation of *Scr*	3.2.1.1
Reverse and forward ncRNAs from *vg* promoter	PRC2 activity modulation	3.2.1.2
sxl promoter transcripts	Regulation of sxl_{Pm} and sxl_{Pe} promoters of the *sxl* gene	3.2.1.4
rga (SIS RNA from *ragena*)	Regulation of *ragena* gene transcripts	3.2.2
7SK	Regulation of RNA polII activity via P-TEFb	3.2.3
CRG (3′UTR overlapping lncRNA from *CASK* gene)	Regulation of RNA polII binding at *CASK* gene	3.2.4
Acal	Trans-acting regulator of JNK signaling	3.2.5
roX1 and roX2	"Paint" the single X chromosome in male somatic cells to organize its chromatin for hyperactivity required for dosage compensation; also involved in differential regulation of genes in heterochromatic regions in males and females	3.2.6
AAGAG repeat RNA	Nuclear matrix constituent	3.3.1
Yar - intergenic RNA from *yellow-achaete* region	Regulation of sleep behavior	3.3.2
Y chromosome transcripts	"Giant" transcripts from Y-chromosome loops in primary spermatocytes required presumably for spermiogenesis	3.3.3
hsrω transcripts	Developmentally expressed and stress induced multiple nuclear and cytoplasmic transcripts from the *hsrω* or *93D* gene; the large nuclear transcripts, together with diverse heterogenous nuclear RNA binding proteins (hnRNPs) and some other proteins, organize the nucleoplasmic omega speckles and modulate a variety of cell regulatory networks	3.3.6
Yar, hsrω-RA, oskar, CR43432	Dual function coding/noncoding RNAs	3.4
circMblRNA, Laccase2 circRNA	Possible sponges/decoys for miRNAs and transcription factors	3.5

3.2 Regulatory lncRNAs Affecting Enhancer/Promoter Activities and Chromatin Organization

A large number of lncRNA genes show tissue and developmental stage-specific expression [81, 86] with a greater proportion being expressed in *Drosophila* nervous systems and gonads [81, 87]. Early developmental stages appear to show greater abundance and diversity of ncRNA species [86]. Further, compared to protein coding genes, the expression profile of lncRNAs is highly temporally restricted since 21% and 42% were found to be significantly upregulated, respectively, at the developmentally critical late embryonic and larval stages [83]. Many of the lncRNAs show conservation across *Drosophila* species, reflecting a purifying selection [88]. The developmental tissue- and stage-specific expression and conservation obviously point to their important regulatory roles. It appears that the large number of uncharacterized lncRNAs may have roles in regulation of expression of their neighboring adjacent genes [89].

3.2.1 lncRNAs from Promoter/Enhancer Regions

The promoter and/or enhancer regions of many genes are transcribed [78, 90–94]. Some of the newly discovered lncRNAs are antisense to protein-coding genes, producing short regulatory RNAs, while many of the pervasive intergenic transcripts seem to actually originate from the newly identified introns [81]. This is significant in view of the location of many enhancers in introns. Transcription of the promoter/enhancer regions is believed to affect their activities. It has been suggested that many of the ncRNAs that overlap the various regulatory elements of other transcripts may not have a specific function except to affect the regulatory regions through transcriptional interference since the act of transcription may displace RNA pol II from the promoter or may dislodge transcription factors or may delay the assembly of pre-initiation complex [95]. On the other hand, in some cases at least, transcription through the promoter/enhancer regions has been shown to specifically enhance the promoter/enhancer activity.

3.2.1.1 Hox Gene Clusters

The promoter/enhancer regions of the Hox gene clusters (*Bithorax-Complex* (*BX-C*) and *Antennapedia-Complex* (*ANTP-C*)) have been extensively studied for their transcriptional activities [88, 96–101]. A well-known feature of the *Hox* gene clusters in diverse taxa is the high degree of conservation of the sequential order of different genes in the cluster in relation to their spatial domains of developmental expression [102–104]. Another common feature of the *Hox*-gene clusters is the production of multiple noncoding transcripts. Thus, as characteristic of vertebrate *Hox* clusters, many intergenic noncoding transcripts arise from *Drosophila BX-C* and *ANTP-C*, exhibiting spatial colinearity in expression and function [101].

The *BX-C* of *Drosophila melanogaster* spans ~310 kb. However, exons for the four genes (*Ultrabithorax* (*Ubx*), *abdominal-A* (*abd-A*), and *Abdominal-B* (*Abd-B*) and one unrelated *glucose transporter 3* (*Glut3*) gene) in this cluster account for only ~16.5 kb, and yet much of the DNA in the *BX-C* is transcribed [100]. The pioneering work by Lewis [102] identified nine genetic domains in the *BX-C*; these were based on locations of mutations affecting different parasegments (PS) in the fly body, starting from the third thoracic (PS5) through the eighth abdominal segment (PS13). Depending upon the locations of mutants that affect a given parasegment, the domains are named as *bithorax* (*bx*), *bithoraxoid* (*bxd*), and *infraabdominal-2* (*iab-2*) through *infraabdominal-8* (*iab-8*). As elegantly shown by Lewis [102], these domains are aligned along the chromosome in the same order as the segments affected by them. Among these "genes" or domains in the *BX-C*, only three, *Ubx*, *abd-A*, and *Abd-B*, are protein coding, while others have regulatory roles and are known to be transcribed. The *bithoraxoid* (*bxd*) domain produces lncRNA [96], while the *iab-4* and *iab-8* produce miRNAs [99]. In situ hybridization studies revealed that the *iab-3* through *iab-8* domains, located between the *Abd-A* and *Abd-B* genes, transcribe first in blastoderm embryos at a time when the gap and pair rule genes establish the segmental addresses of the *BX-C* domains [100, 105–109]. It is significant that the ncRNAs produced by a particular DNA domain are seen specifically in the body segments regulated by that domain. These transcribed regions are typically delimited by insulator elements so that a loss of insulator function permits transcription to proceed through these elements, which is associated with segmental transformations [101, 110]. The 92 kb long iab-8 transcript, produced by the intergenic region between *abd-A* and *Abd- B* genes in neural cells of the eighth abdominal segment at embryonic stage 14, represses *abd-A* through the miR-iab-8 embedded in its intron, and through transcriptional interference because of an overlap of the 3′ end of the *iab-8* and the *abd-A* promoter [111]. It is interesting that the male and female sterility following knocking down of the iab-8 transcripts is due to behavioral phenotypes, which in males involve failure to bend the abdomen during mating, while in females there may be a peristaltic wave disorder so that eggs fail to pass through the oviduct [111].

The *bxd* mutant was first identified by C. B. Bridges in 1919 [88] and is now known to map to a cis-regulatory region of the *BX-C*, which is extensively transcribed into lncRNAs. Despite the extensive studies on bxd lncRNAs, inferences about their roles have been contradictory [88, 100]. While Sanchez-Elsner et al. [112] suggested the *bxd* transcription to activate transcription of the adjacent *Ultrabithorax*, Petruk et al. [98] found this gene's transcriptional activity to repress *Ultrabithorax*. On the other hand, a more recent study manipulated the *bxd* region so that the embryos did not produce the normal bxd lncRNA and found that even in the absence of full-length bxd transcripts, regulation of the *bxd* domain remained unaffected and the flies looked normal. It was, therefore, concluded that the bxd lncRNA has no apparent function [100]. However, since the first exon of the bxd RNA was not affected by the manipulations, it remains possible that the absence of a phenotype may be due to an independent function of that exon or the act of transcription at the enhancer by itself may be necessary/sufficient for the enhancer to function [100]. More studies are needed to ascertain if the bxd transcription is only a "transcriptional noise" or has hitherto unknown mode of action.

It is significant that some of the vertebrate lncRNAs and miRNAs show syntenic conservation in insects and, therefore, the *Hox* cluster ncRNAs may be an ancestral feature [101]. However, their mechanism/s of action and precise functions are as yet unclear.

A novel lincX lncRNA from a region overlapping the distal cis-regulatory elements of *Scr* in the *ftz-Antp* interval at the *ANTP-C* has been found to be associated with activation of *Scr* in cis [113]; these transcripts also have a transvection effect [114], so that their expression on one chromosome affects expression on the homolog in a pairing-dependent manner.

3.2.1.2 *Vestigial* Gene

Some promoter-associated transcripts switch between forward and reverse directions and thereby regulate the enhancer activity [93, 115]. An example is the forward and reverse transcription covering the Polycomb/Trithorax response elements (PRE/TREs) of the *vestigial* (*vg*) gene, which switches the status of the PRE/TRE between silencing and activation, respectively [115]. While in vitro, the reverse as well as the forward ncRNAs inhibit the histone methyltransferase (HMT) activity of Polycomb Repressor Complex 2 (PRC2), only the reverse strand binds with PRC2 in vivo [93, 115]. It has been suggested [115] that the PRC2 may not have inherently different affinities for different ncRNAs but it is the regulated availability of a given ncRNA for interaction with PRC2 that is important for inhibition of its HMT activity. Strand switching of noncoding RNAs is seen at many Polycomb binding sites in *Drosophila* and vertebrate genomes, leading to the proposal [115] that the PRE/TRE may exist in a "neutral" state so that neither they nor their associated genes are transcribed, but in tissues where the gene can transcribe, the PRE/TRE transcription can be switched in forward or reverse mode to either silence or activate the gene through modulation of the PRC2 activity.

3.2.1.3 *Yellow* and *White* Enhancers

Intergenic transcription through the *yellow* and *white* enhancers has been found to suppress enhancer action in transgenic model [90, 92]. It was found that the distances between regulatory elements like promoter and enhancer and the strength of pass-through transcription affect the degree of inhibition of *yellow* and *white* enhancers.

3.2.1.4 *sxl* Promoter Transcripts

The *sxl* gene is the master regulator of sex determination and dosage compensation in *Drosophila*, acting through its two promoters which differentially switch its transcriptional and splicing patterns in males and females [116, 117]. Responding to the X chromosome counting signal, the establishment promoter of *sxl*, sxl_{Pe}, is activated in female but not in male embryos; the Sxl protein from sxl_{Pe} -dependent transcripts affects the splicing of transcripts from the *sxl* maintenance promoter, sxl_{Pm}. This promoter is transcribed in both sexes after the initial expression of sxl_{Pe} decays in females. However, in the absence of Sxl protein in male embryos, sxl_{Pm}-dependent transcripts undergo differential splicing to retain a translation-terminating exon so that male embryos fail to produce the full-length Sxl protein. It is reported [118] that dynamically expressed sense and antisense lncRNAs from regions R1 and R2, upstream of sxl_{Pe}, influence the X chromosome count sensitive sxl_{Pe} promoter. The R2 antisense lncRNA expression is regulated by the X chromosome counting genes and coincides with sxl_{Pe} promoter-dependent transcription. Experiments with transgenic lines revealed that the R1 and R2 lncRNAs can act in trans, having a negative and positive effect, respectively, in females presumably by affecting the Polycomb/Trithorax chromatin marks which in turn influence the time and strength of sxl_{Pe} promoter activity [118].

3.2.2 Stable Intronic Sequence RNAs

Although majority of intron transcripts are generally believed to be degraded after splicing, a few are known to give rise to snoRNAs, Cajal body-specific RNAs, or miRNAs [119, 120]. In addition, a number of intronic transcripts have been reported in diverse taxa to function as stable intronic sequence RNAs or sisRNA to regulate host gene expression or to act as molecular sinks or to regulate translation [120]. A 600 nt long RNA, corresponding to the single ~700 bp intronic region of the *hsrω* lncRNA gene (see Sect. 3.3.6 below) was suggested to be a putative stable intronic RNA in *Drosophila* [121, 122]. The spliced out intron of the *Delta* primary transcript is reported to accumulate near the site of transcription [123]. Functions of these stable intronic RNAs remain unknown. Thirty-four sisRNAs were identified in 0–2 h *Drosophila* embryos as linear or circular molecules; some of them continued to be present in larval and adult stages as well [124]. A more detailed study of transcripts from the *ragena* (*rga*) gene revealed an autoregulatory negative loop in which the stable intronic RNA, sisR-I derived from the ragena pre-mRNA, represses the cis-natural antisense transcript of *ragena* (ASTR); interestingly, ASTR transcripts seem to promote high ragena pre-mRNA synthesis during embryonic development so that repression of ASTR activity by sisR-I results in reduction of ragena pre-mRNA as well as sisR-I itself [124].

3.2.3 7SK RNA Regulating RNA pol II Activity Via P-TEFb

The RNA pol III-transcribed 331 nt long 7SK RNA is relatively well studied in mammalian cells [125, 126]. The more recently identified *Drosophila* 7SK RNA is a 444 nt long snRNA which, like its mammalian counterpart, regulates the availability of the positive transcription elongation factor, P-TEFb, and thereby functions as a negative regulator of transcription [46, 125–128]. The high conservation of the

7SK snRNA and its associated proteins in vertebrates and insects suggests early evolutionary origin of this important regulatory molecule.

The *Drosophila* 7SK RNA sequesters P-TEFb, a cyclin-dependent kinase, and dHEXIM (homolog of mammalian HEXIM1 and HEXIM2) proteins. The dLARP7 (homolog of mammalian La related protein) and the 7SK methyl phosphate capping enzyme (Bin3, homolog of mammalian MePCE) associate with the 7SK RNA and regulate the release of P-TEFb [46, 127–129]. Recent structural studies on mammalian 7SK snRNP suggest that LARP7, essential for function and stability of 7SK RNA, uses its N- as well as C-terminal domains to stabilize the 7SK RNA in a closed structure formed by joining the conserved sequences at the 5′-end with the foot of the 3′ hairpin [130]. Sequestration of P-TEFb by 7SK snRNP blocks phosphorylation of RNA pol II so that the transcriptional elongation is blocked and the RNA pol II may stay as paused polymerase [128, 129]. Contrary to earlier suggestion of inducible recruitment of P-TEFb to gene promoters during transcriptional activation, recent studies suggest that the 7SK snRNP complex containing inactive P-TEFb remains localized, across the genome, to promoter-proximal regions prior to gene activation or may get localized during activation [131]. The strategic recruitment of catalytically inactive but primed P-TEFb to gene promoters and enhancers promotes kinase activation to facilitate release of the paused RNA pol II. It has recently been shown that SRSF2 binds to the promoter-associated nascent RNA and coordinates the release of P-TEFb from the 7SK and consequently the RNA pol II [132]. As expected, disruption of assembly of the 7SK snRNP has very severe effects at early developmental stages in vertebrates as well as insects [127–129].

It remains to be seen if the 7SK RNA-mediated sequestration of P-TEFb is also involved in the global inhibition of transcription observed in stressed cells [46]. Following the dissociation of 7SK snRNP from P-TEFb and HEXIM, it forms a complex with several hnRNPs (hnRNPA1, hnRNAPA2, hnRNPQ1, and hnRNPR). Whether the association of hnRNPs is only to stabilize 7SK RNA or this complex has other functions remains to be understood [128].

3.2.4 CRG: A 3′UTR Overlapping RNA Regulates RNA pol II Binding at the Upstream CASK *Gene*

Many of the lncRNAs overlap with 3′UTRs of protein-coding genes [82]. One such example is the predominantly neural tissue that expressed lncRNA from the *Cask regulatory RNA gene* (*CRG*) which participates in locomotor and climbing activity of *Drosophila* by positively regulating expression of the neighboring *Ca^{2+}/calmodulin-dependent protein kinase* (*CASK*) gene [133]. The CRG RNA is 2672 nt long, non-spliced, and polyadenylated lncRNA that is transcribed in the same direction as the adjacent protein-coding *CASK* gene. The *CRG* shows strong conservation in all the annotated *Drosophila* species' genomes, and although partially overlapping with the 3′UTR of *CASK* gene, the CRG and CASK are transcribed separately with the CRG transcript extending beyond the CASK 3′UTR [133]. *CRG* mutants show reduced *CASK* expression and defective climbing ability similar to the *CASK* mutant phenotype. The *CRG* mutant defects can be rescued by *CASK* overexpression. The CRG RNA is required for recruitment of RNA polymerase II to the CASK promoter and thus for its enhanced expression [133]. Genome-wide and whole-body expression profile in CRG_{D1877} mutants revealed that 491 genes were downregulated, while 329 genes were upregulated in the *CRG* mutant larvae. However, the most obvious phenotype of the *CRG* mutants, the defective locomotor behavior and inefficient climbing activity, is related to the reduced expression of the *CASK* gene in the absence of the CRG RNA [133]. How CRG RNA facilitates RNA pol II binding at the *CASK* promoter remains to be examined.

3.2.5 Acal: A Trans-acting lncRNA That Regulates JNK- Signaling During Embryonic Dorsal Closure

The *acal* gene, located between *lola* and *pcg* genes and producing a 2336 nt long nuclear trans-acting lncRNA, acts as a negative regulator of JNK signaling during the dorsal closure in *Drosophila* embryos. This RNA is conserved in diverse dipteran species and is dynamically expressed and processed into 20–120 nt fragments which appear to be different from other small RNAs like miRNAs [134]. Expression of *acal* is regulated by *raw*, while the acal transcripts negatively trans-regulate *Connector of kinase to AP1* (*Cka*) and *anterior open* (*aop*) genes in the embryonic lateral epidermis and thus restrain JNK activation in the leading edge cells during the dorsal closure stage. Since the acal transcripts show genetic interaction with Polycomb, a critical member of the PRC1 [134], it is possible that this RNA may regulate the PRC1 activity and thereby regulate *Cka* and *aop*. It remains to be seen if the *acal* action is through its processed smaller RNAs or the unprocessed and processed acal RNAs perform different functions.

3.2.6 RNA on X (roX) Binds with Chromatin and Brings About Pan-Chromosomal Remodeling of X Chromosome in Male to Achieve Dosage Compensation

The phenomenon of dosage compensation is associated with chromosomal sex determination where one sex is heterogametic and the other is homogametic for the sex chromosomes. This results in an imbalance in the dosages of the sex chromosome-linked genes in the two sexes and calls for dosage compensation [135, 136]. Eutherian mammals achieve the equalization of expression of X chromosome-linked genes by a random inactivation of one of the X chromosomes in somatic cells of females [15], which is achieved through orchestrated cis action of several lncRNAs, including the better known X-inactive-specific transcript or XIST, produced by the *Xic* locus on the X chromosome [47, 137–143]. On the other hand, dosage compensation in *Drosophila* is achieved through hyperactivation of the single X chromosome in male somatic cells [144, 145], which involves chromatin remodeling along the length of the X chromosome so that its genes transcribe at nearly twice the rate of each of the two X chromosomes in female cells. It is in this pan-chromosomal remodeling that the two lncRNAs, the roX1 and roX2, play crucial roles. The polyadenylated roX1 and roX2 transcripts were initially identified as male-specific lncRNAs that were more abundant in *Drosophila* brain and were later found, by RNA:RNA in situ hybridization, to exclusively "paint" the X chromosome in male polytene nuclei and hence the names *RNA on X1* (*roX1*) and *RNA on X2* (*roX2*) [42, 43, 146, 147]. The dosage compensation complex or DCC includes the roX1 and roX2 lncRNAs and at least five proteins, viz., male-specific lethal-1 (Msl-1, scaffolding protein), Msl-2 (RING finger E3 ubiquitin-protein ligase), Msl-3 (chromodomain protein), Males-absent-on-the-first (Mof, histone acetyl transferase), and Maleless (Mle, DNA/RNA helicase). The DCC "paints" the entire X chromosome in male somatic cells and remodels its chromatin for hyperactivity [47, 142, 147–155]. The male-specific expression of Msl-2 protein keeps the X-linked *roX1* and *roX2* genes active only in males through gene-internal enhancers [156, 157]. The roX RNAs positively autoregulate their expression so that the presence of the newly assembled DCC complex around the *roX* genes sustains transcription and X chromosome-specific spreading in males [158].

The X-linked *roX1* and *roX2* genes are believed to be functionally redundant since deletion of any one of them has little effect although the absence of both the genes causes male-specific lethality. In spite of their functional redundancy, they produce transcripts that are dissimilar in size and sequence. The current annotation of these two genes at Flybase (www.flybase.org) shows that *roX1* produces five transcripts of 3758, 3460, 4421, 3785, and 483 nt, respectively, while the *roX2* gene generates six transcripts of 1368, 693, 652, 602, 562, and 522 nt, respectively. The significance of such multiple transcripts is not clear.

In normal male cells, the roX transcripts specifically begin their association with the single X chromosome at several X chromosomal chromatin entry sites (CES) or high-affinity sequence (HAS) sites which are ~1.5 kb in length and recruit very high levels of DCC [159–161]. The HAS sites contain smaller 21 nt MSL recognition element (MRE) that is necessary for recruitment of the DCC. The MRE motifs, although distributed through the genome, are ~twofold more enriched on the X [160]. The *roX1* and *roX2* gene sites themselves provide the first HAS for their transcripts to bind, in association with protein members of the DCC. From the HAS, the DCC moves along the chromosome till the entire length of the chromosome is "painted." The H3K36me3 active chromatin mark and association of JIL-1, a histone H3 serine 10 kinase, appear as predictive marks for the HAS on male X chromosome [155, 161, 162]. The specificity of DCC binding to X-chromosomal MREs seems to be mediated by chromatin-linked adapter for MSL proteins (CLAMP), a zinc finger protein, which specifically binds with the MRE sequences in the HAS sites, especially on the *roX* loci and the male X chromosome in general [163, 164]. Genomic and evolutionary studies indicate that GA-dinucleotide repeats expanded and accumulated on the X chromosome over evolutionary time, leading to enhanced density of CLAMP-binding sites on the X chromosome and thus driving the evolution of dosage compensation [165]. The siRNAs derived from X-linked long repeat 1.688 satellite sequence elements have also been implicated in promoting the specific association of DCC components with the X chromosome in male via an unknown mechanism [166]. In another study, the 1.688 satellite-related sequences in *D. melanogaster* themselves have been suggested to play a primary role as recognition elements for the DCC and to be responsible for interspecific hybrid incompatibility [167].

The order of assembly of the DCC or association of its different components with the X chromosome is not fully understood, but it is believed that the HAS facilitate binding of Msl-1, Msl-2, and roX transcripts with male X chromosome [147, 148, 154, 155, 168–171]. It has recently [172] been suggested that MSL2 uses its two CXC and proline/basic-residue-rich domains for interaction with complex DNA elements on the X chromosome. The CXC domain binds a novel motif defined by DNA sequence and shape at a subclass of MSL2-binding sites, which are named as PionX (pioneering sites on the X) as they are the first chromosomal sites to be bound during de novo MSL-DCC assembly. This pioneering interaction of the MSL2–MSL1 sub-complex, without the other protein and RNA subunits, with the PionX sites is followed by occupation of the nearby non-PionX HAS sites as the full DCC is assembled after transcription of the roX RNAs.

The functional redundancy of roX1 and roX2, in spite of their different sizes and sequences, is due to the presence of conserved and multiple GU-boxes or roX-boxes in both the transcripts [173–175]. The Mle's helicase activity [169, 176] is suggested to remodel, in ATP-dependent manner, the roX box large stem-loop into two smaller stem loops with which Msl-2 binds, and this is followed by association of Msl-1, Msl-3, and MOF to make the full DCC [175, 177]. The DCC-associated MOF acetylates the H4K16 so that the chromatin assumes a more open conformation to permit the hyper-transcription of X-linked genes in male cells. The various chromatin organizers also play important roles through interaction with the DCC since the male X chromosome assumes a bloated appearance when chromatin remodeler components like Iswi, NURF301, Su(var)3-7 (suppressor of variegation 3-7), and HP1 (heterochromatin protein 1) are downregulated or absent [178–182].

A recent study [70] on evolutionary dynamics of roX RNAs in different *Drosophila* species, diverged during the past ~40 million years, revealed 47 new roX orthologs. Interestingly, transgenic roX orthologs and engineered synthetic lncRNAs are claimed to rescue roX deficiency suggesting that the focal structural repeats (roX boxes) have been maintained during evolution to mediate the roX function [70]. Likewise, the genomic occupancy maps of roX RNAs in four *Drosophila* species revealed the individual binding site to turnover rapidly but remain within nearby chromosomal neighborhoods. Such differential evolutionary conservation and divergence of different parts of the same lncRNA agrees with the suggestion that the rapid evolutionary divergence of base sequence of lncRNAs, which at one time was taken as strong evidence for classifying them as "junk" or "selfish", actually provides a good strategy for rapid evolution of adaptive changes [46, 70]. The divergent evolutionary trajectories of different parts of lncRNAs indicate the critical roles of conserved structural motifs within the less conserved landscape of lncRNA sequences [44, 46, 69, 70, 183].

The roX transcripts have also been found to differentially regulate expression of genes located in heterochromatic regions of chromosomes in the two sexes [184, 185]. Following the observation that loss of HP1 protein differentially affects expression of genes located in heterochromatin regions in the two sexes, it has been found that the roX transcripts together with Msl1, Msl3, and Mle proteins are required for normal expression of autosomal heterochromatin genes in males but not in females [185]. Interestingly, the regions of roX transcripts that are important for dosage compensation and for male sex-limited heterochromatin functioning are separable and Mle, but not Jil-1 kinase, contributes to heterochromatic gene expression [185]. The roX lncRNAs thus participate in two distinct regulatory systems, viz., dosage compensation of X-linked genes and modulation of heterochromatin activity in male flies.

3.3 Other lncRNAs

3.3.1 AAGAG Repeat RNA Is an Essential Component of Nuclear Matrix

Several hundred nucleotide long transcripts from the pericentromeric AAGAG repeats have been found to be critical constituents of the nuclear matrix [186]. Both strands of these satellite DNA sequences are transcribed and are associated with nuclear matrix. The polypurine strand seems to be more abundant. The AAGAG RNA is essential for viability since global or tissue-specific RNAi for these transcripts disrupted nuclear chromatin organization with lethality at embryonic or late larval/pupal stage, respectively. The specific roles of these transcripts are not known, but their association with nuclear matrix suggests that these lncRNAs may help in nuclear architecture [186].

Several other satellite sequences, besides those on the Y chromosome (see later), are also known to be transcribed to produce siRNA, piRNA, and other lncRNAs [187]. Functions of many of them are not yet known.

3.3.2 Yellow-Achaete Intergenic RNA (yar) Regulates Sleep Behavior

The intergenic region between the *yellow* and *achaete* genes produces multiple noncoding cytoplasmic transcripts, named as yar (yellow-achaete intergenic RNA) [188]. The Flybase (http://flybase.org) shows this gene to produce eight alternatively spliced lncRNAs ranging in size from 841 to 1569 nt. The *yar* gene carries motifs in its sequence, including in its promoter, which are conserved across

Drosophila species representing 40–60 million years of evolution. Although the *yar* nulls show normal viability without any visible phenotype, their night time sleep is reduced and fragmented, with sleep rebound following sleep deprivation being diminished [188]. The yar transcripts carry <75 amino acid encoding ORFs, which leaves the possibility that they may produce small functional polypeptides. The cytoplasmic location of yar transcripts also suggests that their regulatory effects may depend upon stabilization or translational regulation of target RNAs [189].

3.3.3 Y Chromosome Transcripts: Essential for Normal Spermiogenesis and Male Fertility

The Y chromosome in *Drosophila melanogaster* contains about 40 Mbp DNA but has only ~13 protein-coding genes [190, 191], which has justified the earlier noted conventional description of this entirely heterochromatic chromosome as "gene desert." As noted earlier, the only obvious function of the Y chromosome in *Drosophila* relates to male fertility as its absence renders the males completely sterile, although without any significant morphological phenotypes [1, 3, 190–192]. The transcriptional activity of the Y chromosome in primary spermatocytes is cytologically visible as lampbrush loops (Fig. 3.1). These loops were estimated to produce unusually large transcripts ranging in size from 260 to 1500 kb [33, 35, 192]. It is interesting that while the Y chromosome remains highly condensed in all somatic cells of males, it opens up in the primary spermatocyte stage and develops chromatin loops of distinctive morphologies along its length [33, 192]. The Y chromosome loops have been studied more extensively in *Drosophila hydei* because of their distinctly larger sizes and uniquely distinctive morphologies (Fig. 3.1). Each of these loops is transcriptionally active, and their distinctive matrices contain variety of proteins produced by other chromosomal genes [34, 192]. Some of these large transcripts include small protein-coding parts, while the bulk of transcripts constitute noncoding introns comprising of a variety of simple and other repetitive sequences and transposable elements [34, 190, 192]. The functional significance of the unusually large transcripts produced by the Y chromosome in primary spermatocytes has remained enigmatic and largely unexamined. Their unusually large size may simply reflect the lack of recombination in the Y chromosome, since it has been argued that intron size is inversely correlated with recombination rate [193]. Some of the satellite sequences in Y chromosome have the potential to form triplex structures and the *kl-3* and *kl-5* loops on Y chromosome in *Drosophila melanogaster* spermatocytes indeed show immunostaining with anti-triplex antibody [191]. Significance of such triplex structure is not yet understood. The large transcripts may provide scaffolds for accumulation of a variety of proteins, which are required at later steps during spermiogenesis [34, 191, 194, 195]. A functional analysis of these unusually large transcripts remains challenging as they mostly contain repetitive and transposable sequences. However, their analysis is expected to provide novel insights about the roles of these giant lncRNAs and the unusual structures associated with them.

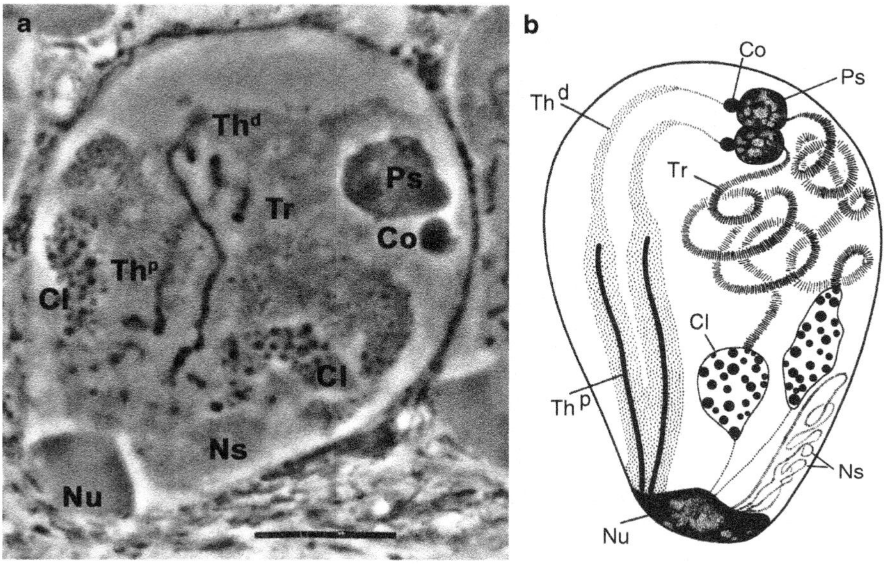

Fig. 3.1 The Y chromosome derived lampbrush loops in the primary spermatocyte nucleus (stage II) of *Drosophila hydei* as seen under phase contrast optics (**a**); a schematic representation of the lampbrush loops is shown in (**b**). All loops are in duplicate, each representing one chromatid of the Y chromosome after replication in the first meiotic prophase. *Cl* clubs, *Co* cones, *Ns* nooses, *Nu* nucleolus (composed of X- and Y- chromosomal components), *Ps* pseudonucleolus, *Th* threads (Th^p and Th^d, distinguished by their associated matrix, are the proximal and distal regions, respectively, of the Th loops) and *Tr* tubular ribbons. Images kindly provided by Prof. W. Hennig

3.3.4 Male-Biased lncRNAs: Roles in Sexual Dimorphism and Species Divergence

A comparison between *D. pseudoobscura* and closely related *D. persimilis* revealed divergent expression of ten novel putative lncRNAs, and, significantly, these were overrepresented among the differentially expressed transcripts in males of the two species [73]. Notwithstanding their differential expression, all these lncRNAs show high sequence conservation in the two species. Significantly, seven of these ten lncRNAs do not show sequence homologies with the other ten *Drosophila* species whose genome sequence is available [73]. Three of the ten lncRNAs were found to be among the top 4% of highly and differentially expressed transcripts in male. Such lncRNAs may be important in sexual dimorphism and species divergence [73]. In a more recent study [196], 1589 intergenic lncRNA loci have been identified in *D. pseudoobscura*, which as in *D. melanogaster*, show prolific expression in male development, more so in testis. Another recent study [197] has reported a large number of testis-specific lncRNAs which affect spermiogenesis in males and show high evolutionary selection.

3.3.5 Stress-Induced *αγ/αβ*-Transcripts of Unknown Functions

Of the five copies of stress-inducible *hsp70* genes in *Drosophila melanogaster*, two are present in the 87A7 cytogenetic region, while three are located at the 87C1 cytogenetic region, with two copies being tandemly repeated and separated from the third reversely oriented copy by a ~38 kb intergenic region. This 38 kb intergenic region includes 10–14 copies of ~1.5 kb αγ/αβ (alpha-gamma/alpha-beta) elements, some of which are transcribed following heat shock as they are immediately downstream of the "gamma" elements homologous to the *hsp70* promoter region [198]. Earlier northern analysis indicated that heat shock induces poly-A+ αγ/αβ-transcripts of 2.5, 1.8, 1.4, and 1.1 kb sizes [199, 200] which do not code for any polypeptide. A series of studies in my laboratory [201–205] indicated a complex interaction between the transcriptional activities of another stress inducible *hsrω* lncRNA gene (see Sect. 3.3.6 and Fig. 3.2e) and the stress inducible *hsp70* and *αγ/αβ* genes located at the 87A and 87C cytogenetic regions. It is interesting that *D. simulans*, a sibling species of *D. melanogaster*, does not carry the *αγ/αβ* sequences in the *hsp70* gene cluster at 87C1 locus [198] and correspondingly, an interaction between transcriptional activities of the *hsrω* locus and *hsp70* genes was not seen in *D. simulans* [202]. Functional significance of the αγ/αβ transcripts remains unknown.

3.3.6 Developmentally Active, Stress Inducible and Essential **hsrω** Gene Modulates Multiple Regulatory Networks

The *heat shock RNA omega* (*hsrω*) or *93D* gene is one of the first identified and extensively studied lncRNA producing gene in *Drosophila* (reviewed in [46, 48, 203–207]). This gene, located at the 93D cytogenetic region of polytene chromosomes, was shown during 1970–1980s to be actively transcribed but without any protein product [38, 41, 208–210]. Interest in this gene was triggered by the finding that it is uniquely induced in larval salivary gland polytene nuclei by a brief in vitro benzamide treatment [211], is developmentally active, is strongly induced (Fig. 3.2a) in cells exposed to thermal or other stresses [208], and is functionally conserved in different species of *Drosophila* [40, 41, 212] and yet does not produce any protein [38, 210]. Subsequently, it was found that this gene is singularly induced by all the amides that were tested [213]. Interestingly, despite the conservation of its functions and of its architecture (two exons, one intron, and tandem repeat units toward 3′ side of the gene) [205], its base sequence shows considerable divergence in different species of *Drosophila* except for the near complete identity in regions around the junctions of its single intron with exon 1 and exon 2, respectively, and a nonamer sequence motif within its unique tandem repeat units [41, 205, 214]. Early studies indicated

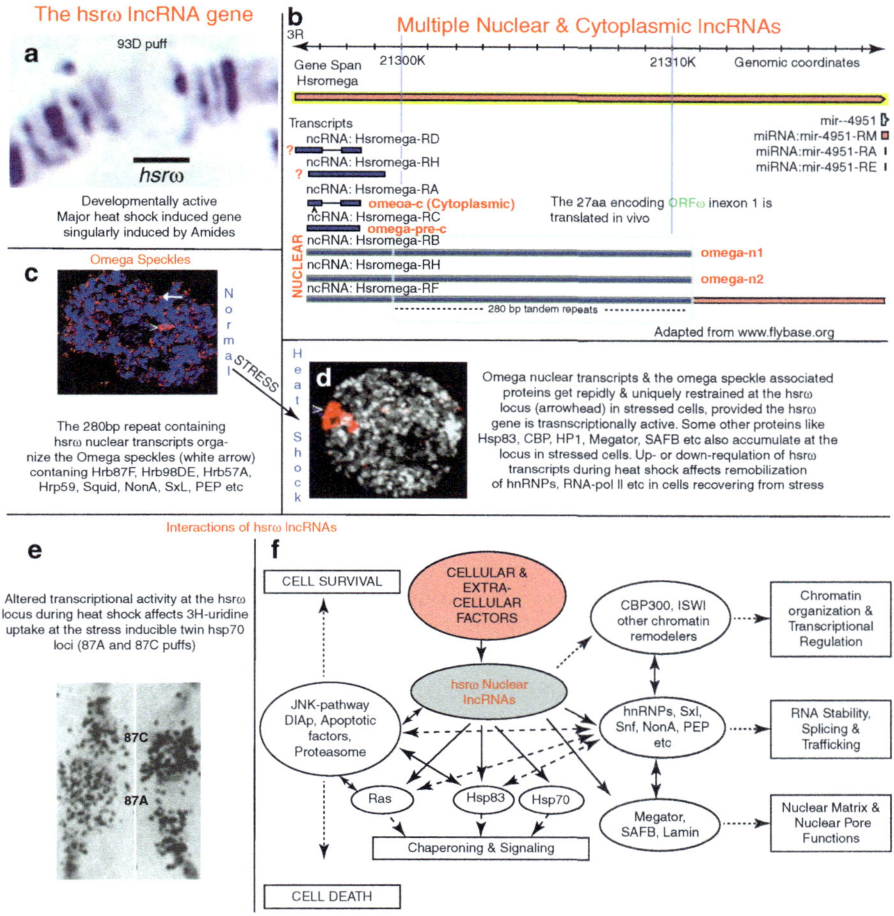

Fig. 3.2 Salient features of the *hsrω* gene, its different nuclear and cytoplasmic transcripts and its known functions/interactions. (**a**) Heat shock-induced 93D puff in polytene nuclei; (**b**) genomic coordinates and different transcripts of the *hsrω* gene as annotated at www.flybase.org; (**c**, **d**). Confocal images of fluorescence in situ hybridization (**c**) of *hsrω* 280 nt repeat riboprobe with nuclear RNA (*red*; DAPI stained DNA blue) and of immunostaining (**d**) for Hrb87F (*red*; DAPI-stained DNA in white) in unstressed (**c**) or heat shocked (**d**) Malpighian tubule nuclei (*arrowheads* in (**c**, **d**) indicate the *hsrω* locus; *arrow* in (**c**) points to one nucleoplasmic omega speckle) (image in (**c**) by Dr. Sonali Sengupta and that in (**d**) by Dr. Anand Singh); (**e**) ^{3}H-uridine autoradiograms of segments of polytene chromosomes showing unequal transcriptional activity at the 87A and 87C loci (87A > 87C in *left* and 87C > 87A in *right panel*) when transcription at the 93D site (not shown) is prevented during heat shock (images adapted from [202]); (**f**) schematic of interactions of *hsrω* nuclear transcripts with different proteins and networks as revealed by various studies (*solid arrows* depict direct interactions, *dashed line arrows* indicate the affected cascading downstream pathways while *dotted-line arrows* indicate the phenotypic consequences). See text for more details

that this gene produces two independently regulated 1.9 kb (omega-pre-c) and >10 kb (omega-n) nuclear lncRNAs, the smaller one of which was spliced to produce a 1.2 kb (omega-c) cytoplasmic transcript [121]. Some early studies indicated that the spliced out 700 nt intron also persisted in cells as a 600 nt RNA [121, 122]. Following recent NGS RNA-sequence data, the current annotation at the Flybase, however, shows this gene to produce seven transcripts, ranging in size from 1.2 to 21 kb and transcribed in same direction (Fig. 3.2b). It is notable that the Hsromega-RD transcript starts within the proximal promoter region of the other Hsromega transcripts (Fig. 3.2b). Little is known so far about the recently annotated Hsromega-RD, Hsromega-RH, and the 21 kb long Hsromega-RF transcripts.

The 280 nt repeats are unique to this locus. In situ localization of the 280 nt repeat units of omega-n transcripts by fluorescence RNA:RNA hybridization [215] led to discovery of the nucleoplasmic omega speckles (Fig. 3.2c) [216]. The 280 nt repeat containing omega-n1 and omega-n2 (a spliced product of omega-n1, [217]) nuclear transcripts have been more extensively studied in relation to their roles in organization and functions of the nucleoplasmic omega speckles [46, 205, 216–222]. Preliminary observations in my laboratory suggest that the 21 kb Hsromega-RF transcripts are also exclusively nuclear and localize, like the omega-n1 and omega-n2 transcripts, at the *hsrω* gene locus and in omega speckles. Omega speckles contain (Fig. 3.2), besides the 280 nt repeat containing large nuclear omega transcripts, a variety of hnRNPs, viz., Hrb87F/Hrp36 (hnRNP A1/A2), Hrb98DE/Hrp38 (hnRNP A), Squid/Hrp40 (hnRNP D), Hrb57A (hnRNP K), Rumpelstiltskin/Hrp59 (hnRNP M), and some other RNA-binding proteins like NonA, Sxl, PEP, etc. [216, 222, 223]. The CBP or P300 (histone acetyltransferase) also shows partial colocalization with omega speckles [224]. It is not yet known if the different omega speckles present in a nucleus are heterogeneous with respect to the pool of associated proteins. The omega-n transcripts as well as the Hrb87F/Hrp36 protein are essential for the assembly of the omega speckles since, in the absence of either of these, omega speckles are not formed [217, 219, 221, 222].

A very unique feature of the omega-n transcripts and omega speckles is that the omega-n transcripts and all the hnRNPs and other proteins that are associated with omega speckles get rapidly and almost exclusively clustered at the *93D* or *hsrω* gene locus following cell stress (Fig. 3.2c) [215, 216, 225]. Active transcription of *hsrω* gene is essential for the stress-induced accumulation of these proteins at its site [222]. In addition to the omega speckle-associated hnRNPs and other proteins, levels of several other proteins like Tpr/Megator, Snf (Sans-fille), SAF B, Hsp83, Ubiquitin Specific Protease-7 (USP7), GMP Synthase (GMPS), ISWI, HP1, etc. also increase at this gene site in stressed cells (reviewed by [48, 205]). It is interesting that like the retention of hsrω-n transcripts in stressed cells at the site of synthesis, the cell stress-induced human sat III lncRNAs also remain confined to the site of transcription and assemble the nuclear stress bodies containing a variety of RNA-binding proteins, CBPs, and heat shock factor [226–228]. In view of such similarities, the nuclear transcripts of *hsrω* gene are considered to be functionally analogous to the human sat III lncRNAs [218]. It may, however, be noted that while the stress-induced global inhibition of transcription following heat shock in human cells requires sat III transcripts [228], the stress-induced remobilization of active RNA pol II on heat shock gene sites in *Drosophila* polytene nuclei is not affected by *hsrω*-null condition [220], indicating that the global inhibition of transcription by heat shock may not be directly mediated by the hsrω lncRNAs.

Unlike heat shock which induces the 93D puff along with the other heat shock puffs, amides singularly induce the *93D* puff. Both these treatments elicit global inhibition of chromosomal transcription and the near exclusive accumulation of hnRNPs, etc. at this gene site [205, 222]. It is interesting, however, that if heat shock and amide treatments are applied together, the 93D puff fails to be induced [203]. Further, the profiles of the *hsrω* transcripts induced by heat shock and amides are different [121], and, unlike after heat shock, Hsp83 does not localize at the 93D puff site [229]. It is also notable that unlike the presence of heat shock elements within a few hundred bp region upstream of the transcription start site [230], the amide-response elements were mapped to be more than 21 kb upstream [231]. The differential expression of *lacZ* reporter gene placed under different lengths of the *hsrω* promoter also indicates the complexity of this gene promoter [232]. Further complexity of this gene promoter is also evident in the fact that the Hsromega-D transcript overlaps the proximal promoter region of the other hsrω transcripts (Fig. 3.2b).

It is believed that the omega speckles function as storage sites for the different hnRNPs and other associated proteins, releasing them as and when required for transcriptional and/or RNA processing activities. Their sequestration at the *hsrω* gene locus in stressed cells is believed to ensure that these proteins remain unavailable for RNA processing and other activities under unfavorable conditions [205, 206, 215, 216, 222]. Chaperones like Hsp83, which accumulate at the *hsrω* gene site in stressed cells [229], may maintain the various accumulated proteins in appropriate conformation so that they can initiate their normal activities as the cells recover from stress. Recent live imaging, FRET and FLIP studies [222], showed that hnRNPs are continuously exchanged between different nuclear compartments and the release of hnRNPs from omega speckles in the nucleoplasm is accompanied by disappearance of the speckle organization, presumably with concurrent breakdown of the associated omega-n transcripts. Therefore, hnRNPs and other proteins move to the *hsrω* gene site in normal and stressed cells in a diffuse rather than in speckled form. Interestingly, however, fully-formed omega speckles emerge from *hsrω* gene locus in unstressed as well as in cells recovering from stress [222]. Chromatin remodelers play significant roles in biogenesis of omega speckles since cells lacking Iswi, the ATPase subunit of several chromatin remodelers [233], fail to generate typical omega speckles [222, 223]. Other preliminary studies in my laboratory (Deoprakash Chaturvedi and Lakhotia, unpublished) suggest that downregulation of several other chromatin remodelers and related proteins like Nurf301, Nurf38, Gcn5, etc. also lead to disappearance of omega speckles even in unstressed cells. It is significant that paraspeckles and nuclear stress bodies, dependent upon the NEAT-1 and satIII lncRNAs, respectively, also require SWI/SNF chromatin remodeling complexes for their assembly [234]. Therefore, it seems that like their roles in chromatin remodeling, these proteins assist in an orderly arrangement of the diverse proteins on the lncRNA scaffolds in different nuclear bodies.

Biophysical considerations suggest that the membraneless nuclear speckles, like the paraspeckles in mammalian cells, may be organized on principles of liquid-liquid phase separation to form droplet organelles [235] as they contain RNA associated with low-complexity domain-containing hnRNPs. It is notable in this context that the variety of hnRNPs associated with omega speckles also show a continuous exchange between the speckle and other compartments; furthermore, even when the hnRNPs and some other proteins get aggregated in hsrω-transcript dependent manner at the *hsrω* locus under conditions of cell stress (Fig. 3.2), some of the hnRNP molecules continue to move in and out of the aggregate [222]. These features suggest that omega speckles are also organized as droplet organelles.

The *hsrω* gene is developmentally active in almost all cells of *Drosophila* [230, 232] and is essential for survival since only about 30% of the *hsrω*-null embryos emerge as weak, short-lived, and relatively poorly fertile adults [217, 236, 237]. Such individuals show very poor thermotolerance as they fail to accumulate the omega-speckle-associated proteins at the *hsrω* gene locus and to redistribute them to their normal locations as the cells recover from the cell stress and, consequently, die during recovery [217, 220–222]. The *hsrω* is also sensitive to cold temperature [238, 239] and fungal infections [240]. Population genetic studies have shown correlation between certain alleles of this gene with latitudinal and altitudinal clines in Australia [241–243]. Nullisomy for the *hsrω* gene is also reported to enhance rates of protein synthesis [237]. Several earlier studies in my laboratory (reviewed in [203, 205]) revealed an intriguing interaction between the transcriptional activity at the *hsrω/93D* puff and that at the 87A and 87C puffs, which are duplicated loci for stress-inducible Hsp70 [198]. As also noted in Sect. 3.3.5 above, whenever the 93D site failed to become active following heat shock, the puffing and ^{3}H-uridine uptake at the twin Hsp70 encoding loci were unequal (Fig. 3.2e). This phenomenon has not been followed further, and, therefore, its basis and significance remain unknown.

Genetic interaction studies have shown that hsrω transcripts interact (Fig. 3.2f), directly or indirectly, with a variety of regulatory pathways including Ras-, Egfr-, and JNK-signaling [224, 244–246]. The ISWI protein physically interacts with the 280 nt repeat units of the hsrω nuclear transcripts [223], while lamin C, DIAP1, and proteasome complex have been found to genetically interact with *hsrω* [217, 224, 245]. Induction of the 93D puff in response to heat shock was found to be affected by levels of β-alanine [247]. cGMP has also been reported to accumulate at the 93D puff in heat-shocked cells [248]. In view of this gene's singular induction with amides, which are potent inhibitors of poly(ADP-ribose) polymerase [249], and ribosylation of hnRNPs being important for their activities [250], it is possible that hsrω transcripts have a role in the poly(ADP-ribose) metabolism. Such interactions with very diverse regulatory pathways (Fig. 3.2f) suggest that lncRNAs like those produced by *hsrω* act as hubs to coordinate different regulatory networks that affect cell's survival or death [46, 48, 205, 251]. Basis for such pleiotropic actions of hsrω and other similar lncRNAs lies in their being associated with a variety of RNA-binding proteins, which themselves are involved in multiple regulatory events and pathways. Regulated sequestration and release of specific RNA-binding proteins by lncRNAs, like the hsrω-n transcripts, is a powerful mode of regulation with subtle as well as far-reaching consequences [46, 48, 205, 206, 215, 218]. Deregulation of such interactions between the lncRNAs and their binding proteins can have severe pathological consequences [252–258]. The 1.2 kb cytoplasmic transcript of the *hsrω* gene may have a role in the global inhibition of translation in heat-shocked cells since its levels are enhanced in stressed cells [121], while *hsrω*-nullisomy is reported to enhance translational activity [237]. The multiplicity of transcripts produced through alternative splicing and/or alternative transcript start or termination sites by many of the lncRNA genes adds to the diversity of their actions.

A homolog of the *hsrω* gene has not yet been identified in other insects although one of the heat shock-induced telomeric Balbiani ring in *Chironomus* species shows properties that are reminiscent of the *hsrω* gene [203, 229, 259, 260].

3.4 Coding-Noncoding (cnc) Bifunctional RNAs

The conventional operative definition of ncRNAs is that they do not have ORF/s, which can encode 75–100 or more amino acids. However, evidence has accumulated in recent years that many small peptides are translated directly from short open-reading frames and have important biological functions [74, 261–265]. For example, the *tarsal-less* (*tal*) gene, which controls tissue folding in *Drosophila*, was earlier identified as lncRNA gene since its ~1.5 kb transcript carries only very short ORFs. These ORFs are now shown [266] to be actually translated into multiple 11 AA peptides to carry out functions of this gene. It is likely that as smaller peptides are more deeply explored, several of the earlier identified ncRNA genes may indeed turn out to encode small peptides. The yar lncRNA (see Sect. 3.3.2 above) may also be a cncRNA if its small polypeptide is found be functional. Revision of the annotation of different transcripts at the Flybase [82] led to reclassification of 63 of the earlier identified lncRNAs as coding genes with 51 of them encoding polypeptides between 30 and 50 amino acids and 12 encoding polypeptides with less than 30 amino acids. Therefore, the bioinformatic identification of ncRNAs based on the absence of "sufficiently long" ORF may not always stand the experimental scrutiny. Ribosomal profiling of putative ncRNAs has in fact revealed actual coding functions of several ncRNAs [261–263, 267, 268]. However, despite their coding potential, many of them also function as ncRNAs, and, therefore, such transcripts have been termed as bifunctional ncRNA or coding-noncoding (cnc) RNAs [74, 262, 267].

As noted in Sect. 3.3.6 and Fig. 3.2, the 1.2 kb cytoplasmic (Hsr-RA or omega-c) transcript of the well-known lncRNA *hsrω* gene includes a short ORF (ORF-omega) that can potentially produce a 27 AA polypeptide. The location of this ORF is conserved in different species of *Drosophila* although length of the ORF shows some variation [41]. In an early study [269], this ORF was shown to be translatable although the translated polypeptide was not detectable. Recent studies (Rashmi Ranjan Sahu and S. C. Lakhotia, unpublished) in my laboratory, using a transgene in which the ORF-omega is fused in frame with downstream GFP indicate that the

GFP-tagged ORF-omega polypeptide is indeed present in unstressed as well as stressed cells. A bioinformatic search indicates the presence of ORF-omega in exon 1 of the *hsrω* gene in different *Drosophila* species, although the length and sequence of the encoded short polypeptide shows little conservation (Eshita Mutt and S. C. Lakhotia, unpublished). Function of the small hsr-omega polypeptide or of the 1.2 kb RNA is not known although in view of its association with mono- or di-ribosomes, and its rapid turnover [121, 269], it has been speculated that the act of translation of this short ORF may serve to monitor the "health" of cell's translational machinery [205]. Since the nuclear transcripts of *hsrω* gene function as lncRNAs, while its cytoplasmic transcript is translated, this gene also may function as bifunctional lncRNA gene.

A large proportion of cytoplasmic lncRNAs in human cells are also reported to be associated with polysomes [270], and many of them with short ORFs are translated [271]. Functions of these small peptides, if any, remain to be known.

Some of the well-known protein-coding genes may also have independent function as lncRNAs [74, 272]. An example in *Drosophila* is the well-known protein-coding gene *oskar*, whose protein product, Oskar, has distinct roles in determination of germ line and differentiation of posterior abdominal segments [273]. The 3′UTR of oskar mRNA, however, also plays independent roles since the absence of this region or mutation therein arrests oogenesis [274]. It appears that this noncoding function of the 3′-UTR of oskar RNA is mediated partly through sequestration of the translational regulator Bruno which binds to Bru response elements in its 3′UTR [74].

Examination of transcriptomic changes in *Drosophila* cells following exposure to ecdysone, the molting hormone in insects, revealed that 4 (CR43432, CR43626, CR45391, and CR45424) of the 11 widely induced genes lacked GO annotations and were listed as lncRNA genes [84]. Unlike the known widespread expression of most lncRNAs in nervous system and in gonads [81, 87], each of these four genes is expressed at high levels in larval salivary glands and fat body. In addition, the CR43432 is also highly expressed, in response to the mid-embryological ecdysone pulse, in 4–14 h embryos. The CR43432 encodes three short, ultra-conserved ORFs and hence constitutes a candidate protein coding gene. Interestingly, the induced levels of this RNA parallel the well-known ecdysone responsive pri (polished rice) RNA, which encodes 11 AA peptides that are critical for ecdysone signal transduction in epidermis [275]. If the CR43432 RNA and its short peptides have independent functions, this gene may also be an example of bifunctional or cncRNA.

The presence of short ORFs in many lncRNAs has implications for evolution of new protein-coding genes. Transcripts of young duplicated, orphan, and lncRNA genes are significantly enriched at specific developmental stages in *D. melanogaster* [276], suggesting that such genes may contribute to origin and evolution of specific phenotypes through evolution into new genes. Examination of evolution, function, and reconstruction of transcriptional histories of six putatively protein-coding de novo genes in *D. melanogaster* [277] revealed that two of them emerged from novel long noncoding RNAs at least five MY prior to evolution of an open reading frame. Interestingly, although these genes are more active in males than females with most strong expression in testes, RNAi experiments revealed their essential requirement in both sexes during metamorphosis, suggesting that protein-coding de novo genes quickly become functionally important in *Drosophila* [277].

It may be mentioned that even the 5′- and 3′-UTRs of many of the typical mRNAs also serve important regulatory functions as RNA. In this respect, even these conventional mRNAs may qualify to be named as bifunctional or cncRNAs. However, the lncRNAs encoding short polypeptides have special significance because of their dual functionality and their potential for evolving into protein-coding genes with novel functions.

3.5 Circular RNAs

Circular RNAs (circRNAs), which have covalently joined 3′ and 5′ ends, are recent additions to the classes of lncRNAs [278–281]. The circRNAs are generated by back-splicing of a downstream splice-donor site with an upstream splice-acceptor site and depending upon choice of the donor and acceptor splice sites, they may include one or more exon/s. If more than one exon is included, the intervening intron(s) also become part of the exon–intron circRNAs or EIciRNAs [279–281]. In a study on *Drosophila* circRNAs, which involved analysis of >100 libraries and >10 billion total RNA-seq reads, >2500 circRNAs were annotated [278]. These were more frequently derived from exons with long flanking introns and had a strong bias for exons containing conserved miRNA sites; many of these circRNAs were common in the nervous system, particularly in older flies [278]. One of the better known circRNA in *Drosophila* is the Muscle blind circular RNA (circMblRNA) whose production is catalyzed by binding of the Mbl splicing factor to its own introns [282]. Unlike circMblRNA, the biogenesis of *Drosophila* Laccase2 circRNA depends upon miniature introns (<150 nt) containing splice sites and inverted repeats, with several hnRNPs and SR-family proteins regulating the back-splicing events [283]. In this respect the *Drosophila* Laccase2 circRNA resembles many of the mammalian cirRNAs that depend upon intronic Alu repeats [279–281]. Circular intronic RNAs (ciRNAs) have also been identified in mammalian cells. Functions of circRNAs are not well understood, but these are believed to function as sponges/decoys for miRNAs and transcription factors, while some of them may also directly regulate transcription [279–281].

3.6 Parallels in Roles of Heterochromatin and ncRNAs

Reexamination of the classical studies on heterochromatin in the light of recent findings about the diverse ncRNAs (short and long ncRNAs) in different organisms, including *Drosophila*, reveals many parallels between them. As noted in the Introduction, the heterochromatic regions of chromosomes remained enigmatic because in spite of their being recognized as "gene deserts," heterochromatin not only persisted through evolution but has done so with remarkably rapid divergence. The varying perceptions about functions of heterochromatin and its varying definitions [1, 9, 13, 14, 284, 285] further contributed to the enigma (see the quote from Cooper at the beginning of Introduction). In recent times, besides the cytological condensed state of chromatin being a parameter for identification as heterochromatin, epigenetic modifications that bring about gene silencing are also taken as marks of heterochromatin [284–286]. Like the enigma of heterochromatin, the ncRNAs also remained enigmatic and, in the absence of knowledge about functions, were considered to be produced by "selfish" or "junk" DNA or as "transcriptional noise." Now that we have begun to explore and appreciate the significance of diverse short and long ncRNAs, it is interesting to note that their varied and sometimes apparently contradictory functional roles are reminiscent of those implied earlier for heterochromatin. Indeed recent studies indicate that heterochromatin and many ncRNAs are functionally and mechanistically interconnected.

Heterochromatin in *Drosophila* chromosomes is present as large pericentromeric blocks of condensed chromatin and as many small regions of intercalary heterochromatin, which are spread across the euchromatin on all chromosomes. These intercalary heterochromatin regions are enriched in repetitive/satellite and transposable element sequences [287] and show ectopic pairing [7] and late-replication, while many such regions also remain under-replicated in polytenized cells [288–290]. The intercalary heterochromatin regions have also been characterized as "forum domains," which are stretches of eukaryotic chromosomal DNA that get excised during spontaneous nonrandom chromosomal fragmentation [291]. Several of the regions under-replicated in endo-replicating nuclei are highly conserved and include important gene clusters like the *BX-C* which are characterized by distant regulator elements and widespread intergenic transcription [290]. In relation to these features of intercalary heterochromatin, it is notable that 20% of the annotated lncRNAs in *D. melanogaster* are associated with the heterochromatic and under-replicated regions [290]. Interestingly, the boundaries of the under-replicated regions and forum domains too are enriched in short ncRNA-encoding sequences and in transcriptionally active but rapidly evolving transposable elements [290, 291]. Late replicating heterochromatic regions, being enriched in conserved "non-coding" elements (lncRNA or retroposed genes), may serve as hotspots and testing ground for evolution of novel genes through expression in testis or even as transcriptional noise [290, 292]. It is notable in this context that the significant roles suggested to be played by heterochromatin and its associated highly repetitive/satellite DNA sequences in speciation and reproductive isolation are paralleled by the propensity for rapid evolution displayed by the various ncRNAs, transposons, etc. The rapid evolution and acquisition of new functions by such DNA sequences play critical roles in reproductive isolation between related species through diverse mechanisms, including hybrid dysgenesis [167, 293–296].

Another interesting property of heterochromatin is "spreading" of the condensed state to neighboring euchromatic loci, and, therefore, the borders of heterochromatin are generally occupied by boundary or insulator elements like Gypsy retroposonderived sequences or the CTCF-binding sites, which are known, at least in some cases, to be regulated by ncRNAs [297]. The interactions between boundary elements and ncRNAs may help maintain the cell-type specific three-dimensional architecture of chromatin in the nucleus, a role that was also implied for heterochromatin [1]. The general inactivity of heterochromatin and its functions in specific cell types are closely related to ncRNA metabolism. The spreading of chromatin organization through facultative heterochromatinization of one of the Xs in somatic cells of female mammals is regulated by panoply of lncRNAs [47, 142, 143]. An opposing role of lncRNAs in determining the state of chromatin condensation and transcriptional activity is exemplified by the roX transcripts in *Drosophila* (see Sect. 3.2.6 above). The siRNA and piRNA pathways regulate transcription and heterochromatin formation [265, 285, 298–300]. Transcription of the heterochromatin-associated retroposons and other small ncRNAs like piRNAs has important roles in maintenance of chromatin state, chromatin stability, hybrid dysgenesis, etc. [295, 301–303].

Transgenerational nongenetic inheritance is another emerging area in which ncRNAs and heterochromatin, as understood in terms of epigenetic modifications that affect chromatin organization and gene activity, also seem to be interconnected [265, 285, 298–300, 304–310].

The various ncRNAs thus display all the "inconsistent galaxy of attributes" ascribed to heterochromatin by Cooper [1] since they act:

(a) *On genes* (regulation of gene activity at various levels, e.g., siRNA, piRNA, various promoter associated lncRNAs)
(b) *Within chromosomes* (cis-regulatory chromatin modifying actions, e.g., roX1 and roX2 transcripts)
(c) *Transchromosomally* (trans-action in gene expression and chromatin organization, e.g., lincX RNA, 7-SK RNA, Acal lncRNA, etc.)
(d) *Metabolically* (acting as tethers, decoys or architectural units for various regulatory molecules and/or nuclear/cytoplasmic bodies, e.g., Y chromosome transcripts, hsrω-nuclear lncRNAs, 3′-UTR of oskar mRNA, circMblRNA, Laccase2 circRNA, etc.)
(e) *On the cells* (various regulatory steps through which the ncRNAs impinge on cellular activities, e.g., all the above examples and yar lncRNA, hsrω-c RNA, etc.)
(f) On development (roles in cell differentiation, sex determination, e.g., promoter/intergenic RNAs from *UBx-C*, *Antp-C* and *sxl* genes, yar, hsrω transcripts, etc.)
(g) In speciation through rapid sequence divergence (repetitive and transposable elements-derived ncRNAs)
(h) In theory as the especial "seat of the unorthodox" noncoding genes

The evolution of biological complexity obviously has been accompanied by increasingly complex regulatory networks, which operate at various levels ranging from basic developmental pathway selections to very subtle and rapid, but effective, moment-to-moment modulations of specific interactions within a cell. It is clear that noncoding RNAs, by virtue of their versatile regulatory and structural roles, play crucial roles in these networks. It is also clear that as we learn more about RNA populations in diverse cells, novel and unexpected modes of their actions would come to light.

It is notable that while the classical cytogeneticists had to undertake rather laborious and painstaking cytological and genetic studies to come out with empirical, but prophetic, inferences about heterochromatin, the currently available tools and reagents permit more direct studies and interpretations.

3.7 Long Reach of ncRNAs Through Multiple Paths

The noncoding DNA, considered "junk" or "selfish" for several decades, is now well established as a critical, and perhaps the most important, component of cell's functional repertoire that produces diverse small and lncRNAs. As indicated by the above discussion on some of the relatively better studied lncRNAs in *Drosophila* (Table 3.1), these entities, like heterochromatin, affect diverse cellular activities in qualitative as well as quantitative manner through widely varying modes. These few examples of lncRNAs already illustrate the diversity of processes and targets that are affected by them through equally diverse modes of actions. The production, in many cases, of multiple lncRNAs by a single gene, adds to the diversity as well as pleiotropy.

With a view to define functional commonality, it has recently been proposed [69] that the lncRNAs may be grouped into three broad categories, viz., (a) that affect cellular architecture by providing scaffolds for cellular substructures like the various nuclear bodies and speckles, (b) that have epigenetic effects in cis or trans on promoters/enhancers of their own or other genes, and (c) that act at translational level through effects on stability or translatability of specific target mRNA/s [69]. However, such categorization of the various lncRNAs may not be mutually exclusive. For example, the lncRNAs like the omega-speckle associated hsrω-n transcripts (see Sect. 3.3.6 above) or the paraspeckle associated Malat-1 or Neat-2 transcripts, besides having an architectural role in organizing the given nuclear speckles, also directly or indirectly affect activity of other genes through regulated release/sequestration of the associated regulatory proteins [46, 311–316]. Some of these effects can have epigenetic consequences. Likewise, some lncRNAs that have translational effects through modulation of stability of own or other mRNAs can also have epigenetic consequences especially if their target mRNAs encode regulatory proteins. Moreover, examples are now available where lncRNA also modulate pre-mRNA splicing [317]. It is clear that as we learn more, most of the lncRNAs would turn out to be pleiotropic because of the network effects that are so common in biology. Such pleiotropic consequences of lncRNAs would defy, as was experienced in past with categorization of heterochromatin in different types, their classification into distinct categories. The propensity of ncRNAs for rapid evolution, like that of the heterochromatin associated DNA sequences, and acquisition of novel functions through changes in small sequence motifs [44–46, 69, 70, 183] also add to the functional flexibility and pleiotropic actions of lncRNAs. Obviously, as ongoing and future studies unravel the intricate network of ncRNAs, it would be increasingly more difficult to apply a reductionist approach to characterize and classify them into discrete functional classes.

Notwithstanding the current widespread excitement about discovery of new lncRNAs and the consequent ever-increasing numbers of publications in recent years, it is obvious that we are beginning to just scratch the tip of huge iceberg, which the noncoding component of any genome is. As the concepts of "junk" and "selfish" DNA themselves become junk and selfish, the genome's "dark matter" [318, 319] would indeed unravel its mysteries so that the diverse roles played by myriads of RNA types in maintaining the characteristic self-organization of biological systems would no longer be considered "unorthodox." The remarkable powers of fly genetics, combined with the newer approaches of genome engineering in vivo [320–325], provide unique opportunities for systematic genetic, molecular, and functional analyses of the dauntingly large number of lncRNAs that are known and remain to be known. While global and large-scale analysis of ncRNA/genes would continue to uncover more such genes in different genomes, application of the power of fly genetics to individual ncRNAs/genes would go a long way in deciphering functions and the underlying mechanism/s of action of specific ncRNA genes. *Drosophila* indeed has a lot to contribute.

Acknowledgments I thank all the former and present researchers in my laboratory who have contributed to the story of the *hsrω* lncRNA gene. Current research on the *hsrω* gene in my laboratory is supported by a grant from the Department of Biotechnology (New Delhi, Govt. of India, grant no. BT/PR6150/COE/34/20/2013). I also thank the Indian National Science Academy (New Delhi) for supporting me through its Senior Scientist program.

References

1. Cooper KW (1959) Cytogenetic analysis of major heterochromatic elements (especially Xh and Y) in Drosophila melanogaster, and the theory of "heterochromatin". Chromosoma 10:535–588
2. Sturtevant A (1913) The linear arrangement of six sex-linked factors in Drosophila, as shown by their mode of association. J Exp Zool 14(1):43–59. doi:10.1002/jez.1400140104
3. Bridges C (1916) Non-disjunction as proof of the chromosome theory of heredity (concluded). Genetics 1:1–52
4. Muller HJ (1914) A gene for the fourth chromosome of *Drosophila*. J Exp Zool 17:325–336
5. Stern C (1927) Ein genetischer und zytologischer Beweis fur Vererbung im Y-Chromosom von *Drosophila melanogaster*. Z Indukt Abstamm Vererb Lehre 44:187–231
6. Stern C (1929) Untersuchungen uber Aberrationen des Y-Chromosoms von *Drosophila melanogaster*. Z Indukt Abstamm Vererb Lehre 51:253–353
7. Slizynski BM (1946) XV. -"Ectopic" pairing and the distribution of heterochromatin in the X-chromosome of salivary gland nuclei of Drosophila melanogaster. Proc Royal Soc Edinburgh B 62(2):114–119. doi:10.1017/S0080455X00009711
8. Heitz E (1928) Das heterochromatin der Moose I. Jb wiss Bot 69:762–818
9. Heitz E (1934) Über α-und β-Heterochromatin sowie Konstanz und Bau der Chromomeren bei *Drosophila*. Biol Zbl 54:588–609

10. Painter T (1934) Salivary chromosomes and the attack on the gene. J Hered 25(12):465–476
11. Beermann W (1952) Chromomerenkonstanz und spezifisehe Modifikationen der Chromosomenstruktur in der Entwieklung und Organdifferenzierung von *Chironomus tentans*. Chromosoma 5:139–198
12. Becker HJ (1959) Die Puffs der Speicheldrüsenchromosomen von *Drosophila melanogaster*. Chromosoma 10(1–6):654–678
13. Brown SW (1966) Heterochromatin. Science 151(3709):417–425
14. Shah V, Lakhotia SC, Rao SRV (1973) Nature of heterochromatin. J Sci Ind Res 32:467–480
15. Lyon MF (1961) Gene action in the X-chromosome of the mouse (Mus musculus L.) Nature 190:372–373
16. Lyon MF (1962) Sex chromatin and gene action in the mammalian X-chromosome. Am J Human Genet 14(2):135
17. Pelling C (1964) Ribonucleic acid synthesis in giant chromosomes. Autoradiographic investigations on Chironomus tentans. Chromosoma 15:71–122
18. Berlowitz L (1965) Correlation of genetic activity, heterochromatization, and RNA metabolism. Proc Natl Acad Sci U S A 53(1):68–73
19. Cooper JEK, Hsu TC (1971) Radiation-induced deletions and translocations of Microtus agrestis sex chromosomes in vivo. Exp Cell Res 67(2):343–351
20. Thomas CA (1971) The genetic organization of chromosomes 3027. AnnRev Genet 5:237–256
21. Gall JG (1981) Chromosome structure and the C-value paradox. J Cell Biol 91(3 Pt 2):3s–14s
22. Edstrom JE (1964) Chromosomal RNA and other nuclear RNA fractions. In: Locke M (ed) The role of chromosomes in development. Academic Press, New York, pp 137–152
23. Soeiro R, Vaughan M, Warner J (1968) The turnover of nuclear DNA-like RNA in HeLa cells. J Cell Biol 39(1):112–118. doi:10.1083/jcb.39.1.112
24. Shearer R, McCarthy B (1967) Evidence for ribonucleic acid molecules restricted to the cell nucleus. Biochemistry 6(1):283–289
25. Goldstein L, Trescott OH (1970) Characterization of RNAs that do and do not migrate between cytoplasm and nucleus. Proc Natl Acad Sci U S A 67(3):1367–1374
26. Weinberg R (1973) Nuclear RNA metabolism. Annu Rev Biochem 42:329–354. doi:10.1146/annurev.bi.42.070173.001553
27. Crick FHC (1958) On protein synthesis. In: Symp Soc Exp Biol, vol 12:138–163
28. Crick F (1970) Central dogma of molecular biology. Nature 227(5258):561–563
29. Mayr E (1970) Populations, species, and evolution. An abridgment of animal species and evolution. Harvard University Press, Cambridge
30. Ohno S (1980) Gene duplication, junk DNA, intervening sequences and the universal signal for their removal. Rev Brasil Genet III 2:99–114
31. Doolittle WF, Sapienza C (1980) Selfish genes, the phenotype paradigm and genome evolution. Nature 284(5757):601–603
32. Orgel LE, Crick FHC, Sapienza C (1980) Selfish DNA. Nature 288(5792):645–646
33. Hess O, Meyer GF (1968) Genetic activities of the Y chromosome in Drosophila during spermatogenesis. Adv Genet 14:171–223
34. Hennig W (1993) Conventional protein coding genes in the Drosophila Y chromosome: is the puzzle of the fertility gene function solved? Proc Natl Acad Sci U S A 90(23):10904–10906
35. Hennig W (1968) Ribonucleic acid synthesis of the Y-chromosome of Drosophila hydei. J Mol Biol 38(2):227–239
36. Lakhotia SC, Jacob J (1974) EM autographic studies on polytene nuclei of *Drosophila melanogaster*. 2. Organization and transcriptive activity of chromocentre. Exp Cell Res 86(2):253–263
37. Carlson M, Brutlag D (1978) A gene adjacent to satellite DNA in *Drosophila melanogaster*. Proc Natl Acad Sci U S A 75(12):5898–5902
38. Lakhotia SC, Mukherjee T (1982) Absence of novel translation products in relation to induced activity of the 93D puff in *Drosophila melanogaster*. Chromosoma 85(3):369–374

39. Peters F, Lubsen N, Walldorf U, Moormann R, Hovemann B (1984) The unusual structure of heat shock locus 2-48B in Drosophila hydei. Mol Gen Genet: MGG 197(3):392–398
40. Walldorf U, Richter S, Ryseck RP, Steller H, Edstrom JE, Bautz EK, Hovemann B (1984) Cloning of heat-shock locus 93D from *Drosophila melanogaster*. EMBO J 3(11):2499–2504
41. Garbe J, Bendena W, Alfano M, Pardue M (1986) A *Drosophila* heat shock locus with a rapidly diverging sequence but a conserved structure. J Biol Chem 261(36):16889–16894
42. Amrein H, Axel R (1997) Genes expressed in neurons of adult male Drosophila. Cell 88(4):459–469
43. Meller V, Wu K, Roman G, Kuroda M, Davis R (1997) roX1 RNA paints the X chromosome of male Drosophila and is regulated by the dosage compensation system. Cell 88(4):445–457
44. Lakhotia S (1996) RNA polymerase II dependent genes that do not code for protein. Indian J Biochem Biophys 33(2):93–102
45. Lakhotia SC (1999) Non-coding RNAs: versatile roles in cell regulation. Curr Sci 77:479–480
46. Lakhotia SC (2012) Long non-coding RNAs coordinate cellular responses to stress. WIREs RNA 3(6):779–796. doi:10.1002/wrna.1135
47. Lakhotia SC (2015) Divergent actions of long noncoding RNAs on X-chromosome remodelling in mammals and *Drosophila* achieve the same end result: dosage compensation. J Genet 94(4):575–584
48. Lakhotia SC (2016) Non-coding RNAs have key roles in cell regulation. Proc Indian Natl Sci Acad 82:1171–1182. doi:10.16943/ptinsa/2016/v82/48404
49. Szymański M, Barciszewska M, Zywicki M, Barciszewski J (2003) Noncoding RNA transcripts. J Appl Genet 44(1):1–19
50. Brosius J, Tiedge H (2004) RNomenclature. RNA Biol 1:81–83
51. Costa F (2005) Non-coding RNAs: new players in eukaryotic biology. Gene 357(2):83–94. doi:10.1016/j.gene.2005.06.019
52. Clark MB, Mattick JS (2011) Long noncoding RNAs in cell biology. Semin Cell Dev Biol 22(4):366–376. doi:10.1016/j.semcdb.2011.01.001
53. Khalil AM, Coller J (2013) Molecular biology of long non-coding RNAs. Springer, New York
54. Roberts TC, Morris KV, Weinberg MS (2014) Perspectives on the mechanism of transcriptional regulation by long non-coding RNAs. Epigenetics 9:13–20
55. Hirose T, Mishima Y, Tomari Y (2014) Elements and machinery of non-coding RNAs: toward their taxonomy. EMBO Rep 15:489–507
56. Liebers R, Rassoulzadegan M, Lyko F (2014) Epigenetic regulation by heritable RNA. PLoS Genet 10:e1004296
57. Jiao AL, Slack FJ (2014) RNA-mediated gene activation. Epigenetics 9(1):27–36
58. Bergmann JH, Spector DL (2014) Long non-coding RNAs: modulators of nuclear structure and function. Curr Opin Cell Biol 26:10–18
59. Böhmdorfer G, Wierzbicki AT (2015) Control of chromatin structure by long noncoding RNA. Trends Cell Biol 25(10):623–632
60. Cech TR, Steitz JA (2014) The noncoding RNA revolution—trashing old rules to forge new ones. Cell 157(1):77–94
61. Rinn JL, Guttman M (2014) RNA and dynamic nuclear organization. Science 345(6202):1240–1241
62. Shibayama Y, Fanucchi S, Magagula L, Mhlanga MM (2014) lncRNA and gene looping: what's the connection? Transcription 5(3):e28658
63. Quinodoz S, Guttman M (2014) Long noncoding RNAs: an emerging link between gene regulation and nuclear organization. Trends Cell Biol 24(11):651–663
64. Fatima R, Akhade VS, Pal D, Rao MRS (2015) Long noncoding RNAs in development and cancer: potential biomarkers and therapeutic targets. Mol Cell Ther 3(1):1–19. doi:10.1186/s40591-015-0042-6
65. Iyer MK, Niknafs YS, Malik R, Singhal U, Sahu A, Hosono Y, Barrette TR, Prensner JR, Evans JR, Zhao S (2015) The landscape of long noncoding RNAs in the human transcriptome. Nat Genet 47(3):199–208

66. Jose AM (2015) Movement of regulatory RNA between animal cells. Genesis 53(7):395–416
67. Quan M, Chen J, Zhang D (2015) Exploring the secrets of long noncoding RNAs. Int J Mol Sci 16(3):5467–5496
68. Chujo T, Yamazaki T, Hirose T (2016) Architectural RNAs (arcRNAs): a class of long noncoding RNAs that function as the scaffold of nuclear bodies. Biochim Biophys Acta 1859(1):139–146
69. Hirose T, Nakagawa S (2016) Clues to long noncoding RNA taxonomy. Biochim Biophys Acta 1859(1):1–2
70. Quinn JJ, Zhang QC, Georgiev P, Ilik IA, Akhtar A, Chang HY (2016) Rapid evolutionary turnover underlies conserved lncRNA–genome interactions. Genes Dev 30(2):191–207
71. Wilk R, Hu J, Blotsky D, Krause HM (2016) Diverse and pervasive subcellular distributions for both coding and long noncoding RNAs. Genes Dev 30(5):594–609. doi:10.1101/gad.276931.115
72. Hiller M, Findeiss S, Lein S, Marz M, Nickel C, Rose D, Schulz C, Backofen R, Prohaska S, Reuter G, Stadler P (2009) Conserved introns reveal novel transcripts in Drosophila melanogaster. Genome Res 19(7):1289–1300. doi:10.1101/gr.090050.108
73. Jiang Z-F, Croshaw DA, Wang Y, Hey J, Machado CA (2011) Enrichment of mRNA-like noncoding RNAs in the divergence of *Drosophila* males. Mol Biol Evol 28(4):1339–1348
74. Sampath K, Ephrussi A (2016) CncRNAs: RNAs with both coding and non-coding roles in development. Development 143(8):1234–1241
75. Inagaki S, Numata K, Kondo T, Tomita M, Yasuda K, Kanai A, Kageyama Y (2005) Identification and expression analysis of putative mRNA-like non-coding RNA in Drosophila. Genes Cells 10(12):1163–1173
76. Tupy J, Bailey A, Dailey G, Evans-Holm M, Siebel C, Misra S, Celniker S, Rubin G (2005) Identification of putative noncoding polyadenylated transcripts in Drosophila melanogaster. Proc Natl Acad Sci U S A 102(15):5495–5500. doi:10.1073/pnas.0501422102
77. Dinger ME, Pang KC, Mercer TR, Mattick JS (2008) Differentiating protein-coding and noncoding RNA: challenges and ambiguities. PLoS Comput Biol 4(11):e1000176. doi:10.1371/journal.pcbi.1000176
78. Young R, Marques A, Tibbit C, Haerty W, Bassett A, Liu J-L, Ponting C (2012) Identification and properties of 1,119 candidate lincRNA loci in the Drosophila melanogaster genome. Genome Biol Evol 4(4):427–442. doi:10.1093/gbe/evs020
79. Will S, Yu M, Berger B (2013) Structure-based whole-genome realignment reveals many novel noncoding RNAs. Genome Res 23(6):1018–1027
80. Vinogradova SV, Soldatov RA, Mironov AA (2013) Genome-wide search for functional noncoding RNA. Mol Biol 47(4):599–604
81. Brown JB, Boley N, Eisman R, May GE, Stoiber MH, Duff MO, Booth BW, Wen J, Park S, Suzuki AM (2014) Diversity and dynamics of the Drosophila transcriptome. Nature 512(7515):393–399
82. Matthews BB, dos Santos G, Crosby MA, Emmert DB, Pierre SES, Gramates LS, Zhou P, Schroeder AJ, Falls K, Strelets V (2015) Gene model annotations for drosophila melanogaster: impact of high-throughput data. G3 5(8):1721–1736
83. Chen B, Zhang Y, Zhang X, Jia S, Chen S, Kang L (2016) Genome-wide identification and developmental expression profiling of long noncoding RNAs during Drosophila metamorphosis. Sci Rep 6:23330. doi:10.1038/srep23330
84. Stoiber M, Celniker S, Cherbas L, Brown B, Cherbas P (2016) Diverse hormone response networks in 41 independent Drosophila cell lines. G3 6(3):683–694
85. Chen M-JM, Chen L-K, Lai Y-S, Lin Y-Y, Wu D-C, Tung Y-A, Liu K-Y, Shih H-T, Chen Y-J, Lin Y-L (2016) Integrating RNA-seq and ChIP-seq data to characterize long noncoding RNAs in Drosophila melanogaster. BMC Genomics 17(1):220. doi:10.1186/s12864-016-2457-0
86. Li Z, Liu M, Zhang L, Zhang W, Gao G, Zhu Z, Wei L, Fan Q, Long M (2009) Detection of intergenic non-coding RNAs expressed in the main developmental stages in Drosophila melanogaster. Nucleic Acids Res 37(13):4308–4314

87. Derrien T, Johnson R, Bussotti G, Tanzer A, Djebali S, Tilgner H, Guernec G, Martin D, Merkel A, Knowles DG (2012) The GENCODE v7 catalog of human long noncoding RNAs: analysis of their gene structure, evolution, and expression. Genome Res 22(9):1775–1789
88. Bonasio R, Shiekhattar R (2014) Regulation of transcription by long noncoding RNAs. Annu Rev Genet 48:433–455
89. Luo S, Lu JY, Liu L, Yin Y, Chen C, Han X, Wu B, Xu R, Liu W, Yan P (2016) Divergent lncRNAs regulate gene expression and lineage differentiation in pluripotent cells. Cell Stem Cell 18(5):637–652. doi:10.1016/j.stem.2016.01.024
90. Erokhin M, Davydova A, Lomaev D, Georgiev P, Chetverina D (2016) The effect of transcription on enhancer activity in Drosophila melanogaster. Russian Jour Genetics 52(1):29–37
91. Wang KC, Chang HY (2011) Molecular mechanisms of long noncoding RNAs. Mol Cell 43(6):904–914
92. Erokhin M, Davydova A, Parshikov A, Studitsky VM, Georgiev P, Chetverina D (2013) Transcription through enhancers suppresses their activity in Drosophila. Epigenetics Chromatin 6(1):1–31. doi:10.1186/1756-8935-6-31
93. Beltran M, García de Herreros A (2016) Antisense non-coding RNAs and regulation of gene transcription. Transcription 7(2):39–43
94. Krishnan J, Mishra RK (2015) Code in the non-coding. Proc Indian Natl Sci Acad 81(3):609–628. doi:10.16943/ptinsa/2015/v81i3/48230
95. Palmer AC, Egan JB, Shearwin KE (2011) Transcriptional interference by RNA polymerase pausing and dislodgement of transcription factors. Transcription 2(1):9–14
96. Lipshitz H, Peattie D, Hogness D (1987) Novel transcripts from the Ultrabithorax domain of the bithorax complex. Genes Dev 1(3):307–322
97. Lewis EB, Knafels JD, Mathog DR, Celniker SE (1995) Sequence analysis of the cis-regulatory regions of the bithorax complex of Drosophila. Proc Natl Acad Sci U S A 92(18):8403–8407
98. Petruk S, Sedkov Y, Riley KM, Hodgson J, Schweisguth F, Hirose S, Jaynes JB, Brock HW, Mazo A (2006) Transcription of bxd noncoding RNAs promoted by trithorax represses Ubx in cis by transcriptional interference. Cell 127(6):1209–1221
99. Marco A (2012) Regulatory RNAs in the light of Drosophila genomics. Brief Funct Genomics 11(5):356–365. doi:10.1093/bfgp/els033
100. Pease B, Borges AC, Bender W (2013) Noncoding RNAs of the Ultrabithorax domain of the Drosophila bithorax complex. Genetics 195(4):1253–1264
101. De Kumar B, Krumlauf R (2016) HOXs and lincRNAs: two sides of the same coin. Sci Adv 2(1):e1501402
102. Lewis EB (1978) A gene complex controlling segmentation in Drosophila. Nature 276(5688):565–570
103. Martinez P, Amemiya CT (2002) Genomics of the HOX gene cluster. Comp Biochem Physiol B Biochem Mol Biol 133(4):571–580
104. Lemons D, McGinnis W (2006) Genomic evolution of Hox gene clusters. Science 313(5795):1918–1922
105. Cumberledge S, Zaratzian A, Sakonju S (1990) Characterization of two RNAs transcribed from the cis-regulatory region of the abd-A domain within the Drosophila bithorax complex. Proc Natl Acad Sci U S A 87(9):3259–3263
106. Sanchez-Herrero E, Akam M (1989) Spatially ordered transcription of regulatory DNA in the bithorax complex of Drosophila. Development 107(2):321–329
107. Bae E, Calhoun VC, Levine M, Lewis EB, Drewell RA (2002) Characterization of the intergenic RNA profile at abdominal-A and Abdominal-B in the Drosophila bithorax complex. Proc Natl Acad Sci U S A 99(26):16847–16852
108. Rank G, Prestel M, Paro R (2002) Transcription through intergenic chromosomal memory elements of the Drosophila bithorax complex correlates with an epigenetic switch. Mol Cell Biol 22(22):8026–8034
109. Maeda RK, Karch F (2009) The bithorax complex of Drosophila: an exceptional Hox cluster. Curr Top Dev Biol 88:1–33

110. Singh NP (2016) Hox genes: let's work together. Proc Indian Natl Sci Acad 82:1229–1236. doi:10.16943/ptinsa/2016/v82/48403
111. Gummalla M, Maeda RK, Alvarez JJC, Gyurkovics H, Singari S, Edwards KA, Karch F, Bender W (2012) abd-A regulation by the iab-8 noncoding RNA. PLoS Genet 8(5):e1002720
112. Sanchez-Elsner T, Gou D, Kremmer E, Sauer F (2006) Noncoding RNAs of trithorax response elements recruit Drosophila Ash1 to Ultrabithorax. Science 311(5764):1118–1123
113. Pettini T, Ronshaugen MR (2016) Transvection and pairing of a Drosophila Hox long noncoding RNA in the regulation of Sex combs reduced. bioRxiv 045617
114. Lewis E (1954) The theory and application of a new method of detecting chromosomal rearrangements in Drosophila melanogaster. Am Nat 88:225–239
115. Herzog VA, Lempradl A, Trupke J, Okulski H, Altmutter C, Ruge F, Boidol B, Kubicek S, Schmauss G, Aumayr K (2014) A strand-specific switch in noncoding transcription switches the function of a Polycomb/Trithorax response element. Nat Genet 46(9):973–981
116. Cline TW, Meyer BJ (1996) Vive la difference: males vs females in flies vs worms. Annu Rev Genet 30(1):637–702
117. Bopp D, Saccone G, Beye M (2013) Sex determination in insects: variations on a common theme. Sex Dev 8(1–3):20–28
118. Mulvey BB, Olcese U, Cabrera JR, Horabin JI (2014) An interactive network of long noncoding RNAs facilitates the Drosophila sex determination decision. Biochim Biophys Acta 1839(9):773–784
119. Hesselberth JR (2013) Lives that introns lead after splicing. WIREs RNA 4(6):677–691
120. Osman I, Tay ML, Pek JW (2016) Stable intronic sequence RNAs (sisRNAs): a new layer of gene regulation. Cell Mol Life Sci 73(18):3507–3519. doi:10.1007/s00018-016-2256-4
121. Bendena W, Garbe J, Traverse K, Lakhotia S, Pardue M (1989) Multiple inducers of the *Drosophila* heat shock locus 93D (*hsr omega*): inducer-specific patterns of the three transcripts. J Cell Biol 108(6):2017–2028
122. Lakhotia SC, Sharma A (1995) RNA metabolism in situ at the 93D heat shock locus in polytene nuclei of Drosophila melanogaster after various treatments. Chromosome Res 3(3):151–161
123. Kopczynski CC, Muskavitch M (1992) Introns excised from the Delta primary transcript are localized near sites of Delta transcription. J Cell Biol 119(3):503–512
124. Pek JW, Osman I, Tay ML-I, Zheng RT (2015) Stable intronic sequence RNAs have possible regulatory roles in Drosophila melanogaster. J Cell Biol 211(2):243–251
125. Wassarman DA, Steitz JA (1991) Structural analyses of the 7SK ribonucleoprotein (RNP), the most abundant human small RNP of unknown function. Mol Cell Biol 11:3432–3445
126. Marz M, Donath A, Verstraete N, Nguyen VT, Stadler PF, Bensaude O (2009) Evolution of 7SK RNA and its protein partners in metazoa. Mol Biol Evol 26(12):2821–2830
127. Nguyen D, Krueger B, Sedore S, Brogie J, Rogers J, Rajendra T, Saunders A, Matera A, Lis J, Uguen P, Price D (2012) The Drosophila 7SK snRNP and the essential role of dHEXIM in development. Nucleic Acids Res 40(12):5283–5297. doi:10.1093/nar/gks191
128. Peterlin BM, Brogie JE, Price DH (2012) 7SK snRNA: a noncoding RNA that plays a major role in regulating eukaryotic transcription. WIREs RNA 3(1):92–103
129. Jennings BH (2013) Pausing for thought: disrupting the early transcription elongation checkpoint leads to developmental defects and tumourigenesis. Bioessays 35(6):553–560
130. Uchikawa E, Natchiar KS, Han X, Proux F, Roblin P, Zhang E, Durand A, Klaholz BP, Dock-Bregeon A-C (2015) Structural insight into the mechanism of stabilization of the 7SK small nuclear RNA by LARP7. Nucleic Acids Res 43(6):3373–3388. doi:10.1093/nar/gkv173
131. McNamara RP, Bacon CW, D'Orso I (2016) Transcription elongation control by the 7SK snRNP complex: releasing the pause. Cell Cycle 15(16):2115–2123. doi:10.1080/15384101.2016.1181241
132. Ji X, Zhou Y, Pandit S, Huang J, Li H, Lin CY, Xiao R, Burge CB, Fu X-D (2013) SR proteins collaborate with 7SK and promoter-associated nascent RNA to release paused polymerase. Cell 153(4):855–868

133. Li M, Wen S, Guo X, Bai B, Gong Z, Liu X, Wang Y, Zhou Y, Chen X, Liu L (2012) The novel long non-coding RNA CRG regulates Drosophila locomotor behavior. Nucleic Acids Res 40(22):11714–11727. doi:10.1093/nar/gks943
134. Ríos-Barrera LD, Gutiérrez-Pérez I, Domínguez M, Riesgo-Escovar JR (2015) acal is a long mon-coding RNA in JNK signaling in epithelial shape changes during Drosophila dorsal closure. PLoS Genet 11(2):e1004927
135. Muller HJ (1932) Further studies on the nature and cause of gene mutations. Int Congr Genet 1:213–255
136. Muller HJ (1950) Evidence of the precision of genetic adaptation. Harvey Lect 43:165–229
137. Brown CJ, Ballabio A, Rupert JL, Lafreniere RG, Grompe M, Tonlorenzi R, Willard HF (1991) A gene from the region of the human X inactivation centre is expressed exclusively from the inactive X chromosome. Nature 349(6304):38–44
138. Brockdorff N, Ashworth A, Kay GF, Cooper P, Smith S, McCabe VM, Norris DP, Penny GD, Patel D, Rastan S (1991) Conservation of position and exclusive expression of mouse Xist from the inactive X chromosome. Nature 351(6324):329–331
139. Spusta S, Goldman M (1999) XISTential wanderings: the role of XIST RNA in X-chromosome inactivation. Curr Sci 77(4):530–538
140. Plath K, Mlynarczyk-Evans S, Nusinow D, Panning B (2002) Xist RNA and the mechanism of X chromosome inactivation. Annu Rev Genet 36:233–278. doi:10.1146/annurev.genet.36.042902.092433
141. Vallot C, Huret C, Lesecque Y, Resch A, Oudrhiri N, Bennaceur A, Duret L, Rougeulle C (2013) XACT, a long non-coding transcript coating the active X chromosome in human pluripotent cells. Epigenetics Chromatin 6(1):1
142. Marchese FP, Huarte M (2014) Long non-coding RNAs and chromatin modifiers: their place in the epigenetic code. Epigenetics 9(1):21–26
143. Briggs SF, Pera RAR (2014) X chromosome inactivation: recent advances and a look forward. Curr Opin Genet Dev 28:78–82
144. Mukherjee AS, Beermann W (1965) Synthesis of ribonucleic acid by the X-chromosomes of Drosophila melanogaster and the problem of dosage compensation. Nature 207(998):785–786
145. Lakhotia SC, Mukherjee AS (1969) Chromosomal basis of dosage compensation in Drosophila. I. Cellular autonomy of hyperactivity of the male X-chromosome in salivary glands and sex differentiation. Genet Res 14(2):137–150
146. Kelley RL (2004) Path to equality strewn with roX. Dev Biol 269(1):18–25
147. Lucchesi JC, Kelly WG, Panning B (2005) Chromatin remodeling in dosage compensation. Annu Rev Genet 39:615–651
148. Lucchesi JC (1998) Dosage compensation in flies and worms: the ups and downs of X-chromosome regulation. Curr Opin Genet Dev 8(2):179–184
149. Gelbart ME, Kuroda MI (2009) Drosophila dosage compensation: a complex voyage to the X chromosome. Development 136(9):1399–1410
150. Georgiev P, Chlamydas S, Akhtar A (2011) Drosophila dosage compensation: males are from Mars, females are from Venus. Fly 5(2):147–154
151. Koya S, Meller V (2011) roX RNAs and genome regulation in Drosophila melanogaster. Prog Mol Subcell Biol 51:147–160. doi:10.1007/978-3-642-16502-3_7
152. Horabin J (2012) Balancing sex chromosome expression and satisfying the sexes. Fly 6(1):26–29. doi:10.4161/fly.18822
153. Ferrari F, Alekseyenko AA, Park PJ, Kuroda MI (2014) Transcriptional control of a whole chromosome: emerging models for dosage compensation. Nat Struct Mol Biol 21(2):118–125
154. Keller CI, Akhtar A (2015) The MSL complex: juggling RNA–protein interactions for dosage compensation and beyond. Curr Opin Genet Dev 31:1–11
155. Lucchesi JC, Kuroda MI (2015) Dosage compensation in Drosophila. Cold Spring Harb Perspect Biol 7(5):a019398
156. Bai X, Larschan E, Kwon S, Badenhorst P, Kuroda M (2007) Regional control of chromatin organization by noncoding roX RNAs and the NURF remodeling complex in Drosophila melanogaster. Genetics 176(3):1491–1499. doi:10.1534/genetics.107.071571

157. Rattner BP, Meller VH (2004) Drosophila male-specific lethal 2 protein controls sex-specific expression of the roX genes. Genetics 166(4):1825–1832
158. Lim CK, Kelley RL (2012) Autoregulation of the Drosophila noncoding roX1 RNA gene. PLoS Genet 8(3):e1002564
159. Sural TH, Peng S, Li B, Workman JL, Park PJ, Kuroda MI (2008) The MSL3 chromodomain directs a key targeting step for dosage compensation of the Drosophila melanogaster X chromosome. Nat Struct Mol Biol 15(12):1318–1325
160. Alekseyenko A, Ho J, Peng S, Gelbart M, Tolstorukov M, Plachetka A, Kharchenko P, Jung Y, Gorchakov A, Larschan E, Gu T, Minoda A, Riddle N, Schwartz Y, Elgin S, Karpen G, Pirrotta V, Kuroda M, Park P (2012) Sequence-specific targeting of dosage compensation in Drosophila favors an active chromatin context. PLoS Genet 8(4):e1002646. doi:10.1371/journal.pgen.1002646
161. Soruco MM, Larschan E (2014) A new player in X identification: the CLAMP protein is a key factor in Drosophila dosage compensation. Chromosome Res 22(4):505–515
162. Jin Y, Wang Y, Johansen J, Johansen KM (2000) JIL-1, a chromosomal kinase implicated in regulation of chromatin structure, associates with the male specific lethal (MSL) dosage compensation complex. J Cell Biol 149(5):1005–1010
163. Larschan E, Soruco MM, Lee O-K, Peng S, Bishop E, Chery J, Goebel K, Feng J, Park PJ, Kuroda MI (2012) Identification of chromatin-associated regulators of MSL complex targeting in Drosophila dosage compensation. PLoS Genet 8(7):e1002830
164. Soruco MM, Chery J, Bishop EP, Siggers T, Tolstorukov MY, Leydon AR, Sugden AU, Goebel K, Feng J, Xia P (2013) The CLAMP protein links the MSL complex to the X chromosome during Drosophila dosage compensation. Genes Dev 27(14):1551–1556
165. Kuzu G, Kaye EG, Chery J, Siggers T, Yang L, Dobson JR, Boor S, Bliss J, Liu W, Jogl G (2016) Expansion of GA dinucleotide repeats increases the density of CLAMP binding sites on the X-chromosome to promote Drosophila dosage compensation. PLoS Genet 12(7):e1006120. doi:10.1371/journal.pgen.1006120
166. Menon DU, Coarfa C, Xiao W, Gunaratne PH, Meller VH (2014) siRNAs from an X-linked satellite repeat promote X-chromosome recognition in Drosophila melanogaster. Proc Natl Acad Sci U S A 111(46):16460–16465. doi:10.1073/pnas.1410534111
167. Gallach M (2015) 1.688 g/cm3 satellite-related repeats: a missing link to dosage compensation and speciation. Mol Ecol 24(17):4340–4347
168. Lyman LM, Copps K, Rastelli L, Kelly RL, Kuroda MI (1997) Drosophila male-specific lethal-4 protein: structure/function analysis and dependence on msl-1 for chromosome association. Genetics 147(4):1743–1753
169. Maenner S, Müller M, Becker PB (2012) Roles of long, non-coding RNA in chromosome-wide transcription regulation: lessons from two dosage compensation systems. Biochimie 94(7):1490–1498
170. Straub T, Zabel A, Gilfillan GD, Feller C, Becker PB (2013) Different chromatin interfaces of the Drosophila dosage compensation complex revealed by high-shear ChIP-seq. Genome Res 23(3):473–485
171. Chery J, Larschan E (2014) X-marks the spot: X-chromosome identification during dosage compensation. Biochim Biophys Acta 1839(3):234–240
172. Villa R, Schauer T, Smialowski P, Straub T, Becker PB (2016) PionX sites mark the X chromosome for dosage compensation. Nature 537:244–248. doi:10.1038/nature19338
173. Kelley RL, Lee O-K, Shim Y-K (2008) Transcription rate of noncoding roX1 RNA controls local spreading of the Drosophila MSL chromatin remodeling complex. Mech Dev 125(11):1009–1019
174. Park S-W, Kang YI, Sypula JG, Choi J, Oh H, Park Y (2007) An evolutionarily conserved domain of roX2 RNA is sufficient for induction of H4-Lys16 acetylation on the Drosophila X chromosome. Genetics 177(3):1429–1437
175. Park S-W, Kuroda MI, Park Y (2008) Regulation of histone H4 Lys16 acetylation by predicted alternative secondary structures in roX noncoding RNAs. Mol Cell Biol 28(16):4952–4962

176. Ilik IA, Quinn JJ, Georgiev P, Tavares-Cadete F, Maticzka D, Toscano S, Wan Y, Spitale RC, Luscombe N, Backofen R (2013) Tandem stem-loops in roX RNAs act together to mediate X chromosome dosage compensation in Drosophila. Mol Cell 51(2):156–173
177. Wutz A (2013) Noncoding RoX RNA remodeling triggers fly dosage compensation complex assembly. Mol Cell 51(2):131–132
178. Deuring R, Fanti L, Armstrong J, Sarte M, Papoulas O, Prestel M, Daubresse G, Verardo M, Moseley S, Berloco M, Tsukiyama T, Wu C, Pimpinelli S, Tamkun J (2000) The ISWI chromatin-remodeling protein is required for gene expression and the maintenance of higher order chromatin structure in vivo. Mol Cell 5(2):355–365
179. Corona D, Clapier C, Becker P, Tamkun J (2002) Modulation of ISWI function by site-specific histone acetylation. EMBO Rep 3(3):242–247. doi:10.1093/embo-reports/kvf056
180. Badenhorst P, Voas M, Rebay I, Wu C (2002) Biological functions of the ISWI chromatin remodeling complex NURF. Genes Dev 16(24):3186–3198. doi:10.1101/gad.1032202
181. Delattre M, Spierer A, Jaquet Y, Spierer P (2004) Increased expression of Drosophila Su (var) 3-7 triggers Su (var) 3-9-dependent heterochromatin formation. J Cell Sci 117(25):6239–6247
182. Spierer A, Seum C, Delattre M, Spierer P (2005) Loss of the modifiers of variegation Su(var)3-7 or HP1 impacts male X polytene chromosome morphology and dosage compensation. J Cell Sci 118(Pt 21):5047–5057. doi:10.1242/jcs.02623
183. Kawaguchi T, Hirose T (2012) Architectural roles of long noncoding RNAs in the intranuclear formation of functional paraspeckles. Front Biosci 17:1729–1746
184. Deng X, Koya SK, Kong Y, Meller VH (2009) Coordinated regulation of heterochromatic genes in Drosophila melanogaster males. Genetics 182(2):481–491
185. Koya SK, Meller VH (2015) Modulation of heterochromatin by male specific lethal proteins and roX RNA in Drosophila melanogaster males. PLoS One 10(10):e0140259
186. Pathak R, Mamillapalli A, Rangaraj N, Kumar R, Vasanthi D, Mishra K, Mishra RK (2013) AAGAG repeat RNA is an essential component of nuclear matrix in Drosophila. RNA Biol 10(4):564–571
187. Pezer Z, Brajković J, Feliciello I, Ugarković D (2011) Transcription of satellite DNAs in insects. Prog Mol Subcell Biol 51:161–178. doi:10.1007/978-3-642-16502-3_8
188. Soshnev A, Ishimoto H, McAllister B, Li X, Wehling M, Kitamoto T, Geyer P (2011) A conserved long noncoding RNA affects sleep behavior in Drosophila. Genetics 189(2):455–468. doi:10.1534/genetics.111.131706
189. Li M, Liu L (2015) Neural functions of long noncoding RNAs in Drosophila. J Comp Physiol A 201(9):921–926
190. Carvalho AB, Koerich LB, Clark AG (2009) Origin and evolution of Y chromosomes: Drosophila tales. Trends Genet 25(6):270–277
191. Piergentili R (2010) Multiple roles of the Y chromosome in the biology of Drosophila melanogaster. Scientific World Journal 10:1749–1767
192. Hennig W, Brand RC, Hackstein J, Hochstenbach R, Kremer H, Lankenau D-H, Lankenau S, Miedema K, Pötgens A (1989) Y chromosomal fertility genes of Drosophila: a new type of eukaryotic genes. Genome 31(2):561–571
193. Carvalho AB, Clark AG (1999) Genetic recombination: intron size and natural selection. Nature 401(6751):344
194. Hennig W (1999) Y chromosomal fertility genes in *Drosophila*. Curr Sci 77:550–552
195. Francisco FO, Lemos B (2014) How do Y-chromosomes modulate genome-wide epigenetic states: genome folding, chromatin sinks, and gene expression. J Genomics 2:94–103. doi:10.7150/jgen.8043
196. Nyberg KG, Machado CA (2016) Comparative expression dynamics of intergenic long noncoding RNAs (lncRNAs) in the genus Drosophila. Genome Biol Evol 8(6):1839–1858. doi:10.1093/gbe/evw116
197. Wen K, Yang L, Xiong T, Di C, Ma D, Wu M, Xue Z, Zhang X (2016) Critical roles of long noncoding RNAs in Drosophila spermatogenesis. Genome Res 26(9):1233–1244. doi:10.1101/gr.199547.115

198. Hellmund D, Serfling E (1984) Structure of hsp70-coding genes in Drosophila. In: Nover L (ed) Heat shock response of eukaryotic cells. Springer, Berlin, pp 23–28
199. Lis J, Neckameyer W, Mirault M-E, Artavanis-Tsakonas S, Lall P, Martin G, Schedl P (1981) DNA sequences flanking the starts of the hsp 70 and αβ heat shock genes are homologous. Dev Biol 83(2):291–300
200. Lengyel JA, Graham ML (1984) Transcription, export and turnover of Hsp70 and alpha beta, two Drosophila heat shock genes sharing a 400 nucleotide 5′ upstream region. Nucleic Acids Res 12(14):5719–5735
201. Sharma A, Lakhotia SC (1995) In situ quantification of hsp70 and alpha-beta transcripts at 87A and 87C loci in relation to hsr-omega gene activity in polytene cells of *Drosophila melanogaster*. Chromosome Res 3:386–393. doi:10.1007/BF00710021
202. Kar Chowdhuri D, Lakhotia SC (1986) Different effects of 93D on 87C heat shock puff activity in *Drosophila melanogaster* and *D. simulans*. Chromosoma 94:279–284. doi:10.1007/BF00290857
203. Lakhotia SC (1987) The 93D heat shock locus in Drosophila: a review. J Genet 66:139–157
204. Lakhotia SC (1989) The 93D heat shock locus of *Drosophila melanogaster*: modulation by genetic and developmental factors. Genome 31(2):677–683
205. Lakhotia SC (2011) Forty years of the 93D puff of Drosophila melanogaster. J Biosci 36(3):399–423
206. Lakhotia SC (2003) The noncoding developmentally active and stress inducible hsr gene of Drosophila melanogaster integrates post-transcriptional processing of other nuclear transcripts. In: Barciszewski J, Erdmann VA (eds) Noncoding RNAs: molecular biology and molecular medicine. Kluwer Academic/Plenum, New York, pp 203–221
207. Pardue ML, Bendena WG, Fini ME, Garbe JC, Hogan NC, Traverse KL (1990) *Hsr-omega*, A novel gene encoded by a *Drosophila* heat shock puff. Biol Bull 179:77–86
208. Mukherjee T, Lakhotia SC (1979) ^{3}H-uridine incorporation in the puff 93D and in chromocentric heterochromatin of heat shocked salivary glands of *Drosophila melanogaster*. Chromosoma 74(1):75–82
209. Ryseck RP, Walldorf U, Hovemann B (1985) Two major RNA products are transcribed from heat-shock locus 93D of Drosophila melanogaster. Chromosoma 93(1):17–20
210. Lengyel JA, Ransom LJ, Graham ML, Pardue ML (1980) Transcription and metabolism of RNA from the Drosophila melanogaster heat shock puff site 93D. Chromosoma 80(3):237–252
211. Lakhotia SC, Mukherjee AS (1970) Activation of a specific puff by benzamide in *D. melanogaster*. Dros Inf Ser 45:108
212. Lakhotia SC, Singh AK (1982) Conservation of the 93D puff of *Drosophila melanogaster* in different species of *Drosophila*. Chromosoma 86:265–278. doi:10.1007/BF00288681
213. Tapadia M, Lakhotia S (1997) Specific induction of the hsr omega locus of Drosophila melanogaster by amides. Chromosome Res 5(6):359–362
214. Garbe J, Bendena W, Pardue M (1989) Sequence evolution of the *Drosophila* heat shock locus *hsr omega*. I. The nonrepeated portion of the gene. Genetics 122(2):403–415
215. Lakhotia S, Ray P, Rajendra T, Prasanth K (1999) The non-coding transcripts of hsr-omega gene in *Drosophila*: do they regulate trafficking and availability of nuclear RNA-processing factors? Curr Sci 77(4):553–563
216. Prasanth K, Rajendra T, Lal A, Lakhotia S (2000) Omega speckles—a novel class of nuclear speckles containing hnRNPs associated with noncoding hsr-omega RNA in *Drosophila*. J Cell Sci 113(Pt 19):3485–3497
217. Mallik M, Lakhotia SC (2011) Pleiotropic consequences of misexpression of the developmentally active and stress-inducible non-coding *hsr omega* gene in *Drosophila*. J Biosci 36(2):265–280. doi:10.1007/s12038-011-9061-x
218. Jolly C, Lakhotia SC (2006) Human sat III and *Drosophila hsr omega* transcripts: a common paradigm for regulation of nuclear RNA processing in stressed cells. Nucleic Acids Res 34(19):5508–5514. doi:10.1093/nar/gkl711
219. Mallik M, Lakhotia S (2009) RNAi for the large non-coding hsr omega transcripts suppresses polyglutamine pathogenesis in *Drosophila* models. RNA Biol 6(4):464–478

220. Lakhotia SC, Mallik M, Singh AK, Ray M (2012) The large noncoding hsromega-n transcripts are essential for thermotolerance and remobilization of hnRNPs, HP1 and RNA polymerase II during recovery from heat shock in *Drosophila*. Chromosoma 121(1):49–70. doi:10.1007/s00412-011-0341-x
221. Singh A, Lakhotia SC (2012) The hnRNP A1 homolog Hrp36 is essential for normal development, female fecundity, omega speckle formation and stress tolerance in Drosophila melanogaster. J Biosci 37(4):659–678
222. Singh AK, Lakhotia SC (2015) Dynamics of hnRNPs and omega speckles in normal and heat shocked live cell nuclei of *Drosophila melanogaster*. Chromosoma 124(3):367–383. doi:10.1007/s00412-015-0506-0
223. Onorati M, Lazzaro S, Mallik M, Ingrassia A, Carreca A, Singh A, Chaturvedi D, Lakhotia S, Corona D (2011) The ISWI chromatin remodeler organizes the hsrω ncRNA-containing omega speckle nuclear compartments. PLoS Genet 7(5):e1002096. doi:10.1371/journal.pgen.1002096
224. Mallik M, Lakhotia SC (2010) Improved activities of CREB binding protein, heterogeneous nuclear ribonucleoproteins and proteasome following downregulation of noncoding hsromega transcripts help suppress poly(Q) pathogenesis in fly models. Genetics 184(4):927–945. doi:10.1534/genetics.109.113696
225. Saumweber H, Symmons P, Kabisch R, Will H, Bonhoeffer F (1980) Monoclonal antibodies against chromosomal proteins of Drosophila melanogaster: establishment of antibody producing cell lines and partial characterization of corresponding antigens. Chromosoma 80(3):253–275
226. Denegri M, Chiodi I, Corioni M, Cobianchi F, Riva S, Biamonti G (2001) Stress-induced nuclear bodies are sites of accumulation of pre-mRNA processing factors. Mol Biol Cell 12:3502–3514
227. Jolly C, Metz A, Govin J, Vigneron M, Turner B, Khochbin S, Vourc'h C (2004) Stress-induced transcription of satellite III repeats. J Cell Biol 164(1):25–33. doi:10.1083/jcb.200306104
228. Goenka A, Sengupta S, Pandey R, Parihar R, Mohanta GC, Mukerji M, Ganesh S (2016) Human satellite-III non-coding RNAs modulate heat shock-induced transcriptional repression. J Cell Sci 129:3541–3552. doi:10.1242/jcs.189803
229. Morcillo G, Diez JL, Carbajal ME, Tanguay RM (1993) HSP90 associates with specific heat shock puffs (hsr omega) in polytene chromosomes of Drosophila and Chironomus. Chromosoma 102(9):648–659
230. Mutsuddi M, Lakhotia SC (1995) Spatial expression of the hsr-omega (93D) gene in different tissues of *Drosophila melanogaster* and identification of promoter elements controlling its developmental expression. Dev Genet 17(4):303–311. doi:10.1002/dvg.1020170403
231. Lakhotia S, Tapadia MG (1998) Genetic mapping of the amide response element(s) of the hsrω locus of Drosophila melanogaster. Chromosoma 107(2):127–135
232. Lakhotia SC, Rajendra TK, Prasanth KV (2001) Developmental regulation and complex organization of the promoter of the non-coding hsr(omega) gene of Drosophila melanogaster. J Biosci 26(1):25–38
233. Toto M, D'Angelo G, Corona DFV (2014) Regulation of ISWI chromatin remodelling activity. Chromosoma 123(1–2):91–102
234. Kawaguchi T, Tanigawa A, Naganuma T, Ohkawa Y, Souquere S, Pierron G, Hirose T (2015) SWI/SNF chromatin-remodeling complexes function in noncoding RNA-dependent assembly of nuclear bodies. Proc Natl Acad Sci U S A 112(14):4304–4309
235. Courchaine EM, Lu A, Neugebauer KM (2016) Droplet organelles? EMBO J 35(15):1603–1612. doi:10.15252/embj.201593517
236. Mohler J, Pardue ML (1982) Deficiency mapping of the 93D heat-shock locus in *Drosophila melanogaster*. Chromosoma 86(4):457–467
237. Johnson T, Cockerell F, McKechnie S (2011) Transcripts from the *Drosophila* heat-shock gene hsr-omega influence rates of protein synthesis but hardly affect resistance to heat knockdown. Mol Genet Genomics 285(4):313–323. doi:10.1007/s00438-011-0610-7
238. Lakhotia SC, Singh AK (1985) Non-inducibility of the 93D heat-shock puff in cold-reared larvae of *Drosophila melanogaster*. Chromosoma 92:48–54. doi:10.1007/BF00327244

239. Purać J, Kojić D, Petri E, Popović ŽD, Grubor-Lajšić G, Blagojević DP (2016) Cold adaptation responses in insects and other arthropods: an "omics" approach. In: Short views on insect genomics and proteomics. Springer, Basel, pp 89–112
240. Lu H-L, Wang JB, Brown MA, Euerle C, Leger RJS (2015) Identification of Drosophila mutants affecting defense to an entomopathogenic fungus. Sci Rep 5:12350. doi:10.1038/srep12350
241. McColl G, Hoffmann AA, McKechnie SW (1996) Response of two heat shock genes to selection for knockdown heat resistance in *Drosophila melanogaster*. Genetics 143(4):1615–1627
242. Collinge JE, Anderson AR, Weeks AR, Johnson TK, McKechnie SW (2008) Latitudinal and cold-tolerance variation associate with DNA repeat-number variation in the hsr-omega RNA gene of *Drosophila melanogaster*. Heredity 101:260–270
243. Clemson AS, Sgrò CM, Telonis-Scott M (2016) Thermal plasticity in Drosophila melanogaster populations from eastern Australia: quantitative traits to transcripts. J Evol Biol 29(12):2447–2463. doi:10.1111/jeb.12969
244. Ray P, Lakhotia SC (1998) Interaction of the non-protein-coding developmental and stress-inducible hsrω gene with Ras genes of *Drosophila melanogaster*. J Biosci 23:377–386
245. Mallik M, Lakhotia SC (2009) The developmentally active and stress-inducible noncoding *hsromega* gene is a novel regulator of apoptosis in *Drosophila*. Genetics 183(3):831–852. doi:10.1534/genetics.109.108571
246. Ray M, Lakhotia SC (2016) Activated Ras expression in eye discs with altered hsrω lncRNA causes JNK-induced Dilp8 secretion and reduces post-pupal ecdysone leading to early pupal death in Drosophila. bioRxiv. doi:10.1101/049882
247. Lakhotia SC, Chowdhuri DK, Burma PK (1990) Mutations affecting β-alanine metabolism influence inducibility of the 93D puff by heat shock in Drosophila melanogaster. Chromosoma 99:296–305. doi:10.1007/BF01731706
248. Spruill WA, Hurwitz DR, Lucchesi JC, Steiner AL (1978) Association of cyclic GMP with gene expression of polytene chromosomes of Drosophila melanogaster. Proc Natl Acad Sci U S A 75(3):1480–1484
249. Wang Y-Q, Wang P-Y, Wang Y-T, Yang G-F, Zhang A, Miao Z-H (2016) An update on Poly (ADP-ribose) polymerase-1 (PARP-1) inhibitors: opportunities and challenges in cancer therapy. J Med Chem 59(21):9575–9598. doi:10.1021/acs.jmedchem.6b00055
250. Ji Y, Tulin AV (2016) Poly (ADP-ribosyl) ation of hnRNP A1 protein controls translational repression in Drosophila. Mol Cell Biol 36(19):2476–2486. doi:10.1128/MCB.00207-16
251. Arya R, Mallik M, Lakhotia SC (2007) Heat shock genes—integrating cell survival and death. J Biosci 32(3):595–610
252. Széll M, Bata-Csörgo Z, Kemény L (2008) The enigmatic world of mRNA-like ncRNAs: their role in human evolution and in human diseases. Semin Cancer Biol 18(2):141–148. doi:10.1016/j.semcancer.2008.01.007
253. Qureshi I, Mehler M (2012) Emerging roles of non-coding RNAs in brain evolution, development, plasticity and disease. Nature Rev Neurosci 13(8):528–541. doi:10.1038/nrn3234
254. Amaral P, Dinger M, Mattick J (2013) Non-coding RNAs in homeostasis, disease and stress responses: an evolutionary perspective. Brief Funct Genomics 12(3):254–278. doi:10.1093/bfgp/elt016
255. Li T, Mo X, Fu L, Xiao B, Guo J (2016) Molecular mechanisms of long noncoding RNAs on gastric cancer. Oncotarget 7(8):8601–8612
256. Lorenzen JM, Thum T (2016) Long noncoding RNAs in kidney and cardiovascular diseases. Nat Rev Nephrol 12(6):360–373
257. Wan P, Su W, Zhuo Y (2016) The role of long noncoding RNAs in neurodegenerative diseases. Mol Neurobiol 54(3):2012–2021. doi:10.1007/s12035-016-9793-6
258. Tripathi BK, Surabhi S, Bhaskar PK, Mukherjee A, Mutsuddi M (2016) The RNA binding KH domain of Spoonbill depletes pathogenic non-coding spinocerebellar ataxia 8 transcripts and suppresses neurodegeneration in Drosophila. Biochim Biophys Acta 1862(9):1732–1741. doi:10.1016/j.bbadis.2016.06.008

259. Nath BB, Lakhotia SC (1991) Search for a Drosophila-93D-like locus in Chironomus and Anopheles. Cytobios 65(260):7–13
260. Martínez-Guitarte J, Morcillo G (2014) Telomeric transcriptome from Chironomus riparius (Diptera), a species with noncanonical telomeres. Insect Mol Biol 23(3):367–380. doi:10.1111/imb.12087
261. Cohen SM (2014) Everything old is new again:(linc) RNAs make proteins! EMBO J 33(9):937–938
262. Pauli A, Valen E, Schier AF (2015) Identifying (non-) coding RNAs and small peptides: challenges and opportunities. Bioessays 37(1):103–112
263. Waterhouse PM, Hellens RP (2015) Plant biology: coding in non-coding RNAs. Nature 520(7545):41–42
264. de Andres-Pablo A, Morillon A, Wery M (2016) LncRNAs, lost in translation or licence to regulate? Curr Genet 63(1):29–33. doi:10.1007/s00294-016-0615-1
265. Butler AA, Webb WM, Lubin FD (2016) Regulatory RNAs and control of epigenetic mechanisms: expectations for cognition and cognitive dysfunction. Epigenomics 8(1):135–151. doi:10.2217/epi.15.79
266. Galindo MI, Pueyo JI, Fouix S, Bishop SA, Couso JP (2007) Peptides encoded by short ORFs control development and define a new eukaryotic gene family. PLoS Biol 5(5):e106
267. Kumari P, Sampath K (2015) cncRNAs: Bi-functional RNAs with protein coding and non-coding functions. Semin Cell Dev Biol 47:40–51. doi:10.1016/j.semcdb.2015.10.024
268. Li H, Hu C, Bai L, Li H, Li M, Zhao X, Czajkowsky DM, Shao Z (2016) Ultra-deep sequencing of ribosome-associated poly-adenylated RNA in early Drosophila embryos reveals hundreds of conserved translated sORFs. DNA Res 23(6):571–580. doi:10.1093/dnares/dsw040
269. Fini ME, Bendena WG, Pardue ML (1989) Unusual behavior of the cytoplasmic transcript of hsr omega: an abundant, stress-inducible RNA that is translated but yields no detectable protein product. J Cell Biol 108(6):2045–2057
270. Carlevaro-Fita J, Rahim A, Guigó R, Vardy LA, Johnson R (2016) Cytoplasmic long noncoding RNAs are frequently bound to and degraded at ribosomes in human cells. RNA 22(6):867–882
271. Ji Z, Song R, Regev A, Struhl K (2015) Many lncRNAs, 5′UTRs, and pseudogenes are translated and some are likely to express functional proteins. Elife 4:e08890
272. Dinger M, Mercer T, Mattick J (2008) RNAs as extracellular signaling molecules. J Mol Endocrinol 40(4):151–159. doi:10.1677/JME-07-0160
273. Ephrussi A, Dickinson LK, Lehmann R (1991) Oskar organizes the germ plasm and directs localization of the posterior determinant nanos. Cell 66(1):37–50
274. Jenny A, Hachet O, Závorszky P, Cyrklaff A, Weston MD, St Johnston D, Erdélyi M, Ephrussi A (2006) A translation-independent role of oskar RNA in early Drosophila oogenesis. Development 133(15):2827–2833
275. Chanut-Delalande H, Hashimoto Y, Pelissier-Monier A, Spokony R, Dib A, Kondo T, Bohère J, Niimi K, Latapie Y, Inagaki S (2014) Pri peptides are mediators of ecdysone for the temporal control of development. Nat Cell Biol 16(11):1035–1044
276. Liu H-Q, Li Y, Irwin DM, Zhang Y-P, Wu D-D (2014) Integrative analysis of young genes, positively selected genes and lncRNAs in the development of Drosophila melanogaster. BMC Evol Biol 14(1):1
277. Reinhardt JA, Wanjiru BM, Brant AT, Saelao P, Begun DJ, Jones CD (2013) De novo ORFs in Drosophila are important to organismal fitness and evolved rapidly from previously non-coding sequences. PLoS Genet 9(10):e1003860
278. Westholm JO, Miura P, Olson S, Shenker S, Joseph B, Sanfilippo P, Celniker SE, Graveley BR, Lai EC (2014) Genome-wide analysis of Drosophila circular RNAs reveals their structural and sequence properties and age-dependent neural accumulation. Cell Rep 9(5):1966–1980
279. Barrett SP, Salzman J (2016) Circular RNAs: analysis, expression and potential functions. Development 143(11):1838–1847. doi:10.1242/dev.128074

280. Chen L-L (2016) The biogenesis and emerging roles of circular RNAs. Nat Rev Mol Cell Biol 17:205–211. doi:10.1038/nrm.2015.32
281. Ebbesen KK, Kjems J, Hansen TB (2016) Circular RNAs: identification, biogenesis and function. Biochim Biophys Acta 1859(1):163–168. doi:10.1016/j.bbagrm.2015.07.007
282. Ashwal-Fluss R, Meyer M, Pamudurti NR, Ivanov A, Bartok O, Hanan M, Evantal N, Memczak S, Rajewsky N, Kadener S (2014) circRNA biogenesis competes with pre-mRNA splicing. Mol Cell 56(1):55–66. doi:10.1016/j.molcel.2014.08.019
283. Kramer MC, Liang D, Tatomer DC, Gold B, March ZM, Cherry S, Wilusz JE (2015) Combinatorial control of Drosophila circular RNA expression by intronic repeats, hnRNPs, and SR proteins. Genes Dev 29(20):2168–2182. doi:10.1101/gad.270421.115
284. Craig JM (2005) Heterochromatin--many flavours, common themes. Bioessays 27(1):17–28. doi:10.1002/bies.20145
285. Wang J, Jia ST, Jia S (2016) New insights into the regulation of heterochromatin. Trends Genet 32(5):284–294. doi:10.1016/j.tig.2016.02.005
286. Lakhotia SC (2004) Epigenetics of heterochromatin. J Biosci 29(3):219–224
287. López-Flores I, Garrido-Ramos M (2012) The repetitive DNA content of eukaryotic genomes. In: Repetitive DNA, vol 7. Karger, Basel, pp 1–28
288. Lakhotia SC, Mukherjee AS (1970) Chromosomal basis of dosage compensation in Drosophila. 3. Early completion of replication by the polytene X-chromosome in male: further evidence and its implications. J Cell Biol 47(1):18–33
289. Zhimulev I, Belyaeva E, Makunin I, Pirrotta V, Volkova E, Alekseyenko A, Andreyeva E, Makarevich G, Boldyreva L, Nanayev R (2003) Influence of the SuUR gene on intercalary heterochromatin in Drosophila melanogaster polytene chromosomes. Chromosoma 111(6):377–398. doi:10.1007/s00412-002-0218-0
290. Makunin IV, Kolesnikova TD, Andreyenkova NG (2014) Underreplicated regions in Drosophila melanogaster are enriched with fast-evolving genes and highly conserved noncoding sequences. Genome Biol Evol 6(8):2050–2060. doi:10.1093/gbe/evu156
291. Tchurikov NA, Kretova OV, Sosin DV, Zykov IA, Zhimulev IF, Kravatsky YV (2011) Genome-wide profiling of forum domains in Drosophila melanogaster. Nucleic Acids Res 39(9):3667–3685. doi:10.1093/nar/gkq1353
292. Kern AD, Barbash DA, Mell JC, Hupalo D, Jensen A (2015) Highly constrained intergenic Drosophila ultraconserved elements are candidate ncrnas. Genome Biol Evol 7(3):689–698. doi:10.1093/gbe/evv011
293. MacIntyre R (2015) Mutation driven evolution. J Hered 106(4):420. doi:10.1093/jhered/esv032
294. Romero-Soriano V, Burlet N, Vela D, Fontdevila A, Vieira C, Guerreiro MPG (2016) Drosophila females undergo genome expansion after interspecific hybridization. Genome Biol Evoution 8(3):556–561
295. Sturm A, Ivics Z, Vellai T (2015) The mechanism of ageing: primary role of transposable elements in genome disintegration. Cell Mol Life Sci 72(10):1839–1847. doi:10.1007/s00018-015-1896-0
296. Oppold A-M, Schmidt H, Rose M, Hellmann SL, Dolze F, Ripp F, Weich B, Schmidt-Ott U, Schmidt E, Kofler R, Hankeln T, Pfenninger M (2016) Chironomus riparius (Diptera) genome sequencing reveals the impact of minisatellite transposable elements on population divergence. bioRxiv doi:10.1101/080721
297. Cohen AL, Jia S (2014) Noncoding RNAs and the borders of heterochromatin. WIREs RNA 5(6):835–847. doi:10.1002/wrna.1249
298. Chen Y-CA, Aravin AA (2015) Non-coding RNAs in transcriptional regulation. Curr Mol Biol Rep 1(1):10–18. doi:10.1007/s40610-015-0002-6
299. Allis CD, Jenuwein T (2016) The molecular hallmarks of epigenetic control. Nat Rev Genet 17(8):487–500. doi:10.1038/nrg.2016.59
300. Khanduja JS, Calvo IA, Joh RI, Hill IT, Motamedi M (2016) Nuclear noncoding RNAs and genome stability. Mol Cell 63(1):7–20. doi:10.1016/j.molcel.2016.06.011

301. Klenov MS, Lavrov SA, Korbut AP, Stolyarenko AD, Yakushev EY, Reuter M, Pillai RS, Gvozdev VA (2014) Impact of nuclear Piwi elimination on chromatin state in Drosophila melanogaster ovaries. Nucleic Acids Res 42(10):6208–6218. doi:10.1093/nar/gku268
302. Romero-Soriano V, Guerreiro MPG (2016) Expression of the retrotransposon Helena reveals a complex pattern of TE Deregulation in Drosophila hybrids. PLoS One 11(1):e0147903. doi:10.1371/journal.pone.0147903
303. Russo J, Harrington AW, Steiniger M (2016) Antisense transcription of retrotransposons in Drosophila: an origin of endogenous small interfering RNA precursors. Genetics 202(1):107–121. doi:10.1534/genetics.115.177196
304. Heard E, Martienssen RA (2014) Transgenerational epigenetic inheritance: myths and mechanisms. Cell 157(1):95–109. doi:10.1016/j.cell.2014.02.045
305. Yan W (2014) Potential roles of noncoding RNAs in environmental epigenetic transgenerational inheritance. Mol Cell Endocrinol 398(1):24–30. doi:10.1016/j.mce.2014.09.008
306. Cortini R, Barbi M, Caré BR, Lavelle C, Lesne A, Mozziconacci J, Victor J-M (2016) The physics of epigenetics. Rev Mod Phys 88(2):025002. doi:10.1103/RevModPhys.88.025002
307. Hanson MA, Skinner MK (2016) Developmental origins of epigenetic transgenerational inheritance. Environ Epigenet 2(1):dvw002. doi:10.1093/eep/dvw002
308. Larriba E, del Mazo J (2016) Role of non-coding RNAs in the transgenerational epigenetic transmission of the effects of reprotoxicants. Int J Mol Sci 17(4):452. doi:10.3390/ijms17040452
309. Schmidt E, Kornfeld J-W (2016) Decoding Lamarck—transgenerational control of metabolism by noncoding RNAs. Pflugers Arch 468(6):959–969. doi:10.1007/s00424-016-1807-8
310. Sharma A (2015) Transgenerational epigenetic inheritance: resolving uncertainty and evolving biology. Biomol Concepts 6(2):87–103. doi:10.1515/bmc-2015-0005
311. Fox AH, Lam YW, Leung AK, Lyon CE, Andersen J, Mann M, Lamond AI (2002) Paraspeckles: a novel nuclear domain. Curr Biol 12(1):13–25
312. Nakagawa S, Hirose T (2012) Paraspeckle nuclear bodies-useful uselessness? Cell Mol Life Sci 69(18):3027–3036. doi:10.1007/s00018-012-0973-x
313. Bernard D, Prasanth KV, Tripathi V, Colasse S, Nakamura T, Xuan Z, Zhang MQ, Sedel F, Jourdren L, Coulpier F, Triller A, Spector DL, Bessis A (2010) A long nuclear-retained non-coding RNA regulates synaptogenesis by modulating gene expression. EMBO J 29(18):3082–3093. doi:10.1038/emboj.2010.199
314. Tripathi V, Ellis JD, Shen Z, Song DY, Pan Q, Watt AT, Freier SM, Bennett CF, Sharma A, Bubulya PA, Blencowe BJ, Prasanth SG, Prasanth KV (2010) The nuclear-retained noncoding RNA MALAT1 regulates alternative splicing by modulating SR splicing factor phosphorylation. Mol Cell 39(6):925–938. doi:10.1016/j.molcel.2010.08.011
315. Miyagawa R, Tano K, Mizuno R, Nakamura Y, Ijiri K, Rakwal R, Shibato J, Masuo Y, Mayeda A, Hirose T, Akimitsu N (2012) Identification of cis- and trans-acting factors involved in the localization of MALAT-1 noncoding RNA to nuclear speckles. RNA 18(4):738–751. doi:10.1261/rna.028639.111
316. Anantharaman A, Jadaliha M, Tripathi V, Nakagawa S, Hirose T, Jantsch MF, Prasanth SG, Prasanth KV (2016) Paraspeckles modulate the intranuclear distribution of paraspeckle-associated Ctn RNA. Sci Rep 6:34043. doi:10.1038/srep34043
317. Villamizar O, Chambers CB, Riberdy JM, Persons DA, Wilber A (2016) Long noncoding RNA Saf and splicing factor 45 increase soluble Fas and resistance to apoptosis. Oncotarget 7(12):13810–13826
318. Kapranov P, Laurent GS (2012) Dark matter RNA: existence, function, and controversy. Front Genet 3:60. doi:10.3389/fgene.2012.00060
319. Clark MB, Choudhary A, Smith MA, Taft RJ, Mattick JS (2013) The dark matter rises: the expanding world of regulatory RNAs. Essays Biochem 54:1–16. doi:10.1042/bse0540001
320. Bassett AR, Liu J-L (2014) CRISPR/Cas9 and genome editing in Drosophila. J Genet Genomics 41(1):7–19

321. Mohr SE, Hu Y, Kim K, Housden BE, Perrimon N (2014) Resources for functional genomics studies in Drosophila melanogaster. Genetics 197(1):1–18
322. Ugur B, Chen K, Bellen HJ (2016) Drosophila tools and assays for the study of human diseases. Dis Model Mech 9(3):235–244
323. Perrimon N, Bonini NM, Dhillon P (2016) Fruit flies on the front line: the translational impact of Drosophila. Dis Model Mech 9(3):229–231
324. Ghosh S, Tibbit C, Liu J-L (2016) Effective knockdown of Drosophila long non-coding RNAs by CRISPR interference. Nucleic Acids Res 44(9):gkw063. doi:10.1093/nar/gkw063
325. Leone S, Santoro R (2016) Challenges in the analysis of long noncoding RNA functionality. FEBS Lett 590(15):2342–2353. doi:10.1002/1873-3468.12308

Chapter 4
Long Noncoding RNAs in the Yeast *S. cerevisiae*

Rachel O. Niederer, Evan P. Hass, and David C. Zappulla

Abstract Long noncoding RNAs have recently been discovered to comprise a sizeable fraction of the RNA World. The scope of their functions, physical organization, and disease relevance remain in the early stages of characterization. Although many thousands of lncRNA transcripts recently have been found to emanate from the expansive DNA between protein-coding genes in animals, there are also hundreds that have been found in simple eukaryotes. Furthermore, lncRNAs have been found in the bacterial and archaeal branches of the tree of life, suggesting they are ubiquitous. In this chapter, we focus primarily on what has been learned so far about lncRNAs from the greatly studied single-celled eukaryote, the yeast *Saccharomyces cerevisiae*. Most lncRNAs examined in yeast have been implicated in transcriptional regulation of protein-coding genes—often in response to forms of stress—whereas a select few have been ascribed yet other functions. Of those known to be involved in transcriptional regulation of protein-coding genes, the vast majority function in *cis*. There are also some yeast lncRNAs identified that are not directly involved in regulation of transcription. Examples of these include the telomerase RNA and telomere-encoded transcripts. In addition to its role as a template-encoding telomeric DNA synthesis, telomerase RNA has been shown to function as a flexible scaffold for protein subunits of the RNP holoenzyme. The flexible scaffold model provides a specific mechanistic paradigm that is likely to apply to many other lncRNAs that assemble and orchestrate large RNP complexes, even in humans. Looking to the future, it is clear that considerable fundamental knowledge remains to be obtained about the architecture and functions of lncRNAs. Using genetically tractable unicellular model organisms should facilitate lncRNA characterization. The acquired basic knowledge will ultimately translate to better understanding of the growing list of lncRNAs linked to human maladies.

R.O. Niederer, Ph.D.
Massachusetts Institute of Technology, Cambridge, MA, USA
e-mail: niederer@mit.edu

E.P. Hass • D.C. Zappulla, Ph.D. (✉)
Department of Biology, Johns Hopkins University, Baltimore, MD, USA
e-mail: ehass1@jhu.edu; zappulla@jhu.edu

M.R.S. Rao (ed.), *Long Non Coding RNA Biology*, Advances in Experimental Medicine and Biology 1008, DOI 10.1007/978-981-10-5203-3_4

Keywords Long noncoding RNA • lncRNA • *Saccharomyces cerevisiae* • Yeast SRG1 • RME2 • PHO84 • CDC28 • GAL10 • pHO • IRT1 • Telomerase RNA • TLC1 • TERRA

Recent advances in high-throughput technologies have revealed pervasive transcription in humans [1–3], as well as in the unicellular yeasts *S. cerevisiae* and *S. pombe*, indicating that at least 75% of their genomes is transcribed [4–6]. Among the transcripts are several classes of noncoding RNA (ncRNA). One class of ncRNAs are those that are longer than 200 nucleotides, which are referred to as long noncoding RNAs, or lncRNAs. While this broad definition includes many classically studied ncRNAs, such as spliceosomal snRNAs and ribosomal rRNAs, discussions of lncRNAs generally focus on the emerging class of transcripts involved in regulation and maintenance of the genome.

Most lncRNAs analyzed so far represent transcripts synthesized by RNA polymerase II, with many in yeast thought to arise from bidirectional promoters [7]. Multiple classes of lncRNAs in yeasts are unstable, including Xrn1 unstable transcripts (XUTs), cryptic unstable transcripts (CUTs), and Nrd1-unterminated transcripts (reviewed in [8]). Often, the function of these transcripts is unknown, although they can regulate gene expression, usually by influencing histone modifications or by interfering with transcription of nearby genes. In fact, transcriptional interference is a widespread phenomenon, occurring in organisms from bacteria to humans.

In unicellular organisms, lncRNAs are extensively used to regulate gene expression in response to environmental conditions. For example, there are additional classes of ncRNAs only expressed under specific circumstances, such as meiosis (MUTs, MeiRNA in *S. pombe*) [9, 10], or in the absence of nonsense-mediated decay (CD-CUTs) [11], underscoring the potential importance of ncRNAs in regulating adaptive changes in yeasts.

4.1 LncRNAs That Are Byproducts of Transcriptional Interference

Transcriptional interference is a phenomenon by which the transcription from one promoter, independent of the produced transcript, interferes with transcription from a second nearby promoter. Examples of transcriptional interference have been found in both bacteria and eukaryotes [12], including several instances in budding yeast. In many cases, the lncRNAs themselves appear to simply be byproducts of transcriptional interference without functions of their own.

One well-studied example of lncRNA transcription-mediated transcriptional interference is repression of the *SER3* gene by the upstream transcript SRG1 (Table 4.1). *SER3* encodes an enzyme required for serine biosynthesis that is

Table 4.1 LncRNA functional classifications in *S. cerevisiae*

		LncRNA examples
Transcriptional regulation	Transcriptional interference	SRG1, IRT1, RME2, pHO-lncRNA
	Gene looping	CDC28 lncRNA
	Cis repression	GAL10 lncRNA
	Trans repression	PHO84 lncRNA
Non-transcriptional functions (e.g., maintaining telomeres)	RNP scaffold	TLC1 (telomerase RNA)
	Telomere regulation	TERRA

strongly repressed when cells are grown in serine-rich medium [13]. The finding that this repression was dependent on the SWI/SNF chromatin-remodeling complex prompted researchers to investigate the elements of the promoter more closely [14]. Chromatin immunoprecipitation (ChIP) experiments revealed that TATA-binding protein (TBP) and RNA polymerase II both bind to a site upstream of the *SER3* TATA box, where there is a second TATA sequence affiliated with the promoter for a noncoding transcript, *SRG1* (*SER3* regulatory gene 1). *SRG1* is transcribed from this promoter on the same strand as *SER3,* and mutation of the *SRG1* TATA box results in *SER3* derepression. When a transcriptional terminator was inserted between the *SRG1* and *SER3* promoters, SRG1 lncRNA-mediated repression was abolished, indicating that transcription across the *SER3* promoter is required for its repression. Furthermore, insertion of the *SRG1* promoter upstream of a different gene (*GAL7*) was sufficient to mediate strong repression of that gene, indicating that *SRG1* transcription, not the SRG1 transcript, is responsible for SRG1-mediated repression.

Subsequent studies revealed that histone modification and nucleosome-remodeling factors play prominent roles in the SRG1-mediated *SER3* transcriptional regulation. *SRG1* transcription was shown to be controlled by the serine-dependent transcriptional activator Cha4, which recruits the SAGA and SWI/SNF complexes to promote *SRG1* transcription [15]. Additionally, *SRG1* transcription exerted its repressive effect on *SER3* by altering nucleosome positioning on the *SER3* promoter [16, 17]. In wild-type cells grown in rich medium, the nucleosome-free region typical of active promoters is absent at the *SER3* promoter. However, if the *SRG1* TATA box is mutated or cells are grown in the absence of serine, a nucleosome-free region opens up at the *SER3* promoter. Deposition of nucleosomes on the *SER3* promoter during *SRG1* transcription was shown to be dependent on Spt6 and Spt16, subunits of the Spt6/Spn1 and FACT complexes, respectively, as well as Spt2—all of which are involved in nucleosome deposition in the wake of transcription. Thus, the current model for *SRG1*-mediated transcriptional interference is that in serine-rich conditions, Cha4, along with SAGA and SWI/SNF, promotes *SRG1* transcription. The act of transcription then results in deposition of nucleosomes across the *SER3* promoter, occluding binding sites for transcription factors in what would otherwise be the *SER3* nucleosome-free region and preventing *SER3* transcription.

The general configuration of lncRNA transcription-mediated transcriptional interference described above—i.e., transcription of an upstream, sense-stranded lncRNA depositing repressive chromatin on a downstream mRNA promoter, as shown in Fig. 4.1a—is not unique to the *SRG1-SER3* system. Recently, expression of the HO endonuclease was found to be regulated by transcription of an upstream lncRNA, termed pHO-lncRNA (Table 4.1), through a nucleosome-repositioning

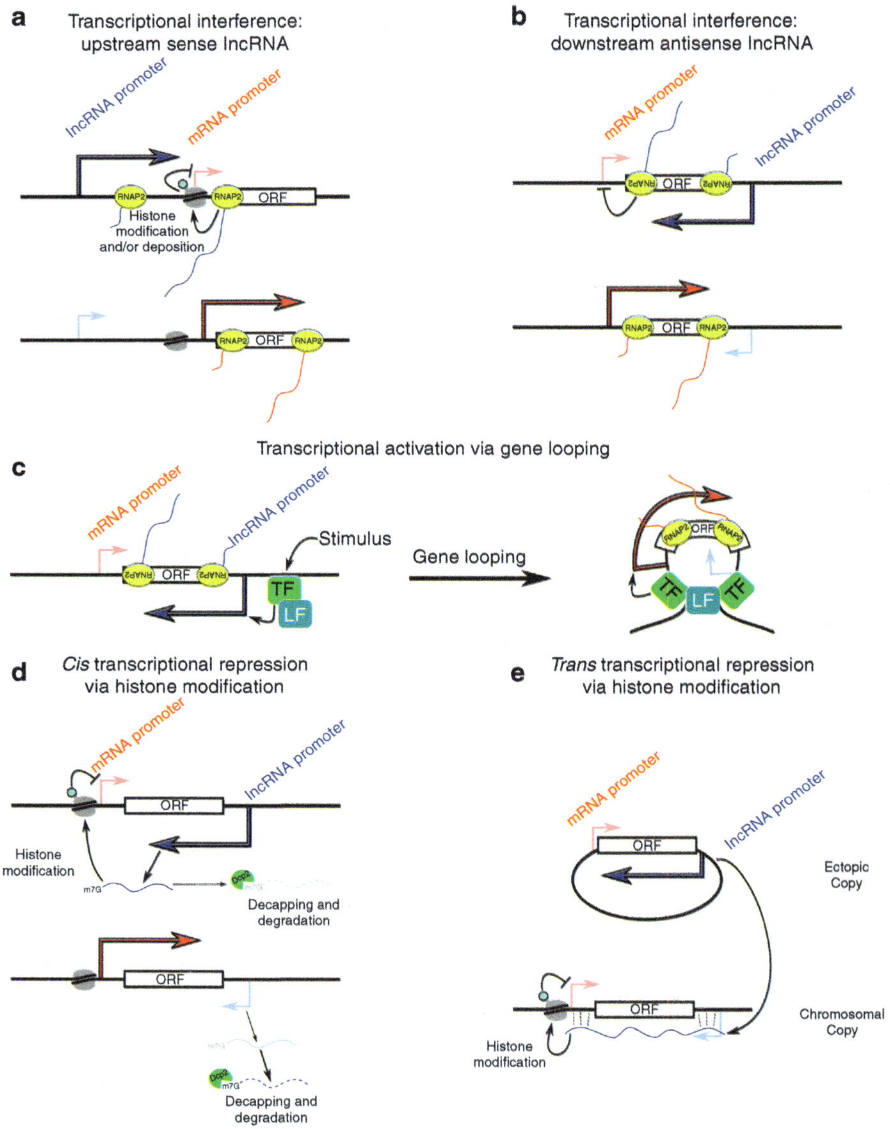

mechanism very similar to *SRG1* [18]. Furthermore, a gene encoding a master regulatory transcription factor for meiosis, *IME1*, is also regulated by transcription of its upstream sense-strand lncRNA, IRT1 (*IME1*-repressive transcript 1), although *IME1* repression is ultimately caused by histone modification rather than nucleosome repositioning [19]. In haploid cells, the meiosis-repressive transcription factor Rme1 activates transcription from the *IRT1* promoter, resulting in transcription across the *IME1* promoter. This transcription recruits the chromatin-modifying factors Set2 and Set3, which deposit repressive histone modifications at the *IME1* promoter, preventing *IME1* transcription, and ultimately meiosis (which would be lethal for haploid cells). However, Rme1 is a haploid-specific gene regulated by the diploid-specific a1/α2 repressor complex. Thus, in diploid cells, *RME1* is repressed, *IRT1* is not transcribed, and the *IME1* promoter remains open for transcription factor binding, allowing the cells to enter meiosis and sporulate when in nutrient-poor conditions.

IME1 is not the only meiotic regulatory gene repressed via lncRNA transcription-mediated transcriptional interference. Transcriptional expression of an RNA methyltransferase also required for the meiotic pathway, *IME4*, is regulated by the noncoding transcript RME2 [20, 21]. Like the *IRT1-IME1* system, RME2 is expressed only in haploid cells and *IME4* only in diploid *MAT*a/α cells. In contrast, *RME2* is transcribed in the antisense orientation relative to *IME4*, and its promoter is located downstream rather than upstream, with the two convergent transcripts

Fig. 4.1 LncRNAs involved in transcriptional regulation. (**a**) Transcriptional interference: upstream sense lncRNA. Transcription of the lncRNA by RNA polymerase II across the mRNA promoter results in deposition of nucleosomes (*gray*) or repressive histone modifications (*cyan circle*) which, in turn, result in occlusion of transcription factor binding sites in the mRNA promoter. When the lncRNA is not transcribed, chromatin at the mRNA promoter remains in an open state, allowing for mRNA transcription. (**b**) Transcriptional interference: downstream antisense lncRNA. Transcription of the lncRNA interferes with and represses mRNA transcription. When lncRNA transcription is turned off (either due to the presence of a transcriptional repressor or the absence of an activator), the mRNA can be transcribed. (**c**) Transcriptional activation via gene looping. An environmental stimulus promotes transcription factor (TF) recruitment to the promoter of a downstream antisense lncRNA. lncRNA transcription and recruitment of looping factors (LF) result in gene looping, stimulating recruitment of the transcription factor to the mRNA promoter and subsequent mRNA transcription. (**d**) *Cis* transcriptional repression via histone modification. When an antisense lncRNA is transcribed, the transcript recruits histone modifiers that deposit repressive chromatin marks at the mRNA promoter. The lncRNA is constantly turned over via a decapping-dependent mechanism, but turnover is outweighed by transcription when the lncRNA is being actively transcribed. However, when lncRNA transcription is downregulated, the small amount of the lncRNA that is present can be cleared from the locus via decapping-dependent degradation, allowing for transcription from the mRNA promoter. (**e**) *Trans* transcriptional repression via histone modification. When ectopically expressed, an antisense lncRNA represses mRNA transcription from the endogenous chromosomal copy via repressive histone modification. This repression in *trans* requires that the lncRNA has homology (*black dotted lines*) with the sequences upstream and downstream of the endogenous gene

overlapping (Fig. 4.1b). A binding site for the diploid-specific a1/α2 repressor complex was found downstream of the *IME4* open reading frame (ORF), and mutating the binding site resulted in ectopic *RME2* transcription and *IME4* downregulation in diploid cells. Inversely, deleting the *RME2* promoter entirely resulted in ectopic *IME4* transcription in haploid cells. Additionally, the transcriptional interference between *RME2* and *IME4* was shown to act strictly in *cis*. Even when two *IME4/RME2* genes were placed next to one another on the chromosome and were mutated such that one would constitutively express *IME4* and the other would constitutively express *RME2*, the latter gene did not repress the former, strongly suggesting that RME2 represses *IME4* through transcriptional interference. However, TBP binding at the *IME4* promoter is not disrupted by *RME2* transcription through it, suggesting a transcriptional interference mechanism distinct from those at the *SER3* and *IME1* promoters. Intriguingly, a 450-bp region within the *IME4* ORF was shown to be required for transcriptional interference by RME2; premature *RME2* termination within this region, deletion of the region, or inversion of the region all result in *IME4* derepression. However, the detailed mechanism for how transcriptional interference works in the *IME4*-RME2 system remains to be fully elucidated.

4.2 Functional lncRNAs

While transcriptional interference may be the best-characterized role of lncRNAs in unicellular organisms, some lncRNAs are known to function on their own. As is the case with lncRNAs broadly, most functional lncRNAs regulate gene expression. This class includes lncRNAs generated in both the sense and antisense direction relative to nearby genes that can utilize both *cis* and *trans* mechanisms [22, 23].

4.2.1 Cis-Acting lncRNAs

One example of a functional lncRNA that acts exclusively in *cis* is the CDC28 lncRNA. Upon osmotic stress, the stress-activated protein kinase Hog1 induces a set of lncRNAs, including an antisense transcript within the *CDC28* gene [24], which encodes a master regulator cyclin-dependent kinase that controls the cell cycle in yeast. Induction of CDC28 promotes efficient reentry into the cell cycle in response to stress. The coordinated induction of both *CDC28* and the antisense lncRNA prompted investigators to examine the relationship between the sense and antisense transcripts. ChIP experiments showed accumulation of Hog1 at both the 5′ and 3′ ends of *CDC28* under conditions of osmostress. However, when transcription of the lncRNA was prevented, Hog1 accumulation was reduced at the 3′ end and completely abolished at the 5′ end of *CDC28*. Expression of the lncRNA from a plasmid failed to promote *CDC28* activation, suggesting the lncRNA functions in *cis*.

Notably, transcription of the CDC28 lncRNA alone is not sufficient to cause *CDC28* induction. When the inducible *GAL1* promoter was used to drive lncRNA expression, sense transcription of *CDC28* was not increased. Conversely, if lncRNA expression was induced using a Hog1-dependent activator, *CDC28* induction was restored. Thus, induction of *CDC28* requires both transcription of the full-length lncRNA and Hog1 association at the 5′ and 3′ ends of the gene. The Hog1 association pattern raised the possibility of gene looping in this region (Fig. 4.1c). In fact, when an essential looping protein, Ssu72 [25], is repressed, induction of *CDC28* is severely reduced. Together these data suggest a model where Hog1 binding at the 3′ end of *CDC28* induces expression of an antisense lncRNA, which then establishes gene looping, facilitating the distribution of Hog1 to both the 3′ and 5′ ends of *CDC28*. Hog1 association then promotes chromatin remodeling and ultimately increased *CDC28* expression.

This pattern of regulation by lncRNA-mediated gene looping isn't limited to *CDC28*. Chromatin conformation capture (3C) experiments show stress-induced looping at the *MMF1* locus, which also exhibits lncRNA-dependent Hog1 accumulation [24]. This general mechanism, schematized in Fig. 4.1c, has previously been observed in human cells [26], suggesting that this may be a conserved function of lncRNAs.

Another well-studied *cis*-acting long noncoding RNA, GAL10 lncRNA, regulates the cellular response to growth on galactose. The set of yeast *GAL* genes, required for growth on galactose-containing medium, have been studied extensively and are considered a model for regulated gene expression in yeast. The GAL10 lncRNA spans two divergently transcribed *GAL* genes, *GAL1* and *GAL10*. Expression of the lncRNA induces deacetylation throughout the transcribed region, and levels of the lncRNA are inversely correlated with the overlapping mRNAs [27]. Interestingly, efficient galactose induction of *GAL1* requires decapping of the *GAL10* lncRNA [28]. Decapping plays a role in the regulation of over 100 lncRNAs, indicating that this may be a widespread phenomenon [28]. This finding is consistent with a model where the lncRNAs themselves play a role in gene regulation (Fig. 4.1d), rather than control being achieved via transcriptional interference. However, this is not a forgone conclusion, since recent studies have shown that lncRNA degradation can be linked to transcription and termination [29], adding additional layers of complexity and nuance to lncRNA-mediated gene regulation.

4.2.2 Trans-*Acting lncRNAs*

While most well-characterized lncRNAs in yeasts act as *cis*-regulators of gene expression, several lncRNAs can mediate changes in gene expression in *trans*. Interestingly, the yeast gene *PHO84* is regulated by antisense lncRNAs that can function both in *cis* and in *trans*. *PHO84* encodes a phosphate transporter and shows decreased expression while *S. cerevisiae* cells age chronologically [23]. Concurrent with the decrease in *PHO84* expression, two antisense lncRNA transcripts spanning the *PHO84* gene show increased expression as cells age. Both transcripts can be considered among the class of unstable transcripts, as their levels appear to be primarily regulated by the nuclear exosome component Rrp6 [23]. Rrp6 protein association with the *PHO84* gene decreases over time, corresponding with the increased levels of both antisense lncRNAs. In addition to stabilization of the antisense lncRNAs, *PHO84* repression also requires the Hda1/2/3 histone deacetylase machinery. As in previous examples given for transcriptional interference, the *PHO84* antisense lncRNAs regulate gene expression by influencing local chromatin. While early reports indicated *cis*-regulation of *PHO84* may proceed via stabilization of the lncRNAs rather than transcriptional interference, more recent work has shown that Rrp6 promotes early termination of the lncRNA rather than degradation [29].

Strikingly, transcriptional expression of an additional copy of *PHO84* in the cell also results in repression of the endogenous gene, indicating the lncRNAs can function in *trans* [22]. While the *cis*-acting mechanism requires histone deacetylation, the *trans*-acting machinery is instead promoted by the Set1 histone methyltransferase. Homology to both the 3′ portion of *PHO84* and an upstream activating sequence (but not the coding sequence itself) is required for *trans*-silencing (Fig. 4.1e). This homology-based *trans*-acting mechanism is particularly striking in *S. cerevisiae,* which lacks the RNAi machinery [30].

The examples given above highlight the tendency of yeast lncRNAs to be involved in cellular responses to environmental and stressful conditions. A prominent role for lncRNAs in cellular reactions to stressful stimuli may be due to the ongoing nature of lncRNA transcription and the lack of a requirement for their translation, which allows lncRNAs to regulate gene expression rapidly and specifically. Both *S. cerevisiae* and *S. pombe* extensively modulate lncRNA gene expression in response to nutrient availability, osmostress, and senescence [31–34]. Senescence-associated lncRNAs have also been identified in human cells [35], suggesting they may be part of a conserved cellular response to senescence. It is not yet known what the roles of the senescence-associated lncRNAs in yeast are, but expression patterns indicate that some may act in *cis* and others in *trans*, highlighting the versatility of lncRNAs in regulating cellular processes [34].

4.3 LncRNAs That Assemble Ribonucleoprotein Complexes: Flexible Scaffolding by Telomerase RNA

LncRNAs, like many canonical noncoding RNAs, can bind to other factors in order to assemble complexes. Such lncRNAs typically have specific domains required to bind proteins, and therefore these portions of the lncRNA transcript tend to be well conserved. This contrasts with other portions of many such lncRNAs, which are apt to change in sequence very rapidly.

The first and most mechanistically characterized lncRNA that scaffolds assembly of a functional lncRNP complex is telomerase RNA. Telomerase is an RNP enzyme that performs a critical reverse transcription reaction to synthesize DNA at the ends of chromosomes to prevent them from eroding due to the "end-replication problem" [35, 36]. Telomerase RNA is considered to be a lncRNA because it does not encode a protein, is greater than 200 nucleotides (except in ciliates), is evolving rapidly in sequence, is typically expressed at a very low level, and is polyadenylated. Telomerase RNA differs from most other lncRNAs in that it is part of an enzyme, although it is probable that some identified lncRNAs will be ultimately discovered to participate in catalytic reactions. Irrespective of whether telomerase RNA should be technically classified as a *bona fide* lncRNA, what has been learned from studying telomerase RNA is directly relevant to understanding the mechanism of function of a significant fraction of lncRNAs (e.g., see reference [42]).

The telomerase RNA in the yeast *S. cerevisiae* has been shown to be a flexible scaffold for protein subunits [37–40]. What does "flexible scaffold" mean in the context of RNA that assembles an RNP complex? One can envision at least three major types of flexibility that an RNA could provide in this context, and RNPs could be built upon flexible scaffold RNAs that have some or all of these properties:

1. *Organizational flexibility*. This form of flexibility refers to the ability of an RNA-protein complex to function in vivo despite relative rearrangement of the subunits' locations in the complex [37, 41]. As one can imagine, "organizational," or "functional," flexibility [38] does not predominate in RNP complexes that require precise positioning of subunits in order to function, such as the ribosome.
2. *Pliability*. This refers to the degree of dynamic bending of the lncRNA in three-dimensional space. To a large extent, pliability of large noncoding RNAs is determined by the extent to which loops and junctions interact (e.g., via formation of pseudoknots) to form stable tertiary structure, as opposed to these single-stranded regions retaining their intrinsically dynamic nature, providing "hinges" between more rigid base-paired helices.
3. *Folded-state heterogeneity*. Multiple native folded states represent a third form of flexibility, since a diverse array of final structures contributes diversity to the ensemble of molecules with the same sequence. This is very likely to impart a range of distances in three-dimensional space between functional elements, such as protein-binding domains, scaffolded by the lncRNA.

Whereas pliability and folded-state heterogeneity of telomerase RNA have yet to be tested extensively, it is clear that *S. cerevisiae* telomerase has pronounced organizational flexibility. This was shown first by repositioning an essential binding site for a critical holoenzyme-specific protein subunit, Est1, on the telomerase RNA, TLC1, and demonstrating that the RNP still functions [37] (Fig. 4.2). Retention of functionality despite protein-subunit repositioning along the RNA in the RNP complex was a striking discovery, given that the critical Est1-interacting site was moved hundreds of nucleotides away from its native location. Such functional tolerance of subunit repositioning would not be expected in RNP complexes with defined higher-order structure, such as the ribosome. Furthermore, the binding site for the Ku

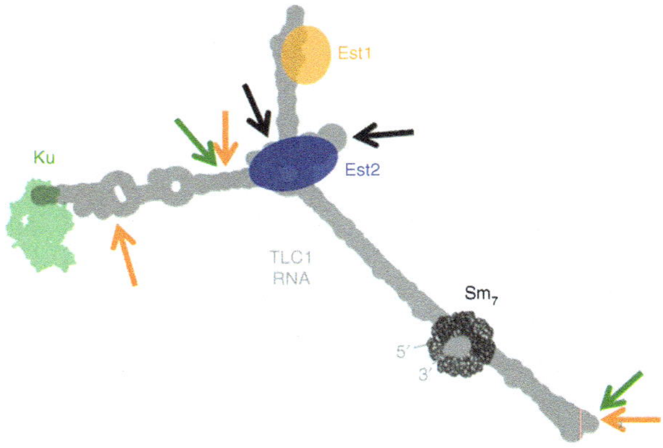

Fig. 4.2 Organizational flexibility of the telomerase RNA in yeast. Regions of the *S. cerevisiae* telomerase RNA, TLC1, that bind to protein subunits of the RNP holoenzyme functioned despite being moved to the indicated positions. Est1 site, *orange* (Zappulla and Cech, *PNAS* 2004). Ku site, *green* (Zappulla et al., *RNA* 2011). Sm site, *black* (Mefford et al., *EMBO* 2013). In the case of moving the Sm-binding site, the ends of TLC1 RNA were also relocated (i.e., a circularly permuted RNA was constructed), since Sm_7 functions in TLC1 3′-end processing

subunit has been relocated with similar results [40], as well as the Sm_7 complex via circular permutation experiments [38] (Fig. 4.2). Overall, these experiments demonstrated that each of the three known holoenzymatic protein subunits of the telomerase RNP can be repositioned with retention of function in vivo, and therefore telomerase RNA has impressive organizational flexibility, with respect to its organization of the RNP holoenzyme complex. In contrast, there is more limited organizational flexibility at the catalytic core of the *S. cerevisiae* telomerase RNA: repositioning of RNA elements in the core caused catalytic inactivation in vitro and loss of function in vivo, although more subtle mutations, such as some circular permutations, still permitted functionality [38].

Flexible scaffold function by telomerase RNA in yeast has provided a paradigm for long noncoding RNAs more generally, including in Xist and other lncRNAs in humans [41, 42]. Thus, fundamental lncRNA functional features discovered and thoroughly tested by studying telomerase RNA in budding yeast can advance understanding and guide future studies of mammalian lncRNAs.

4.4 TERRA lncRNA

Another kind of long noncoding RNA is transcribed from the ends of chromosomes, at telomeres. This telomeric repeat-containing RNA, "TERRA"—not to be confused with the telomerase RNA—was first identified in human cells, but was shortly thereafter also identified in yeast [43–45]. By leveraging the genetic tractability and molecular biology tools in yeast, much has been learned about TERRA, as well as

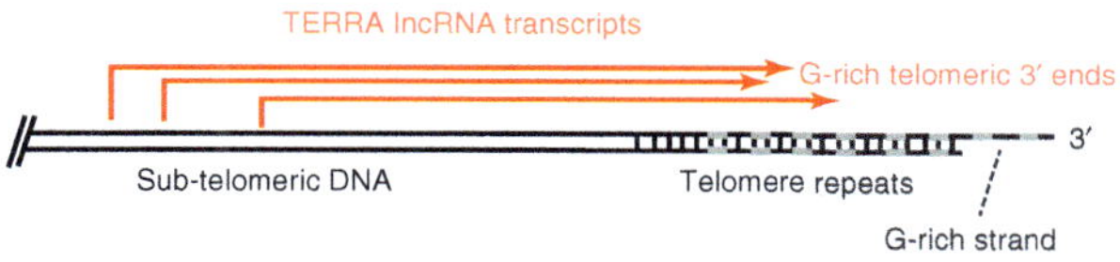

Fig. 4.3 TERRA lncRNAs are transcribed from the subtelomeric region into the telomeric repeats. TERRA RNAs have G-rich 3′ ends, like the chromosomes' 3′ ends, since they are transcribed in the direction indicated, therefore using the C-rich 5′ end of the chromosome as the template during transcription by RNA polymerase II. Diversity of TERRA RNA lengths and sequences stem from use of multiple promoters, as well as differences in the 3′ ends

other lncRNAs encoded by DNA near chromosome ends. TERRA has been found to be a widely conserved telomere-associated lncRNA among eukaryotes. Furthermore, it also has been found to have a role in telomere-length regulation and human diseases such as cancer [46].

TERRA promoters reside in subtelomeric regions, and therefore the 5′ regions of TERRA RNA transcripts are not telomeric repeats, but rather are encoded by sequences internal to the distal telomere repeats, i.e., in subtelomeric loci (see Fig. 4.3) [47, 48]. Since the 3′ ends of chromosomes are G-rich telomeric repeats, and the C-rich strand is the template during TERRA transcription, the TERRA RNA is therefore also G-rich. This permits TERRA to potentially hybridize with the telomerase enzyme's RNA C-rich template that synthesizes telomeric DNA by reverse transcription, and, indeed, it has been demonstrated that TERRA can inactivate telomerase RNPs by this RNA-RNA base-pairing in vitro [49, 50]. TERRA length is heterogeneous as a result of diverse RNA polymerase II promoter positions relative to the telomeres, as well as a variety of 3′ ends that are ultimately formed. A minority of TERRA RNAs are polyadenylated, although in *S. cerevisiae* it has been reported that the vast majority has a poly-A tail [44].

In ascertaining the fundamental functions of TERRA and other telomeric lncRNAs in eukaryotes, as with lncRNAs overall, it will be important to disambiguate the roles of the RNAs per se from the act of transcription. It has been proposed that transcription of telomeres plays a role in remodeling individual telomeres during alternative telomeric epigenetic states, such as in response to stress, DNA damage, etc. Telomere transcription and/or telomeric lncRNAs could also play a role in regulating telomerase recruitment and access, perhaps even guiding telomerase to short telomeres in greatest need of extension.

4.5 Concluding Remarks

Long noncoding RNAs continue to be discovered at a rapid rate, particularly now that their identification is being actively pursued and their roles are being more widely appreciated and studied. The current handful of examples of lncRNAs with ascribed functions will expand considerably, concomitant with the hard work of characterizing lncRNA functions, architecture, mechanisms, binding partners, cellular localization, and life cycles. In the coming years, the list of different functional classes of lncRNAs is apt to grow exponentially. Ultimately, the term "long noncoding RNA" is very likely to be supplanted by categories based on functional classes, as functions are identified. The potential for model organisms—particularly those with facile genetics such as *S. cerevisiae*—to assist in lncRNA characterization is substantial, particularly given the great challenge that characterizing lncRNAs represents.

References

1. Bertone P, Stolc V, Royce TE et al (2004) Global identification of human transcribed sequences with genome tiling arrays. Science 306(5705):2242–2246
2. Guttman M, Amit I, Garber M et al (2009) Chromatin signature reveals over a thousand highly conserved large non-coding RNAs in mammals. Nature 458(7235):223–227
3. Carninci P, Kasukawa T, Katayama S et al (2005) The transcriptional landscape of the mammalian genome. Science 309(5740):1559–1563
4. Nagalakshmi U, Wang Z, Waern K et al (2008) The transcriptional landscape of the yeast genome defined by RNA sequencing. Science 320(5881):1344–1349
5. Jacquier A (2009) The complex eukaryotic transcriptome: unexpected pervasive transcription and novel small RNAs. Nat Rev Genet 10(12):833–844
6. Wilhelm BT, Marguerat S, Watt S et al (2008) Dynamic repertoire of a eukaryotic transcriptome surveyed at single-nucleotide resolution. Nature 453(7199):1239–1243
7. Xu Z, Wei W, Gagneur J et al (2009) Bidirectional promoters generate pervasive transcription in yeast. Nature 457(7232):1033–1037
8. Wu J, Delneri D, O'Keefe RT (2012) Non-coding RNAs in *Saccharomyces cerevisiae*: what is the function? Biochem Soc Trans 40(4):907–911
9. Lardenois A, Liu Y, Walther T et al (2011) Execution of the meiotic noncoding RNA expression program and the onset of gametogenesis in yeast require the conserved exosome subunit Rrp6. Proc Natl Acad Sci U S A 108(3):1058–1063
10. Shichino Y, Yamashita A, Yamamoto M (2014) Meiotic long non-coding meiRNA accumulates as a dot at its genetic locus facilitated by Mmi1 and plays as a decoy to lure Mmi1. Open Biol 4(6):140022
11. Toesca I, Nery CR, Fernandez CF et al (2011) Cryptic transcription mediates repression of subtelomeric metal homeostasis genes. PLoS Genet 7(6):e1002163
12. Shearwin KE, Callen BP, Egan JB (2005) Transcriptional interference—a crash course. Trends Genet 21(6):339–345
13. Martens JA, Winston F (2002) Evidence that Swi/Snf directly represses transcription in *S. cerevisiae*. Genes Dev 16(17):2231–2236
14. Martens JA, Laprade L, Winston F (2004) Intergenic transcription is required to repress the *Saccharomyces cerevisiae SER3* gene. Nature 429(6991):571–574
15. Martens JA, Wu PY, Winston F (2005) Regulation of an intergenic transcript controls adjacent gene transcription in *Saccharomyces cerevisiae*. Genes Dev 19(22):2695–2704
16. Hainer SJ, Pruneski JA, Mitchell RD et al (2011) Intergenic transcription causes repression by directing nucleosome assembly. Genes Dev 25(1):29–40
17. Thebault P, Boutin G, Bhat W et al (2011) Transcription regulation by the noncoding RNA SRG1 requires Spt2-dependent chromatin deposition in the wake of RNA polymerase II. Mol Cell Biol 31(6):1288–1300

18. Yu Y, Yarrington RM, Chuong EB et al (2016) Disruption of promoter memory by synthesis of a long noncoding RNA. Proc Natl Acad Sci U S A 113(34):9575–9580
19. van Werven FJ, Neuert G, Hendrick N et al (2012) Transcription of two long noncoding RNAs mediates mating-type control of gametogenesis in budding yeast. Cell 150(6): 1170–1181
20. Hongay CF, Grisafi PL, Galitski T et al (2006) Antisense transcription controls cell fate in *Saccharomyces cerevisiae*. Cell 127(4):735–745
21. Gelfand B, Mead J, Bruning A et al (2011) Regulated antisense transcription controls expression of cell-type-specific genes in yeast. Mol Cell Biol 31(8):1701–1709
22. Camblong J, Beyrouthy N, Guffanti E et al (2009) Trans-acting antisense RNAs mediate transcriptional gene cosuppression in *S. cerevisiae*. Genes Dev 23(13):1534–1545
23. Camblong J, Iglesias N, Fickentscher C et al (2007) Antisense RNA stabilization induces transcriptional gene silencing via histone deacetylation in *S. cerevisiae*. Cell 131(4):706–717
24. Nadal-Ribelles M, Sole C, Xu Z et al (2014) Control of Cdc28 CDK1 by a stress-induced lncRNA. Mol Cell 53(4):549–561
25. Ansari A, Hampsey M (2005) A role for the CPF 3′-end processing machinery in RNAP II-dependent gene looping. Genes Dev 19(24):2969–2978
26. Lai F, Orom UA, Cesaroni M et al (2013) Activating RNAs associate with mediator to enhance chromatin architecture and transcription. Nature 494(7438):497–501
27. Houseley J, Rubbi L, Grunstein M et al (2008) A ncRNA modulates histone modification and mRNA induction in the yeast *GAL* gene cluster. Mol Cell 32(5):685–695
28. Geisler S, Lojek L, Khalil AM et al (2012) Decapping of long noncoding RNAs regulates inducible genes. Mol Cell 45(3):279–291
29. Castelnuovo M, Rahman S, Guffanti E et al (2013) Bimodal expression of *PHO84* is modulated by early termination of antisense transcription. Nat Struct Mol Biol 20(7):851–858
30. Aravind L, Watanabe H, Lipman DJ et al (2000) Lineage-specific loss and divergence of functionally linked genes in eukaryotes. Proc Natl Acad Sci U S A 97(21):11319–11324
31. Bumgarner SL, Neuert G, Voight BF et al (2012) Single-cell analysis reveals that noncoding RNAs contribute to clonal heterogeneity by modulating transcription factor recruitment. Mol Cell 45(4):470–482
32. Huber F, Bunina D, Gupta I et al (2016) Protein abundance control by non-coding antisense transcription. Cell Rep 15(12):2625–2636
33. Leong HS, Dawson K, Wirth C et al (2014) A global non-coding RNA system modulates fission yeast protein levels in response to stress. Nat Commun 5:3947
34. Niederer RO, Papadopoulos N, Zappulla DC (2016) Identification of novel noncoding transcripts in telomerase-negative yeast using RNA-seq. Sci Rep 6:19376
35. Olovnikov AM (1973) A theory of marginotomy. The incomplete copying of template margin in enzymic synthesis of polynucleotides and biological significance of the phenomenon. J Theor Biol 41(1):181–190
36. Greider CW, Blackburn EH (1985) Identification of a specific telomere terminal transferase activity in *Tetrahymena* extracts. Cell 43:405–413
37. Zappulla DC, Cech TR (2004) Yeast telomerase RNA: a flexible scaffold for protein subunits. Proc Natl Acad Sci U S A 101(27):10024–10029
38. Mefford MA, Rafiq Q, Zappulla DC (2013) RNA connectivity requirements between conserved elements in the core of the yeast telomerase RNP. EMBO J 32(22):2980–2993
39. Lebo KJ, Zappulla DC (2012) Stiffened yeast telomerase RNA supports RNP function *in vitro* and *in vivo*. RNA 18(9):1666–1678
40. Zappulla DC, Goodrich KJ, Arthur JR et al (2011) Ku can contribute to telomere lengthening in yeast at multiple positions in the telomerase RNP. RNA 17(2):298–311
41. Zappulla DC, Cech TR (2006) RNA as a flexible scaffold for proteins: yeast telomerase and beyond. Cold Spring Harb Symp Quant Biol 71:217–224
42. Guttman M, Donaghey J, Carey BW et al (2011) lincRNAs act in the circuitry controlling pluripotency and differentiation. Nature 477(7364):295–300

43. Azzalin CM, Reichenbach P, Khoriauli L et al (2007) Telomeric repeat containing RNA and RNA surveillance factors at mammalian chromosome ends. Science 318(5851):798–801
44. Luke B, Panza A, Redon S et al (2008) The Rat1p 5′ to 3′ exonuclease degrades telomeric repeat-containing RNA and promotes telomere elongation in *Saccharomyces cerevisiae*. Mol Cell 32(4):465–477
45. Azzalin CM, Lingner J (2015) Telomere functions grounding on TERRA firma. Trends Cell Biol 25(1):29–36
46. Barthel FP, Wei W, Tang M et al (2017) Systematic analysis of telomere length and somatic alterations in 31 cancer types. Nat Genet 49(3):349–357
47. Nergadze SG, Farnung BO, Wischnewski H et al (2009) CpG-island promoters drive transcription of human telomeres. RNA 15(12):2186–2194
48. Pfeiffer V, Lingner J (2012) TERRA promotes telomere shortening through exonuclease 1-mediated resection of chromosome ends. PLoS Genet 8(6):e1002747
49. Redon S, Reichenbach P, Lingner J (2010) The non-coding RNA TERRA is a natural ligand and direct inhibitor of human telomerase. Nucleic Acids Res 38(17):5797–5806
50. Schoeftner S, Blasco MA (2008) Developmentally regulated transcription of mammalian telomeres by DNA-dependent RNA polymerase II. Nat Cell Biol 10(2):228–236

Chapter 5
Long Noncoding RNAs in Plants

Hsiao-Lin V. Wang and Julia A. Chekanova

Abstract The eukaryotic genomes are pervasively transcribed. In addition to protein-coding RNAs, thousands of long noncoding RNAs (lncRNAs) modulate key molecular and biological processes. Most lncRNAs are found in the nucleus and associate with chromatin, but lncRNAs can function in both nuclear and cytoplasmic compartments. Emerging work has found that many lncRNAs regulate gene expression and can affect genome stability and nuclear domain organization both in plant and in the animal kingdom. Here, we describe the major plant lncRNAs and how they act, with a focus on research in *Arabidopsis thaliana* and our emerging understanding of lncRNA functions in serving as molecular sponges and decoys, functioning in regulation of transcription and silencing, particularly in RNA-directed DNA methylation, and in epigenetic regulation of flowering time.

Keywords Plant lncRNAs • Noncoding RNAs • Epigenetics • Exosome • FLC • Transcriptional regulation

5.1 Introduction

In eukaryotes, transcriptome studies showed that >90% of the genome is transcribed and a myriad of transcripts corresponds to noncoding RNAs (ncRNAs) [1, 2], including long ncRNAs (lncRNAs), which are classically >200 nt long and have no discernable coding potential [3–5]. Plant genomes produce tens of thousands of lncRNAs from intergenic, intronic, or coding regions. RNA Pol II transcribes most lncRNAs (from the sense or antisense strands); plants also have Pol IV and Pol V, the two plant-specific RNA polymerases that can produce lncRNAs [6, 7]. Majority of described up-to-date plant lncRNAs are polyadenylated, while in yeast and mammals, there are many non-polyadenylated lncRNAs as well [8]. However, there are

H.-L.V. Wang • J.A. Chekanova (✉)
School of Biological Sciences, University of Missouri-Kansas City,
Kansas City, MO 64110, USA
e-mail: chekanovaj@gmail.com

M.R.S. Rao (ed.), *Long Non Coding RNA Biology*, Advances in Experimental Medicine and Biology 1008, DOI 10.1007/978-981-10-5203-3_5

several well-studied important functional non-polyadenylated lncRNAs [9–11]; and the recent work in *Arabidopsis* found that abiotic stress induced the production of hundreds of non-polyadenylated lncRNAs [12–14].

Most lncRNAs can be broadly classified based on their relationships to protein-coding genes: (1) long intergenic ncRNAs (lincRNAs) (Fig. 5.1A); (2) lncRNAs produced from introns (incRNAs), which can be transcribed in any orientation relative to coding genes (Fig. 5.1B); and (3) antisense RNAs and natural antisense transcripts (NATs), which are transcribed from the antisense strand of genes (Fig. 5.1C and D) [15]. Various types of lncRNAs are also transcribed near transcription start sites (TSSs) and transcription termination sites (TTSs) or from enhancer regions (eRNAs) (Fig. 5.1G) and splice sites. For example, yeast produces cryptic unstable transcripts (CUTs) and stable unannotated transcripts (SUTs) from around TSSs [16], Xrn1-sensitive XUTs [17], and Nrd1-dependent NUTs [18, 19], and mammalian cells produce PROMPTs and upstream antisense RNAs (uaRNAs) [20] and others (Fig. 5.1E and F).

Information about TSS-proximal lncRNAs in plants remains scant. However, recent analyses of nascent RNA from *Arabidopsis* seedlings obtained using a combination of global nuclear run-on sequencing (GRO-seq), 5′ GRO-seq, and RNA-seq did not detect upstream antisense TSS-proximal ncRNAs [21]. These data suggest a possibility that divergent transcription is lacking in *Arabidopsis* (and likely maize), in contrast to the situation in many other eukaryotes, indicating that eukaryotic promoters might be not inherently bidirectional. In *Arabidopsis* TSS-proximal lncRNAs that were observed in the RNA exosome-deficient lines include the upstream noncoding transcripts (UNTs), which are transcribed as sense RNAs and are colinear with the 5′ ends of the associated protein-coding gene, extending into the first intron. The UNTs resemble yeast CUTs and mammalian PROMPTs [1].

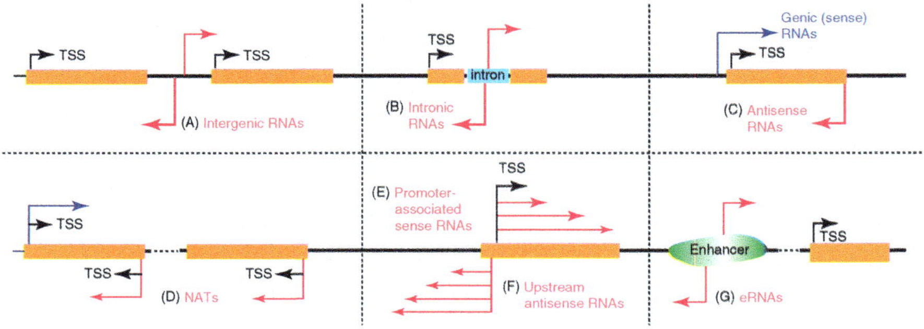

Fig. 5.1 Classification of lncRNAs based on their relationship to protein-coding genes. *Orange boxes* correspond to the protein-coding genes and *pink lines* correspond to lncRNAs. *Arrows* indicate the direction of transcription. Each panel depicts a subtype of lncRNAs: intergenic or long intergenic noncoding RNAs (lincRNAs) (*A*), intronic RNAs (*B*), antisense RNAs (*C*), natural antisense transcripts (NATs) (*D*), promoter-proximal sense (*E*) and upstream antisense RNAs (*F*), eRNAs (*G*)

The exosome-sensitive enhancer RNAs (eRNAs) produced from enhancer regions make up a large proportion of non-polyadenylated lncRNAs in mammalian cells (Fig. 5.1G) [8]. However, information about plant enhancers has only recently started to emerge. An analysis of chromatin signatures predicted over 10,000 plant intergenic enhancers [22]. However, their potential roles as transcriptional enhancers in vivo will require follow-up experiments, and eRNAs have not yet been reported in plants.

5.2 Recent Advances in Studying Plant lncRNAs

Mammalian lncRNAs are by far the best-studied. However, in recent years, identification of plant lncRNAs has largely caught up with mammalian field. The plant databases where the information on lncRNAs can be found are summarized in Table 5.1.

Table 5.1 List of plant lncRNA databases

Database	Descriptions/features	Website	Ref
The Arabidopsis Information Resource (TAIR)	Comprehensive database of *Arabidopsis thaliana* genome, including annotated genome sequences (TAIR10), gene structures, and transcriptome data for coding and nonprotein-coding loci. TAIR has multiple analytical tools: interactive genome browser, BLAST, motif analysis, bulk data retrieval, and a chromosome map tool	https://www.arabidopsis.org/	[23]
Araport11	A comprehensive database based on Arabidopsis Col-0 version 11 (Araport11) includes additional coding and noncoding annotations compared to TAIR10, such as lincRNAs, NATs, and other ncRNAs	https://www.araport.org/	[24]
Plant long noncoding RNA database (PLncDB)	This database includes a curated list of >13,000 lincRNAs identified using RNA-seq and tiling array and their organ-specific expression and the differential expression in RdDM mutants. PLncDB has a genome browser for viewing the association of various epigenetic markers	http://chualab.rockefeller.edu/gbrowse2/homepage.html	[3]
Green Non-coding Database (GREENC)	GREENC has >120,000 annotated lncRNAs from 37 plant species and algae. The user can access the coding potential and folding energy for each lncRNA	http://greenc.sciencedesigners.com/wiki/Main_Page	[25]
NONCODE v4.0	NONCODE includes >500,000 lncRNAs from 16 species. *Arabidopsis* is the only plant species, as NONCODE focuses on non-plant species, including human and mouse	http://www.noncode.org/index.php	[26]

(continued)

Table 5.1 (continued)

Database	Descriptions/features	Website	Ref
CANTATAdb	CANTATAdb contains >45,000 plant lncRNAs from ten model plant species. In addition to tissue-specific expressions and coding potential, each lncRNA is also evaluated based on potential roles in splicing regulation and miRNA modulations	http://cantata.amu.edu.pl/	[27]
Plant ncRNA database (PNRD)	PNRD has >25,000 ncRNAs of 11 different types and from 150 plant species. It also includes analytical tools, such as an miRNA predictor, coding potential calculator, and customized genome browser	http://structuralbiology.cau.edu.cn/PNRD/	[28]
Plant Natural Antisense Transcripts DataBase (PlantNATsDB)	A database for natural antisense transcripts (NATs) from 70 plant species, associated gene information, small RNA expression, and GO annotation	http://bis.zju.edu.cn/pnatdb/	[29]

An examination of >200 transcriptome data sets in *Arabidopsis* identified ~40,000 candidate lncRNAs; these included NATs (>30,000) and lincRNAs (>6000) [3, 4, 30]. Most of the lincRNAs did not produce smRNAs, and, like mammalian lncRNAs, the lincRNA transcript levels were 30–60-fold lower than that of transcript levels of the associated mRNA. Work in *Arabidopsis* found that NAT pairs, lncRNAs transcribed from opposite strands, occur widely: ~70% of protein-coding loci in *Arabidopsis* produce candidate NAT pairs 200–12,370 nt long (average length of 731 nt) [4]. Some NAT pairs show complete overlap (~60%), but others have complementary segments at their 5′ or the 3′ ends.

The expression levels of many lincRNAs differ significantly depending on the tissue and also change during stress; this indicates that lncRNAs undergo dynamic regulation and act in regulation of development and stress responses [30]. The expression levels of many NATs also are tissue-specific and change in response to biotic or abiotic stresses. For example, a recent study identified ~1400 NATs that respond to light; of the NAT pairs, about half respond in the same direction, and half respond in opposite directions. For the light-responsive NATs, the associated genes also showed peaks of histone acetylation; the acetylation levels changed with the changes in NAT expression in response to light [4].

Among the lncRNAs, *Arabidopsis* and rice have intermediate-sized ncRNAs (im-ncRNAs), which are 50–300 nt long [31, 32] and originate from 5′ UTRs, coding regions, and introns. The genes associated with 5′ UTR im-ncRNAs tended to have higher expression and H3K4me3 and H3K9ac histone marks, which are associated with transcriptional activation. Plants that have reduced levels of some im-ncRNAs showed developmental phenotypes or detectable molecular changes [31].

While we continue to gain better understandings of the mechanisms of lncRNA action, the mechanisms that regulate lncRNAs in plants remain limited. Like all transcripts, lncRNAs undergo transcriptional level regulation and regulation that affects lncRNA biogenesis, processing, and turnover. One of the players in this regulation is the exosome complex, which plays a major role in regulating the quantity, quality, and processing of various transcripts, including lncRNAs. The exosome complex is a conserved machinery with 3′–5′ exoribonuclease activity that consists of nine-subunit core associated with its enzymatic subunits, Rrp44 and Rrp6. The depletion of the *Arabidopsis* exosome allowed identification of a number of *Arabidopsis* ncRNAs as well as the genomic regions where the exosome is involved in their metabolism [1].

5.3 Molecular Functions of Plant lncRNAs

lncRNAs are present at low levels and show little sequence conservation compared with mRNAs; therefore, early studies questioned their importance and necessity and also suggested that lncRNAs might result from transcriptional noise. Indeed, considerable debate remains about the functionality of lncRNAs. However, evidence has emerged in recent years to indicate that many lncRNAs function in a large number of diverse molecular processes in eukaryotic cells; these include the regulation of yeast mating type [33, 34] and modulation of embryonic stem cell pluripotency and various diseases [35]. In plants, lncRNAs function in gene silencing, flowering time control, organogenesis in roots, photomorphogenesis in seedlings, abiotic stress responses, and reproduction [5, 11–14, 36–40].

For their effects on gene regulation, lncRNAs act at multiple levels and with simple or complex mechanisms. lncRNAs can act in *cis* or *trans*, function by sequence complementarity to RNA or DNA, and be recognized via specific sequence motifs or secondary/tertiary structures (Fig. 5.2a). At the most simple level, lncRNAs can serve as precursors to smRNAs (Fig. 5.2b), as in the case of RNA Pol IV transcripts [6, 42–46]. Some lncRNAs keep regulatory proteins or microRNAs from interacting with their DNA or RNA targets by acting as decoys that mimic the targets (Fig. 5.2c). Some of the plant examples include the *Arabidopsis* microRNA target mimics IPS1 lncRNA and the decoy ASCO-lncRNA [38, 47].

In animal systems, some lncRNAs directly affect Pol II and its associated transcriptional machinery by promoting phosphorylation of transcription factors (TFs) regulating their DNA-binding activity [48]. Many lncRNAs affect different processes related to transcription, including the initiation and elongation of transcripts, by affecting the pausing of RNA Pol II. Other lncRNAs act as scaffolds to recruit enzymes that remodel chromatin and thus alter chromatin structure and nuclear organization (Fig. 5.2d) (reviewed in [49]). Examples of plant lncRNAs that regulate transcription have started to emerge; for example, HID1 binds to the promoter of PIF3 gene to downregulate its expression [39]. However, no plant lncRNAs have yet been implicated in regulation of transcription elongation or Pol II pausing.

Different types of lncRNAs associate with chromatin and act as scaffolds that allow the assembly of complexes of chromatin-modifying enzymes. Recruitment of

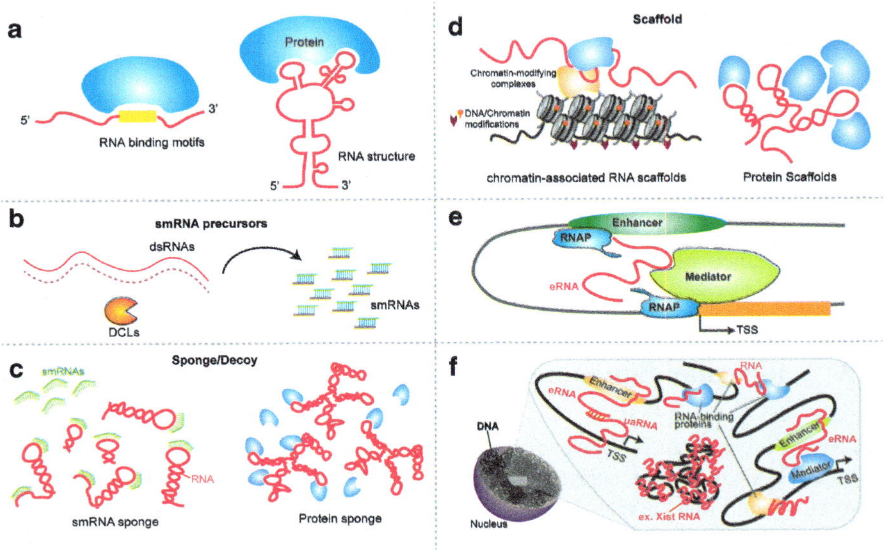

Fig. 5.2 Example lncRNAs and the mechanisms of their action. (**a**) Specific sequence motifs or secondary structures could be required for lncRNA function. (**b**) lncRNAs, specifically, the double-stranded transcripts, can serve as precursors to smRNAs in the RNA interference (RNAi) pathway. (**c**) lncRNAs can function as scaffolds for the recruitment of chromatin-modifying factors or as a platform for assembly of protein complexes. (**d**) lncRNAs can function as molecular sponges or decoys for smRNAs and also act as decoys to titrate away RNA-binding proteins. (**e**) The eRNAs, which are expressed from enhancers, are regulated by the exosome and can interact with other regions of DNA, such as enhancers or promoters, affecting the topology of the local DNA and thus altering gene expression. Adapted from [41]. (**f**) lncRNAs that interact with several chromatin-remodeling proteins and chromatin regions could affect higher-order nuclear structure

these proteins can require small RNAs or not. For example, the siRNA-directed DNA methylation (RdDM) pathway, which occurs specifically in plants, requires small RNAs [37]. Other lncRNAs can recruit complexes of enzymes that remodel chromatin but do not require smRNAs. The mechanism that provides targeting specificity for these lncRNAs remains to be discovered. Work in mammalian systems showed that lncRNAs can interact with proteins of the Trithorax group and activate transcription via trimethylation of histone H3K4 [50]. Other lncRNAs interact with proteins that modify histones with repressive marks, such as Polycomb Repressive Complex 2 (PRC2), to repress transcription via methylation of histone H3K27 [51]. The best-studied RNAi-independent pathway that relies on lncRNAs interacting with Polycomb is epigenetic regulation via histone modifications and expression of *Arabidopsis FLOWERING LOCUS C* (*FLC*).

Additional examples include enhancer RNAs (eRNAs), shown to be involved in regulation of transcription initiation. Enhancers are regulatory genomic regions that are shown to be involved in transcriptional regulation through targeting promoters of protein-coding genes in a tissue-specific and developmental manner as well as modulating spatial organization of the genome [52]. Work in mammalian systems has shown that exosome-sensitive eRNAs function in activation of transcription, consistent with the enhancer function. Some eRNAs act in *cis* to recruit complexes of coactivator proteins that form chromosome loops that connect the enhancer with its promoter, thus activating gene expression (Fig. 5.2e) [41, 53]. However, no eRNAs have not been identified in plants yet. The exosome function of resolving R-loops, which are RNA-DNA triplexes, might also reduce genomic instability in the regions expressing eRNAs [41]. R-loops form during transcription and can persist in regions that are divergently transcribed [54]. These results suggest that the exosome modulates the interactions among the key elements that regulate gene expression and the organization of the nucleus.

The examples of the well-studied plant lncRNAs with established functions and mechanisms of action are listed in Table 5.2.

Table 5.2 List of plant lncRNAs

lncRNAs	Description and function	References
ASCO-lncRNA	Functions in lateral root development in *Arabidopsis*. Regulator of alternative splicing. Works as a decoy lncRNA	[38]
IPS1	Functions in regulating phosphate balance and phosphate starvation response in *Arabidopsis*. Competes with *PHO2* mRNA for interaction with miR399 and acts as non-cleavable miRNA target	[47]
HID1	Functions in regulation of photomorphogenesis in *Arabidopsis* seedlings. *Trans*-acting lncRNA (236 nt) acts by associating with the *PIF3* promoter and represses its transcription. Evolutionary conserved in land plants	[39]
COOLAIR	Functions in regulation of flowering in *Arabidopsis* in both vernalization and autonomous pathways. Modulates *FLC* expression by multiple mechanisms	[55]
COLDAIR	Functions in regulation of flowering in *Arabidopsis* in the vernalization pathway. Associates with Polycomb to mediate silencing of *FLC* and affects chromatin looping at *FLC* in response to vernalization	[9]
COLDWRAP	Functions in regulation of flowering in *Arabidopsis* in the vernalization pathway. Participates in and coordinates vernalization-mediated Polycomb silencing of the FLC. Also affects formation of an intragenic chromatin loop that represses *FLC*	[11]
ASL	Functions in regulation of flowering in the autonomous pathway in *Arabidopsis*. AtRRP6L regulates ASL to modulate H3K27me3 levels.	[10]
APOLO	Functions in regulation of auxin signaling outputs in *Arabidopsis*. Participates in chromatin loop dynamics. Affects formation of a chromatin loop in the *PID* promoter region	[56]

(continued)

Table 5.2 (continued)

lncRNAs	Description and function	References
Pol IV transcripts	Technically shorter in length than the standard lncRNAs. Function in silencing of transposons (TEs) and repeats in RdDM pathway. Serve as precursors for siRNAs in RdDM pathway	[57–59]
Pol V transcripts	Function in silencing TEs and repeats in RdDM pathway. Serve as a scaffold lncRNAs for assembly of siRNAs and proteins in RdDM pathway	[60]
ENOD40	Functions in regulation of symbiotic interactions between leguminous plants and soil bacteria in *Medicago truncatula*. Suggested to function in re-localization of proteins in plants	[38, 61]
LDMAR	Regulates photoperiod-sensitive male sterility in rice by affecting DNA methylation in the LDMAR promoter region. The precise mechanism of LDMAR function and the interaction between LDMAR and siRNAs remain to be clarified	[62]

5.4 Plant lncRNAs Functioning as Molecular Sponges and Decoys

Work in *Arabidopsis* identified lncRNAs that compete with microRNAs (miRNAs) or mimic the targets of miRNAs; similar function was also identified in animal systems. For example, the IPS1 lncRNA plays a role in regulating phosphate balance and uptake by competing for binding the *PHO2* mRNA. PHO2 negatively regulates phosphate transporters and is itself downregulated by miR399 cleavage of its mRNA; IPS1 serves as mimic that cannot be cleaved by miR399 due to the mismatch but can titrate off miR399 [47]. Bioinformatics approaches also have predicted many additional miRNA target mimics in *Arabidopsis*, but the functions of many of these remain to be deciphered [63].

The *Arabidopsis* ASCO-lncRNA functions as decoy and regulates plant root development. ASCO-RNA competes with the binding of nuclear speckle RNA-binding proteins (NSRs), regulators of alternative splicing, to their targets; hijacking the NSRs changes the splicing patterns of NSR-regulated mRNA targets resulting in the production of alternative splice isoforms and leading to switch of developmental fates in plant roots (Fig. 5.3) [38].

5.5 Plant lncRNAs Functioning in Regulation of Transcription and Silencing

5.5.1 Regulation of PIF3 Transcription by HID1 im-ncRNA

One of the interesting *Arabidopsis* lncRNAs, HIDDEN TREASURE 1 (HID1), also classified in original study as im-ncRNA with a length of 236 nt, is involved in the regulation of transcription of the transcription factor PIF3, a member of

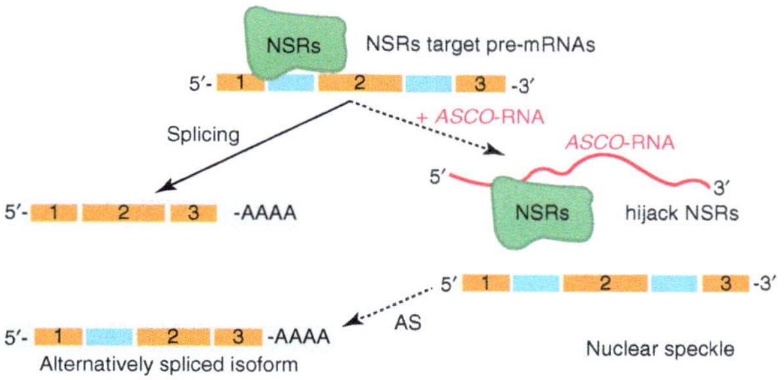

Fig. 5.3 Plant lncRNAs can affect the expression of proteins that regulate alternative splicing. The ASCO-lncRNA functions as a decoy that competes with mRNAs for binding to NSR splicing regulators

"phytochrome-interacting factors" (PIFs), a family of basic helix-loop-helix (bHLH) transcription factors [39]. HID1 is evolutionarily conserved in land plants and functions in *trans* as a component of an RNA-protein complex. It interacts with the promoter region of PIF3 and suppresses PIF3 transcription. The HID1 imncRNA is among rare examples of lncRNAs for which it was shown that its function requires its secondary structure. The secondary structure of HID1 in *Arabidopsis* and rice shows substantial conservation and expression of OsHID1 could complement the *Arabidopsis hid1* mutant phenotype, indicating its importance in regulation of photomorphogenesis in seedlings.

5.5.2 Role of lncRNAs in RdDM

In plants, lncRNAs also function in epigenetic silencing, acting via siRNA-dependent DNA methylation (RdDM) (Fig. 5.4). RdDM in plants has similar mechanisms to gene silencing mediated by siRNAs in *S. pombe* [64–67]. RdDM primarily functions to repress transcription of transposons and repetitive sequences and requires RNA Pol IV and Pol V, two plant-specific RNA polymerases [6], and perhaps some involvement of RNA Pol II [68]. RNA Pol IV produces ncRNAs that serve as templates for 24 nt siRNAs, and RNA Pol V transcribes lncRNAs, which act as scaffolds that the AGO-siRNA complex recognizes through sequence complementarity (reviewed in [37]). In *Arabidopsis*, most siRNAs are generated by Pol IV; however, Pol V and Pol II can also make siRNA templates, suggesting additional complexity involved in siRNA biogenesis [69–72].

Identification of the Pol IV- and particularly Pol V-produced lncRNAs has remained challenging until recently [57–60]. One of the recent genome-wide studies identified Pol IV/RDR2-dependent transcripts (P4RNAs) from thousands of *Arabidopsis* loci. Interestingly, these P4RNAs are transcribed mainly from intergenic regions; 65% of the P4RNAs overlapped with transposable elements or repeats, and 9% of the RNAs overlapped with genes [57]. The Pol IV/RDR2-dependent transcripts are non-polyadenylated and produced from the sense and antisense DNA strands. Surprisingly, instead of a 5′ triphosphate, the P4RNAs have a monophosphate [57].

Until very recently Pol V transcripts eluded detection on the genome-wide scale due to the very low levels of their accumulation, which made them difficult to detect using RNA-seq. Based on the analysis of the individual transcripts, Pol V lncRNAs are

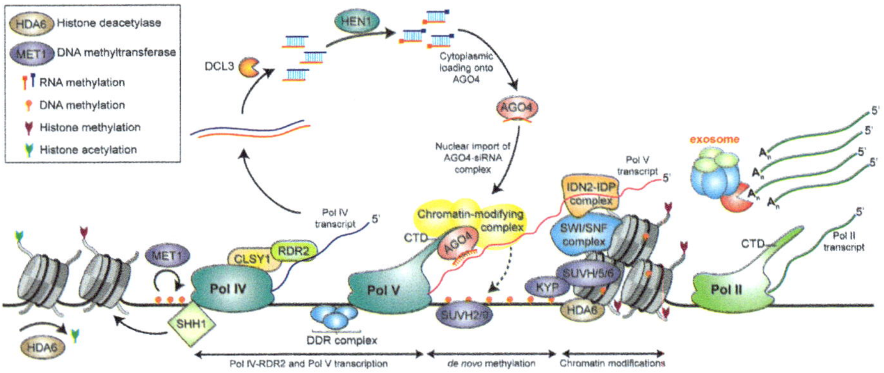

Fig. 5.4 LncRNAs participating in the RdDM pathway. Transcripts produced by Pol IV are precursors for 24 nt siRNA; transcripts produced by Pol V are scaffolds and siRNA targets. SHH1 reads the H3K9me status of chromatin and recruits Pol IV; then the chromatin-remodeling protein CLSY1 assists in the passage of Pol IV [73]. Pol IV transcripts are transcribed by RDR2 into double-stranded RNAs (dsRNAs) before they are processed by DCL3 into 24 nt siRNAs and stabilized by methylation at the 3′ end by HEN1. These siRNAs associate with AGO and return to the nucleus as a part of the AGO-siRNA complex, which targets nascent Pol V scaffold transcripts. Pol V is recruited by SUVH2 or SUVH9 to its target genomic loci marked by DNA methylation [74], and Pol V transcription is facilitated by the DDR complex [75].The IDN2-IDP complex binds to Pol V scaffold RNAs and interacts with the SWI/SNF complex, which adjusts the position of nucleosomes [76]. The AGO4-siRNA complex interacts with Pol V; in this interaction, the siRNA in the complex base pairs with the transcript produced by Pol V to target a chromatin-modifying complex that catalyzes de novo methylation at the genomic loci. Then, the silencing mediated by DNA methylation is further amplified by methylation of histone H3K9 by KYP, SUVH5, and SUVH6 (reviewed in [37]). The silencing of solo LTRs requires the exosome, which does not act via siRNAs and DNA methylation. Rather, the exosome interacts with transcripts from a nearby scaffold-producing region and acts in silencing the solo LTR by altering chromatin structure via H3K9 histone methylation, suggesting this may function in parallel with RdDM

non-polyadenylated and either tri-phosphorylated or capped at the 5′ ends [6]. Recent genome-wide study using RIP-seq identified 4502 individual Pol V-associated transcripts [60]. It was previously annotated that the Pol V-transcribed regions have an average length of 689 nt. Surprisingly, it was found that experimentally identified Pol V lncRNAs are shorter than previously annotated, with their median size ranging from 196 to 205 nt yet spanning the entire region. This data suggested that Pol V might not transcribe the entire regions continuously but is possibly controlled by internal promoters situated within the annotated regions that lead to active Pol V transcription.

Unlike RNA polymerases I, II, and III, which use specific sequence elements that identify their promoters, no specific DNA sequence elements were found in Pol V-transcribed regions. Instead internal repressive chromatin modifications appeared to control Pol V transcription and contribute to initiation by internal promoters. Interestingly, Pol V produces lncRNAs bidirectionally on annotated Pol V transcripts with no correlations in strand preference. However, despite Pol V that transcribes both strands of DNA, a subset of Pol V transcripts on transposons was found to be enriched on one strand in a way that indicated that limited strand preference of

Pol V in these loci may be involved in determining boundaries of heterochromatin on transposons.

Previous genome-wide studies using ChIP-seq identified Pol V-associated genomic regions and found Pol V may also function in pathways other than the RdDM pathway [6, 75–80]. About 75% of Pol V-occupied genomic sites are transposons and repetitive sequences that also have 24 nt siRNAs and high levels of DNA methylation, indicating that Pol V induces RdDM at these sites. The other 25% of Pol V-associated sites include many protein-coding genes that have lower methylation levels and do not associate with siRNAs. This indicates that Pol V may also function in other silencing pathways [77]. Pol II also can produce scaffold transcripts that recruit siRNAs bound by AGO [68]. However, it remains unclear how Pol II targets specific intergenic loci and how Pol II interacts with Pol IV and Pol V.

Interestingly, the exosome also appears to play some role in silencing of these regions. A genome-wide study that identified exosome targets found many polyadenylated substrates of the exosome complex that corresponded to ncRNAs from centromeric regions, repetitive sequences, and other siRNA-producing loci and undergo RdDM-mediated silencing [1]. However, when we explored the connection between the two silencing pathways, RdDM and the exosome in *Arabidopsis*, we found that mutants of the core exosome subunits only produce a small effect on smRNAs [81]. This differs from results found in studies of the exosome in fission yeast, as in this system, the exosome prevents RNAs from spuriously entering into smRNA pathways [65]. Instead, less H3K9me2 was observed at several loci controlled by RdDM in exosome-deficient lines. The exosome interacts genetically with RNA Pol V and physically associates with polyadenylated Pol II transcripts from the regions generating Pol V scaffold RNAs [81]. These observations indicate that the exosome functions in lncRNA metabolism or processing in scaffold-generating regions. The exosome may also mediate the interactions among Pol II, Pol V, and Pol IV, modulating transcriptional repression. One outstanding question is whether and how the exosome (possibly acting through lncRNAs) contributes to silencing of loci via fine-tuning histone modifications and if the same mechanism of action can be observed genome wide.

However, *Arabidopsis* exosome subunits have diverse functions [1]. The additional enzymatic subunit, AtRRP6L1, is independent of the exosome core functions [10]. Mutations in AtRRP6L1 effect siRNA metabolism and DNA methylation [82]. Therefore, the exosome and the additional enzymatic subunits played an important role in regulation of ncRNAs, including siRNAs, in the RdDM pathway.

5.6 lncRNAs in the Regulation of Flowering

Because of the importance of flowering time regulation for plant adaptation to different latitudes, the lncRNAs that regulate flowering are among the best-studied functional plant lncRNAs. Work in *Arabidopsis* has shown that these lncRNAs regulate the initiation of flowering by modulating the expression of *FLOWERING LOCUS C (FLC)*, which encodes a MADS-box transcription factor. FLC represses downstream genes required for flowering and thus negatively regulates flowering, acting in a dose-dependent manner. FLC functions in the vernalization pathway, which modulates flowering time in response to prolonged low temperature, and in the autonomous pathway, which modulates flowering time independently of environmental factors [83].

The regulation of flowering time involves epigenetic silencing of *FLC*, mainly via modification of histones. Repression of *FLC* requires PRC2, which is recruited to *FLC* and methylates histone H3K27. Alteration of chromatin, particularly changes in histone modifications that remove H3K4me3, H3K36me3, and H2Bub1 and replace those modifications with H3K27me3, epigenetically represses *FLC* expression (reviewed in [36]).

The lncRNAs COLDAIR, COLDWRAP, and COOLAIR are transcribed from *FLC* and function in *FLC* epigenetic silencing (Fig. 5.5) [9, 11, 84]. Vernalization induces transient transcription of COLDAIR, a 5′ capped, non-polyadenylated lncRNA, transcribed from *FLC* intron 1, in the same direction as *FLC* (Fig. 5.5). CURLY LEAF (CLF), a homolog of mammalian EZH2 (an enzymatic component of PRC2), binds to COLDAIR, and knockdown of COLDAIR decreases CLF and

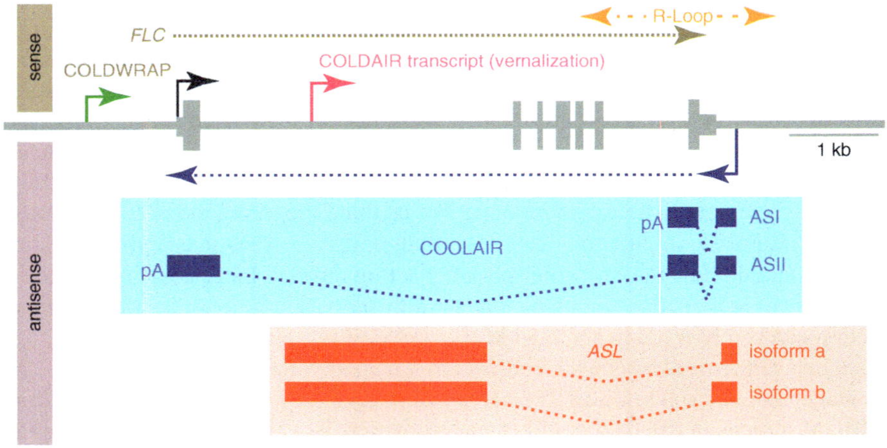

Fig. 5.5 Regulatory lncRNAs produced from the *FLC* locus. Diagram of the *FLC* locus [84]. The FLC transcriptional start site is indicated by *black arrow*, and the *vertical bars* indicate exons in the *FLC* sense transcript. During vernalization, the COLDAIR lncRNA (*pink*) is transcribed in the sense direction, starting in the first intron of *FLC*. Another sense lncRNA, COLDWARP, is transcribed from the repressed promoter of FLC (*green*). The COOLAIR (*blue*) and ASL (*red*) lncRNA transcripts are transcribed from the indicated start sites (*purple arrow*) in the antisense direction; both result from alternative polyadenylation at poly(A) site either in the sense promoter region or intron 6. The ASL lncRNA also undergoes alternative splicing. *Blue boxes* indicate the exon of COOLAIR; *red boxes* indicate the exons of AS I and II; *dotted lines* indicate the spliced regions. ASL covers *FLC* intron I. *Yellow dotted lines* indicate the R-loops, in the COOLAIR promoter region, and repress COOLAIR transcription

H3K27me3 enrichment at *FLC* in response to cold. This thus hampers the repression of *FLC* during vernalization and indicates that COLDAIR's repression of *FLC* is essential for the vernalization response [9]. Previous work suggested that PRC2 recruitment to *FLC* requires COLDAIR for the initiation of epigenetic silencing, analogous to the functions of the mammalian lncRNAs HOTAIR and Xist [51]. However, mammalian PRC2 shows high-affinity binding to unrelated RNAs; therefore, other factors, in addition to lncRNAs, may provide the specificity that targets PRC2 to *FLC* [85].

An additional Polycomb-interacting lncRNA, cold of winter-induced noncoding RNA from the promoter (COLDWRAP), was identified to be expressed from the upstream promoter region of *FLC* locus and shown to function in repression of *FLC* (Fig. 5.5) [11]. COLDWRAP is a 316 nt lncRNA that is transcribed in the sense direction with its transcription start located 225 nt upstream from the *FLC* mRNA. COLDAIR and COLDWRAP both have 5′ caps, but most transcripts of COLDWRAP appear to be non-polyadenylated. Interestingly, association of the Polycomb complex with COLDWRAP appears to be specific, as native CLF binds significantly to the sense strand of COLDWRAP but only weakly to the antisense strand. In addition, the 5′ half of COLDWRAP and several stable secondary structures identified in this region are needed for RNA-protein interactions. Importantly, COLDWRAP working in a cooperative manner with COLDAIR is necessary for vernalization-mediated FLC silencing. COLDWRAP functions to retain Polycomb at the FLC promoter through the formation of a repressive intragenic chromatin loop forming a stable repressive chromatin structure.

The COOLAIR is a set of lncRNAs transcribed from the 3′ end of *FLC* in the antisense direction, which are alternatively spliced and polyadenylated, proximal AS I and distal AS II [55]. In response to cold, the locus first produces COOLAIR, then COLDAIR, before H3K27me3 accumulates; therefore, initial studies indicated that COOLAIR may act early in vernalization [55]. However, knockdown of COOLAIR did not affect the vernalization response [86]. Rather, COOLAIR increases the rate of *FLC* transcriptional repression during vernalization, and its function does not require PRC2 or H3K27me3 [36, 87]. The COOLAIR knockdown desynchronized the change from H3K36me to H3K27me3 in *FLC*; therefore, this switch at *FLC* may require COOLAIR or transcription in the antisense direction [87].

COOLAIR represses *FLC* in the vernalization and autonomous pathways. In the autonomous pathway, COOLAIR 3′ end processing affects the *FLC* chromatin [84]. The autonomous pathway factors FCA, FY, and FPA, along with the polyadenylation cleavage factors CstF64 and CstF77, and the spliceosome component PRP8, favor the production of AS I by increasing usage of the proximal COOLAIR polyadenylation site [84, 88, 89]. This increases levels of the FLOWERING LOCUS D (FLD) histone demethylase at *FLC* leading to H3K4me2 demethylation of *FLC* [90].

Unraveling the functional importance of transcription of COOLAIR and the functions of COOLAIR transcripts remains challenging. Since it is difficult to determine whether it is the COOLAIR transcription, COOLAIR transcripts, or both that are functionally important, the secondary RNA structure of COOLAIR was recently determined experimentally [91]. It was found that even despite the relatively low sequence identity between *Arabidopsis* and evolutionarily divergent Brassicaceae species, the structures showed remarkable evolutionary conservation. This conservation applied to multi-helix junctions and through covariation of a non-contiguous DNA sequence. The observed conservation of COOLAIR lncRNA structure in the Brassicaceae indicates that the COOLAIR lncRNA itself is very likely to function in regulation of *FLC*, although the process of antisense transcription from *FLC* may also affect *FLC* regulation.

Recent work also discovered the Antisense Long (ASL) transcript in early-flowering *Arabidopsis* ecotypes that do not require vernalization for flowering [10]. In contrast to the other lncRNAs transcribed from *FLC*, ASL does not get polyadenylated, although it is alternatively spliced. The ASL transcript is >2000 nucleotides long and is transcribed from the antisense strand, starting at the same promoter as COOLAIR. The 5′ regions of COOLAIR and ASL overlap, but ASL spans intron 1 (important for maintenance of *FLC* silencing) and includes the COLDAIR region, which is transcribed in the sense direction. The ASL transcript physically associates with the *FLC* locus and H3K27me3 [10], suggesting that ASL and COOLAIR play different roles in *FLC* silencing and perhaps in the maintenance of H3K27me3.

It is interesting that the exosome again is involved in the regulation of the antisense transcript and does so in a surprising way. Two of the exosome components, RRP6-Like (RRP6L) proteins, are involved in lncRNA-mediated regulation of flowering. RRP6, one of the catalytic subunits, has both core-complex-dependent and core-complex-independent functions [92, 93]. In *Arabidopsis*, RRP6L1 and RRP6L2 regulate COOLAIR and ASL expression or processing in the exosome core-complex-independent way [10]. Mutations of RRP6L also derepress *FLC*; this delays flowering. The AS I and II downregulation observed in RRP6Ls multiple mutants resembled the patterns that occur in CstF64 and CstF77 mutants, which are 3′ end processing factors [10, 84], indicating that COOLAIR 3′ end processing may require RRP6Ls.

Very surprisingly, emerging work indicates that RRP6Ls have a major role in regulation of the synthesis or biogenesis of ASL, as RRP6Ls mutants lack (or have minuscule amounts of) ASL transcript. This result finding is unexpected because RRP6 functions as a 3′–5′ exoribonuclease and RRP6 mutants generally fail to degrade or process certain RNAs; thus, these mutants usually overaccumulate certain RNAs. However, recent work found that the abundance of many yeast mRNAs also decreased in the *rrp6Δ* mutants [19]. Similarly, in humans, inactivation of the RRP6 homolog also causes a dramatic decrease in Xist levels [94].

Another function of RRP6Ls involves affecting the epigenetic marks at *FLC*; mutants of RRP6L have decreased H3K27me3 levels and decreased density of nucleosomes at *FLC*. These mutants therefore show increased expression of *FLC* and delayed flowering. RRP6L1 physically interacts with the ASL RNA and with chromatin at *FLC*; this indicates that RRP6Ls may regulate ASL to maintain H3K27me3 levels at *FLC*. Therefore, RRP6Ls regulate *FLC* lncRNAs, and their regulation of various antisense RNAs may affect *FLC* silencing [10].

R-loops that form over the COOLAIR promoter region affect COOLAIR transcription, although effects of R-loop formation on *FLC* expression are not fully unclear [95]. Failure of the termination of transcription can often produce R-loops

[96], which can recruit the exosome co-transcriptionally through the noncanonical pathway for 3′ end processing [19]. Work in mammals showed that RRP6 can resolve deleterious R-loops [41]; thus, plant RRP6Ls may affect both the processing and expression of antisense transcripts from *FLC* in a similar manner.

In mammalian systems, lncRNAs have key roles in molding the three-dimensional organization of the nucleus (Fig. 5.2f) [97–99]. In plants, emerging research is beginning to reveal the role of lncRNAs in architecture of the nucleus, and some RNA studies also indicate that lncRNAs may have similar roles in 3-D nuclear architecture in plants and animals. Several studies have also addressed genome organization using Hi-C approach in *Arabidopsis* [100–104]. The RdDM pathway likely also affects the higher-order structure of chromatin by acting with MORC proteins. In *Arabidopsis*, MORC6 may have ATPase activity and interact with the DDR complex component DMS3; the action of this complex may be analogous to that of mammalian cohesin-like proteins that function in inactivation of the X-chromosome in mice. Consistent with this, MORC1 and MORC6 mutant plants have de-condensed pericentromeric heterochromatin [105]. The promoter and 3′ terminator of *FLC* form gene loops [106, 107], and COLDAIR and COLDWRAP lncRNAs participate in this process [11]. *FLC* alleles also undergo long-distance interactions, clustering during vernalization-mediated epigenetic silencing. This interaction requires VRN5 and VERNALIZATION 2, two PRC2 *trans*-acting factors [108]. However, we lack information on how lncRNAs function in long-distance interactions of the chromatin at *FLC*. As illustrated by *FLC*, plant lncRNAs carry out diverse, varied, and important functions. Our understanding of lncRNA functions continues to emerge as new studies uncover the mechanisms controlling lncRNA transcription and processing.

5.7 Concluding Remarks

The recent discovery that genomes undergo pervasive transcription opened many questions on the functions of these RNAs. Since then, studies in the various kingdoms of eukaryotes have broadened our understanding of the biogenesis and functions of various lncRNAs. However, although various studies have identified and classified many categories of lncRNAs, the functions of lncRNAs, and how they carry out these functions, remain to be discovered. Work in plants identifying lncRNAs systematically has caught up with work in other systems. Plant studies have also discovered lncRNA functions in controlling flowering time and RdDM-mediated silencing of genes. However, many other lncRNAs remain to be examined. The regulation of plant lncRNA synthesis and biogenesis also will require further work to elucidate. Understanding the mechanisms that control plant lncRNA expression and biogenesis will require integration of bioinformatics, genetic, and biochemical data to provide a complete understanding of lncRNA function and biology. A complete understanding of the various facets of plant lncRNAs will reciprocally advance our understanding of lncRNAs in other species.

References

1. Chekanova JA, Gregory BD, Reverdatto SV, Chen H, Kumar R, Hooker T, Yazaki J, Li P, Skiba N, Peng Q, Alonso J, Brukhin V, Grossniklaus U, Ecker JR, Belostotsky DA (2007) Genome-wide high-resolution mapping of exosome substrates reveals hidden features in the Arabidopsis transcriptome. Cell 131(7):1340–1353. doi:10.1016/j.cell.2007.10.056
2. Kapranov P, Cheng J, Dike S, Nix DA, Duttagupta R, Willingham AT, Stadler PF, Hertel J, Hackermüller J, Hofacker IL, Bell I, Cheung E, Drenkow J, Dumais E, Patel S, Helt G, Ganesh M, Ghosh S, Piccolboni A, Sementchenko V, Tammana H, Gingeras TR (2007) RNA maps reveal new RNA classes and a possible function for pervasive transcription. Science 316(5830):1484–1488. doi:10.1126/science.1138341
3. Jin J, Liu J, Wang H, Wong L, Chua N-H (2013) PLncDB: plant long non-coding RNA database. Bioinformatics 29(8):1068–1071. doi:10.1093/bioinformatics/btt107
4. Wang H, Chung PJ, Liu J, Jang I-C, Kean MJ, Xu J, Chua N-H (2014) Genome-wide identification of long noncoding natural antisense transcripts and their responses to light in Arabidopsis. Genome Res 24(3):444–453. doi:10.1101/gr.165555.113
5. Zhang Y-C, Liao J-Y, Li Z-Y, Yu Y, Zhang J-P, Li Q-F, Qu L-H, Shu W-S, Chen Y-Q (2014) Genome-wide screening and functional analysis identify a large number of long noncoding RNAs involved in the sexual reproduction of rice. Genome Biol 15(12):512. doi:10.1186/s13059-014-0512-1
6. Wierzbicki AT, Haag JR, Pikaard CS (2008) Noncoding transcription by RNA polymerase Pol IVb/Pol V mediates transcriptional silencing of overlapping and adjacent genes. Cell 135(4):635–648. doi:10.1016/j.cell.2008.09.035
7. Li L, Eichten SR, Shimizu R, Petsch K, Yeh C-T, Wu W, Chettoor AM, Givan SA, Cole RA, Fowler JE, Evans MM, Scanlon MJ, Yu J, Schnable PS, Timmermans MC, Springer NM, Muehlbauer GJ (2014) Genome-wide discovery and characterization of maize long noncoding RNAs. Genome Biol 15(2):R40. doi:10.1186/gb-2014-15-2-r40
8. Andersson R, Gebhard C, Miguel-Escalada I, Hoof I, Bornholdt J, Boyd M, Chen Y, Zhao X, Schmidl C, Suzuki T, Ntini E, Arner E, Valen E, Li K, Schwarzfischer L, Glatz D, Raithel J, Lilje B, Rapin N, Bagger FO, Jørgensen M, Andersen PR, Bertin N, Rackham O, Burroughs AM, Baillie JK, Ishizu Y, Shimizu Y, Furuhata E, Maeda S, Negishi Y, Mungall CJ, Meehan TF, Lassmann T, Itoh M, Kawaji H, Kondo N, Kawai J, Lennartsson A, Daub CO, Heutink P, Hume DA, Jensen TH, Suzuki H, Hayashizaki Y, Müller F, FANTOM C, Forrest ARR, Carninci P, Rehli M, Sandelin A (2014) An atlas of active enhancers across human cell types and tissues. Nature 507(7493):455–461. doi:10.1038/nature12787
9. Heo JB, Sung S (2011) Vernalization-mediated epigenetic silencing by a long intronic noncoding RNA. Science 331(6013):76–79. doi:10.1126/science.1197349
10. Shin J-H, Chekanova JA (2014) Arabidopsis RRP6L1 and RRP6L2 function in FLOWERING LOCUS C silencing via regulation of antisense RNA synthesis. PLoS Genet 10(9):e1004612. doi:10.1371/journal.pgen.1004612
11. Kim D-H, Sung S (2017) Vernalization-triggered intragenic chromatin loop formation by long noncoding RNAs. Dev Cell 40(3):302–312.e4. doi:10.1016/j.devcel.2016.12.021
12. Di C, Yuan J, Wu Y, Li J, Lin H, Hu L, Zhang T, Qi Y, Gerstein MB, Guo Y, Lu ZJ (2014) Characterization of stress-responsive lncRNAs in Arabidopsis thaliana by integrating expression, epigenetic and structural features. Plant J 80(5):848–861. doi:10.1111/tpj.12679
13. Yuan J, Zhang Y, Dong J, Sun Y, Lim BL, Liu D, Lu ZJ (2016) Systematic characterization of novel lncRNAs responding to phosphate starvation in Arabidopsis thaliana. BMC Genomics 17(1):655. doi:10.1186/s12864-016-2929-2
14. Li S, Yamada M, Han X, Ohler U, Benfey PN (2016) High-resolution expression map of the Arabidopsis root reveals alternative splicing and lincRNA regulation. Dev Cell 39(4):508–522. doi:10.1016/j.devcel.2016.10.012
15. Mattick JS, Rinn JL (2015) Discovery and annotation of long noncoding RNAs. Nat Struct Mol Biol 22(1):5–7. doi:10.1038/nsmb.2942

16. Xu Z, Wei W, Gagneur J, Perocchi F, Keller C, Camblong J, Guffanti E, Stutz F, Huber W, Steinmetz LM (2009) Bidirectional promoters generate pervasive transcription in yeast. Nature 457(7232):1033–1037. doi:10.1038/nature07728
17. van Dijk EL, Chen CL, d'Aubenton-Carafa Y, Gourvennec S, Kwapisz M, Roche V, Bertrand C, Silvain M, Legoix-Né P, Loeillet S, Nicolas A, Thermes C, Morillon A (2011) XUTs are a class of Xrn1-sensitive antisense regulatory non-coding RNA in yeast. Nature 475(7354):114–117. doi:10.1038/nature10118
18. Schulz D, Schwalb B, Kiesel A, Baejen C, Torkler P, Gagneur J, Soeding J, Cramer P (2013) Transcriptome surveillance by selective termination of noncoding RNA synthesis. Cell 155(5):1075–1087. doi:10.1016/j.cell.2013.10.024
19. Fox MJ, Gao H, Smith-Kinnaman WR, Liu Y, Mosley AL (2015) The exosome component Rrp6 is required for RNA polymerase II termination at specific targets of the Nrd1-Nab3 pathway. PLoS Genet 11(2):e1004999. doi:10.1371/journal.pgen.1004999
20. Flynn RA, Almada AE, Zamudio JR, Sharp PA (2011) Antisense RNA polymerase II divergent transcripts are P-TEFb dependent and substrates for the RNA exosome. Proc Natl Acad Sci 108(26):10460–10465. doi:10.1073/pnas.1106630108
21. Hetzel J, Duttke SH, Benner C, Chory J (2016) Nascent RNA sequencing reveals distinct features in plant transcription. Proc Natl Acad Sci 113(43):12316–12321. doi:10.1073/pnas.1603217113
22. Zhu B, Zhang W, Zhang T, Liu B, Jiang J (2015) Genome-wide prediction and validation of intergenic enhancers in Arabidopsis using open chromatin signatures. Plant Cell 27(9):2415–2426. doi:10.1105/tpc.15.00537
23. Lamesch P, Berardini TZ, Li D, Swarbreck D, Wilks C, Sasidharan R, Muller R, Dreher K, Alexander DL, Garcia-Hernandez M, Karthikeyan AS, Lee CH, Nelson WD, Ploetz L, Singh S, Wensel A, Huala E (2012) The Arabidopsis Information Resource (TAIR): improved gene annotation and new tools. Nucleic Acids Res 40(Database issue):D1202–D1210. doi:10.1093/nar/gkr1090
24. Cheng C-Y, Krishnakumar V, Chan AP, Thibaud-Nissen F, Schobel S, Town CD (2017) Araport11: a complete reannotation of the Arabidopsis thaliana reference genome. Plant J 89(4):789–804. doi:10.1111/tpj.13415
25. Paytuví Gallart A, Hermoso Pulido A, Anzar Martínez de Lagrán I, Sanseverino W, Aiese Cigliano R (2015) GREENC: a Wiki-based database of plant lncRNAs. Nucleic Acids Res. doi:10.1093/nar/gkv1215. gkv1215
26. Zhao Y, Li H, Fang S, Kang Y, Wu W, Hao Y, Li Z, Bu D, Sun N, Zhang MQ, Chen R (2016) NONCODE 2016: an informative and valuable data source of long non-coding RNAs. Nucleic Acids Res 44(D1):D203–D208. doi:10.1093/nar/gkv1252
27. Szcześniak MW, Rosikiewicz W, Makałowska I (2016) CANTATAdb: a collection of plant long non-coding RNAs. Plant Cell Physiol 57(1):e8–e8. doi:10.1093/pcp/pcv201
28. Yi X, Zhang Z, Ling Y, Xu W, Su Z (2015) PNRD: a plant non-coding RNA database. Nucleic Acids Res 43(Database issue):D982–D989. doi:10.1093/nar/gku1162
29. Chen D, Yuan C, Zhang J, Zhang Z, Bai L, Meng Y, Chen L-L, Chen M (2012) PlantNATsDB: a comprehensive database of plant natural antisense transcripts. Nucleic Acids Res 40(Database issue):D1187–D1193. doi:10.1093/nar/gkr823
30. Liu J, Jung C, Xu J, Wang H, Deng S, Bernad L, Arenas-Huertero C, Chua N-H (2012) Genome-wide analysis uncovers regulation of long intergenic noncoding RNAs in Arabidopsis. Plant Cell 24(11):4333–4345. doi:10.1105/tpc.112.102855
31. Wang Y, Wang X, Deng W, Fan X, Liu T-T, He G, Chen R, Terzaghi W, Zhu D, Deng XW (2014) Genomic features and regulatory roles of intermediate-sized non-coding RNAs in Arabidopsis. Mol Plant 7(3):514–527. doi:10.1093/mp/sst177
32. Liu T-T, Zhu D, Chen W, Deng W, He H, He G, Bai B, Qi Y, Chen R, Deng XW (2013) A global identification and analysis of small nucleolar RNAs and possible intermediate-sized non-coding RNAs in Oryza sativa. Mol Plant 6(3):830–846. doi:10.1093/mp/sss087
33. van Werven FJ, Neuert G, Hendrick N, Lardenois A, Buratowski S, van Oudenaarden A, Primig M, Amon A (2012) Transcription of two long noncoding RNAs mediates mating-type control of gametogenesis in budding yeast. Cell 150:1170–1181. doi:10.1016/j.cell.2012.06.049

34. Zofall M, Yamanaka S, Reyes-Turcu FE, Zhang K, Rubin C, Grewal SIS (2012) RNA elimination machinery targeting meiotic mRNAs promotes facultative heterochromatin formation. Science 335(6064):96–100. doi:10.1126/science.1211651
35. Flynn RA, Chang HY (2014) Long noncoding RNAs in cell-fate programming and reprogramming. Cell Stem Cell 14(6):752–761. doi:10.1016/j.stem.2014.05.014
36. Berry S, Dean C (2015) Environmental perception and epigenetic memory: mechanistic insight through FLC. Plant J 83(1):133–148. doi:10.1111/tpj.12869
37. Matzke MA, Mosher RA (2014) RNA-directed DNA methylation: an epigenetic pathway of increasing complexity. Nat Rev Genet 15(6):394–408. doi:10.1038/nrg3683
38. Bardou F, Ariel F, Simpson CG, Romero-Barrios N, Laporte P, Balzergue S, Brown JWS, Crespi M (2014) Long noncoding RNA modulates alternative splicing regulators in Arabidopsis. Dev Cell 30(2):166–176. doi:10.1016/j.devcel.2014.06.017
39. Wang Y, Fan X, Lin F, He G, Terzaghi W, Zhu D, Deng XW (2014) Arabidopsis noncoding RNA mediates control of photomorphogenesis by red light. Proc Natl Acad Sci 111(28):10359–10364. doi:10.1073/pnas.1409457111
40. Wang D, Qu Z, Yang L, Zhang Q, Liu Z-H, Do T, Adelson DL, Wang Z-Y, Searle I, Zhu J-K (2017) Transposable elements (TEs) contribute to stress-related long intergenic noncoding RNAs in plants. Plant J. doi:10.1111/tpj.13481
41. Pefanis E, Wang J, Rothschild G, Lim J, Kazadi D, Sun J, Federation A, Chao J, Elliott O, Liu Z-P, Economides AN, Bradner JE, Rabadan R, Basu U (2015) RNA exosome-regulated long non-coding RNA transcription controls super-enhancer activity. Cell 161(4):774–789. doi:10.1016/j.cell.2015.04.034
42. Zilberman D, Cao X, Jacobsen SE (2003) ARGONAUTE4 control of locus-specific siRNA accumulation and DNA and histone methylation. Science 299(5607):716–719. doi:10.1126/science.1079695
43. Xie Z, Johansen LK, Gustafson AM, Kasschau KD, Lellis AD, Zilberman D, Jacobsen SE, Carrington JC (2004) Genetic and functional diversification of small RNA pathways in plants. PLoS Biol 2(5):E104. doi:10.1371/journal.pbio.0020104.sg002
44. Zheng X, Zhu J, Kapoor A, Zhu J-K (2007) Role of Arabidopsis AGO6 in siRNA accumulation, DNA methylation and transcriptional gene silencing. EMBO J 26(6):1691–1701
45. Gao Z, Liu H-L, Daxinger L, Pontes O, He X, Qian W, Lin H, Xie M, Lorkovic ZJ, Zhang S, Miki D, Zhan X, Pontier D, Lagrange T, Jin H, Matzke AJM, Matzke M, Pikaard CS, Zhu J-K (2010) An RNA polymerase II- and AGO4-associated protein acts in RNA-directed DNA methylation. Nature 465(7294):106–109. doi:10.1038/nature09025
46. Matzke M, Kanno T, Daxinger L, Huettel B, Matzke AJ (2009) RNA-mediated chromatin-based silencing in plants. Curr Opin Cell Biol 21(3):367–376. doi:10.1016/j.ceb.2009.01.025
47. Franco-Zorrilla JM, Valli A, Todesco M, Mateos I, Puga MI, Rubio-Somoza I, Leyva A, Weigel D, García JA, Paz-Ares J (2007) Target mimicry provides a new mechanism for regulation of microRNA activity. Nat Genet 39(8):1033–1037. doi:10.1038/ng2079
48. Wang P, Xue Y, Han Y, Lin L, Wu C, Xu S, Jiang Z, Xu J, Liu Q, Cao X (2014) The STAT3-binding long noncoding RNA lnc-DC controls human dendritic cell differentiation. Science 344(6181):310–313. doi:10.1126/science.1251456
49. Bonasio R, Shiekhattar R (2014) Regulation of transcription by long noncoding RNAs. Annu Rev Genet 48:433–455. doi:10.1146/annurev-genet-120213-092323
50. Wang KC, Yang YW, Liu B, Sanyal A, Corces-Zimmerman R, Chen Y, Lajoie BR, Protacio A, Flynn RA, Gupta RA, Wysocka J, Lei M, Dekker J, Helms JA, Chang HY (2011) A long noncoding RNA maintains active chromatin to coordinate homeotic gene expression. Nature 472(7341):120–124. doi:10.1038/nature09819
51. Tsai MC, Manor O, Wan Y, Mosammaparast N, Wang JK, Lan F, Shi Y, Segal E, Chang HY (2010) Long noncoding RNA as modular scaffold of histone modification complexes. Science 329(5992):689–693. doi:10.1126/science.1192002
52. Li W, Notani D, Rosenfeld MG (2016) Enhancers as non-coding RNA transcription units: recent insights and future perspectives. Nat Rev Genet 17(4):207–223. doi:10.1038/nrg.2016.4

53. Lai F, Orom UA, Cesaroni M, Beringer M, Taatjes DJ, Blobel GA, Shiekhattar R (2013) Activating RNAs associate with mediator to enhance chromatin architecture and transcription. Nature 494(7438):497–501. doi:10.1038/nature11884
54. Skourti-Stathaki K, Proudfoot NJ (2014) A double-edged sword: R loops as threats to genome integrity and powerful regulators of gene expression. Genes Dev 28(13):1384–1396. doi:10.1101/gad.242990.114
55. Swiezewski S, Liu F, Magusin A, Dean C (2009) Cold-induced silencing by long antisense transcripts of an Arabidopsis polycomb target. Nature 462(7274):799–802. doi:10.1038/nature08618
56. Ariel F, Jegu T, Latrasse D, Romero-Barrios N, Christ A, Benhamed M, Crespi M (2014) Noncoding transcription by alternative RNA polymerases dynamically regulates an auxin-driven chromatin loop. Mol Cell 55(3):383–396. doi:10.1016/j.molcel.2014.06.011
57. Li S, Vandivier LE, Tu B, Gao L, Won SY, Li S, Zheng B, Gregory BD, Chen X (2015) Detection of Pol IV/RDR2-dependent transcripts at the genomic scale in Arabidopsis reveals features and regulation of siRNA biogenesis. Genome Res 25(2):235–245. doi:10.1101/gr.182238.114
58. Blevins T, Podicheti R, Mishra V, Marasco M, Wang J, Rusch D, Tang H, Pikaard CS (2015) Identification of Pol IV and RDR2-dependent precursors of 24 nt siRNAs guiding de novo DNA methylation in Arabidopsis. eLife 4:e09591. doi:10.7554/eLife.09591
59. Zhai J, Bischof S, Wang H, Feng S, Lee T-F, Teng C, Chen X, Park SY, Liu L, Gallego-Bartolome J, Liu W, Henderson IR, Meyers BC, Ausin I, Jacobsen SE (2015) A one precursor one siRNA model for Pol IV-dependent siRNA biogenesis. Cell 163(2):445–455. doi:10.1016/j.cell.2015.09.032
60. Böhmdorfer G, Sethuraman S, Rowley MJ, Krzyszton M, Rothi MH, Bouzit L, Wierzbicki AT (2016) Long non-coding RNA produced by RNA polymerase V determines boundaries of heterochromatin. eLife 5:1325. doi:10.7554/eLife.19092
61. Campalans A, Kondorosi A, Crespi M (2004) Enod40, a short open reading frame-containing mRNA, induces cytoplasmic localization of a nuclear RNA binding protein in Medicago truncatula. Plant Cell 16(4):1047–1059. doi:10.1105/tpc.019406
62. Ding J, Lu Q, Ouyang Y, Mao H, Zhang P, Yao J, Xu C, Li X, Xiao J, Zhang Q (2012) A long noncoding RNA regulates photoperiod-sensitive male sterility, an essential component of hybrid rice. Proc Natl Acad Sci 109(7):2654–2659. doi:10.1073/pnas.1121374109
63. Wu H-J, Wang Z-M, Wang M, Wang X-J (2013) Widespread long noncoding RNAs as endogenous target mimics for microRNAs in plants. Plant Physiol 161(4):1875–1884. doi:10.1104/pp.113.215962
64. Bühler M, Haas W, Gygi SP, Moazed D (2007) RNAi-dependent and -independent RNA turnover mechanisms contribute to heterochromatic gene silencing. Cell 129(4):707–721. doi:10.1016/j.cell.2007.03.038
65. Bühler M, Spies N, Bartel DP, Moazed D (2008) TRAMP-mediated RNA surveillance prevents spurious entry of RNAs into the Schizosaccharomyces pombe siRNA pathway. Nat Struct Mol Biol 15(10):1015–1023. doi:10.1038/nsmb.1481
66. Reyes-Turcu FE, Zhang K, Zofall M, Chen E, Grewal SIS (2011) Defects in RNA quality control factors reveal RNAi-independent nucleation of heterochromatin. Nat Struct Mol Biol 18(10):1132–1138. doi:10.1038/nsmb.2122
67. Holoch D, Moazed D (2015) RNA-mediated epigenetic regulation of gene expression. Nat Rev Genet 16(2):71–84. doi:10.1038/nrg3863
68. Zheng B, Wang Z, Li S, Yu B, Liu JY, Chen X (2009) Intergenic transcription by RNA polymerase II coordinates Pol IV and Pol V in siRNA-directed transcriptional gene silencing in Arabidopsis. Genes Dev 23(24):2850–2860. doi:10.1101/gad.1868009
69. Zhang X, Henderson IR, Lu C, Green PJ, Jacobsen SE (2007) Role of RNA polymerase IV in plant small RNA metabolism. Proc Natl Acad Sci 104(11):4536–4541. doi:10.1073/pnas.0611456104
70. Lee T-F, Gurazada SGR, Zhai J, Li S, Simon SA, Matzke MA, Chen X, Meyers BC (2012) RNA polymerase V-dependent small RNAs in Arabidopsis originate from small, intergenic loci including most SINE repeats. Epigenetics 7(7):781–795. doi:10.4161/epi.20290

71. You W, Lorkovic ZJ, Matzke AJM, Matzke M (2013) Interplay among RNA polymerases II, IV and V in RNA-directed DNA methylation at a low copy transgene locus in Arabidopsis thaliana. Plant Mol Biol 82(1–2):85–96. doi:10.1007/s11103-013-0041-4
72. Sasaki T, Lee T-F, Liao W-W, Naumann U, Liao J-L, Eun C, Huang Y-Y, Fu JL, Chen P-Y, Meyers BC, Matzke AJM, Matzke M (2014) Distinct and concurrent pathways of Pol II- and Pol IV-dependent siRNA biogenesis at a repetitive trans-silencer locus in Arabidopsis thaliana. Plant J 79(1):127–138. doi:10.1111/tpj.12545
73. Law JA, Du J, Hale CJ, Feng S, Krajewski K, Palanca AMS, Strahl BD, Patel DJ, Jacobsen SE (2013) Polymerase IV occupancy at RNA-directed DNA methylation sites requires SHH1. Nature 498(7454):385–389. doi:10.1038/nature12178
74. Liu Z-W, Shao C-R, Zhang C-J, Zhou J-X, Zhang S-W, Li L, Chen S, Huang H-W, Cai T, He X-J (2014) The SET domain proteins SUVH2 and SUVH9 are required for Pol V occupancy at RNA-directed DNA methylation loci. PLoS Genet 10(1):e1003948. doi:10.1371/journal.pgen.1003948
75. Zhong X, Hale CJ, Law JA, Johnson LM, Feng S, Tu A, Jacobsen SE (2012) DDR complex facilitates global association of RNA polymerase V to promoters and evolutionarily young transposons. Nat Struct Mol Biol 19(9):870–875. doi:10.1038/nsmb.2354
76. Zhu Y, Rowley MJ, Böhmdorfer G, Wierzbicki AT (2013) A SWI/SNF chromatin-remodeling complex acts in noncoding RNA-mediated transcriptional silencing. Mol Cell 49(2):298–309. doi:10.1016/j.molcel.2012.11.011
77. Wierzbicki AT, Cocklin R, Mayampurath A, Lister R, Rowley MJ, Gregory BD, Ecker JR, Tang H, Pikaard CS (2012) Spatial and functional relationships among Pol V-associated loci, Pol IV-dependent siRNAs, and cytosine methylation in the Arabidopsis epigenome. Genes Dev 26(16):1825–1836. doi:10.1101/gad.197772.112
78. Wierzbicki AT, Ream TS, Haag JR, Pikaard CS (2009) RNA polymerase V transcription guides ARGONAUTE4 to chromatin. Nat Genet 41(5):630–634. doi:10.1038/ng.365
79. Zheng Q, Rowley MJ, Böhmdorfer G, Sandhu D, Gregory BD, Wierzbicki AT (2012) RNA polymerase V targets transcriptional silencing components to promoters of protein-coding genes. Plant J. doi:10.1111/tpj.12034
80. Böhmdorfer G, Rowley MJ, Kuciński J, Zhu Y, Amies I, Wierzbicki AT (2014) RNA-directed DNA methylation requires stepwise binding of silencing factors to long non-coding RNA. Plant J 79(2):181–191. doi:10.1111/tpj.12563
81. Shin J-H, Wang H-LV, Lee J, Dinwiddie BL, Belostotsky DA, Chekanova JA (2013) The role of the Arabidopsis exosome in siRNA-independent silencing of heterochromatic loci. PLoS Genet 9(3):e1003411. doi:10.1371/journal.pgen.1003411.s007
82. Zhang H, Tang K, Qian W, Duan C-G, Wang B, Zhang H, Wang P, Zhu X, Lang Z, Yang Y, Zhu J-K (2014) An Rrp6-like protein positively regulates noncoding RNA levels and DNA methylation in Arabidopsis. Mol Cell 54(3):418–430. doi:10.1016/j.molcel.2014.03.019
83. Amasino RM, Michaels SD (2010) The timing of flowering. Plant Physiol 154(2):516–520. doi:10.1104/pp.110.161653
84. Liu F, Marquardt S, Lister C, Swiezewski S, Dean C (2009) Targeted 3′ processing of antisense transcripts triggers Arabidopsis FLC chromatin silencing. Science 327(5961):94–97. doi:10.1126/science.1180278
85. Davidovich C, Wang X, Cifuentes-Rojas C, Goodrich KJ, Gooding AR, Lee JT, Cech TR (2015) Toward a consensus on the binding specificity and promiscuity of PRC2 for RNA. Mol Cell 57(3):552–558. doi:10.1016/j.molcel.2014.12.017
86. Helliwell CA, Robertson M, Finnegan EJ, Buzas DM, Dennis ES (2011) Vernalization-repression of Arabidopsis FLC requires promoter sequences but not antisense transcripts. PLoS One 6(6):e21513. doi:10.1371/journal.pone.0021513
87. Csorba T, Questa JI, Sun Q, Dean C (2014) Antisense COOLAIR mediates the coordinated switching of chromatin states at FLC during vernalization. Proc Natl Acad Sci 111(45):16160–16165. doi:10.1073/pnas.1419030111
88. Hornyik C, Terzi LC, Simpson GG (2010) The Spen family protein FPA controls alternative cleavage and polyadenylation of RNA. Dev Cell 18(2):203–213. doi:10.1016/j.devcel.2009.12.009

89. Marquardt S, Raitskin O, Wu Z, Liu F, Sun Q, Dean C (2014) Functional consequences of splicing of the antisense transcript COOLAIR on FLC transcription. Mol Cell 54(1):156–165. doi:10.1016/j.molcel.2014.03.026
90. Liu F, Quesada V, Crevillen P, Bäurle I, Swiezewski S, Dean C (2007) The Arabidopsis RNA-binding protein FCA requires a lysine-specific demethylase 1 homolog to downregulate FLC. Mol Cell 28(3):398–407. doi:10.1016/j.molcel.2007.10.018
91. Hawkes EJ, Hennelly SP, Novikova IV, Irwin JA, Dean C, Sanbonmatsu KY (2016) COOLAIR antisense RNAs form evolutionarily conserved elaborate secondary structures. Cell Rep 16(12):3087–3096. doi:10.1016/j.celrep.2016.08.045
92. Callahan KP, Butler JS (2008) Evidence for core exosome independent function of the nuclear exoribonuclease Rrp6p. Nucleic Acids Res 36(21):6645–6655. doi:10.1093/nar/gkn743
93. Kiss DL, Andrulis ED (2010) Genome-wide analysis reveals distinct substrate specificities of Rrp6, Dis3, and core exosome subunits. RNA 16(4):781–791. doi:10.1261/rna.1906710
94. Ciaudo C, Bourdet A, Cohen-Tannoudji M, Dietz HC, Rougeulle C, Avner P (2006) Nuclear mRNA degradation pathway(s) are implicated in Xist regulation and X chromosome inactivation. PLoS Genet 2(6):e94. doi:10.1371/journal.pgen.0020094
95. Sun Q, Csorba T, Skourti-Stathaki K, Proudfoot NJ, Dean C (2013) R-loop stabilization represses antisense transcription at the Arabidopsis FLC locus. Science 340(6132):619–621. doi:10.1126/science.1234848
96. Skourti-Stathaki K, Kamieniarz-Gdula K, Proudfoot NJ (2014) R-loops induce repressive chromatin marks over mammalian gene terminators. Nature 516(7531):436–439. doi:10.1038/nature13787
97. Hacisuleyman E, Goff LA, Trapnell C, Williams A, Henao-Mejia J, Sun L, McClanahan P, Hendrickson DG, Sauvageau M, Kelley DR, Morse M, Engreitz J, Lander ES, Guttman M, Lodish HF, Flavell R, Raj A, Rinn JL (2014) Topological organization of multichromosomal regions by the long intergenic noncoding RNA Firre. Nat Struct Mol Biol 21(2):198–206. doi:10.1038/nsmb.2764
98. Engreitz JM, Pandya-Jones A, McDonel P, Shishkin A, Sirokman K, Surka C, Kadri S, Xing J, Goren A, Lander ES, Plath K, Guttman M (2013) The Xist lncRNA exploits three-dimensional genome architecture to spread across the X chromosome. Science 341(6147):1237973. doi:10.1126/science.1237973
99. Quinodoz S, Guttman M (2014) Long noncoding RNAs: an emerging link between gene regulation and nuclear organization. Trends Cell Biol 24(11):651–663. doi:10.1016/j.tcb.2014.08.009
100. Feng S, Cokus SJ, Schubert V, Zhai J, Pellegrini M, Jacobsen SE (2014) Genome-wide Hi-C analyses in wild-type and mutants reveal high-resolution chromatin interactions in Arabidopsis. Mol Cell 55(5):694–707. doi:10.1016/j.molcel.2014.07.008
101. Grob S, Schmid MW, Grossniklaus U (2014) Hi-C analysis in Arabidopsis identifies the knot, a structure with similarities to the flamenco locus of Drosophila. Mol Cell 55(5):678–693. doi:10.1016/j.molcel.2014.07.009
102. Wang C, Liu C, Roqueiro D, Grimm D, Schwab R, Becker C, Lanz C, Weigel D (2015) Genome-wide analysis of local chromatin packing in Arabidopsis thaliana. Genome Res 25(2):246–256. doi:10.1101/gr.170332.113
103. Liu C, Weigel D (2015) Chromatin in 3D: progress and prospects for plants. Genome Biol 16(1):170. doi:10.1186/s13059-015-0738-6
104. Liu C, Wang C, Wang G, Becker C, Zaidem M, Weigel D (2016) Genome-wide analysis of chromatin packing in Arabidopsis thaliana at single-gene resolution. Genome Res. doi:10.1101/gr204032.116. gr204032116
105. Moissiard G, Cokus SJ, Cary J, Feng S, Billi AC, Stroud H, Husmann D, Zhan Y, Lajoie BR, McCord RP, Hale CJ, Feng W, Michaels SD, Frand AR, Matteo P, Dekker J, Kim JK, Jacobsen SE (2012) MORC family ATPases required for heterochromatin condensation and gene silencing. Science 336(6087):1448–1451. doi:10.1126/science.1221472
106. Crevillen P, Sonmez C, Wu Z, Dean C (2013) A gene loop containing the floral repressor FLC is disrupted in the early phase of vernalization. EMBO J 32(1):140–148. doi:10.1038/emboj.2012.324

107. Jegu T, Latrasse D, Delarue M, Hirt H, Domenichini S, Ariel F, Crespi M, Bergounioux C, Raynaud C, Benhamed M (2014) The BAF60 subunit of the SWI/SNF chromatin-remodeling complex directly controls the formation of a gene loop at FLOWERING LOCUS C in Arabidopsis. Plant Cell 26(2):538–551. doi:10.1105/tpc.113.114454
108. Rosa S, De Lucia F, Mylne JS, Zhu D, Ohmido N, Pendle A, Kato N, Shaw P, Dean C (2013) Physical clustering of FLC alleles during polycomb-mediated epigenetic silencing in vernalization. Genes Dev 27(17):1845–1850. doi:10.1101/gad.221713.113

Chapter 6
Long Noncoding RNAs in Mammalian Development and Diseases

Parna Saha, Shreekant Verma, Rashmi U. Pathak, and Rakesh K. Mishra

Abstract Following analysis of sequenced genomes and transcriptome of many eukaryotes, it is evident that virtually all protein-coding genes have already been discovered. These advances have highlighted an intriguing paradox whereby the relative amount of protein-coding sequences remain constant but nonprotein-coding sequences increase consistently in parallel to increasing evolutionary complexity. It is established that differences between species map to nonprotein-coding regions of the genome that surprisingly is transcribed extensively. These transcripts regulate epigenetic processes and constitute an important layer of regulatory information essential for organismal development and play a causative role in diseases. The noncoding RNA-directed regulatory circuit controls complex characteristics. Sequence variations in noncoding RNAs influence evolution, quantitative traits, and disease susceptibility. This chapter presents an account on a class of such noncoding transcripts that are longer than 200 nucleotides (long noncoding RNA—lncRNA) in mammalian development and diseases.

Keywords lncRNAs • Evolution of complexity • Epigenetic modifications • Imprinting • Chromosome inactivation • Body patterning • Nuclear architecture • Cellular differentiation

6.1 Introduction

Recent technical advancements in high-throughput sequencing have revealed that a majority of eukaryotic genome is pervasively transcribed. Large-scale analysis (ENCODE project) has shown that ~75% of human genome is transcribed in various cell lines [1]. Why a cell spends so much of its resources on RNA production has captured the imagination of scientific community. Many of these RNAs are long

P. Saha • S. Verma • R.U. Pathak (✉) • R.K. Mishra (✉)
CSIR-Centre for Cellular and Molecular Biology, Uppal Road, Hyderabad 500007, India
e-mail: rashmi@ccmb.res.in; mishra@ccmb.res.in

M.R.S. Rao (ed.), *Long Non Coding RNA Biology*, Advances in Experimental Medicine and Biology 1008, DOI 10.1007/978-981-10-5203-3_6

transcripts with no apparent protein-coding potential. Initially these transcripts were discounted as artifacts and were thought to be the result of expression "noise" rather than expression "choice." But many features of these transcripts indicate a definite and important role. For example, transcription of lncRNAs is initiated from conserved promoters. Many of the transcripts are alternatively spliced and display predicted structures. They are dynamically expressed during differentiation and disease in a cell- and tissue-specific manner. However, the main argument posed against a functional role is the lack of primary sequence conservation among these transcripts. But studies have shown that lack of conservation does not necessarily mean lack of function.

lncRNA molecules are involved in diverse biological processes like genomic imprinting, dosage compensation, epigenetic and transcriptional regulation, chromosome conformation, cell cycle regulation, stem cell differentiation/reprogramming, and allosteric enzymatic activity [2]. The structure and biogenesis of lncRNAs is very similar to that of mRNAs. Like mRNAs, they are transcribed by RNA polymerase II from genomic loci in the epigenetic context similar to protein-coding genes. They are 5′-capped and spliced and commonly have a polyadenylated tail. However, unlike mRNAs, they may undergo alternative forms of processing at 3′-end. For example, an RNase P-assisted cleavage at 3′-end results in a lncRNA with stable 3′-terminal RNA triplex structure, instead of a polyadenylated tail [3]. Although lncRNAs lack coding capacity, they possess the intriguing ability to adopt a secondary/tertiary structure that may relate to their function. Depending on their position and direction of transcription in relation to protein-coding genes, lncRNAs may be classified as antisense, intergenic, intronic, bidirectional, processed, or pseudogene transcripts [4, 5]. Mechanism of action of lncRNAs is also very diverse. They may regulate genes in *cis* (i.e., in close proximity to site of transcription) or in *trans* (at a distance from transcription site) [6]. They may act as *scaffolds* to bring a group of proteins into spatial proximity, as *guides* to recruit proteins to DNA, as *decoys* to titrate away proteins, or as *enhancer RNAs* involved in chromosomal looping in enhancer-like manner [2]. Some lncRNAs are precursor to smaller regulatory RNAs, like miRNA or piwi RNAs or they may bind to complimentary RNAs to affect their turnover [7].

Mostly the lncRNAs are expressed at low levels in a highly tissue-specific manner, so much so that their expression profiles are important markers for disease or developmental state [8]. Many a times they are found next to protein-coding genes that are under tight transcriptional control, and often their expression pattern correlates with tissue differentiation, development, and disease [9]. The widespread dysregulation of lncRNA expression in human diseases and the finding that many lncRNAs are enriched for SNPs that associate with human traits/diseases have highlighted the need to understand the functional contribution of these RNAs [10, 11]. However, the study of lncRNAs using model organisms is confounded by the fact that these RNAs exhibit poor primary sequence conservation. Exons of lncRNA evolve much faster than protein-coding gene sequence and most lncRNAs are lineage specific [8, 12, 13].These RNAs rather show conservation along genomic

Table 6.1 Number of noncoding and coding transcripts in different organisms

Organism	Genome size (Mb)	lncRNAs[a]
Human	3300	141,353
Mouse	2800	117,405
D. melanogaster	120	54,819
A. thaliana	135	3853
C. elegans	100	3271
S. cerevisiae	12.5	61

[a]www.noncode.org

position (synteny), short sequence motifs, or secondary structure [14, 15]. Because of more likely structural than sequence conservation, functionality of lncRNAs could be organized into modular domains similar to proteins organized into functional motifs.

Understandably organismal complexity correlates better with the expression repertoire of lncRNA than with that of protein-coding genes (Table 6.1) [16]. This presents a pressing need to explore the functional relevance of such transcripts in the context of evolution of developmental mechanisms. In most vertebrates, exhaustive annotation of lncRNA is still not available, primarily due to incomplete genome sequences and partial annotation of protein-coding genes. Further, majority of the annotated lncRNAs remain functionally uncharacterized and only a small fraction have been explored for their biological relevance. In this chapter we give an overview of some of the characterized mammalian lncRNAs and their etiology in human diseases.

6.2 Diverse Function of lncRNAs in Mammalian Gametogenesis and Development

A large number of mammalian lncRNAs (mostly in human and mice) have been discovered in recent genome-wide expression studies. They have been found to play important role in almost all stages of mammalian development, i.e., gametogenesis, embryogenesis (during preimplantation stages as well as in placenta), body axis patterning, pre-/postnatal tissue development, and organogenesis. In diploid organisms, most genes are expressed from both alleles, but some are expressed from only one allele in a parent of origin-specific manner. Genomic imprinting and X-chromosome inactivation (XCI) are two such phenomena that lead to monoallelic expression of genes. These phenomena come into play during gametogenesis/embryonic development and have lncRNAs as a key player in the process. Similarly, spatiotemporally coordinated embryonic expression of Hox genes leads to body axis patterning in bilaterians. Epigenetic features and lncRNAs bring about this coordination of Hox gene expression. Several studies point to functional role of lncRNAs in mammalian development.

6.2.1 lncRNA in Genomic Imprinting

In mammals, some genes are epigenetically "imprinted" mainly by DNA methylation during gametogenesis by a process called "genomic imprinting." This results in allele-specific expression of either maternally or paternally inherited genes in developing embryo. The imprinting process happens during early gametogenesis and approximately 1% of mammalian protein-coding genes get imprinted. Initial cues to the phenomenon came from early experiments where nuclear transfer in mouse zygotes reconstructed from two maternal pronuclei (gynogenones) or from two paternal pronuclei (androgenones) failed to develop, while zygotes carrying one paternal pronucleus and maternal pronucleus were able to develop [17, 18]. Later, genome-wide studies and deletion and transgenic approaches led to the identification of several imprinted genes most of which are essential and have been implicated in developmental process.

To date, more than 150 imprinted genes in mouse and about half that number has been identified in humans. Most imprinted genes are organized in clusters that contain three or more genes. The size of the cluster spans from a few kilobases to several megabases on different chromosomes [19, 20] (www.mousebook.org). An imprinting control region (ICR) controls gene expression in each imprinted cluster. ICRs are rich in CpG dinucleotides and carry parental allele-specific germline-derived DNA-methylated regions (gDMR). This pattern of gDMR is maintained throughout development [21, 22]. The allele-specific expression of imprinted genes in a cluster in daughter cells after subsequent cell divisions is conferred and managed by histone modifications, insulators, and higher-order chromatin organizations [19, 20]. Surprisingly most imprinted clusters identified have one or more associated lncRNA that have been found to be inherently essential to the allele-specific expression. In general, lncRNAs show reciprocal parental allele-specific expression when compared to the imprinted genes in a cluster. ICRs are mostly located in or near the promoter of lncRNA. Further, by many overexpression and deletion experiments, it has been confirmed that lncRNA regulates imprinting of the locus in *cis* or in *trans* or both.

lncRNAs known to be involved in genomic imprinting are listed below (Table 6.2), and the mechanisms of action of two relatively better understood examples are discussed here. Although the imprinting-associated lncRNAs do not employ a common mechanism for epigenetic control, they do offer valuable insights into the biology of lncRNAs in general. Interestingly, most imprinted lncRNAs are relatively conserved at functional as well as sequence level between mice and humans. This makes genomic imprinting an attractive model system to study lncRNA-dependent epigenetic mechanisms during human development and diseases using mouse models.

6.2.1.1 H19

H19 gene encodes for a 2.3-kb lncRNA. It is among one of the first discovered and widely investigated imprinted genes in mammals. In mouse, *H19* is present along with insulin-like growth factor (*Igf2*) gene at distal segment of chromosome 7. This

Table 6.2 lncRNAs involved in mammalian genomic imprinting

Imprinted cluster	lncRNA and expression (M or P)	Type of lncRNA	*Cis*-silencing function	Genes imprinted and expression (M or P)	Refs
Igf2	*H19* (M)	Intergenic	Yes	*Igf2* (P)	[23, 24]
Kcnq1	*Kcnq1ot1*	Antisense	Yes	*Kcnq1*, *Cdkn1c*, *Slc22a18*, *Phlda2*, *Ascl2*, *Cd81*, *Tssc4*, *Tspan32*, *Osbpl5* (M)	[25]
Igf2r	*Airn*(P)	Antisense	Yes	*Igf2r*, *Slc22a2*, *Slc22a3* (M)	[26]
Pws/As	*UBE3A-ATS*	Antisense	Yes	*MAGEL2* (P), *NDN* (P), *SNRPN* (P),	[27–29]
	Ipw	Intergenic	n.d.	*SNORD115* (P),	
	Pwcr1	n.d.	Yes	*SNORD116* (P), *UBEA3A* (M)	
Dlk1	*Gtl2* (M)	Antisense	n.d.	*DLK1* (P), *DIO3* (P), *RTL1*(P)	[30–32]
	Rtl1as (M)	Antisense	Yes		
	Rian (M)	Intergenic	n.d.		
	Mirg (M)	Intergenic	n.d.		
Gnas	*Nespas*	Antisense	Yes	*Gnas* (M), *Nesp* (M)	[33, 34]
	Exon1A	Antisense	n.d.		

M maternal, *P* paternal, *n.d.* not determined

region is syntenic to the locus 11p15.5 in human [23, 35]. A differentially methylated ICR, which lies in between the two genes, regulates mutually exclusive mono-allelic expression of *H19* and *Igf2* at the locus. A common enhancer located downstream of *H19* drives the expression of both the genes. The ICR and *H19* promoter are methylated in paternal allele. On the maternal allele, the un-methylated ICR binds to an architectural known as CCCTC-binding factor (CTCF) responsible for long-range chromatin interactions and chromatin looping. CTCF further triggers recruitment of cohesin to ICR, resulting in higher-order chromatin conformation that restricts the enhancer access to *Igf2* promoter. A methylated, thus unoccupied, ICR on the paternal chromosome on the other hand poses no restriction, and enhancer interacts with the *Igf2* promoter driving its expression (Fig. 6.1a) [36, 37].

Although expression of *H19* is mostly studied in relation to imprinting of *H19–Igf2* locus, studies have been carried out to understand the functions of *H19* lncRNA. *H19* knockout mice are viable and fertile with growth defects and reduced muscle regeneration capacities [35, 38]. For example, *H19* deletion on the maternally inherited chromosome led to an increase in *Igf2* expression and increased body weight that could be rescued by deletion of one *Igf2* allele. Although *H19* is highly expressed during embryogenesis, it is effective only in specific cell lineages. Apart from imprinting, deletion/overexpression of *H19* affects embryonic growth. This is because *H19* is part of an imprinted gene network (IGN), which consists of 16 co-expressing imprinted genes that include many growth regulators such as *Igf2*, *Igf2r*,

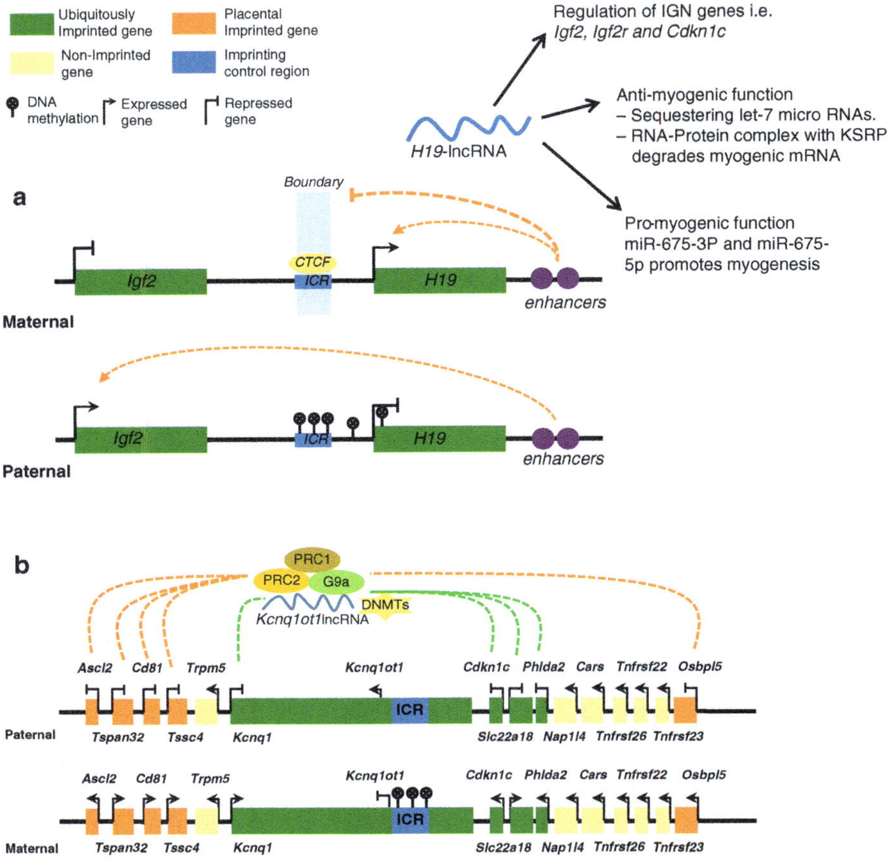

Fig. 6.1 Imprinted regulation of genes of *Igf2* and *Kcnq1* cluster by *H19* and *Kcnq1ot1* lncRNAs, respectively. (**a**) *H19* and *Igf2* show reciprocally exclusive mono-allelic expression from maternal and paternal loci, respectively. A differentially methylated ICR between the two genes and downstream enhancers regulates parent of origin-specific expression of both the genes. The ICR and *H19* promoter are methylated in paternal chromosome that represses *H19* expression, while enhancer interacts with the *Igf2* promoter driving its expression. On the maternal chromosome, the ICR is un-methylated and CTCF is bound to it. CTCF triggers recruitment of cohesin to ICR and higher-order chromatin conformation that restricts the enhancer access to *Igf2* promoter. *H19* lncRNA also involves in regulation of genes of imprinting gene networks (IGN). *H19* lncRNA shows dual function of anti-myogenic and pro-myogenic during mesenchymal stem cells (C2C12) differentiation into myocytes. (**b**) At *Kcnq1* cluster *Kcnq1ot1* expresses paternally, while all the imprinted protein-coding genes are maternally expressed. These imprinted genes are of two types—(1) placenta-specific imprinted genes (PIGs: *Ascl2*, *Cd81*, *Tssc4*, *Tspan32*, *Osbpl5*) which show imprinted silencing only in placental tissues and (2) ubiquitously imprinted genes (UIGs: *Kcnq1*, *Cdkn1c*, *Slc22a18*, *Phlda2*) which show imprinted silencing in both placental and embryonic tissues. The promoter for the *Kcnq1ot1* coincides with the differentially methylated ICR (*Kcnq1* ICR/ KVDMR1). The maternal allele-specific methylation of *Kcnq1ot1* promoter restricts the expression of lncRNA from maternal chromosome. Paternally expressed *Kcnq1ot1* lncRNA interacts with modifiers of chromatin (EZH2 and G9a) and DNA (DNMT1) that bind paternal alleles in *cis* and silence the imprinted genes by establishing higher-order chromatin compartment enriched in repressive histone and DNA modifications

and *Cdkn1c*. The noncoding RNA acts in *trans* to bring about its effect on IGN. It interacts with methyl-CpG-binding protein (MBD-1) that methylates the DMRs of IGN members like *Igf2*, *Slc38a4*, and *Peg1* via H3K9 methyltransferase [39]. This function of *H19* lncRNA that leads to establishment of H3K9me3-associated repressive chromatin occurs on both the parental alleles and is related to embryonic growth regulation.

After embryogenesis, *H19* expresses at low levels in all tissues and at a very high level in muscle. During postnatal tissue differentiation, *H19* has been implicated in contrasting pro- and anti-myogenic functions. Using mouse multipotent mesenchymal cells (C2C12 cells), it has been shown that depletion of *H19* accelerates muscle differentiation suggesting an anti-myogenic function. Two different mechanisms have been suggested for this function. In one of the studies, *H19* from human as well as mice is reported to carry conserved binding sites of Let-7 microRNAs, a pro-myogenic factor. It thus acts as a competing endogenous RNA (CeRNA), a natural sponge that sequesters Let-7 and controls its level. In another study, *H19* has been shown to interact with an RNA processing protein known as K homology-type splicing regulatory protein (KSRP). The resulting RNA–protein complex facilitates interaction between exosome and labile transcripts of protein "myogenin" promoting its degradation and eventually restricting the differentiation of C2C12 into myocytes. However, in contrast to the above findings, a pro-myogenic function of *H19* has been reported that is mediated by two microRNAs, miR-675-3p and miR-675-5p, originating from the exon1 of the *H19* transcript. *H19* along with miR-675-3p/miR-675-5p induces C2C12 differentiation into myocytes. Downregulation of *H19* or blocking the action of miR-675-3p/miR-675-5p prevents C2C12 differentiation [40–42]. The apparently contrasting functions can be reconciled with possible mechanisms that inhibit the primary role of *H19* which is to prevent myogenesis. Once its function needs to be changed, the RNA gets degraded or processed in a way that miRNAs from its exon1 are generated to eventually promote myogenesis.

In conclusion, *H19* lncRNA is an epigenetic regulator of transcription. It executes its activity by behaving as a CeRNA, miRNA precursor, or scaffold to recruit proteins. It is involved in multitude of biological processes like imprinting, growth, differentiation, and myogenesis [43].

6.2.1.2 Kcnq1ot1

Kcnq1ot1 is a 91-kb-long noncoding RNA that maps to *Kcnq1* gene in antisense orientation. The imprinted cluster approximately ~1 Mb in length and encompassing 12 genes is present at the distal end of the seventh chromosome in mouse. Its human orthologue is located on chromosome 11p15.5 [25, 44]. Promoter for the *Kcnq1ot1* gene lies in the tenth intron of *Kcnq1* host gene and coincides with the differentially methylated ICR (*Kcnq1* ICR/KVDMR1). The maternal allele-specific methylation of *Kcnq1ot1* promoter restricts the expression of lncRNA from paternal chromosome in antisense direction with respect to host gene. All the imprinted protein-coding genes are maternally expressed. These imprinted genes are of two types—(1) placenta-specific imprinted genes (PIGs: *Ascl2*, *Cd81*, *Tssc4*, *Tspan32*, *Osbpl5*) which show imprinted silencing only in placental tissues and (2) ubiquitously imprinted genes (UIGs: *Kcnq1*, *Cdkn1c*, *Slc22a18*, *Phlda2*) which show imprinted silencing in both placental and embryonic tissues [25, 45, 46] (Fig. 6.1b).

The antisense *Kcnq1ot1* RNA is required for silencing of both UIGs and PIGs. Paternal silencing is lost when *Kcnq1ot1* promoter is deleted or a prematurely truncated RNA is produced [47, 48]. Interestingly, *Kcnq1ot1* also employs lineage-specific mechanism of action as after initiating imprinting of UIGs as well as PIGs, it is involved in the maintenance of silencing at UIGs alone [49]. To unravel the mechanism of action of the lncRNA, biochemical and genetic studies have been carried out in cells and transgenic mice models. The studies show that *Kcnq1ot1* lncRNA interacts with modifiers of chromatin (EZH2 and G9a) and DNA (DNMT1) to recruit them in *cis* to silence the imprinted genes [44, 46, 50]. The allelic silencing is achieved by establishment of higher-order chromatin compartment enriched in repressive histone modifications such as H3K27me3, H3K9me2, and H2AK119ub [50, 51]. While silencing of UIGs is controlled by repressive histone modifications and maintained by methylation of somatic DMRs, silencing of PIGs is controlled by repressive histone modification only. Recently, *Kcnq1ot1* lncRNA has been shown to mediate targeting of the entire repressed loci to distinct perinucleolar repressive compartment by virtue of a conserved 890-bp repeat-containing domain present at its 5′-end.

6.2.2 lncRNA in Dosage Compensation

In higher eukaryotes, the number of sex chromosomes differs between the two sexes. Organisms have evolved different strategies to compensate for this discrepancy by adjusting gene expression levels. To equalize transcription level of genes present on sex chromosome, the chromatin structure is modulated epigenetically. The epigenetic mechanism on one extreme leads to inactivation of one of the X chromosome in females (as observed in mammals) and on the other extreme leads to twofold higher expression of genes on the single X chromosome in males (as observed in *Drosophila*). The curiously opposite ways lead to equal level of expression of sex chromosome-associated genes in different organisms.

In female mammals, the epigenetic process of X-chromosome inactivation (XCI) regulates gene dosage of extra X chromosome. Initially the phenomena was noticed by Murray Barr in 1949 when he observed that female cat cells possess a condensed subnuclear structure which is now called as "Barr body" in his honor. Later studies demonstrated that the Barr body is nothing but a condensed X chromosome which is also transcriptionally silent [52–54]. Later a 17-kb noncoding murine transcript *Xist* was discovered that initiated the fascinating era of lncRNA biology [55]. Further discovery of its 40-kb-long antisense transcript *Tsix* highlighted the fact that untranslated RNAs dominate the regulation of XCI [56]. The process of XCI is similar to genomic imprinting as the silenced genes are clustered, are influenced by a long-distance master control region, and are associated with multiple lncRNAs.

In eutherian mammals, the process of XCI occurs in two different ways. During early embryogenesis, paternal X chromosome is inactivated in preimplanted embryos. As the embryo reaches blastula stage and gets implanted into the uterus, the outer blastular cells (future placenta) retain paternal XCI, while imprinting is erased from inner cell mass (future embryo). As these inner blastular cells (epiblasts) differentiate, either of the parental chromosome has an equal chance of inactivation (random XCI). The eutherian mammalian female is thus essentially a mosaic, with randomly active paternal/maternal X chromosome. In marsupials, however, the choice of inactivation is always fixed to paternal X chromosome.

Random XCI is a coordinated stepwise process that results in silencing of ~1000 genes along the inactive X chromosome. The process is controlled by X-inactivation center *(Xic)* that codes for lncRNA with regulatory properties. The lncRNAs from *Xic* work in *cis* as well as in *trans*. In the first step which has been referred to as "counting," X chromosome-to-autosome ratio (X:A) is measured. XCI is initiated in female cells where X:A = 1 and is blocked in male cells where the ratio is 0.5 [57, 58]. Molecular details of this measurement remain elusive, but *trans*-activity of the two lncRNAs, *Tsix* and *Xite*, is implicated in the process [59]. The next step results in random "choice" of one of the X chromosome to remain active (Xa) and the other one to get inactivated (Xi). The final "sensing" is a permissive state for XCI, similar to the initial counting step but distinct from it. Eventually the Xi is epigenetically marked by repressive chromatin and DNA methylation to a transcriptionally inert state, while Xa remains open for transcription.

The *Xic* is a 100–200-kb region with at least seven lncRNA genes of which six have been shown to have specific function during XCI (Table 6.3, Fig. 6.2). Prior to initiation of XCI, the lncRNAs *Xist*, *Tsix*, and *Xite* are expressed from both the Xs at low levels. Mutually exclusive selection of Xa and Xi necessitates interchromosomal interaction and robust feedback mechanism. The 5′-end of *Tsix* gene binds to the

Table 6.3 lncRNAs involved in XCI

lncRNA	Functions	Refs
Xist	Initiation and spreading of XCI on Xi	[66]
Tsix	Negative regulation of *Xist*, dosage sensor (measurement of X:A ratio), and X-chromosome pairing for choice of Xi/Xa	[59]
Xite	Positive regulation of *Tsix*, dosage sensor (measurement of X:A ratio), and X-chromosome pairing for choice of Xi/Xa	[59, 67, 68]
DXPas34	Involve in dual function as an enhancer and a repressor of *Tsix*, counting, and X-chromosome pairing for choice of Xi/Xa	[69]
Tsx	Negatively regulates *Xist* and positively regulates *Tsix*	[70]
Linx	Co-expresses with *Tsix* and potentially involved in positive regulation of *Tsix*	[71]
RepA	Play role in upregulation of *Xist* by recruitment of PRC2 and altering the chromatin structure at *Xist* promoter on Xi	[72]
Jpx/Enox	Activates *Xist* upregulation by evicting CTCF binding to *Xist* promoter on Xi	[72, 73]
Ftx	Positive regulator of *Xist*	[74]

protein CTCF that leads to a brief transient contact of the two Xs at *Xic*. However, a role for the two lncRNA transcripts (*Tsix/Xite*) is also envisaged in the process as inhibiting Pol II activity results in abrogation of X–X pairing suggesting that the pairing requires new transcription. The contact of the two Xs results in establishment of an asymmetry and choice of Xi and Xa [60, 61]. The process of establishment of asymmetry is not clear, but it has been postulated that the proximity of Xs directs irreversible shift of proteins (Oct4 and CTCF) from one allele (future Xi) to the other (future Xa) [61–63]. Once chosen, *Tsix* is expressed in an allele-specific manner from

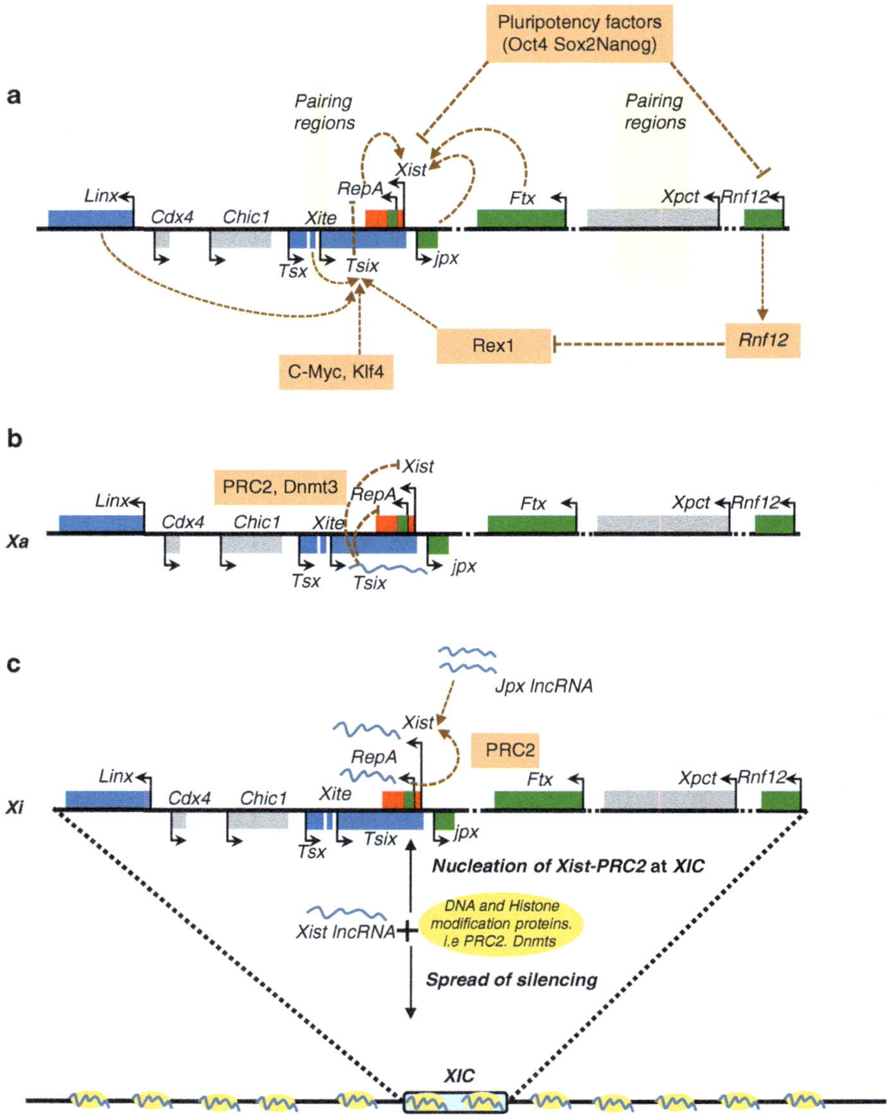

the Xa. As *Tsix* is antisense to *Xist*, its expression results in removal of latter from Xa in *cis*. lncRNAs *Xite* and *DXPas34* positively regulate expression of *Tsix* from Xa. Finally to seal the active state permanently, *Tsix* lncRNA directs DNA methylation (by Dnmt3a) at the *Xist* promoter resulting in stable silencing of *Xist* allele on Xa. On the other hand, removal of transcription factors (Oct4 and CTCF) from *Tsix/Xite* promoter on Xi causes a drop in transcription of these lncRNAs. In the absence of *Tsix*, *Xist* lncRNA being transcribed from Xi becomes abundant and coats it in *cis*. *Xist*-coated Xi becomes enriched for repressive chromatin marks (H3K27me3), but *Xist* transcription continues unabated in otherwise heterochromatic environment. One of the first changes that follows depletion of *Tsix* on Xi is enrichment of polycomb repressive complex (PRC2) on *Xist* promoter. This is brought about by *RepA* lncRNA that recruits PRC2 to 5′-end of *Xist*. This creates a heterochromatic patch at *Xist* promoter, essential and stimulatory for its expression [64, 65].

In conclusion, the lncRNAs coded by the *Xic* perform versatile functions to coordinate the process of XCI. They can uniquely define an address in the genome as they remain tethered to their locus of transcription and guide regulatory mechanisms in *cis*. They can also function as transcript-level regulator by RNAi (*Xist–Tsix* pair) or act as tethers and guides to recruit chromatin modifiers (*RepA* recruitment of PRC2, *Xist* recruitment of Dnmt3a).

6.2.3 Linc-ing RNAs to Body Patterning

Hox genes are the important group of transcriptional factors encoding genes, arranged in clusters in the bilaterian genomes. They define the body axis patterning through precisely coordinated spatiotemporal expression in the developing

Fig. 6.2 lncRNA-mediated X-chromosome inactivation. (**a**) X-inactivation center (Xic) of mouse encompasses ~500 kb regions of X chromosome that has several lncRNA loci (*Xist*, *Tsix*, *Jpx*, Ftx, *RepA*, *Xite*, *Tsx*, *Linx*) as well as protein-coding genes (*gray*). lncRNAs are involved in positive (*RepA*, *Jpx*, and *Ftx—green*) or negative (*Xite*, *Tsx*, *Linx—blue*) regulation of *Xist* by activating *Tsix*, an antisense lncRNA and negative regulator of *Xist*. Other than lncRNA at the Xic, the protein-coding *Rnf12* gene (*green*) which encodes an ubiquitin ligase is also known to promote *Xist* upregulation. Pluripotency factors (Oct4, Sox2, Nanog, C-Myc, Klf4, and Rex1) are thought to block *Xist* expression directly or indirectly through *Tsix* activation. During early embryogenesis before implantation, both the X chromosomes are active, and both express the *Tsix* lncRNA, which negatively regulates the *Xist* lncRNA. (**b**) At the onset of XCI during development or ESC differentiation, several events such as decrease in pluripotency factor levels (OCT2, NANOG, SOX2, and REX1), chromosome pairing (involves pairing region), increase in *Xist* activator expression (*jpx*, *Ftx*, and *RepA*), and induction of mono-allelic expression of *Tsix* facilitate coordinated induction of *Xist* upregulation at XIC of random chosen future Xi chromosome from one of the two X chromosomes. At the Xic of the second X chromosome (Xa), Tsix expression is maintained and proposed to be regulated by its *cis* activator lncRNAs (*DXPas34*, *Tsx*, *xite*, *Linx*) which restrict *Xist* expression from Xa. (**c**) Upregulation of *Xist* initiates XIC by coating Xi chromosome in *cis* at Xic that spread to all over the Xi chromosome. Xist coating to the entire chromosome is accompanied by recruitments of DNA and histone modifiers which direct the series of epigenetic modification that progressively silence most of the X-linked genes

embryo—a hallmark of Hox genes. While invertebrates have one set of Hox genes, vertebrates typically have four sets referred to as HoxA, HoxB, HoxC, and HoxD clusters due to the events of genome duplication during the course of evolution [75]. More recently many kinds of regulatory RNAs that are involved in Hox gene regulation have been discovered. Large body of work in the last decade has discovered and studied the noncoding RNAs in the Hox clusters, and most of them have been found to be long intergenic noncoding (linc)RNAs. The expression of these intergenic transcripts correlates with transcription of neighboring Hox genes. Very often these lincRNAs show syntenic or positional conservation between mouse and humans suggesting a common function. Here we discuss the current understanding of the role and mechanism of action of some these lincRNAs in different mammalian Hox clusters. Table 6.4 enlists various lincRNAs involved in regulation of mammalian Hox genes.

Table 6.4 lncRNAs involved in regulation of mammalian Hox genes

lncRNA	Expressed in	Transcribed from	Function	Refs
HoxA				
Halr1 and *Halr1-os*	Human, mouse ESCs	Region between *Hoxa1* and *Skap2*, ~50 kb from the 3′ end of *Hoxa1*	Retinoic acid-dependent regulation of HoxA genes	[76, 77]
Haunt (HoxA upstream noncoding transcript)	Human ESC, NPC, and NSC	40 kb upstream of HoxA cluster	Attenuates enhancer–promoter contacts acting as RA-dependent brake during ESC differentiation	[78]
HOTAIRM1 (Hox antisense intergenic RNA myeloid 1)	Human, mouse, and other mammals	Transcribed antisense to *Hoxa1* from a shared promoter between *Hoxa1* and *Hoxa2*	Specific to myeloid lineage and involved in granulocyte maturation	[79, 80]
HOXA-AS2 (HoxA cluster antisense RNA 2)	Promyelocytic leukemic cells and neutrophils	Isoforms of 339 to 2045 nucleotides from intergenic region between *Hoxa3* and *Hoxa4*	Induced by IFNγ in PMNs and TNF in NB4 cells for negative regulation of ATRA-induced TRAIL (TNF-related apoptosis-inducing ligand) during myeloid differentiation	[81]
HOXA11-AS	Human and mouse	Antisense transcript from promoter of *Hoxa11*	In gametogenesis. Knockdown results in male and female sterility due to uterine defects and failure of testes to descend from abdomen, respectively	[82]

(continued)

Table 6.4 (continued)

lncRNA	Expressed in	Transcribed from	Function	Refs
HIT18844	Human abdominal tissues like the colon, uterus, and prostate	265 bp from 5′-end of HOXA cluster, ~1.8 kb from Hoxa13	Possess ultra-conserved short stretch that results in secondary structural motif. Function unknown	[83]
HOTTIP (HoxA transcript at the distal tip)	Human and mouse	3.7-kb polyadenylated transcript starting ~330b upstream of *Hoxa13*	Spatiotemporally controls expression of 5′ HoxA genes	[84]
HoxB				
Hobbit1 (HoxB intergenic transcript)	Human and mouse	Intergenic region between *Hoxb4* and *Hoxb5*	Retinoic acid-dependent regulation of HoxB genes	[85, 86]
HoxC				
HOTAIR (Hox transcript antisense intergenic RNA)	Human	2158-bp, polyadenylated transcript from intergenic region between *Hoxc11* and *Hoxc12*	Regulates expression of HoxD genes by acting as a molecular scaffold for binding of LSD1/CoREST/REST complex	[87–89]
HoxD				
Hotdog and *Tog* (transcript from telomeric desert of HoxD cluster and twin of Hotdog)	Mouse	Telomeric region downstream of HoxD cluster	Specific to development cecum, regulation of HoxD genes	[90]
HOXD-AS1	Human	Intergenic between *Hoxd1* and *Hoxd3*	Retinoic acid-induced cell differentiation	[91]

6.2.3.1 HoxA Cluster

Heater is one of the well-studied lncRNA loci in HoxA cluster. The coding potential of this region was discovered during analysis of RNA deep sequencing data in mouse ES cells (ESCs) [76]. Heater region encodes for two lincRNAs—*halr1(linc-Hoxa1)* and *halr1os1*. Studies on the *linc-Hoxa1* (transcribed in opposite direction to *Hoxa1* and is 12 kb long) in mouse ESCs revealed that the *linc-Hoxa1* has three isoforms of which the isoform 1 is most abundant. *Hoxa1* and *linc-Hoxa1* are sensitive to retinoic acid (RA) and single transcript counting showed that their levels are antagonistic to each other. Indeed knockdown of *linc-Hoxa1* increased the level of *Hoxa1* mRNA, but the result was not reproduced when siRNA was used. To solve this mystery, the investigators checked for the levels of *linc-Hoxa1* in the nucleus and cytoplasm using RNA FISH under both experimental conditions. The number of sites of active transcription of *linc-Hoxa1* decreased by the use of antisense oligonucleotides but not siRNAs, indicating the repression of *Hoxa1* by *linc-Hoxa1* occurs only at the site of its transcription (not by overall cellular abundance) and requires the proximity of these two genes in *cis* (neighboring *Hoxa2* levels were unaffected). Another interesting observation was that only when *linc-Hoxa1* is <10 molecules *Hoxa1* transcripts are detectable highlighting the subtlety of transcriptional regulation by these RNAs. In summary, in the absence of RA, *Hoxa1* adopts a conformation that is physically proximal to *linc-Hoxa1*. Such conformation results in repression of*Hoxa1* transcription by abrogating the binding of RA receptors to retinoic acid response elements (RAREs). When RA is present, it binds to RAREs in the *Hoxa1* promoter and pulls it out from the regulation of *linc-Hoxa1*. This finely orchestrated regulation also requires the presence of protein factor purine-rich element-binding protein B (PURB) that binds to *linc-Hoxa1* as shown in Fig. 6.3a [77]. Thus the Heater region through its multiple RAREs regulates the effect of retinoids on the noncoding transcripts that in turn fine-tunes neighboring Hox gene expression.

Another noncoding transcript known as *HOTAIRM1* was identified in HoxA cluster in human peripheral blood neutrophils of myeloid lineage, hence the name [79]. *HOTAIRM1* is not conserved across species in terms of sequence, but similarly localizing transcripts are present in other species. *HOTAIRM1* preferentially associates with CpG islands near the TSS(s) in all mammalian species. Knockdown of transcript results in lowering of expression of *Hoxa1*, *Hoxa4*, and subtly *Hoxa5*, but not *Hoxa9*, *Hoxa10*, and *Hoxa11* (*cis* effect). *HOTAIRM1* knockdown in all-trans retinoic acid (ATRA)-induced human promyelocytic leukemic cells (NB4) also showed *trans* effect by abrogating G/S cell cycle progression by interfering with CD11b, CD18, and β2 integrin signaling pathways that are involved in granulocyte maturation [80].

The discovery of *HOTTIP* was spurred by the observation that 5′-end of the HoxA cluster in anatomically distal cells (foreskin and foot fibroblast) shows broad H3K4me3 peaks and abundant chromatin interactions in contrary to the H3K27me3 marks and no long-range interactions in proximal cells (lung fibroblast). *HOTTIP* has the presence of bivalent histone marks (H3K4me3 and H3K27me3) indicating its poised regulatory function at the diametrically opposite chromatin domain at 5′-end of HoxA cluster as compared to 3′-end. Knockdown of *HOTTIP* reduced the level of transcripts from distal genes *Hoxa13*, *Hoxa11*, and in lesser severity for more proximal genes *Hoxa10–Hoxa7*. Depletion of *HOTTIP* in distal limb bud of chicken embryos resulted in bending and shortening of distal bony elements. There was an increase in H3K27me3 and overall decrease in H3K4me3 marks over the 5′-end of HoxA cluster. These observations indicate that *HOTTIP* promotes transcription of 5′ *HOXA* genes in *cis*, in a proximity-dependent fashion in distal tissues through the deposition of H3K4me3 marks. *HOTTIP* acts as a regulatory switch at the distal end of HoxA cluster by interacting with WDR5 to recruit MLL complex that activates 5′*HOXA* genes. *HOTTIP* is an elegant example of how noncoding transcript couples 3D genome organization with chromatin landscape to spatiotemporally coordinate the developmental pattern [84] Fig. 6.3b.

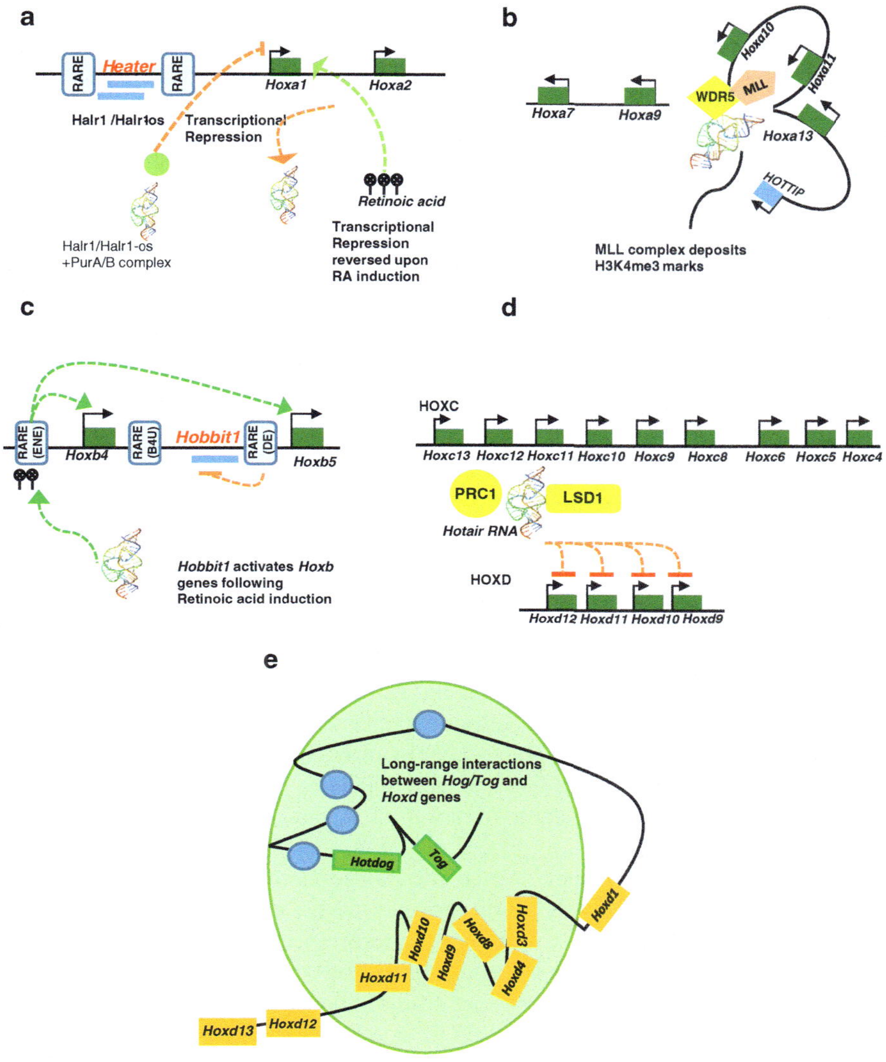

Fig. 6.3 Schematic showing the mechanism of action of Hox lincRNAs. (**a**) *Heater* region encodes two lncRNAs *halr1* and *halr1-os*. In the absence of retinoic acid, *halr1/halr1-os* binds to PURB and transcriptionally represses *Hoxa1*. When retinoic acid is present, it binds to RAREs (retinoic acid response elements) upstream of *Hoxa1* and prevents *halr1* from repressing *Hoxa1* expression. (**b**) *HOTTIP* is a regulatory RNA at the distal end of HoxA cluster that interacts with WDR5 to recruit MLL complex proteins to activate 5′ *Hoxa* genes. (**c**) *Hobbit1* regulates the expression of *Hoxb* genes in the presence of retinoic acid that acts on RAREs present in the regulatory regions. (**d**) *HOTAIR* is transcribed from HoxC cluster and regulates the expression of *Hoxd* genes by acting as a dual molecular scaffold that recruits PRC2 and LSD1/REST/coREST complexes at posterior *Hoxd* genes. (**e**) *Hotdog* and *twin of hotdog*, transcribed from the telomeric desert downstream of HoxD cluster, fold the *Hoxd3*–*Hoxd11* and the enhancers (*blue circles*) into an active TAD to regulate the long-range interactions necessary for proper gene expression during cecum development

6.2.3.2 HoxB Cluster

Hobbit1, the only prominent lncRNA from HoxB cluster, is a noncoding transcript that activates gene expression following RA induction. Figure 6.3c shows early neural enhancer (ENE) and distal element-retinoic acid response element (DE-RARE) enhancer modulate the expression of *Hoxb* genes in neural crest during rostral expansion maintaining the distinct domains of anterior and posterior *Hoxb* genes [85]. DE-RARE can modulate RA response of *Hobbit1* and anteriorize the 5′ *Hoxb* gene expression. This exemplifies the cross talk of *cis* regulatory DE-RARE with noncoding RNA *Hobbit1* and subsequently the neighboring *Hoxb* genes in response to the developmental cues (RA gradient) during organogenesis [86].

6.2.3.3 HoxC Cluster

HOTAIR is the most widely studied lncRNAs from Hox clusters (HoxC cluster). It was discovered in a microarray study of human Hox loci in primary fibroblast with 11 different positional identities. Knockdown of *HOTAIR* had no effect on HoxC cluster genes on Chr12 but severely affected *HoxD* cluster genes spanning 40 kb on Chr2. This was a remarkable example of noncoding transcripts acting in *trans* [87]. Chromatin immunoprecipitation studies revealed that depletion of *HOTAIR* lowered the levels of H3K27me3 and Suz12 over the HoxD cluster. Later studies showed that *HOTAIR* interacts with Ezh2 (T345) and thus recruits the PRC2 complex at the HoxD [92]. Not only this, *HOTAIR* also acts a dual molecular scaffold, as its 5′-end binds PRC2 complex, while the 3′-end binds LSD1/CoREST/REST complex. Thus it tethers the two complexes to coordinate H3K27 methylation (deposition of repressive marks) and H3K4 demethylation (removal of activation marks) as depicted in Fig. 6.3d [88].

Schoroderet and colleagues deleted the mouse *Hotair* region along with the HoxC cluster with surprisingly no discernable in vivo effects concluding that the long noncoding RNAs have evolved to perform species-specific function [89]. However, in an equally surprising report that followed this study, Li et al. showed that precise *Hotair* conditional knockout results in homeotic transformation of spine and metacarpal carpal bones. Interestingly depletion of *Hotair* also affects many other non-Hox genes including those from imprinted loci like *Dlk1-Meg3* and *Igf2-H19* [93]. Reanalysis of Schoroderet results indicated that *HoxCΔ* resulted in upregulation of all other *Hox* genes and removed genes that function antagonistic to *Hotair* leading to compensation of the deletion. The drastic difference in results of *HoxCΔ* knockout and *Hotair* knockdown on using different experimental approaches highlights the need of fine-scale experimentation to study intricate regulation of Hox lincRNAs.

HOTAIR mis-expression has also been implicated in many cancers [94–100] as it has been reported to have multiple protein partners like proteins involved in cytoskeletal and respiratory chain function [101].

6.2.3.4 HoxD Cluster

In HoxD cluster, *Hotdog* and *Tog* (*Hotdog*, lncRNA from telomeric desert of HoxD cluster; *Tog*, twin of *Hotdog* in opposite direction) are the noncoding transcripts specific to developing cecum. They arise from the telomeric region downstream of HoxD cluster. In cecum, *Hoxd9*, *Hoxd10*, and *Hoxd11* are highly expressed, while *Hoxd12* and *Hoxd13* are repressed. These two lncRNAs were discovered while trying to understand how closely spaced genes in HoxD cluster maintain a distinct chromatin/expression domain. Chromosome conformation capture and chromatin immunoprecipitation studies within the HoxD cluster showed that *Hotdog* and *Tog* fold the expressed Hox genes along with their enhancers and the telomeric desert in an active topological domain (H3K4me1/3 marked) as in Fig. 6.3e. The domain allows their precise expression during cecum budding. The repressed Hox genes remain out of the active domain. When the telomeric gene desert was separated from the HoxD cluster by chromosome inversion, the HoxD cluster genes were silenced, but the noncoding transcripts were still produced. Deletion of region from *Hoxd9* to *Hoxd11* abrogates this spatial configuration and abolishes *Hotdog/Tog* transcription as well. These results suggest a model of long-range enhancer sharing, and *Hotdog/Tog* are the example of how noncoding transcripts coordinate long-range interactions connecting distal regulatory elements [90].

6.2.4 lncRNA in Tissue Development and Organogenesis

Embryonic stem cells make cell-fate choices by gene regulatory programs. Terminal differentiation of cells results in patterning of tissues. The functionality of tissue throughout the life is maintained by adult stem cells. Functional studies have revealed that lncRNAs play an active role in gene regulation at virtually every stage of progression of differentiation of ESCs, namely, cell cycle, pluripotency, differentiation, cell survival, apoptosis, etc. They coordinate exit from pluripotency to terminal differentiation. After tissue differentiation, they have emerged as an important class of regulator for maintenance of adult stem cells [102].

Differentiation of skin is a regulated multistep process. Skin differentiation is well characterized at molecular level, to the extent that skin tissue can be regenerated ex vivo and grafted in vivo. It, thus, provides a robust system, where role of lncRNAs could be investigated. Two lncRNAs, *ANCR* and *TINCR*, expressed in epidermal tissue progenitor cells, play a crucial role in epidermal differentiation. They exhibit antagonistic function, where *TINCR* (terminal differentiation-induced noncoding RNA) promotes differentiation, and *ANCR* (antidifferentiation noncoding RNA) inhibits differentiation [103, 104]. *TINCR* is a cytoplasmic lncRNA expressed at low levels in epidermal progenitor cells. Its expression is induced during differentiation. A 25-nucleotide region in *TINCR* (*TINCR*-Box) interacts with a RNA-binding protein staufen1 (STAU1). The resulting *TINCR*-STAU1 complex mediates stabilization of many mRNAs that encode for proteins required for differentiation of keratinocytes. Accordingly, downregulation of TINCR in human squamous cells leads to carcinoma. lncRNA *ANCR* on the other hand enforces undifferentiated cell state within the epidermis.

A few lncRNAs are responsible for maintenance of two different states. For example, lincRNA known as *TUNA* (a.k.a. *megamind* in zebrafish) is a highly conserved noncoding RNA expressed in cells of neural lineages in the adult brain, spinal cord, and eyes. It has been observed that under different cellular contexts, by virtue of its unique protein partners, *TUNA* may maintain pluripotency of ESCs or, in a contrasting role, coordinate neural lineage commitment. This is possible because *TUNA* operates through multiple molecular mechanisms at transcriptional or posttranscriptional level [105].

Recently generation of KO animal models have been used to elucidate the role of lncRNA in tissue patterning. For example, deletion of complete lncRNA *Hotair* in mouse led to skeletal deformities and homeotic transformation including abnormalities in the wrist and spine [93]. The *Mdgt* KO mice showed abnormally low body weight and slower growth. KO of *Fendrr* or *Peril* led to peri-/postnatal lethality of the animal [106]. Apart from generation of KO models, extensive characterization of ESCs has revealed the role of many lncRNAs involved in enforcement of pluripotency in these cells. The common mechanisms of action employed by these noncoding RNAs though remain the same. A vast majority of them act via interacting with chromatin modifiers including the readers, writers, and erasers of histone marks. A subset acts as competing endogenous RNA by "sponging" out miRNAs. Table 6.5 lists lncRNAs involved in mammalian tissue development and organogenesis.

6.3 lncRNA in Nuclear Architecture and Chromosome Structure Maintenance

Genomes are hierarchically folded into complex higher-order structure that gives rise to chromatin fiber, chromosomal domains, and condensed chromosomes during cell division. In interphase nuclei, chromosomes occupy distinct territory that can be defined as the nuclear space taken up by the particular chromosome. During cell division, chromatin further gets compacted into distinct X-shaped condensed chromosome, where centromere and telomere play an important role to maintain its integrity. The higher-order organization of chromatin is directly linked to gene regulation, and any defect in the organization perturbs gene expression causing diseases. Several diseases arise as a result of aberrant chromosome numbers (aneuploidy) and chromosome instability during cell division. The role of specific proteins in chromatin organization is well established and lncRNAs have now emerged as new players in this domain. lncRNAs have been found to be an integral part of this global phenomenon of higher-order chromatin organization/modulation and chromosome structure maintenance. In this section we discuss the role played by lncRNAs in nuclear organization.

Table 6.5 List of lncRNA involved in tissue development and organogenesis

lncRNA	Expressed in	Function	Refs
Skin development			
TINCR	Human	Epidermal differentiation by posttranscriptional mechanism	[104]
ANCR	Human	Suppresses differentiation pathway to maintain epidermal adult stem cell	[103]
Blood development			
LincRNA-EPS	Mouse Hematopoietic organs	Prevents apoptosis in erythroid differentiation	[107]
alncRNA-EC1	Mouse Fetal liver erythroid cell	Regulates erythropoiesis (enhancer-associated RNA)	[108]
alncRNA-EC7	Mouse Fetal liver erythroid cell	Regulates expression of *Band3* (enhancer-associated RNA)	[108]
DLEU2	Human and mouse	Regulates erythropoiesis and B cell maturation Represses of *SPRYD7* gene	[108, 109]
elncRNA-EC1	Mouse	Involved in erythroblast differentiation	[108]
lincRNA-EC9	Mouse	Involved in erythroblast differentiation	[108]
Eye development			
Tug1	Human, mouse, dog, cow, and rat Retinal cells	Involved in cone photoreceptor specification Associates with PRC2	[110]
RNCR2 (MIAT/Gomafu)	Mouse Retinal cells	Involved in retinal cell specification	[111]
Six3	Mouse Eye and retinal cells	Involved in neural specification in ES cells of the retina and eye (promoter-associated RNA)	[112]
Vax2os	Human, other primates, and mouse Retina (outer neuroblastic layer)	Involved in retina cell specification Regulates cell cycle progression of photoreceptor progenitor cells in ventral retina	[113, 114]
Cardiac development			
aHIF	Human	Associated with cardiac pathology (Hypoxia-inducible factor 1A antisense RNA)	[115]
Kcnq1ot	Human	Involved in cardiogenesis Regulates chromatin reorganization at imprinted loci	[116]
ANRIL	Human	Involved in atherosclerosis, carcinomas, and inflammatory response Interacts with CBX7 of PRC1 complex (antisense noncoding RNA in the INK4 locus)	[117, 118]

(continued)

Table 6.5 (continued)

lncRNA	Expressed in	Function	Refs
SENCR	Human	Regulation of endothelial differentiation from pluripotent cells Controls the angiogenic capacity of human umbilical vascular endothelial cells (HUVEC) (Cytoplasmic lncRNA)	[119, 120]
LIPCAR	Human	Biomarker for myocardial infarction (Mitochondrial lncRNA)	[121]
CARL	Human	Inhibits anoxia-mediated mitochondrial fission and apoptosis Acts as mir-539 sponge (Cardiac apoptosis-related lncRNA)	[122]
Mhrt (Myheart)	Human Adult heart	Protects against cardiomyopathy A chromatin-remodeler and antagonizes Brg1 (myosin heavy-chain-associated RNA transcript)	[105]
MIAT	Human	Regulates diabetes mellitus-induced microvascular dysfunction Regulates expression of vascular endothelial growth factor and miR-150-5p (myocardial infarction-associated transcript)	[123–126]
Braveheart	Mouse Cardiac cells	Regulates cardiovascular lineage commitment Epigenetic regulator that interact with Suz12	[127]
Fendrr	Mouse	Involved in differentiation of multiple mesenchyme-derived tissues Associates with PRC2	[128, 129]
Immunological development			
TMEVPG1	Human and mouse peripheral blood lymphocytes (NK+ cells, CD4+ and CD8+ T lymphocytes)	Involved in immunity modulation	[130–132]
NeST	Mouse	Involved in immunity modulation Regulates IFNγ transcription	[133]
lncDC	Human	Controls dendritic cell differentiation Binds to STAT3	[134]
ZFAT-AS	Human CD19+ B cell	Regulates B cell function and implicated in autoimmune thyroid disease	[135, 136]

(continued)

Table 6.5 (continued)

lncRNA	Expressed in	Function	Refs
THRIL	Human	Regulates TNFα expression, immune response, and inflammation (heterogeneous nuclear ribonucleoprotein L-related immune-regulatory long noncoding RNA)	[42]
PCAER	Human	Modulates immune response Prevents binding of p50 subunit of repressive NF-κB complex to COX-2 promoter (p50-associated COX-2 extragenic RNA)	[137]
KIR-AS	Human Hematopoietic progenitors Myeloid precursors	Controls gene expression in progenitor cells (killer cell immunoglobulin-like receptor—antisense)	[138]
PRINS	Primates	Keratinocyte stress response and psoriasis pathogenesis (Psoriasis-susceptibility-related RNA gene induced by stress)	[139]
Neuronal development			
AK055040	Human	Involved in neuronal differentiation Interacts with polycomb group proteins (Promoter-associated RNA)	[140]
AK091713	Human	Involved in neurogenesis Overlaps with miRNAs like Mir125B and LET7A	[140]
AK124648	Human ES cells	Involved in promoting pluripotency and neuronal differentiation (promoter-associated RNA)	[140]
CDKN2B-AS/ ANRIL	Human Many cell types	Involved in atherosclerosis, carcinomas, and inflammatory response Interacts with CBX7 of PRC1 complex (antisense noncoding RNA in the INK4 locus/CDKN2B antisense RNA)	[117, 118]
BACE1-AS	Human and mouse	Positive regulator of BACE1 expression target for anisomycin-mediated suppression of ovarian stem cell cancer	[141]
BC1/BC200	Human Nervous system—dendrites	Regulates synaptogenesis Interacts with FMRP and translational machineries to regulate spatially restricted synaptic turnover	[142]
BDNF-AS	Human Nervous system—neurites	Controls development of neurite elaboration (Natural antisense transcript)	[142]

(continued)

Table 6.5 (continued)

lncRNA	Expressed in	Function	Refs
CDRas1	Human and zebrafish	miRNA decoy and circular RNA	[143]
Cyano	Zebrafish	miRNA (miR-7) decoy	[144]
DALI	Human Neuroblastoma cells	Controls neural differentiation by direct interaction with POU3F and DNMT1	[145]
Dlx1AS	Human	Controls neural differentiation (enhancer-associated RNA)	[146–148]
Evf2/Dlx6AS	Mouse Central nervous system	GABAergic interneuron specification Interacts with transcription factor DLX and methyl-CpG-binding protein MECP2 to epigenetically regulate expression of interneuron lineage genes	[149, 150]
GDNF-OS	Human	Transcriptional regulator of *GDNF* (promoter-associated RNA)	[142, 151]
GOMAFU	Human and mouse Dividing neural stem cells and neurons	Inhibits amacrine cell specification and Muller glia differentiation Interacts with splicing factors to regulate alternative splicing of several neuronal genes	[152, 153]
Kcna2AS	Human	Involved in causation of pain and hypersensitivity Inhibits Kcna2 expression that leads to decreased voltage-gated potassium current and increased membrane potential	[154]
Linc-Brn1a and *LincBrn1b*	Mouse Neural stem cell	Differentiation of neural stem cells and cortical neuron development Regulates basal cortical progenitor turnover	[106]
Linc00299	Human All tissues, predominantly brain	Involved in neurodevelopment, particularly brain development	[155]
Linc00237	Human	Causes MOMO (macrosomia, obesity, macrocephaly, and ocular abnormalities) syndrome	[156]
Peril	Human Brain and spinal cord	Controls the cell cycle, energy metabolism, and immune response genes Transcribed from 110 kb downstream of Sox2	[106]
MSN1PAS	Human	Involved in synapse development	[157]
MALAT1	Human and mouse Neurons	Involved in synaptogenesis and synapse formation Recruits splicing proteins to transcription sites	[158, 159]

(continued)

Table 6.5 (continued)

lncRNA	Expressed in	Function	Refs
Megamind	Human, mouse, zebrafish	Involved in brain morphogenesis and eye development	[144]
Neat1	Human and mouse	Induces of paraspeckle formation (Architectural lincRNA)	[160]
Pantr2/BRN1B	Human and mouse Brain	Regulation of differentiation of delaminating neural progenitor cells	[161]
Paupar	Neuroblastoma cells	Interacts with Pax6 to regulate cell cycle and differentiation	[162]
PNKY	Human and mouse Brain	Regulates neural stem cell turnover Balances self-renewal and differentiation of neural stem cells by regulation splicing regulator PTBP1	[163]
RMST	Mouse ESCs	Involved in neural differentiation Recruits Sox2 to neurogenesis-promoting genes	[140]
Sox2dt	Mouse Brain	Regulates Sox2 expression in neurogenic regions of the brain (enhancer-associated RNA)	[164]
SIX3OS	Human Eye	Specification of photoreceptors, bipolar cells, and Muller glia through SIX3 target genes	[112]
TUG1	Human Retinal cells	Retinal cell-type specification and proliferation	[110]
TUNA	Mouse and zebrafish	Recruits RNA-binding proteins (NCL, PTBP1, and hnRNP-K) to neural gene promoters	[165]
utNgn1	Mouse	Involved in neocortical development Regulates transcription of neurogenin (enhancer RNA transcript)	[166]
VAX2OS1	Human Retinal cells	Involved in retinal cell-type specification and proliferation	[114]
Skeletal muscle development			
Yam-1	Mouse Myoblasts	Regulator of myogenesis Muscle-associated lincRNA positively regulated by YY1 and represses muscle differentiation genes like myogenin (YY1-associated muscle lncRNA)	[167]
Linc-MD1	Human and mouse Myoblasts	Regulator of myogenesis Competing endogenous RNA that acts as a sponge for miR-133	[168]

(continued)

Table 6.5 (continued)

lncRNA	Expressed in	Function	Refs
Other organogenesis			
Mdgt	Mouse	Involved in embryonic development Transcribed from a region close to Hoxd1	[106]
Manr, linc-Cox2	Human and mouse Lungs	Involved in organogenesis of lung	[169]
FIRRE	Human	Controls topological organization of multiple chromosomal region Tethers inactive X chromosome to nucleolus	[170, 171]

6.3.1 Interphase Chromatin

That RNA is a significant component of nuclear architecture is known for the past four decades [172]. Early studies have conceptualized nuclear matrix (NuMat—a skeletal framework in the nucleus) as a scaffold predominantly made up of RNA and protein components. NuMat serves as a platform for virtually all nuclear processes, namely, DNA replication, repair, RNA transcription, and splicing. Moreover it is suggested that NuMat plays a fundamental role regulating gene expression [173]. NuMat is sensitive to RNase, indicating critical role of RNA in formation of the structure. Studies in recent years have shown that repeat-containing lncRNAs are involved in building up of the nucleo-skeleton. This phenomenon appears to be conserved across species. For example, *AAGAG* repeat-containing lncRNA is an important component of *Drosophila* NuMat, which when depleted leads to lethality at larval stages [174]. Similarly purine-rich *GAA* repeat-containing lncRNA was found in mammalian cells [175]. Another study provides compelling evidence that RNA transcribed from *LINE-1* interspersed repeats form a significant component of interphase chromatin in human cells. Interestingly interspersed repeat sequences, which account for almost half of human genome, were abundantly transcribed, and the repeat lncRNAs were found associated with euchromatin. Adapter protein SAF-A, by virtue of its DNA- as well as RNA-binding domains, links the *LINE-1*lncRNA to chromatin. The lncRNA stably associates with chromatin and its removal leads to aberrant chromatin distribution and condensation [176]. Thus *LINE-1* lncRNA and other lncRNA species directly associate with chromatin to add to its stability and functionality from a new class of lncRNAs known as chromatin-associated RNAs (caRNAs). However, studies on the role of the NuMat RNA in chromatin architecture and caRNAs are limited, and further explorations are needed to unravel the mechanistic details of the process.

6.3.2 Euchromatin/Heterochromatin

Based on its transcriptional property, interphase chromatin can be distinguished as euchromatin and heterochromatin. Euchromatin is loosely packed, replicates early, and is permissive to transcription, while heterochromatin is compact, replicates later, and is refractive to transcription. These chromatin states are epigenetically marked by differentially modified histones and DNA methylation. Several lncRNAs are known that mediate these epigenetic changes by recruiting chromatin remodeling complexes to specific loci. For example, the lncRNA *HOTAIR* (described in Sect. 6.2.3) originates in HoxC but silences transcription at HoxD locus in *trans* by recruiting polycomb remodeling complex PRC2 to induce silent chromatin. Very recently, a novel lncRNA CAT7 has been identified in human neuronal cells, responsible for fine-tuning stable gene silencing by guiding PRC1 activity [177]. Other remodeling complexes like MLL and G9a methyltransferases are similarly directed by their associated lncRNAs. Thus a small repertoire of chromatin remodeling complexes with little DNA-binding specificity can be directed to a large number of genomic loci, in a spatially/temporally regulated manner, by the virtue of lncRNA molecules that act as guides. Even at constitutive heterochromatic regions (centromere and telomere), lncRNAs play a role in directing heterochromatin organization.

Apart from heterochromatin, lncRNAs regulate the functionality of euchromatin as well. It is now established that to function effectively, regulatory elements like enhancers, promoters, and boundaries are transcribed. The initial cues to the finding came from a pioneer study where using tiling microarrays the authors found that transcripts arise from beginning and end of protein-coding genes [178]. Follow-up studies confirmed bidirectional transcription from CpG-rich, nucleosome-depleted regions at gene promoters using a separate pre-initiation complex. Such transcription generates transcription start site-associated RNAs (TSSa-RNAs). These TSSa-RNAs regulate transcription initiation events [179, 180]. Such TSSa-RNAs are not only responsible for turning genes on, but at times they are responsible for causing gene-specific repression also. For example, in human cell lines, DNA damage induces the expression of lncRNA from cyclin D1 promoter, which modulates the levels of RNA-binding protein known as translocated in liposarcoma (TLS). Protein TLS in turn modulates histone acetyl-transferase activity at the loci to silence the neighborhood [181]. Similarly enhancers are transcribed in cells where they are supposed to remain active, and this strategy is used for regulation of key developmental genes [182]. According to a report, in human ESCs, ~19% of lncRNAs are enhancer RNAs (eRNAs) [183]. Interestingly, eRNAs are also often bidirectionally transcribed. Coming to another important class of regulatory elements known as boundaries/insulators, a classical example of lncRNA involved in creating a functional boundary comes from the imprinted H19/IGF2 locus, described in detail in Sect. 6.2.1. While boundaries are technically described as elements that, when present in between, prevent enhancer to promoter cross talk, insulators separate two distinct epigenetically modified chromatin domains, and when present at the junction, they may restrict the spread of heterochromatin into euchromatin. Most of mammalian boundaries and insulators are known to bind to protein CTCF [184]. CTCF further recruits cohesin complex to the loci, and this loading of cohesin is essential for insulator function. Some lncRNAs are known to act as scaffolds that stabilize interaction of CTCF along with other factors to boundaries/insulators. For example, a DEAD-box RNA helicase (p68) associates with lncRNA known as steroid receptor RNA activator (*SRA*). This complex then recruits CTCF to execute insulator function at H19/IGF2 locus [185]. Many a time, transcription of tRNA genes (tRNAs are also lncRNAs) results in establishment of a boundary [186, 187]. In yet another example, tissue-specific transcription of a retrotransposon repeat at murine growth hormone locus leads to establishment of a boundary that blocks the influence of neighboring repressive chromatin [188]. From these examples it thus becomes evident that lncRNAs function as master regulators that control the functionality of euchromatic regulatory elements.

6.3.3 Genomic Stability

In addition to the lncRNAs mentioned above that directly interact with chromatin, there are other noncoding transcripts that are indirectly involved in maintaining the genomic stability. One such lncRNA known as *NORAD/LINC00657* (noncoding RNA activated by DNA damage) is a highly conserved and abundant transcript present in cytoplasm of human cells (more than 300–1000 copies per cell) [189]. The lncRNA was initially identified for inducing p53-mediated response to DNA damage in mouse and human cells [76]. Later investigations found that targeted inactivation of *NORAD* triggers changes in ploidy level and results in variable chromosome numbers in karyotypically stable human cells. This suggested *NORAD* to have a role in chromosomal stability. *NORAD* has conserved binding sites that sequester the PUMILIO proteins (PUM1/2). PUMILIO are RNA-binding proteins that induce chromosomal instability by repressing mitosis, DNA repair, and DNA replication factors. Interestingly, in the human brain, expression of *NORAD* decreases with increasing age. These studies indicate multiple roles of *NORAD* by alternative mechanisms that are yet to be identified. However, the discovery of *NORAD*–PUMILIO genomic stability pathway has attracted scientific community's attention to explore other unknown lncRNAs involved in genomic stability, maintenance, and their link to chromosomal abnormalities.

6.3.4 Nuclear Compartmentalization

Eukaryotic nucleus is very well compartmentalized at structural and functional level by mechanisms that are conserved across species. The mammalian nucleus contains discrete subnuclear bodies that carry out specific functions [190]. A distinguishing feature of the nuclear bodies that differentiates them from conventional cytoplasmic organelles is that a lipid membrane does not delimit them. Their structural integrity is entirely maintained by protein–RNA and protein–protein interactions. The nuclear bodies are highly dynamic as they assemble/disassemble during every cell division. They are rapidly formed as a response to specific cellular triggers [191]. Many nuclear bodies form around the site of transcription of lncRNAs. For example, nucleolus forms around site of rRNA transcription and stress bodies form around transcribing satellite III repeats. The lncRNA transcripts at these loci act as templates to assemble RNA-binding proteins that in turn result in the formation of nuclear bodies. In addition, lncRNAs can function as architectural element away from where they are transcribed. For example, the lncRNA *NEAT1* is a polyadenylated nuclear-retained transcript, essential for the formation of paraspeckles [160]. Its causal role in the formation of paraspeckles is proven as in the absence of its transcription in human ESCs, paraspeckles are not formed [192]. Similarly, over the last decade, many noncoding RNAs like *MALAT1* (for nuclear speckles), *TUG1* (for polycomb bodies), U-snRNAs (for Cajal bodies), and U7-snRNA (for histone body locus) have been found that play an architectural role in nuclear body formation.

6.3.5 Topological Domains

Chromosome conformation capture techniques have revealed that distally located DNA elements come in close proximity in three-dimensional nuclear space. Such contacts are cell and context specific with functional consequences. These contacts define chromatin loops that provide topological framework for co-regulated genes, commonly known as topological domains (TADs) [193]. Latest developments show that lncRNAs play a vital role in chromatin looping. A remarkable example of lncRNAs involved organization of genome into TADs is that of *Firre* (functional intergenic repeating RNA element) in humans and mouse. *Firre* is expressed from a macrosatellite locus in mouse and contains several cohesin- and CTCF-binding sites required for its functionality. Repeat domains in *Firre* through its interaction with hnRNPU (a nuclear matrix component protein) localize across a 5-Mb TAD on X chromosome. By serving as a scaffold, *Firre* mediates intra-chromosomal bridges to define the TADs. Thus *Firre* plays an architectural role in organizing the X chromosome in TADs that have similar expression state [170]. This lncRNA also mediates the X-chromosome tethering to nucleolar surface where the repressive state is maintained through H3K27 methylation. Obviously and interestingly enough *Firre* itself escapes X-inactivation.

6.3.6 Centromere

Centromeres are specialized structures for proper segregation and equal partitioning of chromosomes during cell division. Centromere is functionally divided into two distinct domains, the core domain which specifies kinetochore formation and its flanking pericentric heterochromatin. DNA at the pericentric heterochromatic region is rich in α-satellite repeats. Core centromeric domain is characterized by the presence of histone H3 variant CENP-A. The pericentromeric heterochromatin contains H3K9 and DNA methylation and associates with heterochromatic protein HP1. Observations suggest that lncRNA transcribed from the satellite repeats lead to heterochromatin establishment as well as proper kinetochore assembly.

Involvement of lncRNA in organization of centromere can be seen across all phyla, from plants, yeast, and invertebrates to vertebrates [194–196]. For example, maize centromeric repeats transcribed from both strands yields 900 nucleotide long transcripts that bind to CENP-A ortholog CENH3 [197]. Frog centromeric repeat (Fcr1) noncoding RNA of ~175 nucleotides is required for normal Aurora-B (kinase) localization to centromere and kinetochore formation in *Xenopus* [198].

In mouse cell lines (MS5 and C2C12), transcripts of up to 4 kb from minor satellite repeats at centromere have been detected under normal physiological condition. These transcripts accumulate during stress or differentiation. Forced accumulation or ectopic expression of the transcripts cause impaired centromeric function and chromosome segregation defects [194]. Human centromeric α-satellite repetitive DNA is transcribed by RNA polymerase II to produce noncoding RNAs of variable size containing repetitive unit 171-bp nucleotides [195, 199]. These α-satellite transcripts have functional importance in chromosome stability and centromere regulation as the RNA is essential for localization of the proteins CENP-C and INCENP to centromere. The centromeric proteins are sequestered in nucleolus during interphase and relocated to centromere during mitosis. That RNA is responsible for proper localization of these proteins is confirmed by RNaseA treatment where RNA depletion abrogates their nucleolar and centromeric localization [195]. Further human studies have found that knockdown of α-satellite induces abnormal mitosis and formation of "grape-shaped" nuclei. The α-satellite transcripts recruit CENP-A and its chaperone HJURP into centromeric chromatin. In addition, α-satellite transcripts can also interact with Shugoshin (Sgo1)—a cohesin protein chaperone that binds and protects cohesin at inner centromere. It is an essential effector for maintaining centromeric cohesion, which if lost prematurely may result in mitotic disruption [200, 201].

Although the role of centromeric proteins in kinetochore assembly and chromosome segregation is well established, the α-satellite transcribed repeat lncRNAs have now emerged as new players in this domain. As these transcripts arise from pericentromeric heterochromatin, their abnormal accumulation reflects derepression of heterochromatin. Such a scenario is indicative of disease and stress. Higher abundance of α-satellite transcripts has been reported in pancreatic and epithelial cancers. Whether it is the cause or a consequence of global heterochromatin, derepression during cancer is a matter still under investigation [202].

6.3.7 Telomeres

Apart from centromeres, telomeres are special structures at chromosome ends that are vital for its integrity and stability during cell division. Telomeres have been termed as the cellular clocks that determine the replicative lifespan of normal somatic cells. This is because cellular senescence is associated with a gradual shortening of telomere length. Telomere shortening results because of limitations of semiconservative DNA replication machinery that cannot fully replicate the end of a linear DNA. A specific ribonucleoprotein complex containing the enzyme telomerase (TERT) is required for DNA replication at the chromosome ends for telomere length homeostasis. TERT is a reverse transcriptase, which elongates telomeric DNA using associated RNA molecule as a template [203]. Majority of human cancer cells possesses active telomerase in contrast to normal somatic cells that have undetectable telomerase activity [204–206]. The RNA component of the telomerase complex is a lncRNA known as *TERC* (telomerase RNA component) that serves as the template for telomeric repeat synthesis and scaffold for assembly of associated factors. *TERC* knockout mice show short telomere, chromosomal instability, and premature aging suggesting its important role [207].

In addition to *TERC*, another novel lncRNA, *TERRA* (telomeric repeat-containing RNA), has been identified in mammals that are transcribed from sub-telomeric regions by RNA polymerase II and have variable lengths ranging between 100 and >9000 nucleotides [208, 209]. *TERRA* molecules play critical role in telomere maintenance as they regulate telomerase activity and heterochromatin formation at chromosome ends. One of the roles of *TERRA* is to recruit proteins including H3K9me3, HP1, and chromatin remodeling factors to promote heterochromatin formation at chromosome ends [210]. The other role envisaged for *TERRA* is in proper capping of the chromosome ends by binding to shelter in proteins (TRF1 and TRF2). Association of *TERRA* to telomeres is not only because of RNA–protein interaction, but recent evidences show that *TERRA* transcripts base pair with template DNA forming RNA–DNA hybrids known as R-loops that are important for telomere stability [211]. As *TERRA* participates in capping, it prevents activation of DNA damage response (DDR) at chromosome ends. In replicating cells lacking telomerase, telomere shortens with every cell division, eliciting a DDR that results in cellular senescence. *TERRA* actively prevents DDR by recruiting lysine-specific demethylase (LSD1) and chromatin remodeling factors to the telomere [212].

Disruption of nuclear organization correlates with diseased states and in some cases the lncRNA has been found to be the aberrant molecule. For example, in an autosomal dominant disease known as facioscapulohumeral muscular dystrophy (FSHD), loss of lncRNA *DBE-T* results in topological reorganization of the locus derepressing several genes [213].

6.4 lncRNA Etiology in Human Diseases and Disorders

As lncRNAs express in a precisely regulated pattern that is related to development/function, it makes sense that their mis-regulation or mutation would cause disease/disorder. Cancer, which poses a big challenge toward community health in the twenty-first century, is still unconquered. Most cancers arise due to somatic/germline mutations that result in loss of cellular homeostasis. Recent evidences suggest that most of these mutations lie in genomic regions that lack protein-coding capacity but express lncRNAs. Indeed genome-wide association studies (GWAS) and comparative transcriptomic studies have associated lncRNAs with cancer as well as several other diseases. Almost half of all traits associated with SNPs in GWAS occur in intergenic sequences and only a small fraction lie in exons [214]. Another study found that lincRNAs are more than fivefold enriched for SNPs than non-expressed intergenic regions which indicates the functional significance of these lncRNAs [215].

The growing awareness of lncRNA regulatory mechanisms and the mechanism of action of lncRNA themselves offers useful therapeutic targets. It is possible to manipulate the levels of these lncRNAs in vivo, using interventions such as treatment with antisense oligonucleotides (ASO). One such example can be seen in a study where in a murine model of Angelman syndrome, ASO-based silencing of disease causing lncRNA *Ube3a-ATS* leads to activation of Ube3a. This restoration of Ube3a activity caused recovery from cognitive deficits associated with the syndrome [216]. Such use of ASO in treatment of a diseased murine model is a step forward toward use of lncRNA-based therapies in treatment of challenging diseases like cancer. To this end, exploratory results obtained using cancer cell lines, mouse models, and nonhuman primates have been very promising. Table 6.6 shows a list of lncRNA associated with various diseases and disorders. We have excluded lncRNAs associated with cancers as they are discussed in a separate chapter in the book.

Table 6.6 lncRNAs associated with human diseases/disorders/syndromes

Disease/disorder/syndrome	lncRNA	Refs
Immune system diseases and syndromes		
Systemic lupus erythematosus	*FNDC1, TAGP, SOD2, WTAP, ACAT2*	[217]
Rheumatoid arthritis	*Multiple lincRNAs*	[218]
Kawasaki disease	*THRIL*	[42]
Thyroid disease	*SAS-ZFAT*	[219]
Sézary syndrome	*SeCATs*	[220]
Skin diseases		
Psoriasis	*PRINS (psoriasis-susceptibility-related RNA)*	[139]
Melanoma	*BANCR, SPRY4-IT1*	[221, 222]
Developmental disorders and syndromes		
FSHD (facioscapulohumeral muscular dystrophy) syndrome	*DBE-T*	[213]
Brachydactyly Type E	*DA125492*	[223]
Immunodeficiency, centromeric region instability, facial anomalies, dyskeratosis congenital, aplastic anemia, idiopathic pulmonary fibrosis	*TERRA*	[224]
Pseudohypoparathyroidism, McCune–Albright syndrome	*NESP-AS*	[34]
Transient neonatal diabetes mellitus	*HYMAI*	[225]
Klinefelter's syndrome	*XIST*	[226]

(continued)

Table 6.6 (continued)

Disease/disorder/syndrome	lncRNA	Refs
Neurodevelopmental disorders, syndromes, and neural diseases		
Fragile X syndrome	*FRM4 (FMR1-AS1), BC1*	[227, 228]
Schizophrenia	*BDNF-AS, Gomafu, DISC-2, Evf2*	[152, 229]
Prader–Willi syndrome	*SNORD116 (HBII-85), C/D box cluster, ZNF127AS*	[230, 231]
Angelman syndrome	*UBE3A-AS*	[232]
Autism spectrum disorders	*ST7OT1, ST7OT2, ST7OT3, PTCHD1AS1, PTCHD1AS2, PTCHD1AS3, SHANK2-AS, BDNF-AS, MSNP1-AS*	[157, 233, 234]
Rett's syndrome	*AK087060, AK081227*	[235]
Microphthalmia 3 syndrome	*SOX2OT*	[164]
2p15-p16.1 microdeletion syndrome	*FLJ16341*	[236]
Down syndrome	*NRON*	[237]
Alzheimer's disease	*BACE1-AS*	[238]
Beckwith–Wiedemann syndrome	*H19 and KCNQ1OT1 l*	[239, 240]
Silver–Russell syndrome	*H19*	[239]
McCune–Albright syndrome	*NESP-AS*	[34]
Neuropathic pain	*KCNA2-AS*	[154]
Cardiac diseases and disorders		
Heart failure	*Mhrt*	[105]
Cardiac hypertrophy	*CHRF, Novlnc6*	[241]
Myocardial infarction	*MIAT, LIPCAR*	[121, 123]
Spectrum of cardiac disorders	*FENDRR, Braveheart, CARL, KCNQ1OT, MALAT1*	[105]
Blood and circulatory system disorders and syndromes		
HELLP (hemolysis, elevated liver enzymes, and low platelets) syndrome	*HELLP*	[242]
Atheromatosis and atherosclerosis	*ANRIL*	[243]

6.5 Concluding Remarks

Soon after the announcement of human genome sequence, it became clear that protein-coding region constitutes only a tiny minority of the whole genome. Nonprotein-coding DNA increases with increasing evolutionary complexity, and surprisingly most of these sequences are transcribed. Investigations have revealed that many classes of lncRNAs are transcribed from ~75% of human genome that was previously regarded as nonfunctional "selfish DNA" and part of evolutionary junk-yard. Interestingly the protein toolkit of organisms has remained the same through billion years of evolution. For example, human and mice share 99% of their protein-coding genes. The phenotypic diversity appears to have been achieved primarily by modular use of a subset of the proteome. Thus spatial and temporal control of gene expression is instrumental in driving evolutionary diversity and lncRNAs have been found to play a key role in the process. This even challenges the central dogma where RNA was just thought to be a passive messenger between DNA and proteins and has brought regulatory role played by lncRNAs to center stage.

Deep transcriptomic analyses have begun to rediscover the RNA world and its relation with organismal complexity. Present evidences argue that evolutionary complexity results due to interactions of few and fairly similar proteins whose expression is spatially and temporally controlled by regulatory RNA network. The primary basis of higher complexity thus lies in the variation and expansion of this regulatory network. The regulatory network represented largely by lncRNAs is more plastic than the protein-coding sequences that are constrained by strict structure–function relationship. Any sequence variation in protein-coding region (mutations) can be toxic and thus deleterious, giving rise to severely compromised phenotype. But sequence variation in regulatory regions is often tolerated with mild consequences and no discernable phenotype. This "mutation" versus "variation" in nature provides the raw material for evolution.

Higher eukaryotes employ RNA-mediated regulatory mechanisms to control a plethora of molecular mechanisms. In the nucleus they regulate gene activity via chromatin remodeling, epigenetic processes, RNA transcription, splicing and processing, etc. In cytosol they can effectively control RNA translation, RNA stability, and signaling. They virtually are the primary control axis of differentiation, development, and diseases, and to find out the basis for complex human diseases, it is essential that all lncRNAs are identified, their expression pattern is unraveled, and mechanism of action is elucidated. A deeper transcriptomic analysis of different cells under physiologic and pathologic conditions may pave the way to understand complex human diseases and, thereby, help to improve the quality of human life.

References

1. Djebali S, Davis CA, Merkel A, Dobin A, Lassmann T, Mortazavi A, Tanzer A, Lagarde J, Lin W, Schlesinger F et al (2012) Landscape of transcription in human cells. Nature 489(7414):101–108
2. Rinn JL, Chang HY (2012) Genome regulation by long noncoding RNAs. Annu Rev Biochem 81:145–166
3. Wilusz JE, JnBaptiste CK, LY L, Kuhn CD, Joshua-Tor L, Sharp PA (2012) A triple helix stabilizes the 3′ ends of long noncoding RNAs that lack poly(A) tails. Genes Dev 26(21):2392–2407
4. Peschansky VJ, Wahlestedt C (2014) Non-coding RNAs as direct and indirect modulators of epigenetic regulation. Epigenetics 9(1):3–12
5. Mattick JS, Rinn JL (2015) Discovery and annotation of long noncoding RNAs. Nat Struct Mol Biol 22(1):5–7
6. Guil S, Esteller M (2012) Cis-acting noncoding RNAs: friends and foes. Nat Struct Mol Biol 19(11):1068–1075
7. Wilusz JE, Sunwoo H, Spector DL (2009) Long noncoding RNAs: functional surprises from the RNA world. Genes Dev 23(13):1494–1504
8. Derrien T, Johnson R, Bussotti G, Tanzer A, Djebali S, Tilgner H, Guernec G, Martin D, Merkel A, Knowles DG et al (2012) The GENCODE v7 catalog of human long noncoding RNAs: analysis of their gene structure, evolution, and expression. Genome Res 22(9):1775–1789

9. Batista PJ, Chang HY (2013) Long noncoding RNAs: cellular address codes in development and disease. Cell 152(6):1298–1307
10. Wapinski O, Chang HY (2011) Long noncoding RNAs and human disease. Trends Cell Biol 21(6):354–361
11. Du Z, Fei T, Verhaak RG, Su Z, Zhang Y, Brown M, Chen Y, Liu XS (2013) Integrative genomic analyses reveal clinically relevant long noncoding RNAs in human cancer. Nat Struct Mol Biol 20(7):908–913
12. Cabili MN, Trapnell C, Goff L, Koziol M, Tazon-Vega B, Regev A, Rinn JL (2011) Integrative annotation of human large intergenic noncoding RNAs reveals global properties and specific subclasses. Genes Dev 25(18):1915–1927
13. Hezroni H, Koppstein D, Schwartz MG, Avrutin A, Bartel DP, Ulitsky I (2015) Principles of long noncoding RNA evolution derived from direct comparison of transcriptomes in 17 species. Cell Rep 11(7):1110–1122
14. Diederichs S (2014) The four dimensions of noncoding RNA conservation. Trends Genet 30(4):121–123
15. Mortimer SA, Kidwell MA, Doudna JA (2014) Insights into RNA structure and function from genome-wide studies. Nat Rev Genet 15(7):469–479
16. Mattick JS, Taft RJ, Faulkner GJ (2010) A global view of genomic information--moving beyond the gene and the master regulator. Trends Genet 26(1):21–28
17. Surani MA, Barton SC, Norris ML (1984) Development of reconstituted mouse eggs suggests imprinting of the genome during gametogenesis. Nature 308(5959):548–550
18. McGrath J, Solter D (1984) Inability of mouse blastomere nuclei transferred to enucleated zygotes to support development in vitro. Science 226(4680):1317–1319
19. Kelsey G, Feil R (2013) New insights into establishment and maintenance of DNA methylation imprints in mammals. Philos Trans R Soc Lond Ser B Biol Sci 368(1609):20110336
20. Wan LB, Bartolomei MS (2008) Regulation of imprinting in clusters: noncoding RNAs versus insulators. Adv Genet 61:207–223
21. Kota SK, Feil R (2010) Epigenetic transitions in germ cell development and meiosis. Dev Cell 19(5):675–686
22. Smallwood SA, Kelsey G (2012) De novo DNA methylation: a germ cell perspective. Trends Genet 28(1):33–42
23. Bartolomei MS, Zemel S, Tilghman SM (1991) Parental imprinting of the mouse H19 gene. Nature 351(6322):153–155
24. Zhang Y, Tycko B (1992) Monoallelic expression of the human H19 gene. Nat Genet 1(1):40–44
25. Pandey RR, Mondal T, Mohammad F, Enroth S, Redrup L, Komorowski J, Nagano T, Mancini-Dinardo D, Kanduri C (2008) Kcnq1ot1 antisense noncoding RNA mediates lineage-specific transcriptional silencing through chromatin-level regulation. Mol Cell 32(2):232–246
26. Sleutels F, Zwart R, Barlow DP (2002) The non-coding air RNA is required for silencing autosomal imprinted genes. Nature 415(6873):810–813
27. de los Santos T, Schweizer J, Rees CA, Francke U (2000) Small evolutionarily conserved RNA, resembling C/D box small nucleolar RNA, is transcribed from PWCR1, a novel imprinted gene in the Prader-Willi deletion region, which is highly expressed in brain. Am J Hum Genet 67(5):1067–1082
28. Wevrick R, Francke U (1997) An imprinted mouse transcript homologous to the human imprinted in Prader-Willi syndrome (IPW) gene. Hum Mol Genet 6(2):325–332
29. Chamberlain SJ, Brannan CI (2001) The Prader-Willi syndrome imprinting center activates the paternally expressed murine Ube3a antisense transcript but represses paternal Ube3a. Genomics 73(3):316–322
30. Seitz H, Youngson N, Lin SP, Dalbert S, Paulsen M, Bachellerie JP, Ferguson-Smith AC, Cavaille J (2003) Imprinted microRNA genes transcribed antisense to a reciprocally imprinted retrotransposon-like gene. Nat Genet 34(3):261–262
31. Hatada I, Morita S, Obata Y, Sotomaru Y, Shimoda M, Kono T (2001) Identification of a new imprinted gene, Rian, on mouse chromosome 12 by fluorescent differential display screening. J Biochem 130(2):187–190
32. Lin SP, Youngson N, Takada S, Seitz H, Reik W, Paulsen M, Cavaille J, Ferguson-Smith AC (2003) Asymmetric regulation of imprinting on the maternal and paternal chromosomes at the Dlk1-Gtl2 imprinted cluster on mouse chromosome 12. Nat Genet 35(1):97–102

33. Liu J, Yu S, Litman D, Chen W, Weinstein LS (2000) Identification of a methylation imprint mark within the mouse Gnas locus. Mol Cell Biol 20(16):5808–5817
34. Williamson CM, Ball ST, Dawson C, Mehta S, Beechey CV, Fray M, Teboul L, Dear TN, Kelsey G, Peters J (2011) Uncoupling antisense-mediated silencing and DNA methylation in the imprinted Gnas cluster. PLoS Genet 7(3):e1001347
35. Poirier F, Chan CT, Timmons PM, Robertson EJ, Evans MJ, Rigby PW (1991) The murine H19 gene is activated during embryonic stem cell differentiation in vitro and at the time of implantation in the developing embryo. Development 113(4):1105–1114
36. Kurukuti S, Tiwari VK, Tavoosidana G, Pugacheva E, Murrell A, Zhao Z, Lobanenkov V, Reik W, Ohlsson R (2006) CTCF binding at the H19 imprinting control region mediates maternally inherited higher-order chromatin conformation to restrict enhancer access to Igf2. Proc Natl Acad Sci U S A 103(28):10684–10689
37. Lewis A, Murrell A (2004) Genomic imprinting: CTCF protects the boundaries. Curr Biol 14(7):R284–R286
38. Gabory A, Ripoche MA, Le Digarcher A, Watrin F, Ziyyat A, Forne T, Jammes H, Ainscough JF, Surani MA, Journot L et al (2009) H19 acts as a trans regulator of the imprinted gene network controlling growth in mice. Development 136(20):3413–3421
39. Monnier P, Martinet C, Pontis J, Stancheva I, Ait-Si-Ali S, Dandolo L (2013) H19 lncRNA controls gene expression of the imprinted gene network by recruiting MBD1. Proc Natl Acad Sci U S A 110(51):20693–20698
40. Kallen AN, Zhou XB, Xu J, Qiao C, Ma J, Yan L, Lu L, Liu C, Yi JS, Zhang H et al (2013) The imprinted H19 lncRNA antagonizes let-7 microRNAs. Mol Cell 52(1):101–112
41. Dey BK, Pfeifer K, Dutta A (2014) The H19 long noncoding RNA gives rise to microRNAs miR-675-3p and miR-675-5p to promote skeletal muscle differentiation and regeneration. Genes Dev 28(5):491–501
42. Giovarelli M, Bucci G, Ramos A, Bordo D, Wilusz CJ, Chen CY, Puppo M, Briata P, Gherzi R (2014) H19 long noncoding RNA controls the mRNA decay promoting function of KSRP. Proc Natl Acad Sci U S A 111(47):E5023–E5028
43. Kanduri C (2016) Long noncoding RNAs: lessons from genomic imprinting. Biochim Biophys Acta 1859(1):102–111
44. Redrup L, Branco MR, Perdeaux ER, Krueger C, Lewis A, Santos F, Nagano T, Cobb BS, Fraser P, Reik W (2009) The long noncoding RNA Kcnq1ot1 organises a lineage-specific nuclear domain for epigenetic gene silencing. Development 136(4):525–530
45. Smilinich NJ, Day CD, Fitzpatrick GV, Caldwell GM, Lossie AC, Cooper PR, Smallwood AC, Joyce JA, Schofield PN, Reik W et al (1999) A maternally methylated CpG island in KvLQT1 is associated with an antisense paternal transcript and loss of imprinting in Beckwith-Wiedemann syndrome. Proc Natl Acad Sci U S A 96(14):8064–8069
46. Mohammad F, Mondal T, Guseva N, Pandey GK, Kanduri C (2010) Kcnq1ot1 noncoding RNA mediates transcriptional gene silencing by interacting with Dnmt1. Development 137(15):2493–2499
47. Fitzpatrick GV, Soloway PD, Higgins MJ (2002) Regional loss of imprinting and growth deficiency in mice with a targeted deletion of KvDMR1. Nat Genet 32(3):426–431
48. Mancini-Dinardo D, Steele SJ, Levorse JM, Ingram RS, Tilghman SM (2006) Elongation of the Kcnq1ot1 transcript is required for genomic imprinting of neighboring genes. Genes Dev 20(10):1268–1282
49. Mohammad F, Pandey GK, Mondal T, Enroth S, Redrup L, Gyllensten U, Kanduri C (2012) Long noncoding RNA-mediated maintenance of DNA methylation and transcriptional gene silencing. Development 139(15):2792–2803
50. Terranova R, Yokobayashi S, Stadler MB, Otte AP, van Lohuizen M, Orkin SH, Peters AH (2008) Polycomb group proteins Ezh2 and Rnf2 direct genomic contraction and imprinted repression in early mouse embryos. Dev Cell 15(5):668–679
51. Umlauf D, Goto Y, Cao R, Cerqueira F, Wagschal A, Zhang Y, Feil R (2004) Imprinting along the Kcnq1 domain on mouse chromosome 7 involves repressive histone methylation and recruitment of Polycomb group complexes. Nat Genet 36(12):1296–1300

52. Barr ML, Bertram EG (1949) A morphological distinction between neurones of the male and female, and the behaviour of the nucleolar satellite during accelerated nucleoprotein synthesis. Nature 163(4148):676
53. Ohno S, Kaplan WD, Kinosita R (1959) Formation of the sex chromatin by a single X-chromosome in liver cells of Rattus norvegicus. Exp Cell Res 18:415–418
54. Lyon MF (1961) Gene action in the X-chromosome of the mouse (Mus musculus L). Nature 190:372–373
55. Borsani G, Tonlorenzi R, Simmler MC, Dandolo L, Arnaud D, Capra V, Grompe M, Pizzuti A, Muzny D, Lawrence C et al (1991) Characterization of a murine gene expressed from the inactive X chromosome. Nature 351(6324):325–329
56. Lee JT, Lu N (1999) Targeted mutagenesis of Tsix leads to nonrandom X inactivation. Cell 99(1):47–57
57. Kay GF, Barton SC, Surani MA, Rastan S (1994) Imprinting and X chromosome counting mechanisms determine Xist expression in early mouse development. Cell 77(5):639–650
58. Clerc P, Avner P (1998) Role of the region 3′ to Xist exon 6 in the counting process of X-chromosome inactivation. Nat Genet 19(3):249–253
59. Lee JT (2005) Regulation of X-chromosome counting by Tsix and Xite sequences. Science 309(5735):768–771
60. Bacher CP, Guggiari M, Brors B, Augui S, Clerc P, Avner P, Eils R, Heard E (2006) Transient colocalization of X-inactivation centres accompanies the initiation of X inactivation. Nat Cell Biol 8(3):293–299
61. Xu N, Tsai CL, Lee JT (2006) Transient homologous chromosome pairing marks the onset of X inactivation. Science 311(5764):1149–1152
62. Xu N, Donohoe ME, Silva SS, Lee JT (2007) Evidence that homologous X-chromosome pairing requires transcription and Ctcf protein. Nat Genet 39(11):1390–1396
63. Donohoe ME, Silva SS, Pinter SF, Xu N, Lee JT (2009) The pluripotency factor Oct4 interacts with Ctcf and also controls X-chromosome pairing and counting. Nature 460(7251):128–132
64. Sun BK, Deaton AM, Lee JT (2006) A transient heterochromatic state in Xist preempts X inactivation choice without RNA stabilization. Mol Cell 21(5):617–628
65. Zhao J, Sun BK, Erwin JA, Song JJ, Lee JT (2008) Polycomb proteins targeted by a short repeat RNA to the mouse X chromosome. Science 322(5902):750–756
66. Penny GD, Kay GF, Sheardown SA, Rastan S, Brockdorff N (1996) Requirement for Xist in X chromosome inactivation. Nature 379(6561):131–137
67. Morey C, Navarro P, Debrand E, Avner P, Rougeulle C, Clerc P (2004) The region 3′ to Xist mediates X chromosome counting and H3 Lys-4 dimethylation within the Xist gene. EMBO J 23(3):594–604
68. Ogawa Y, Lee JT (2003) Xite, X-inactivation intergenic transcription elements that regulate the probability of choice. Mol Cell 11(3):731–743
69. Cohen DE, Davidow LS, Erwin JA, Xu N, Warshawsky D, Lee JT (2007) The DXPas34 repeat regulates random and imprinted X inactivation. Dev Cell 12(1):57–71
70. Anguera MC, Ma W, Clift D, Namekawa S, Kelleher RJ III, Lee JT (2011) Tsx produces a long noncoding RNA and has general functions in the germline, stem cells, and brain. PLoS Genet 7(9):e1002248
71. Nora EP, Lajoie BR, Schulz EG, Giorgetti L, Okamoto I, Servant N, Piolot T, van Berkum NL, Meisig J, Sedat J et al (2012) Spatial partitioning of the regulatory landscape of the X-inactivation centre. Nature 485(7398):381–385
72. Tian D, Sun S, Lee JT (2010) The long noncoding RNA, Jpx, is a molecular switch for X chromosome inactivation. Cell 143(3):390–403
73. Sun S, Del Rosario BC, Szanto A, Ogawa Y, Jeon Y, Lee JT (2013) Jpx RNA activates Xist by evicting CTCF. Cell 153(7):1537–1551
74. Chureau C, Chantalat S, Romito A, Galvani A, Duret L, Avner P, Rougeulle C (2011) Ftx is a non-coding RNA which affects Xist expression and chromatin structure within the X-inactivation center region. Hum Mol Genet 20(4):705–718
75. McGinnis W, Krumlauf R (1992) Homeobox genes and axial patterning. Cell 68(2):283–302

76. Guttman M, Amit I, Garber M, French C, Lin MF, Feldser D, Huarte M, Zuk O, Carey BW, Cassady JP et al (2009) Chromatin signature reveals over a thousand highly conserved large non-coding RNAs in mammals. Nature 458(7235):223–227
77. Maamar H, Cabili MN, Rinn J, Raj A (2013) Linc-HOXA1 is a noncoding RNA that represses Hoxa1 transcription in cis. Genes Dev 27(11):1260–1271
78. Yin Y, Yan P, Lu J, Song G, Zhu Y, Li Z, Zhao Y, Shen B, Huang X, Zhu H et al (2015) Opposing roles for the lncRNA haunt and its genomic locus in regulating HOXA gene activation during embryonic stem cell differentiation. Cell Stem Cell 16(5):504–516
79. Zhang X, Lian Z, Padden C, Gerstein MB, Rozowsky J, Snyder M, Gingeras TR, Kapranov P, Weissman SM, Newburger PE (2009) A myelopoiesis-associated regulatory intergenic non-coding RNA transcript within the human HOXA cluster. Blood 113(11):2526–2534
80. Zhang X, Weissman SM, Newburger PE (2014) Long intergenic non-coding RNA HOTAIRM1 regulates cell cycle progression during myeloid maturation in NB4 human promyelocytic leukemia cells. RNA Biol 11(6):777–787
81. Zhao H, Zhang X, Frazao JB, Condino-Neto A, Newburger PE (2013) HOX antisense lincRNA HOXA-AS2 is an apoptosis repressor in all trans retinoic acid treated NB4 promyelocytic leukemia cells. J Cell Biochem 114(10):2375–2383
82. Hsieh-Li HM, Witte DP, Weinstein M, Branford W, Li H, Small K, Potter SS (1995) Hoxa 11 structure, extensive antisense transcription, and function in male and female fertility. Development 121(5):1373–1385
83. Sasaki YT, Sano M, Kin T, Asai K, Hirose T (2007) Coordinated expression of ncRNAs and HOX mRNAs in the human HOXA locus. Biochem Biophys Res Commun 357(3):724–730
84. Wang KC, Yang YW, Liu B, Sanyal A, Corces-Zimmerman R, Chen Y, Lajoie BR, Protacio A, Flynn RA, Gupta RA et al (2011) A long noncoding RNA maintains active chromatin to coordinate homeotic gene expression. Nature 472(7341):120–124
85. Ahn Y, Mullan HE, Krumlauf R (2014) Long-range regulation by shared retinoic acid response elements modulates dynamic expression of posterior Hoxb genes in CNS development. Dev Biol 388(1):134–144
86. De Kumar B, Parrish ME, Slaughter BD, Unruh JR, Gogol M, Seidel C, Paulson A, Li H, Gaudenz K, Peak A et al (2015) Analysis of dynamic changes in retinoid-induced transcription and epigenetic profiles of murine Hox clusters in ES cells. Genome Res 25(8):1229–1243
87. Rinn JL, Kertesz M, Wang JK, Squazzo SL, Xu X, Brugmann SA, Goodnough LH, Helms JA, Farnham PJ, Segal E et al (2007) Functional demarcation of active and silent chromatin domains in human HOX loci by noncoding RNAs. Cell 129(7):1311–1323
88. Tsai MC, Manor O, Wan Y, Mosammaparast N, Wang JK, Lan F, Shi Y, Segal E, Chang HY (2010) Long noncoding RNA as modular scaffold of histone modification complexes. Science 329(5992):689–693
89. Schorderet P, Duboule D (2011) Structural and functional differences in the long non-coding RNA hotair in mouse and human. PLoS Genet 7(5):e1002071
90. Delpretti S, Montavon T, Leleu M, Joye E, Tzika A, Milinkovitch M, Duboule D (2013) Multiple enhancers regulate Hoxd genes and the Hotdog LncRNA during cecum budding. Cell Rep 5(1):137–150
91. Yarmishyn AA, Batagov AO, Tan JZ, Sundaram GM, Sampath P, Kuznetsov VA, Kurochkin IV (2014) HOXD-AS1 is a novel lncRNA encoded in HOXD cluster and a marker of neuroblastoma progression revealed via integrative analysis of noncoding transcriptome. BMC Genomics 15(Suppl 9):S7
92. Kaneko S, Li G, Son J, CF X, Margueron R, Neubert TA, Reinberg D (2010) Phosphorylation of the PRC2 component Ezh2 is cell cycle-regulated and up-regulates its binding to ncRNA. Genes Dev 24(23):2615–2620
93. Li L, Liu B, Wapinski OL, Tsai MC, Qu K, Zhang J, Carlson JC, Lin M, Fang F, Gupta RA et al (2013) Targeted disruption of Hotair leads to homeotic transformation and gene derepression. Cell Rep 5(1):3–12
94. Ge XS, Ma HJ, Zheng XH, Ruan HL, Liao XY, Xue WQ, Chen YB, Zhang Y, Jia WH (2013) HOTAIR, a prognostic factor in esophageal squamous cell carcinoma, inhibits WIF-1 expression and activates Wnt pathway. Cancer Sci 104(12):1675–1682

95. Battistelli C, Cicchini C, Santangelo L, Tramontano A, Grassi L, Gonzalez FJ, de Nonno V, Grassi G, Amicone L, Tripodi M (2016) The snail repressor recruits EZH2 to specific genomic sites through the enrollment of the lncRNA HOTAIR in epithelial-to-mesenchymal transition. Oncogene 36(7):942–955
96. Berrondo C, Flax J, Kucherov V, Siebert A, Osinski T, Rosenberg A, Fucile C, Richheimer S, Beckham CJ (2016) Expression of the long non-coding RNA HOTAIR correlates with disease progression in bladder cancer and is contained in bladder cancer patient urinary exosomes. PLoS One 11(1):e0147236
97. Lee M, Kim HJ, Kim SW, Park SA, Chun KH, Cho NH, Song YS, Kim YT (2016) The long non-coding RNA HOTAIR increases tumour growth and invasion in cervical cancer by targeting the notch pathway. Oncotarget 7(28):44558–44571
98. Luo ZF, Zhao D, Li XQ, Cui YX, Ma N, CX L, Liu MY, Zhou Y (2016) Clinical significance of HOTAIR expression in colon cancer. World J Gastroenterol 22(22):5254–5259
99. Milevskiy MJ, Al-Ejeh F, Saunus JM, Northwood KS, Bailey PJ, Betts JA, McCart Reed AE, Nephew KP, Stone A, Gee JM et al (2016) Long-range regulators of the lncRNA HOTAIR enhance its prognostic potential in breast cancer. Hum Mol Genet 25(15):3269–3283
100. Heubach J, Monsior J, Deenen R, Niegisch G, Szarvas T, Niedworok C, Schulz WA, Hoffmann MJ (2015) The long noncoding RNA HOTAIR has tissue and cell type-dependent effects on HOX gene expression and phenotype of urothelial cancer cells. Mol Cancer 14:108
101. Zheng P, Xiong Q, Wu Y, Chen Y, Chen Z, Fleming J, Gao D, Bi L, Ge F (2015) Quantitative proteomics analysis reveals novel insights into mechanisms of action of long noncoding RNA Hox transcript antisense intergenic RNA (HOTAIR) in HeLa cells. Mol Cell Proteomics 14(6):1447–1463
102. Flynn RA, Chang HY (2014) Long noncoding RNAs in cell-fate programming and reprogramming. Cell Stem Cell 14(6):752–761
103. Kretz M, Webster DE, Flockhart RJ, Lee CS, Zehnder A, Lopez-Pajares V, Qu K, Zheng GX, Chow J, Kim GE et al (2012) Suppression of progenitor differentiation requires the long noncoding RNA ANCR. Genes Dev 26(4):338–343
104. Kretz M, Siprashvili Z, Chu C, Webster DE, Zehnder A, Qu K, Lee CS, Flockhart RJ, Groff AF, Chow J et al (2013) Control of somatic tissue differentiation by the long non-coding RNA TINCR. Nature 493(7431):231–235
105. Han P, Li W, Lin CH, Yang J, Shang C, Nurnberg ST, Jin KK, Xu W, Lin CY, Lin CJ et al (2014) A long noncoding RNA protects the heart from pathological hypertrophy. Nature 514(7520):102–106
106. Sauvageau M, Goff LA, Lodato S, Bonev B, Groff AF, Gerhardinger C, Sanchez-Gomez DB, Hacisuleyman E, Li E, Spence M et al (2013) Multiple knockout mouse models reveal lincRNAs are required for life and brain development. eLife 2:e01749
107. Atianand MK, Hu W, Satpathy AT, Shen Y, Ricci EP, Alvarez-Dominguez JR, Bhatta A, Schattgen SA, McGowan JD, Blin J et al (2016) A long noncoding RNA lincRNA-EPS acts as a transcriptional brake to restrain inflammation. Cell 165(7):1672–1685
108. Alvarez-Dominguez JR, Hu W, Yuan B, Shi J, Park SS, Gromatzky AA, van Oudenaarden A, Lodish HF (2014) Global discovery of erythroid long noncoding RNAs reveals novel regulators of red cell maturation. Blood 123(4):570–581
109. Klein U, Lia M, Crespo M, Siegel R, Shen Q, Mo T, Ambesi-Impiombato A, Califano A, Migliazza A, Bhagat G et al (2010) The DLEU2/miR-15a/16-1 cluster controls B cell proliferation and its deletion leads to chronic lymphocytic leukemia. Cancer Cell 17(1):28–40
110. Young TL, Matsuda T, Cepko CL (2005) The noncoding RNA taurine upregulated gene 1 is required for differentiation of the murine retina. Curr Biol 15(6):501–512
111. Rapicavoli NA, Poth EM, Blackshaw S (2010) The long noncoding RNA RNCR2 directs mouse retinal cell specification. BMC Dev Biol 10:49
112. Rapicavoli NA, Poth EM, Zhu H, Blackshaw S (2011) The long noncoding RNA Six3OS acts in trans to regulate retinal development by modulating Six3 activity. Neural Dev 6:32
113. Krol J, Krol I, Alvarez CP, Fiscella M, Hierlemann A, Roska B, Filipowicz W (2015) A network comprising short and long noncoding RNAs and RNA helicase controls mouse retina architecture. Nat Commun 6:7305

114. Meola N, Pizzo M, Alfano G, Surace EM, Banfi S (2012) The long noncoding RNA Vax2os1 controls the cell cycle progression of photoreceptor progenitors in the mouse retina. RNA 18(1):111–123
115. Vausort M, Wagner DR, Devaux Y (2014) Long noncoding RNAs in patients with acute myocardial infarction. Circ Res 115(7):668–677
116. Korostowski L, Sedlak N, Engel N (2012) The Kcnq1ot1 long non-coding RNA affects chromatin conformation and expression of Kcnq1, but does not regulate its imprinting in the developing heart. PLoS Genet 8(9):e1002956
117. Yap KL, Li S, Munoz-Cabello AM, Raguz S, Zeng L, Mujtaba S, Gil J, Walsh MJ, Zhou MM (2010) Molecular interplay of the noncoding RNA ANRIL and methylated histone H3 lysine 27 by polycomb CBX7 in transcriptional silencing of INK4a. Mol Cell 38(5):662–674
118. Holdt LM, Beutner F, Scholz M, Gielen S, Gabel G, Bergert H, Schuler G, Thiery J, Teupser D (2010) ANRIL expression is associated with atherosclerosis risk at chromosome 9p21. Arterioscler Thromb Vasc Biol 30(3):620–627
119. Bell RD, Long X, Lin M, Bergmann JH, Nanda V, Cowan SL, Zhou Q, Han Y, Spector DL, Zheng D et al (2014) Identification and initial functional characterization of a human vascular cell-enriched long noncoding RNA. Arterioscler Thromb Vasc Biol 34(6):1249–1259
120. Boulberdaa M, Scott E, Ballantyne M, Garcia R, Descamps B, Angelini GD, Brittan M, Hunter A, McBride M, McClure J et al (2016) A role for the long noncoding RNA SENCR in commitment and function of endothelial cells. Mol Ther 24(5):978–990
121. Kumarswamy R, Bauters C, Volkmann I, Maury F, Fetisch J, Holzmann A, Lemesle G, de Groote P, Pinet F, Thum T (2014) Circulating long noncoding RNA, LIPCAR, predicts survival in patients with heart failure. Circ Res 114(10):1569–1575
122. Wang K, Long B, Zhou LY, Liu F, Zhou QY, Liu CY, Fan YY, Li PF (2014) CARL lncRNA inhibits anoxia-induced mitochondrial fission and apoptosis in cardiomyocytes by impairing miR-539-dependent PHB2 downregulation. Nat Commun 5:3596
123. Ishii N, Ozaki K, Sato H, Mizuno H, Saito S, Takahashi A, Miyamoto Y, Ikegawa S, Kamatani N, Hori M et al (2006) Identification of a novel non-coding RNA, MIAT, that confers risk of myocardial infarction. J Hum Genet 51(12):1087–1099
124. Jiang Q, Shan K, Qun-Wang X, Zhou RM, Yang H, Liu C, Li YJ, Yao J, Li XM, Shen Y et al (2016) Long non-coding RNA-MIAT promotes neurovascular remodeling in the eye and brain. Oncotarget 7(31):49688–49698
125. Liao J, He Q, Li M, Chen Y, Liu Y, Wang J (2016) LncRNA MIAT: myocardial infarction associated and more. Gene 578(2):158–161
126. Yan B, Yao J, Liu JY, Li XM, Wang XQ, Li YJ, Tao ZF, Song YC, Chen Q, Jiang Q (2015) lncRNA-MIAT regulates microvascular dysfunction by functioning as a competing endogenous RNA. Circ Res 116(7):1143–1156
127. Klattenhoff CA, Scheuermann JC, Surface LE, Bradley RK, Fields PA, Steinhauser ML, Ding H, Butty VL, Torrey L, Haas S et al (2013) Braveheart, a long noncoding RNA required for cardiovascular lineage commitment. Cell 152(3):570–583
128. Grote P, Wittler L, Hendrix D, Koch F, Wahrisch S, Beisaw A, Macura K, Blass G, Kellis M, Werber M et al (2013) The tissue-specific lncRNA Fendrr is an essential regulator of heart and body wall development in the mouse. Dev Cell 24(2):206–214
129. Grote P, Herrmann BG (2013) The long non-coding RNA Fendrr links epigenetic control mechanisms to gene regulatory networks in mammalian embryogenesis. RNA Biol 10(10):1579–1585
130. Collier SP, Collins PL, Williams CL, Boothby MR, Aune TM (2012) Cutting edge: influence of Tmevpg1, a long intergenic noncoding RNA, on the expression of Ifng by Th1 cells. J Immunol 189(5):2084–2088
131. Li H, Hao Y, Zhang D, Fu R, Liu W, Zhang X, Xue F, Yang R (2016) Aberrant expression of long noncoding RNA TMEVPG1 in patients with primary immune thrombocytopenia. Autoimmunity 49(7):496–502
132. Wang J, Peng H, Tian J, Ma J, Tang X, Rui K, Tian X, Wang Y, Chen J, Lu L et al (2016) Upregulation of long noncoding RNA TMEVPG1 enhances T helper type 1 cell response in patients with Sjogren syndrome. Immunol Res 64(2):489–496

133. Gomez JA, Wapinski OL, Yang YW, Bureau JF, Gopinath S, Monack DM, Chang HY, Brahic M, Kirkegaard K (2013) The NeST long ncRNA controls microbial susceptibility and epigenetic activation of the interferon-gamma locus. Cell 152(4):743–754
134. Wang P, Xue Y, Han Y, Lin L, Wu C, Xu S, Jiang Z, Xu J, Liu Q, Cao X (2014) The STAT3-binding long noncoding RNA lnc-DC controls human dendritic cell differentiation. Science 344(6181):310–313
135. Archer K, Broskova Z, Bayoumi AS, Teoh JP, Davila A, Tang Y, Su H, Kim IM (2015) Long non-coding RNAs as master regulators in cardiovascular diseases. Int J Mol Sci 16(10):23651–23667
136. Sigdel KR, Cheng A, Wang Y, Duan L, Zhang Y (2015) The emerging functions of long noncoding RNA in immune cells: autoimmune diseases. J Immunol Res 2015:848790
137. Krawczyk M, Emerson BM (2014) p50-associated COX-2 extragenic RNA (PACER) activates COX-2 gene expression by occluding repressive NF-kappaB complexes. eLife 3:e01776
138. Wright PW, Huehn A, Cichocki F, Li H, Sharma N, Dang H, Lenvik TR, Woll P, Kaufman D, Miller JS et al (2013) Identification of a KIR antisense lncRNA expressed by progenitor cells. Genes Immun 14(7):427–433
139. Szell M, Danis J, Bata-Csorgo Z, Kemeny L (2016) PRINS, a primate-specific long noncoding RNA, plays a role in the keratinocyte stress response and psoriasis pathogenesis. Pflugers Arch 468(6):935–943
140. Ng SY, Johnson R, Stanton LW (2012) Human long non-coding RNAs promote pluripotency and neuronal differentiation by association with chromatin modifiers and transcription factors. EMBO J 31(3):522–533
141. Modarresi F, Faghihi MA, Patel NS, Sahagan BG, Wahlestedt C, Lopez-Toledano MA (2011) Knockdown of BACE1-AS nonprotein-coding transcript modulates beta-amyloid-related hippocampal neurogenesis. Int J Alzheimers Dis 2011:929042
142. Modarresi F, Faghihi MA, Lopez-Toledano MA, Fatemi RP, Magistri M, Brothers SP, van der Brug MP, Wahlestedt C (2012) Inhibition of natural antisense transcripts in vivo results in gene-specific transcriptional upregulation. Nat Biotechnol 30(5):453–459
143. Zhang Y, Sun L, Xuan L, Pan Z, Li K, Liu S, Huang Y, Zhao X, Huang L, Wang Z et al (2016) Reciprocal changes of circulating long non-coding RNAs ZFAS1 and CDR1AS predict acute myocardial infarction. Sci Rep 6:22384
144. Ulitsky I, Shkumatava A, Jan CH, Sive H, Bartel DP (2011) Conserved function of lincRNAs in vertebrate embryonic development despite rapid sequence evolution. Cell 147(7):1537–1550
145. Chalei V, Sansom SN, Kong L, Lee S, Montiel JF, Vance KW, Ponting CP (2014) The long non-coding RNA Dali is an epigenetic regulator of neural differentiation. eLife 3:e04530
146. Dinger ME, Amaral PP, Mercer TR, Pang KC, Bruce SJ, Gardiner BB, Askarian-Amiri ME, Ru K, Solda G, Simons C et al (2008) Long noncoding RNAs in mouse embryonic stem cell pluripotency and differentiation. Genome Res 18(9):1433–1445
147. Mercer TR, Qureshi IA, Gokhan S, Dinger ME, Li G, Mattick JS, Mehler MF (2010) Long noncoding RNAs in neuronal-glial fate specification and oligodendrocyte lineage maturation. BMC Neurosci 11:14
148. Kraus P, Sivakamasundari V, Lim SL, Xing X, Lipovich L, Lufkin T (2013) Making sense of Dlx1 antisense RNA. Dev Biol 376(2):224–235
149. Feng J, Bi C, Clark BS, Mady R, Shah P, Kohtz JD (2006) The Evf-2 noncoding RNA is transcribed from the dlx-5/6 ultraconserved region and functions as a dlx-2 transcriptional coactivator. Genes Dev 20(11):1470–1484
150. Berghoff EG, Clark MF, Chen S, Cajigas I, Leib DE, Kohtz JD (2013) Evf2 (Dlx6as) lncRNA regulates ultraconserved enhancer methylation and the differential transcriptional control of adjacent genes. Development 140(21):4407–4416
151. Airavaara M, Pletnikova O, Doyle ME, Zhang YE, Troncoso JC, Liu QR (2011) Identification of novel GDNF isoforms and cis-antisense GDNFOS gene and their regulation in human middle temporal gyrus of Alzheimer disease. J Biol Chem 286(52):45093–45102
152. Barry G, Briggs JA, Vanichkina DP, Poth EM, Beveridge NJ, Ratnu VS, Nayler SP, Nones K, Hu J, Bredy TW et al (2014) The long non-coding RNA Gomafu is acutely regulated in response to neuronal activation and involved in schizophrenia-associated alternative splicing. Mol Psychiatry 19(4):486–494

153. Takahashi S, Ohtsuki T, SY Y, Tanabe E, Yara K, Kamioka M, Matsushima E, Matsuura M, Ishikawa K, Minowa Y et al (2003) Significant linkage to chromosome 22q for exploratory eye movement dysfunction in schizophrenia. Am J Med Genet B Neuropsychiatr Genet 123B(1):27–32
154. Zhao X, Tang Z, Zhang H, Atianjoh FE, Zhao JY, Liang L, Wang W, Guan X, Kao SC, Tiwari V et al (2013) A long noncoding RNA contributes to neuropathic pain by silencing Kcna2 in primary afferent neurons. Nat Neurosci 16(8):1024–1031
155. Talkowski ME, Maussion G, Crapper L, Rosenfeld JA, Blumenthal I, Hanscom C, Chiang C, Lindgren A, Pereira S, Ruderfer D et al (2012) Disruption of a large intergenic noncoding RNA in subjects with neurodevelopmental disabilities. Am J Hum Genet 91(6):1128–1134
156. PY V, Toutain J, Cappellen D, Delrue MA, Daoud H, El Moneim AA, Barat P, Montaubin O, Bonnet F, Dai ZQ et al (2012) A homozygous balanced reciprocal translocation suggests LINC00237 as a candidate gene for MOMO (macrosomia, obesity, macrocephaly, and ocular abnormalities) syndrome. Am J Med Genet A 158A(11):2849–2856
157. Kerin T, Ramanathan A, Rivas K, Grepo N, Coetzee GA, Campbell DB (2012) A noncoding RNA antisense to moesin at 5p14.1 in autism. Sci Transl Med 4(128):128ra140
158. Bernard D, Prasanth KV, Tripathi V, Colasse S, Nakamura T, Xuan Z, Zhang MQ, Sedel F, Jourdren L, Coulpier F et al (2010) A long nuclear-retained non-coding RNA regulates synaptogenesis by modulating gene expression. EMBO J 29(18):3082–3093
159. Zhang B, Arun G, Mao YS, Lazar Z, Hung G, Bhattacharjee G, Xiao X, Booth CJ, Wu J, Zhang C et al (2012) The lncRNA Malat1 is dispensable for mouse development but its transcription plays a cis-regulatory role in the adult. Cell Rep 2(1):111–123
160. Clemson CM, Hutchinson JN, Sara SA, Ensminger AW, Fox AH, Chess A, Lawrence JB (2009) An architectural role for a nuclear noncoding RNA: NEAT1 RNA is essential for the structure of paraspeckles. Mol Cell 33(6):717–726
161. Goff LA, Groff AF, Sauvageau M, Trayes-Gibson Z, Sanchez-Gomez DB, Morse M, Martin RD, Elcavage LE, Liapis SC, Gonzalez-Celeiro M et al (2015) Spatiotemporal expression and transcriptional perturbations by long noncoding RNAs in the mouse brain. Proc Natl Acad Sci U S A 112(22):6855–6862
162. Vance KW, Sansom SN, Lee S, Chalei V, Kong L, Cooper SE, Oliver PL, Ponting CP (2014) The long non-coding RNA Paupar regulates the expression of both local and distal genes. EMBO J 33(4):296–311
163. Ramos AD, Andersen RE, Liu SJ, Nowakowski TJ, Hong SJ, Gertz CC, Salinas RD, Zarabi H, Kriegstein AR, Lim DA (2015) The long noncoding RNA Pnky regulates neuronal differentiation of embryonic and postnatal neural stem cells. Cell Stem Cell 16(4):439–447
164. Amaral PP, Neyt C, Wilkins SJ, Askarian-Amiri ME, Sunkin SM, Perkins AC, Mattick JS (2009) Complex architecture and regulated expression of the Sox2ot locus during vertebrate development. RNA 15(11):2013–2027
165. Lin N, Chang KY, Li Z, Gates K, Rana ZA, Dang J, Zhang D, Han T, Yang CS, Cunningham TJ et al (2014) An evolutionarily conserved long noncoding RNA TUNA controls pluripotency and neural lineage commitment. Mol Cell 53(6):1005–1019
166. Onoguchi M, Hirabayashi Y, Koseki H, Gotoh Y (2012) A noncoding RNA regulates the neurogenin1 gene locus during mouse neocortical development. Proc Natl Acad Sci U S A 109(42):16939–16944
167. Lu L, Sun K, Chen X, Zhao Y, Wang L, Zhou L, Sun H, Wang H (2013) Genome-wide survey by ChIP-seq reveals YY1 regulation of lincRNAs in skeletal myogenesis. EMBO J 32(19):2575–2588
168. Cesana M, Cacchiarelli D, Legnini I, Santini T, Sthandier O, Chinappi M, Tramontano A, Bozzoni I (2011) A long noncoding RNA controls muscle differentiation by functioning as a competing endogenous RNA. Cell 147(2):358–369
169. Carpenter S, Aiello D, Atianand MK, Ricci EP, Gandhi P, Hall LL, Byron M, Monks B, Henry-Bezy M, Lawrence JB et al (2013) A long noncoding RNA mediates both activation and repression of immune response genes. Science 341(6147):789–792

170. Hacisuleyman E, Goff LA, Trapnell C, Williams A, Henao-Mejia J, Sun L, McClanahan P, Hendrickson DG, Sauvageau M, Kelley DR et al (2014) Topological organization of multichromosomal regions by the long intergenic noncoding RNA Firre. Nat Struct Mol Biol 21(2):198–206
171. Yang F, Deng X, Ma W, Berletch JB, Rabaia N, Wei G, Moore JM, Filippova GN, Xu J, Liu Y et al (2015) The lncRNA Firre anchors the inactive X chromosome to the nucleolus by binding CTCF and maintains H3K27me3 methylation. Genome Biol 16:52
172. Berezney R, Coffey DS (1974) Identification of a nuclear protein matrix. Biochem Biophys Res Commun 60(4):1410–1417
173. Nickerson J (2001) Experimental observations of a nuclear matrix. J Cell Sci 114(Pt 3):463–474
174. Pathak RU, Mamillapalli A, Rangaraj N, Kumar RP, Vasanthi D, Mishra K, Mishra RK (2013) AAGAG repeat RNA is an essential component of nuclear matrix in Drosophila. RNA Biol 10(4):564–571
175. Zheng R, Shen Z, Tripathi V, Xuan Z, Freier SM, Bennett CF, Prasanth SG, Prasanth KV (2010) Polypurine-repeat-containing RNAs: a novel class of long non-coding RNA in mammalian cells. J Cell Sci 123(Pt 21):3734–3744
176. Hall LL, Carone DM, Gomez AV, Kolpa HJ, Byron M, Mehta N, Fackelmayer FO, Lawrence JB (2014) Stable C0T-1 repeat RNA is abundant and is associated with euchromatic interphase chromosomes. Cell 156(5):907–919
177. Ray MK, Wiskow O, King MJ, Ismail N, Ergun A, Wang Y, Plys AJ, Davis CP, Kathrein K, Sadreyev R et al (2016) CAT7 and cat7l long non-coding RNAs tune polycomb repressive complex 1 function during human and zebrafish development. J Biol Chem 291(37):19558–19572
178. Kapranov P, Cheng J, Dike S, Nix DA, Duttagupta R, Willingham AT, Stadler PF, Hertel J, Hackermuller J, Hofacker IL et al (2007) RNA maps reveal new RNA classes and a possible function for pervasive transcription. Science 316(5830):1484–1488
179. Venters BJ, Pugh BF (2013) Genomic organization of human transcription initiation complexes. Nature 502(7469):53–58
180. Guenther MG, Levine SS, Boyer LA, Jaenisch R, Young RA (2007) A chromatin landmark and transcription initiation at most promoters in human cells. Cell 130(1):77–88
181. Wang X, Arai S, Song X, Reichart D, Du K, Pascual G, Tempst P, Rosenfeld MG, Glass CK, Kurokawa R (2008) Induced ncRNAs allosterically modify RNA-binding proteins in cis to inhibit transcription. Nature 454(7200):126–130
182. Kim TK, Hemberg M, Gray JM, Costa AM, Bear DM, Wu J, Harmin DA, Laptewicz M, Barbara-Haley K, Kuersten S et al (2010) Widespread transcription at neuronal activity-regulated enhancers. Nature 465(7295):182–187
183. Sigova AA, Mullen AC, Molinie B, Gupta S, Orlando DA, Guenther MG, Almada AE, Lin C, Sharp PA, Giallourakis CC et al (2013) Divergent transcription of long noncoding RNA/mRNA gene pairs in embryonic stem cells. Proc Natl Acad Sci U S A 110(8):2876–2881
184. Ong CT, Corces VG (2014) CTCF: an architectural protein bridging genome topology and function. Nat Rev Genet 15(4):234–246
185. Yao H, Brick K, Evrard Y, Xiao T, Camerini-Otero RD, Felsenfeld G (2010) Mediation of CTCF transcriptional insulation by DEAD-box RNA-binding protein p68 and steroid receptor RNA activator SRA. Genes Dev 24(22):2543–2555
186. Ebersole T, Kim JH, Samoshkin A, Kouprina N, Pavlicek A, White RJ, Larionov V (2011) tRNA genes protect a reporter gene from epigenetic silencing in mouse cells. Cell Cycle 10(16):2779–2791
187. Raab JR, Chiu J, Zhu J, Katzman S, Kurukuti S, Wade PA, Haussler D, Kamakaka RT (2012) Human tRNA genes function as chromatin insulators. EMBO J 31(2):330–350
188. Lunyak VV, Prefontaine GG, Nunez E, Cramer T, BG J, Ohgi KA, Hutt K, Roy R, Garcia-Diaz A, Zhu X et al (2007) Developmentally regulated activation of a SINE B2 repeat as a domain boundary in organogenesis. Science 317(5835):248–251

189. Lee S, Kopp F, Chang TC, Sataluri A, Chen B, Sivakumar S, Yu H, Xie Y, Mendell JT (2016) Noncoding RNA NORAD regulates genomic stability by sequestering PUMILIO proteins. Cell 164(1–2):69–80
190. Misteli T (2007) Beyond the sequence: cellular organization of genome function. Cell 128(4):787–800
191. Dundr M, Misteli T (2010) Biogenesis of nuclear bodies. Cold Spring Harb Perspect Biol 2(12):a000711
192. Chen LL, Carmichael GG (2009) Altered nuclear retention of mRNAs containing inverted repeats in human embryonic stem cells: functional role of a nuclear noncoding RNA. Mol Cell 35(4):467–478
193. Dekker J, Rippe K, Dekker M, Kleckner N (2002) Capturing chromosome conformation. Science 295(5558):1306–1311
194. Bouzinba-Segard H, Guais A, Francastel C (2006) Accumulation of small murine minor satellite transcripts leads to impaired centromeric architecture and function. Proc Natl Acad Sci U S A 103(23):8709–8714
195. Wong LH, Brettingham-Moore KH, Chan L, Quach JM, Anderson MA, Northrop EL, Hannan R, Saffery R, Shaw ML, Williams E et al (2007) Centromere RNA is a key component for the assembly of nucleoproteins at the nucleolus and centromere. Genome Res 17(8):1146–1160
196. Ohkuni K, Kitagawa K (2011) Endogenous transcription at the centromere facilitates centromere activity in budding yeast. Curr Biol 21(20):1695–1703
197. Topp CN, Zhong CX, Dawe RK (2004) Centromere-encoded RNAs are integral components of the maize kinetochore. Proc Natl Acad Sci U S A 101(45):15986–15991
198. Blower MD (2016) Centromeric transcription regulates aurora-B localization and activation. Cell Rep 15(8):1624–1633
199. Black BE, Cleveland DW (2011) Epigenetic centromere propagation and the nature of CENP-a nucleosomes. Cell 144(4):471–479
200. Quenet D, Dalal Y (2014) A long non-coding RNA is required for targeting centromeric protein a to the human centromere. eLife 3:e03254
201. Liu H, Qu Q, Warrington R, Rice A, Cheng N, Yu H (2015) Mitotic transcription installs Sgo1 at centromeres to coordinate chromosome segregation. Mol Cell 59(3):426–436
202. Ting DT, Lipson D, Paul S, Brannigan BW, Akhavanfard S, Coffman EJ, Contino G, Deshpande V, Iafrate AJ, Letovsky S et al (2011) Aberrant overexpression of satellite repeats in pancreatic and other epithelial cancers. Science 331(6017):593–596
203. Greider CW, Blackburn EH (1985) Identification of a specific telomere terminal transferase activity in Tetrahymena extracts. Cell 43(2 Pt 1):405–413
204. Smogorzewska A, de Lange T (2004) Regulation of telomerase by telomeric proteins. Annu Rev Biochem 73:177–208
205. Lundblad V (1998) Telomerase catalysis: a phylogenetically conserved reverse transcriptase. Proc Natl Acad Sci U S A 95(15):8415–8416
206. Hug N, Lingner J (2006) Telomere length homeostasis. Chromosoma 115(6):413–425
207. Blasco MA, Lee HW, Hande MP, Samper E, Lansdorp PM, DePinho RA, Greider CW (1997) Telomere shortening and tumor formation by mouse cells lacking telomerase RNA. Cell 91(1):25–34
208. Azzalin CM, Reichenbach P, Khoriauli L, Giulotto E, Lingner J (2007) Telomeric repeat containing RNA and RNA surveillance factors at mammalian chromosome ends. Science 318(5851):798–801
209. Schoeftner S, Blasco MA (2008) Developmentally regulated transcription of mammalian telomeres by DNA-dependent RNA polymerase II. Nat Cell Biol 10(2):228–236
210. Arnoult N, Van Beneden A, Decottignies A (2012) Telomere length regulates TERRA levels through increased trimethylation of telomeric H3K9 and HP1alpha. Nat Struct Mol Biol 19(9):948–956
211. Flynn RL, Cox KE, Jeitany M, Wakimoto H, Bryll AR, Ganem NJ, Bersani F, Pineda JR, Suva ML, Benes CH et al (2015) Alternative lengthening of telomeres renders cancer cells hypersensitive to ATR inhibitors. Science 347(6219):273–277
212. Porro A, Feuerhahn S, Delafontaine J, Riethman H, Rougemont J, Lingner J (2014) Functional characterization of the TERRA transcriptome at damaged telomeres. Nat Commun 5:5379

213. Cabianca DS, Casa V, Bodega B, Xynos A, Ginelli E, Tanaka Y, Gabellini D (2012) A long ncRNA links copy number variation to a polycomb/trithorax epigenetic switch in FSHD muscular dystrophy. Cell 149(4):819–831
214. Hindorff LA, Sethupathy P, Junkins HA, Ramos EM, Mehta JP, Collins FS, Manolio TA (2009) Potential etiologic and functional implications of genome-wide association loci for human diseases and traits. Proc Natl Acad Sci U S A 106(23):9362–9367
215. Hangauer MJ, Vaughn IW, McManus MT (2013) Pervasive transcription of the human genome produces thousands of previously unidentified long intergenic noncoding RNAs. PLoS Genet 9(6):e1003569
216. Meng L, Ward AJ, Chun S, Bennett CF, Beaudet AL, Rigo F (2015) Towards a therapy for Angelman syndrome by targeting a long non-coding RNA. Nature 518(7539):409–412
217. Shi L, Zhang Z, AM Y, Wang W, Wei Z, Akhter E, Maurer K, Costa Reis P, Song L, Petri M et al (2014) The SLE transcriptome exhibits evidence of chronic endotoxin exposure and has widespread dysregulation of non-coding and coding RNAs. PLoS One 9(5):e93846
218. Muller N, Doring F, Klapper M, Neumann K, Schulte DM, Turk K, Schroder JO, Zeuner RA, Freitag-Wolf S, Schreiber S et al (2014) Interleukin-6 and tumour necrosis factor-alpha differentially regulate lincRNA transcripts in cells of the innate immune system in vivo in human subjects with rheumatoid arthritis. Cytokine 68(1):65–68
219. Shirasawa S, Harada H, Furugaki K, Akamizu T, Ishikawa N, Ito K, Ito K, Tamai H, Kuma K, Kubota S et al (2004) SNPs in the promoter of a B cell-specific antisense transcript, SAS-ZFAT, determine susceptibility to autoimmune thyroid disease. Hum Mol Genet 13(19):2221–2231
220. Lee CS, Ungewickell A, Bhaduri A, Qu K, Webster DE, Armstrong R, Weng WK, Aros CJ, Mah A, Chen RO et al (2012) Transcriptome sequencing in Sezary syndrome identifies Sezary cell and mycosis fungoides-associated lncRNAs and novel transcripts. Blood 120(16):3288–3297
221. Li R, Zhang L, Jia L, Duan Y, Li Y, Bao L, Sha N (2014) Long non-coding RNA BANCR promotes proliferation in malignant melanoma by regulating MAPK pathway activation. PLoS One 9(6):e100893
222. Khaitan D, Dinger ME, Mazar J, Crawford J, Smith MA, Mattick JS, Perera RJ (2011) The melanoma-upregulated long noncoding RNA SPRY4-IT1 modulates apoptosis and invasion. Cancer Res 71(11):3852–3862
223. Maass PG, Rump A, Schulz H, Stricker S, Schulze L, Platzer K, Aydin A, Tinschert S, Goldring MB, Luft FC et al (2012) A misplaced lncRNA causes brachydactyly in humans. J Clin Invest 122(11):3990–4002
224. Maicher A, Kastner L, Luke B (2012) Telomeres and disease: enter TERRA. RNA Biol 9(6):843–849
225. Temple IK, Shield JP (2002) Transient neonatal diabetes, a disorder of imprinting. J Med Genet 39(12):872–875
226. McHugh CA, Chen CK, Chow A, Surka CF, Tran C, McDonel P, Pandya-Jones A, Blanco M, Burghard C, Moradian A et al (2015) The Xist lncRNA interacts directly with SHARP to silence transcription through HDAC3. Nature 521(7551):232–236
227. Ladd PD, Smith LE, Rabaia NA, Moore JM, Georges SA, Hansen RS, Hagerman RJ, Tassone F, Tapscott SJ, Filippova GN (2007) An antisense transcript spanning the CGG repeat region of FMR1 is upregulated in premutation carriers but silenced in full mutation individuals. Hum Mol Genet 16(24):3174–3187
228. Khalil AM, Faghihi MA, Modarresi F, Brothers SP, Wahlestedt C (2008) A novel RNA transcript with antiapoptotic function is silenced in fragile X syndrome. PLoS One 3(1):e1486
229. Merelo V, Durand D, Lescallette AR, Vrana KE, Hong LE, Faghihi MA, Bellon A (2015) Associating schizophrenia, long non-coding RNAs and neurostructural dynamics. Front Mol Neurosci 8:57
230. Sahoo T, del Gaudio D, German JR, Shinawi M, Peters SU, Person RE, Garnica A, Cheung SW, Beaudet AL (2008) Prader-Willi phenotype caused by paternal deficiency for the HBII-85 C/D box small nucleolar RNA cluster. Nat Genet 40(6):719–721
231. Jong MT, Gray TA, Ji Y, Glenn CC, Saitoh S, Driscoll DJ, Nicholls RD (1999) A novel imprinted gene, encoding a RING zinc-finger protein, and overlapping antisense transcript in the Prader-Willi syndrome critical region. Hum Mol Genet 8(5):783–793

232. Runte M, Huttenhofer A, Gross S, Kiefmann M, Horsthemke B, Buiting K (2001) The IC-SNURF-SNRPN transcript serves as a host for multiple small nucleolar RNA species and as an antisense RNA for UBE3A. Hum Mol Genet 10(23):2687–2700
233. Noor A, Whibley A, Marshall CR, Gianakopoulos PJ, Piton A, Carson AR, Orlic-Milacic M, Lionel AC, Sato D, Pinto D et al (2010) Disruption at the PTCHD1 locus on Xp22.11 in autism spectrum disorder and intellectual disability. Sci Transl Med 2(49):49ra68
234. Vincent JB, Petek E, Thevarkunnel S, Kolozsvari D, Cheung J, Patel M, Scherer SW (2002) The RAY1/ST7 tumor-suppressor locus on chromosome 7q31 represents a complex multi-transcript system. Genomics 80(3):283–294
235. Petazzi P, Sandoval J, Szczesna K, Jorge OC, Roa L, Sayols S, Gomez A, Huertas D, Esteller M (2013) Dysregulation of the long non-coding RNA transcriptome in a Rett syndrome mouse model. RNA Biol 10(7):1197–1203
236. Hancarova M, Simandlova M, Drabova J, Mannik K, Kurg A, Sedlacek Z (2013) A patient with de novo 0.45 Mb deletion of 2p16.1: the role of BCL11A, PAPOLG, REL, and FLJ16341 in the 2p15-p16.1 microdeletion syndrome. Am J Med Genet A 161A(4):865–870
237. Arron JR, Winslow MM, Polleri A, Chang CP, Wu H, Gao X, Neilson JR, Chen L, Heit JJ, Kim SK et al (2006) NFAT dysregulation by increased dosage of DSCR1 and DYRK1A on chromosome 21. Nature 441(7093):595–600
238. Faghihi MA, Modarresi F, Khalil AM, Wood DE, Sahagan BG, Morgan TE, Finch CE, St Laurent G III, Kenny PJ, Wahlestedt C (2008) Expression of a noncoding RNA is elevated in Alzheimer's disease and drives rapid feed-forward regulation of beta-secretase. Nat Med 14(7):723–730
239. Eggermann T (2009) Silver-Russell and Beckwith-Wiedemann syndromes: opposite (epi) mutations in 11p15 result in opposite clinical pictures. Horm Res 71(Suppl 2):30–35
240. Wevrick R, Kerns JA, Francke U (1994) Identification of a novel paternally expressed gene in the Prader-Willi syndrome region. Hum Mol Genet 3(10):1877–1882
241. Ounzain S, Micheletti R, Beckmann T, Schroen B, Alexanian M, Pezzuto I, Crippa S, Nemir M, Sarre A, Johnson R et al (2015) Genome-wide profiling of the cardiac transcriptome after myocardial infarction identifies novel heart-specific long non-coding RNAs. Eur Heart J 36(6):353–368a
242. van Dijk M, Visser A, Buabeng KM, Poutsma A, van der Schors RC, Oudejans CB (2015) Mutations within the LINC-HELLP non-coding RNA differentially bind ribosomal and RNA splicing complexes and negatively affect trophoblast differentiation. Hum Mol Genet 24(19):5475–5485
243. Burd CE, Jeck WR, Liu Y, Sanoff HK, Wang Z, Sharpless NE (2010) Expression of linear and novel circular forms of an INK4/ARF-associated non-coding RNA correlates with atherosclerosis risk. PLoS Genet 6(12):e1001233

Chapter 7
Long Noncoding RNAs in Cancer and Therapeutic Potential

Arun Renganathan and Emanuela Felley-Bosco

Abstract Long noncoding RNAs (lncRNAs) are the major elements of the mammalian transcriptome that is emerging as a central player controlling diverse cellular mechanisms. Most of the well-studied lncRNAs so far are found to be crucial in regulating cellular processes such as cell cycle, growth, and apoptosis that ensure homeostasis. Owing to their location and distribution in the genome, lncRNAs influence the transcription of a wide range of proteins directly or indirectly by transcriptional and posttranscriptional alterations, which opens up the "LncRNA-cancer paradigm" in a context-dependent manner, i.e., either oncogenic or tumor suppressive. Thus, this chapter is a consolidation of lncRNA association in exhibiting or suppressing the typical cancer hallmarks such as continuous proliferation, surpassing apoptosis, genomic instability, drug resistance, invasion, and metastasis studied till date. In addition, special focus has been given on the efficient application of lncRNAs as potential targets for therapeutics that holds a great promise for future cancer therapy.

Keywords Cancer • Natural antisense transcripts • Antisense oligonucleotides • Chemoresistance • Epigenetic regulation • Therapeutic target • Gene regulation • ceRNA

A. Renganathan
Molecular Biology and Genetics Unit, Jawaharlal Nehru Centre for Advanced Scientific Research, Jakkur, Bangalore, India

E. Felley-Bosco (✉)
Laboratory of Molecular Oncology, Division of Thoracic Surgery, University Hospital Zürich, Zürich, Switzerland
e-mail: emanuela.felley-bosco@usz.ch

M.R.S. Rao (ed.), *Long Non Coding RNA Biology*, Advances in Experimental Medicine and Biology 1008, DOI 10.1007/978-981-10-5203-3_7

7.1 Introduction

Advanced sequencing methods and arrays identified that the human genome is composed of only 2% of genes that encodes for proteins while the major portion of the genome is transcribed without any immediate definite purpose [1]. These noncoding RNAs (ncRNAs) have been classified based on their size into small ncRNAs—those smaller than 200 nucleotides such as small interfering RNA (siRNA), microRNA (miRNA), piwi-interacting RNA (piRNA)—and long ncRNAs, those larger than 200 nucleotides [2, 3]. Long noncoding RNAs (lncRNAs) constitute the major class of ncRNAs as the mammalian transcriptome has revealed that there are nearly three times as many lncRNA genes as protein-coding genes [4, 5], and there has been a steep rise of research focus on the lncRNAs recently owing to their influence in several biological processes [6, 7].

Though a large portion is yet to be explored, the lncRNAs identified to date are classified as stand-alone lncRNAs (located in sequence space without overlapping coding genes), natural antisense transcripts (transcription from the antisense DNA strand of annotated transcription units), pseudogenes (replica of genes that have lost their coding capacity due to mutations), long intronic ncRNAs (long transcripts within introns), and divergent transcripts/promoter-associated transcripts/enhancer RNAs (transcriptional by-products that maintain the environment of open chromatin, or enhance or regulate biological processes) [8]. Gene annotations of lncRNAs show poorly defined boundaries, such as the absence of typical transcription initiation and termination hallmarks that distinguished them from coding regions [9–11].

Owing to their loci in the genome, lncRNAs are expected to play crucial roles in regulating the gene expression during differentiation and development that positively or negatively regulate or maintain cellular homeostasis [12, 13]. Though we are yet to completely understand the importance of all the lncRNAs known so far, most of them are identified to be associated with the transcription and post-transcription of coding regions, especially as competent RNAs regulating miRNA levels and mRNA stability and translation [14]. Since the identification of HOX antisense intergenic RNA (*HOTAIR*)—the lncRNA that interacts with the chromatin and represses transcription of human HOX genes regulating development [15]—extensive studies revealed that some lncRNAs regulated chromatin organization, and their deregulation heavily contributed to several pathological conditions especially cancer [16, 17]. Several lncRNAs have been identified to be involved in cell cycle alterations, evasion of apoptosis, and metastasis causing various cancers including lung, liver, prostate, breast, and ovarian cancers [18–20]. Searching Pubmed (https://www.ncbi.nlm.nih.gov/pubmed) using "lncRNA and cancer" as keywords revealed that the number of published studies has increased dramatically in the last 6 years (Fig. 7.1).

The analysis of lncRNAs as novel drivers of tumorigenesis holds a strong platform for the development of anticancer treatments targeting RNA molecules.

Hence, this chapter focuses on the functional perspective of possible lncRNA targets and consolidates their association in different genetic and cell signaling pathways that contribute toward cancer development and progression.

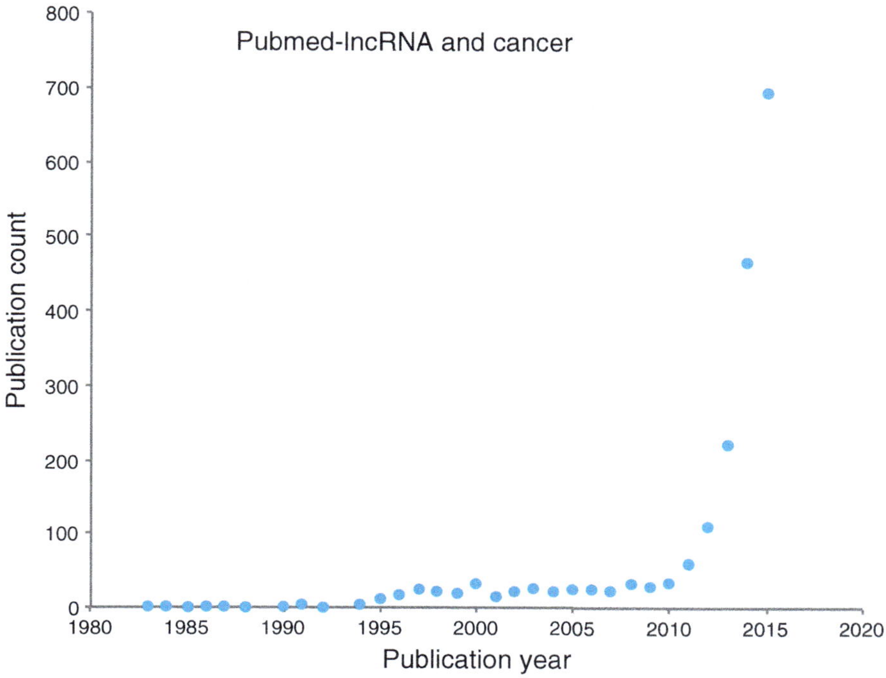

Fig. 7.1 Publication counts in Pubmed using "lncRNA and Cancer" as keywords

7.2 Long Noncoding RNA-Mediated Regulatory Mechanisms

LncRNAs operate through different modes of action (Table 7.1). Transcription from an upstream promoter of lncRNA can negatively or positively affect expression of the downstream gene by inhibiting RNA polymerase II recruitment or inducing chromatin remodeling, respectively. For example, lncRNAs from SINEs (a class of retrotransposon) block transcription of heat-shock genes by binding to Pol II to prevent formation of preinitiation complexes [21, 22].

Histone modifications and DNA methylation are essential for stable repression of genes and have been associated in cancer with deregulation of lncRNA expression. LncRNAs can modulate gene transcription through epigenetic modulation by guiding chromatin-modifying complexes to target genomic DNA loci. For example, lncRNAs such as *HOTAIR* and X-inactive specific transcript (*Xist*)/*RepA* associate with the chromatin-modifying factors EZH2 in PRC2 (a key methyl transferase) and the Pc/Cbx family proteins in PRC1 (proteins that bind to trimethylated H3K27) [8, 23]. *Xist* interacts with EZH2 and SUZ12 of PRC2 complex via repeat A region (*RepA*) for the epigenetic repression of specific genes on the X-chromosome, and *HOTAIR* recruits PRC2 complex at a different locus (HOXD) to suppress specific genes [15, 24].

Table 7.1 Long noncoding RNA-mediated regulatory mechanisms

LNC RNA	Function	Mechanisms
Gene expression control		
Regulation of mRNA transcription		
All	Increased or decreased expression of nearby genes	Modification of PolII recruitment
XIST	X inactivation	Chromatin-mediated repression
HOTAIR	Repression at the HOXD locus	Chromatin-mediated repression
HOTTIP	Activation at the HOXA locus	Chromatin-mediated activation
ANRIL	Repression at the INK4bARF-INK4a locus	Chromatin-mediated repression
AIRN	Imprinting at the IGF2R cluster	Chromatin-mediated repression, transcription interference
GAS5	Repression of glucocorticoid receptor-mediated transcription	DNA mimicry
NRON	Repression of NFAT-mediated transcription	Inhibition of transcription factor nucleocytoplasmic shuttling
Modulation of mRNA posttranscriptional regulatory pathways		
HULC	Downregulation of miRNA-mediated repression	Sequestration of miRNA
PTENP1 pseudogene	Upregulation of PTEN	Sequestration of miRNA
Regulation of mRNA processing		
MALAT1	Ser/Arg splicing factor regulation	Scaffolding of subnuclear domains
Regulation of proteins		
Regulation of protein activity		
GAS5	Repression of glucocorticoid receptor-mediated transcription	DNA mimicry
SAMMSON	Interaction with p32	Regulation of mitochondrial biogenesis
NRON	Repression of NFAT-mediated transcription	Inhibition of transcription factor nucleocytoplasmic shuttling
Organization of protein complexes		
HOTAIR	Repression at the HOXD locus	Recruitment of PRC2
ANRIL	Repression at the INK4b–ARF–INK4a locus	Recruitment of PRC1 and PRC2
TERC	Addition of telomeric repeats to the ends of chromosomes	Organizational scaffold for telomerase components and template for repeat addition
NEAT1	Assembly of paraspeckles	Nucleation of subnuclear domains

An antisense transcript is able to hybridize to the overlapping sense transcript and block recognition of the splice sites by the spliceosome, thus resulting in an alternatively spliced transcript. In this context, the lncRNA metastasis-associated lung adenocarcinoma transcript 1 (*MALAT1*) and *Gomafu/MIAT* have been associated with alternate splicing and hindering the spliceosome formation, respectively

[25, 26]. Antisense transcripts also affect the alternative polyadenylation site selection that influences the mRNA stability [27]. Certain lncRNAs such as natural antisense from the 3′-UTR of inducible nitric oxide synthase have been found to stabilize its sense counterpart by aiding in the recruitment of stabilizing factors [28]. Alternatively, hybridization of the sense and antisense transcripts can allow Dicer to generate endogenous siRNAs.

By binding to specific protein partners, a noncoding transcript can modulate the activity of the protein, serve as a structural component that allows a larger RNA-protein complex to form, or alter where the protein localizes in the cell. A recent example is SAMMSON, a lncRNA expressed in cutaneous melanoma, which, by binding a protein called p32, allows its localization in mitochondria and normal mitochondria biogenesis. Depletion of SAMMSON leads to stress associated with accumulation of mitochondrial peptide precursors and mitochondrial import defects and, consequently, p53-independent apoptosis [29].

Apart from chromatin remodeling, lncRNAs exhibit classic transcriptional regulation by acting as decoys. Decoys for transcription factors may be like lncRNA promoter of CDKN1A antisense DNA damage activated RNA (*PANDA*) sequestering NF-YA away from its pro-apoptotic target genes or lncRNA growth arrest-specific 5 (*Gas5*) *RNA* competing for binding with transcription factors. Alternatively, by binding to protein, lncRNA can influence cellular localization as it is the case of lncRNA-noncoding repressor of NFAT (*NRON*) lncRNA that prevents NFAT transfer into the nucleus hindering its interaction with the import in family of nuclear transport proteins [8].

Finally, lncRNAs can be processed to yield small RNAs, such as miRNAs, piRNAs, and other less well-characterized classes of small transcripts.

7.3 Long Noncoding RNAs: A Cancer Association

Though the knowledge of identification, function, and deregulation of lncRNAs is still in its infancy, studies have documented the oncogenic and tumor suppressive functions of lncRNAs in several common cancer types. In this section, we document the involvement of lncRNAs at different levels in transcription and translation in cancer-associated alterations.

7.3.1 Transcriptional Variations

Dysregulated lncRNAs have been largely associated with disease progression due to their influence on genes nearby (*cis*) or distant (*trans-regulation*) in addition to acting as enhancers [30]. The most widely studied ncRNA *H19* is one of the best examples for lncRNA involved in transcriptional regulation as aberrant *H19* expression, observed in several solid tumors such as hepatocellular carcinoma, shows direct activation of c-MYC-driven downregulation of P53, positively influencing cell proliferation and clonogenicity [31, 32]. The lncRNA colorectal cancer associated transcript 1-long isoform (CCAT-1 L, also known as CARLo-5), transcribed from an upstream super enhancer locus of *Myc*, functions as an enhancer RNA regulating Myc transcription [33, 34].

Some lncRNAs can associate with transcription factors (TF), and the dysregulatory lncRNA-TF-gene triplets have been associated with tumorigenesis [35]. For example, recent studies on glioblastoma-associated lncRNAs have shown involvement of specific lncRNA-TF-gene triplets such as *HOTAIR*-MXI1-CD58/PRKCE and *HOTAIR*-ATF5-NCAM1 in enhancing their target gene expression and contributing to glioblastoma prognosis [35]. LncRNAs HOXA transcript located at the 5′ end of the HOXA cluster (*HOTTIP*) and HOX antisense intergenic RNA myeloid 1 (*HOTAIRM1*) are other important lncRNAs known to be associated with myeloma and leukemia. Enhanced *HOTTIP* and *HOTAIRM1* resulted in upregulation of myeloid differentiation and activation of HOXA genes associated with MLL-gene rearrangements causing leukemia [36, 37].

LncRNAs have also been shown to transactivate the steroid receptors. For example, SRA (steroid receptor RNA activator) ncRNA has been observed to alter estrogen/progesterone receptor status contributing to breast carcinogenesis [38].

PANDA, which as mentioned above represses pro-apoptotic genes, showed p53-dependent induction after DNA damage [39]. Alu-mediated p21 transcriptional regulator (APTR) has been identified to repress the transcription of CDK inhibitor p21 by recruiting PRC2 complex to the p21 promoter [40].

7.3.2 Posttranscriptional Modifications

Apart from transcriptional regulation, lncRNAs are also widely implicated in the posttranscriptional regulation of mRNAs such as splicing, transport, translation, and degradation.

As mentioned above, MALAT1 is known to be involved in splicing events and functions in the regulation of alternate splicing and modulates the activity of spliceosome components contributing to tumorigenesis; it is, for example, necessary for correct splicing of B-Myb, a transcription factor involved in G2/M transition [41].

Natural antisense transcripts (NAT) may regulate splicing of overlapping sense transcripts through base pairing such as the alternative processing of Zeb2/Sip1, a transcriptional repressor of E-cadherin. The natural antisense prevents the splicing of pre-mRNA IRES, leading to increase in the levels of Zeb2 protein and decrease in E-cadherin mRNA and protein in different cancer types [42].

Some lncRNAs are involved in stabilizing and promoting the translation of mRNAs by extended base pairing with them. 5′aHIF-1α and 3′aHIF-1α are two antisense transcripts involved in different regulatory mechanisms. The 3′aHIF-1α, which lacks a 5′cap and a poly (A+) tail, is implicated in increasing instability of

HIF-1α mRNA, while 5′ aHIF-1α, which has a 5′ cap and a poly(A+) tail, has complex and diverse functions [43]. Both transcripts serve as a marker of poor prognosis of breast cancer [44]. LncRNA Staufen1-mediated mRNA decay is another critical way of decreasing the stability of mRNAs [45].

7.3.3 Decoy Elements and Competitive Endogenous RNA

Some lncRNAs function as competitive endogenous RNA and decoys by binding miRNA sponges and reducing their inhibitory effect on their natural targets [46], and lncRNA sponges are identified to be involved in cancer progression. For example, the lncRNAs, highly upregulated in liver cancer (HULC) and papillary thyroid carcinoma susceptibility candidate 3 (PTCSC3), are examples of miRNA-lncRNA competitive interactions involved in hepatocellular carcinoma and thyroid cancer, respectively. The HULC transcript contains miR-372-binding sites, and its overexpression reduced miR-372 expression and activity in the liver cancer cell line Hep3B [47] and acts as a "molecular decoy." In thyroid cancers, PTCSC3 is downregulated, and PTCSC3 overexpression leads to a decrease in miR-574-5p, causing growth inhibition, cell-cycle arrest, and increased apoptosis [48]. Similarly, urothelial cancer associated 1 (UCA1) upregulates a potent oncogene ERBB4 by binding miR-193-3p in lung cancer [49]. In gastric cancer, maternally expressed 3 (MEG3) upregulates Bcl-2 by sequestering miR-181-a [50].

LncRNAs may follow other strategies to act as decoys, and some can modify the phosphorylation of proteins masking thereby other motifs, like long intergenic noncoding RNA for kinase activation (LINK-A) and HIF1a [51] where LINK-A-induced phosphorylation of HIF1a masks the hydroxylation site responsible for HIF1a degradation, thereby leading to activation of HIF1α transcriptional programs under normoxic conditions. In addition, some lncRNAs such as noncoding RNA activated by DNA damage (NORAD) act as sponges for a whole set of proteins, such as the PUMILIO family that would drive chromosomal instability by repressing mitotic, DNA repair, and DNA replication factors at alternate condition [52].

7.4 Long Noncoding RNAs: Involvement in Cancer Hallmarks

In the past, cancer drivers have been searched within protein-coding genes residing in recurrent alterations in cancer genomes. However, the lack of protein coding genes in cancer-associated genetic alterations and the fact that only 2% of the human genome is translated into proteins together with the finding that about 70% of the human genome is transcribed into RNA are strong arguments supporting a role for noncoding RNAs in tumor development.

Although only Xist and Malat1, described in details below, are the two lncRNAs for which a clear genetic link with tumorigenesis has been established [53, 54], in this section we summarize experimental evidence indicating how some specific lncRNAs are involved in the different hallmarks of cancer. According to Lnc2 Cancer (http://www.bio-bigdata.com/lnc2cancer/), which is a manually curated database [55] that provides comprehensive experimentally supported associations between lncRNA and human cancer, there are 579 human lncRNAs linked to cancer. Therefore we decided to select lncRNA for which involvement was supported by functional studies using RNAi and *in vitro* short hairpin-mediated knockdown. For RNAi, we selected genes where at least three independent studies had been reported, and we will illustrate based on our own experience why such stringency is necessary. Since the database was last updated Jan 26, 2016, we then added lncRNAs selected with the same criteria and published after this date. We illustrate the involvement of these different lncRNAs in the different hallmarks of cancer [56, 57]. Indeed cancer is a complex disease rising from altered intracellular regulatory networks and intercellular communication. Intracellular signaling networks are modulated in cancer to sustain proliferation, impair cytostatic and differentiation signals, enhance viability, and promote motility. As pointed out in some recent reviews [58–60], most of the hallmarks of cancer, as described by Hanahan and Weinberg in 2000, are modulated by the activity of multiple lncRNAs (Fig. 7.2).

7.4.1 Sustaining Proliferative Signaling

Through one or the other mechanisms previously mentioned, several lncRNAs sustain proliferative signaling. For example, multiple lines of evidence are now implicating lncRNAs in Myc-driven cancers. Amplification of the 8q24 locus is a well-characterized oncogenic event in many types of human malignancies resulting in *MYC* amplification. In a mouse model of Myc oncogenesis, single-copy amplification of *Myc* alone was insufficient to enhance tumor formation, whereas amplification of a multi-gene segment encompassing *Myc* and the lncRNA plasmacytoma variant translocation 1 (*Pvt1*) promoted efficient tumor development [61]. Co-amplification of *PVT1* and *MYC* increased Myc protein levels, while depletion of *PVT1* in Myc-driven human colon cancer cells impaired proliferation.

Another example is the prostate-specific lncRNA prostate cancer gene expression marker 1 (PCGEM1), also residing at the 8q24 locus, which binds to Myc and enhances Myc's transcriptional activation of several genes involved in various metabolic processes required for growth of prostate cancer cells [62], while in gastric carcinoma myc mRNA is stabilized by gastric carcinoma high expressed transcript 1 (GHET1) [63].

Epigenetic modifications can promote cancer growth. For example, lncRNA HOTTIP interacts with the WDR5/MLL complex, which enhances histone H3 lysine 4 trimethylation to activate the expression of multiple 5′ *HOXA* genes [36]

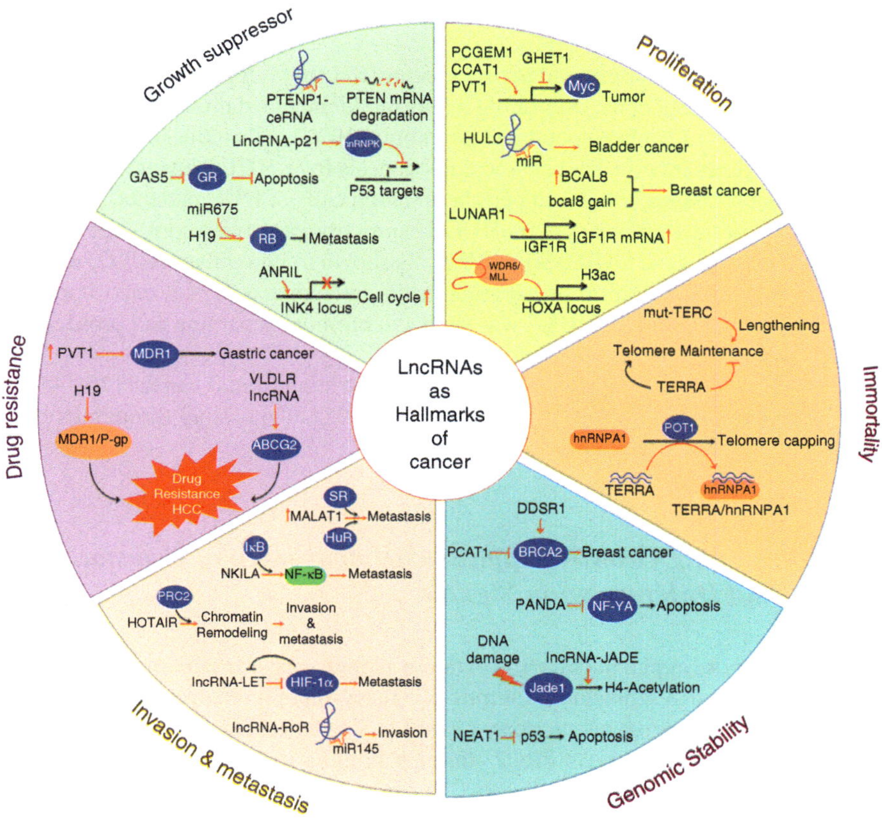

Fig. 7.2 Cancer hallmarks and associated lncRNAs

and increases HOXA13 promoting growth, gemcitabine resistance, and metastasis in pancreatic cancer [64].

Alterations in DNA elements that regulate the 3D organization of chromatin have been suggested to result in aberrant promoter-enhancer interactions that drive cancer development. NOTCH1-orchestrated events bring together the lncRNA leukemia-induced noncoding activator RNA (LUNAR1) and the enhancer of *IGF1R*, which encodes a receptor for insulin-like growth factor that is essential for the survival of T-ALL cells. Silencing of LUNAR1 leads to inhibition of the growth of T-ALL cells due to the loss of IGF1R expression and loss of trophic signals [65].

Within the 19 lncRNAs identified as having cancer-associated genomic alterations and correlated with patient survival, Yan et al. 2015 found that silencing breast cancer-associated lncRNA8 (BCAL8) significantly reduced the proliferation of breast cancer cells *in vitro* and *in vivo*. Decreased survival was significantly associated with higher expression of *BCAL8* RNA and genomic gain of the *BCAL8* gene.

Higher expression of BCAL8 RNA was also significantly correlated with poor clinical outcome in other cancers [66].

HULC lncRNA is conserved across species [67], and it localizes within the ribosomes in liver cancer cells, suggesting it may modulate translational activity. Indeed, silencing of HULC in hepatocellular carcinoma (HCC) cells induced global mRNA changes in genes involved in hepatocarcinogenesis [67]. HULC levels are increased in liver cancer tissues but also in peripheral blood cells of HCC patients, suggesting modulation of the immune system. HULC expression positively correlates with that of hepatitis virus gene x (HBVx). HBVx was shown to upregulate HULC, which in turn promotes proliferation of hepatoma cells by suppressing the oncosuppressor p18 [68]. In other studies, HULC was shown to promote hepatoma cell proliferation by modulation of lipid metabolism [69].

Linc00152 promotes cell proliferation in gastric and renal cancer [70]. UCA1 promotes the growth *in vivo* of bladder cancer cells [71] possibly through its property of acting as sponge for different miR [72].

7.4.2 Evading Growth Suppressors/Impairment of Cytostatic and Differentiation Signals

Several lncRNAs impair the function of tumor suppressors. For example, lncRNAH19 serves as a microRNA precursor for miR-675 that promotes oncogenesis by targeting Rb [73, 74]. The oncogenic property of H19 is also attributed to its full-length processed transcript that targets PRC2 (through binding to EZH2, the histone lysine methyltransferase component of PRC2) to genes that promote cancer metastasis [75].

INK4b-ARF-INK4a encodes three tumor suppressor proteins, p15(INK4b), p14(ARF), and p16(INK4a), and its transcription is a key requirement for replicative or oncogene-induced senescence and constitutes an important barrier for tumor growth. The antisense noncoding RNA in the INK4 locus (ANRIL) gene is transcribed in the antisense orientation of the INK4b-ARF-INK4a gene cluster and induces silencing of p15INK4b through heterochromatin formation, and elevated ANRIL expression is associated with low p15INK4b expression in leukemic cells [76].

Some lncRNAs regulate expression of tumor suppressors by influencing various parts of transcription and translation. lncRNA-p21, which is induced by DNA damage in a p53-dependent manner, interacts with hnRNPK, a transcriptional repressor, to regulate p21 (CDKN1A) and arrest the cell cycle in a p21-dependent manner [77, 78]. Interestingly, transcriptional activation of p21 tumor suppressor depends on the posttranscriptional silencing of a p21-specific antisense transcript [79] which allows epigenetically silencing of the sense transcript.

MEG3 binds to and activates p53-dependent transcription of a subset of p53-regulated genes [80].

Poliseno and colleagues proposed a regulatory role for lncRNAs in the binding of miRNAs and revealed a mechanism that has significant implications in cancer biology [81] after observing that the lncRNAPTENP1 (phosphatase and tensin homolog (PTEN) pseudogene) acts as a molecular sponge for miRNAs that target PTEN mRNA for degradation. They called this class of lncRNAs competitive endogenous RNAs (ceRNAs). The PTENP1-sequestered miRNAs also target other tumor suppressor genes, including CDKN1A, and silencing PTENP1 downregulates CDKN1A and increased proliferation in PTEN-null cells. Consistent with the role of PTENP1 to act as a sponge, deletion of PTENP1 was associated with decreased PTEN expression levels in a cohort of colon tumor samples.

Prostate cancer gene 3 (PCA3) is an antisense intronic lncRNA that downregulates an as yet unrecognized tumor suppressor gene, a human homolog of the *Drosophila prune* gene, *PRUNE2*, through a process that involves RNA editing mediated by a supramolecular complex containing adenosine deaminase acting on RNA family members [82].

The selective advantage of tumor cells is driven also by tolerance of nutrient stress and, in some cancers, preservation of an undifferentiated tumor cell population. LncRNA Gas5 is induced in cells arrested by nutrient deprivation or withdrawal of growth factors. A specific form of mature Gas5 transcript blocks glucocorticoid-responsive gene expression by binding to the DNA binding domain of the glucocorticoid receptor (GR) and acting as a decoy [83, 84]. This blockade of GR decreases expression of, e.g., the cellular inhibitor of apoptosis [84], thereby enhancing apoptosis under stressed conditions. However, we [85] and others (reviewed in [86]) have observed that cells can express various transcripts at the same time, not all of them being able to decoy GR. Thereby the effect on GR may depend on a specific and equilibrate balance of splicing. The presence of different variants makes it also difficult to interpret some published data based on quantitative polymerase chain reaction detecting some but not all variants or functional studies using gene silencing targeting regions not conserved in all transcripts.

Several other lncRNAs have been implicated in apoptosis such as SPRY4 intron transcript 1 (SPRY4-IT1) [87]. BC200, also called brain cytoplasmic RNA 1 (BCYRN1), is transcriptionally induced by estrogen in breast cancer cells, and it prevents apoptosis by modulating alternative splicing of a member of the Bcl-2 family, Bcl-x [88].

7.4.3 *Enabling Replicative Immortality*

In contrast to normal cells that are able to pass through only a limited number of cell division cycles, tumor cells show nearly unlimited replication. Replication potential is limited by the length of chromosome ends, the telomeres. Tumor cells have found two ways to circumvent the loss of telomeres: about 90% of all human cancers express a specialized enzyme, called telomerase, which is able to add telomeric repeats to the end of the chromosomes.

The telomerase consists of a protein component, a reverse transcriptase named telomerase reverse transcriptase (TERT), and an RNA primer, also known as telomerase RNA component (TERC). Single nucleotide polymorphisms at the TERC locus are associated with telomere lengthening and an increased risk of developing high-grade glioma [89]. Furthermore, TERC copy-number gain strongly predicts the progression of premalignant oral cavity neoplasms to invasive cancer [90].

The telomeric repeat containing RNA (TERRA) transcribed from subtelomeric and telomeric DNA sequences exerts both telomerase-dependent and telomerase-independent effects on telomere maintenance [91]. One role for TERRA involves its dynamic regulation during the cell cycle, which regulates the exchange of single-strand DNA binding protein replication protein A by protection of telomere 1 (POT-1) and, thus, telomere capping [92].

7.4.4 Maintenance of Genomic Stability

Recent reports have also identified a role for lncRNAs in the maintenance of genome stability. Several lncRNAs have been observed as essential for DNA repair by homologous recombination: prostate cancer-associated transcript 1 (PCAT1) and DNA damage-sensitive RNA1 (DDSR1). PCAT1 posttranslationally inhibits BRCA2 [93], while DDSR1 is suggested to interact with BRCA1 [94].

lncRNA-JADE is induced after DNA damage in an ataxia-telangiectasia mutated-dependent manner. LncRNA-JADE transcriptionally activates Jade1, a key component in the human acetylase binding to ORC1 histone acetylation complex. Consequently, lncRNA-JADE induces histone H4 acetylation in the DNA damage response. Markedly higher levels of lncRNA-JADE have been observed in human breast tumors in comparison with normal breast tissues, and knockdown of lncRNA-JADE inhibits breast tumor growth *in vivo* [95].

Genetic evidence proved that nuclear enriched abundant transcript 1 (NEAT1) is engaged in a negative feedback loop with p53 and thereby modulates cancer formation in mice by dampening oncogene-dependent activation of p53. Silencing Neat1 expression in mice, which prevents paraspeckle formation, sensitized preneoplastic cells to DNA-damage-induced cell death and impaired skin tumorigenesis. Consistent with this finding, NEAT1 targeting sensitized established human cancer cells to both chemotherapy and p53 reactivation therapy [96].

PANDA, which is one of the lncRNAs induced from the p21 promoter after DNA damage, inhibits DNA-damage-induced apoptosis by binding to the transcription factor NF-YA and blocking its recruitment to pro-apoptotic genes [39].

7.4.5 Invasion and Metastasis

Most cancer deaths are caused by metastasis rather than the primary tumors. Several lncRNAs have been associated with invasion and metastasis. For example, overexpression of MALAT1, an evolutionarily conserved, abundant nuclear lncRNA, was found to predict a high risk of metastatic progression in patients with early-stage non-small cell lung cancer [97]. While MALAT1 loss of function in mouse revealed that it is a nonessential gene in development or for adult normal tissue homeostasis [98, 99], depletion of MALAT1 in lung carcinoma cells impairs cellular motility *in vitro* and metastasis in mice [100], suggesting that MALAT1 overexpression in cancer may drive gain-of-function phenotypes not observed during normal tissue development or homeostasis, not only mediated by regulation of alternative splicing as mentioned above but also possibly through interaction with HuR [101], a member of the ELAVL family, which has been reported to contribute to the stabilization of ARE-containing mRNAs.

LncRNAs also mediate metastasis programs through chromatin deregulation. Overexpression of the HOTAIR in breast cancer enforces a mesenchymal cellular phenotype which promotes breast cancer metastasis by reprogramming the chromatin landscape genome-wide via recruitment of PRC2, a histone H3 lysine 27 methylase involved in developmental gene silencing and cancer progression [102].

The lncRNA second chromosome locus associated with prostate-1 (SChLAP1), associated with poor prognosis and metastatic prostate cancer progression [103], promotes prostate cancer invasion and metastasis by disrupting the metastasis-suppressing activity of the SWI/SNF complex [104].

Recent identification of metastasis-suppressing lncRNAs has opened a new perspective on a link between the tumor microenvironment and lncRNA modulation of the metastasis phenotype.

The lncRNA NF-KappaB interacting lncRNA (NKILA), which is induced by nuclear factor kB (NFkB) in response to inflammatory signaling, mediates a negative feedback loop suppressing NF-kB signaling by binding the cytoplasmic NF-kB/IkB complex and preventing IkB phosphorylation, NF-kB release, and nuclear localization [51].NKILA suppression in human breast cancer is linked to metastatic dissemination and poor prognosis.

The lncRNA low expression in tumor (LET) connects the hypoxia response to metastasis. Hypoxia-induced histone deacetylase 3 suppresses the LET promoter, decreasing LET expression and facilitating NF90 accumulation since binding of LET to NF90 drives its degradation. This event contributes to hypoxia-induced cellular invasion [105].

Multiple lncRNAs increase invasiveness of cancer cells and facilitate metastasis. Examples of these are lncRNA regulator of reprogramming (RoR) in breast cancer [106] and lncRNA activated by transforming growth factor-β (ATB) in HCC. lncRNA-RoR likely serves as a "sponge" for miR-145 that is important for regulation of ADP-ribosylation factor 6, a protein involved in invasion of breast cancer cells [106]. Transforming growth factor-β was found to induce the expression of lncRNA-ATB in HCC cells, which facilitated epithelial to mesenchymal transition (EMT), cellular invasion, and organ colonization by HCC cells [107]. lncRNA-ATB competitively binds miR-200 to activate the expression of ZEB1 and ZEB2 during EMT, while interactions with interleukin-11 mRNA enhance Stat3 signaling to promote metastasis.

Profiling EGF-induced changes in expression of lncRNAs in mammary epithelial cells, which mirrors gene expression patterns in breast cancer patients, led to the identification of a subset of 11 EGF-regulated lncRNAs, the expression patterns of which could be used to predict survival time of breast cancer patients. *In vitro* studies of the selected lncRNAs identified LncRNA inhibiting metastasis (LINC01089/LIMT), an EGF-downregulated lncRNA, as a regulator of mammary cell migration and invasion. Correspondingly, animal studies have shown that depletion of LIMT enhances metastasis formation *in vivo*. Downregulation of LIMT characterizes breast cancer patients diagnosed with either basal-like or HER2-enriched tumors [108].

RNA pulldown after UV crosslinking, coupled to RNA immunoprecipitation experiments, demonstrated the direct and specific interaction between HOTAIR and Snail and validated *in vivo* the binding between HOTAIR and EZH2. The interaction among Snail, HOTAIR, and EZH2 on epithelial genes was found instrumental for the execution of the EMT [109].

7.4.6 Drug Resistance

The increase in the ability of cancer cells to repair DNA damage helps them to bypass the cytotoxicity of chemotherapeutics; therefore any of the lncRNA involved in genomic stability mentioned above can modulate drug sensitivity.

However, there are other mechanisms of chemoresistance that can be modulated by lncRNA (reviewed in [110]).

Enhanced drug efflux caused by overexpression of ATP-binding cassette (ABC) transporters is one of the main mechanisms of drug resistance. The members of ABC transporters regulate the absorption, distribution, and clearance of xenobiotics. In hepatocellular carcinoma, upregulation of H19 and VLDLR lncRNAs induces the expression of MDR1/P-glycoprotein (P-gp) and ABCG2, respectively [111, 112]. In gastric cancers, overexpression of PVT1 caused an elevation in the expression levels of MDR1 [113].

7.5 lncRNAs as Therapeutic Targets

LncRNA biology has already suggested many promising therapeutic targets in cancer. In this section, we will review several strategies to inhibit the function of oncogenic lncRNA or modulate their epigenetic effect. We also mention how specific expression pattern of lncRNA expression can be exploited for therapeutic purposes. Although for the time being there is only one trial involving lncRNAs (www.clinicaltrials.gov, NCT02641847) and they are not targeted for therapy but are investigated for validation of signature in breast cancer, there are no doubts that this will change in the near future.

As mentioned in previous sections, several lncRNAs work as natural antisense transcripts (NATs) to genes of therapeutic interest. NATs are RNA transcripts which overlap a sense protein-coding gene and often act to regulate the associated loci through the recruitment of histone-modifying complexes and induction of transcriptional silencing. Targeting these NATs with antisense oligonucleotides, also known as antagoNATs [114], results in loss of this epigenetic silencing and consequently transcription activation of the sense gene.

To select the best methodology to use for inhibiting lncRNA, a prior knowledge of lncRNA cellular localization is critical to achieve robust lncRNA modulation [115]. Indeed, while several strategies have been successfully employed to deplete lncRNAs, small interfering RNAs, upon loading in the RISC complex in the cytoplasm, can efficiently deplete cytoplasmic lncRNAs but have variable success in targeting predominantly nuclear lncRNAs. Antisense oligonucleotides (ASOs, described in details below) can, on the other hand, robustly deplete the transcripts regardless of their cellular localization.

Another important information is provided by genetic engineering of mice deficient in a given lncRNA. As indicated in a previous section, MALAT-1 knockout mice show a minimal phenotype, indicating that toxicity resulting from disruption of MALAT-1 would be unlikely.

A number of nucleic acid chemical modifications have been developed in order to impart drug-like properties on nucleic acids used to target lncRNA by increasing the target binding affinity, reduce clearance, increase nuclease stability, and improve pharmacokinetic properties [116, 117]. For example, 2′-O-(2-methoxy) ethyl oligonucleotides increase the life span of the oligonucleotide in the complex milieu of nucleases within the cell and reduce degradation products, which may also have effects on the cells [118].

7.5.1 Antisense Oligonucleotides (ASOs)

ASOs are short DNA sequences complementary to RNA of interest. Typically, one of two ASO designs is used: gapmers and mixmers. Gapmers are short oligonucleotide molecules that consist of RNA-based flanking sequences and an internal DNA "gap" region. Binding to a complementary RNA target results in the formation of an RNA-DNA heteroduplex, which is a substrate for cleavage by the enzyme RNase [119]. RNase H is primarily located in the nucleus, and so gapmer technology is ideal for targeting noncoding RNA transcripts with nuclear functions.

Conversely, mixmers consist of alternating nucleic acid chemistries such as locked nucleic acid (LNA) in combination with other types of monomers, typically DNA but also RNA or 2′-OMe-RNA monomers. LNA comprises a class of RNA analogues in which the furanose ring of the ribose sugar is chemically locked in an RNA-mimicking conformation by the introduction of a O2',C4'-methylene linkage providing higher thermal stability. Mixmers are designed to sterically block association of the target transcript with other nucleic acids or ribonucleoproteins [120]. Mixmers are designed so that they do not contain strings of consecutive DNA nucleotides and are therefore not RNaseH competent. Mixmers could be used to block direct association between a pseudogene transcript and its cognate parental mRNA or to inhibit the binding of proteins, such as epigenetic remodeling complexes.

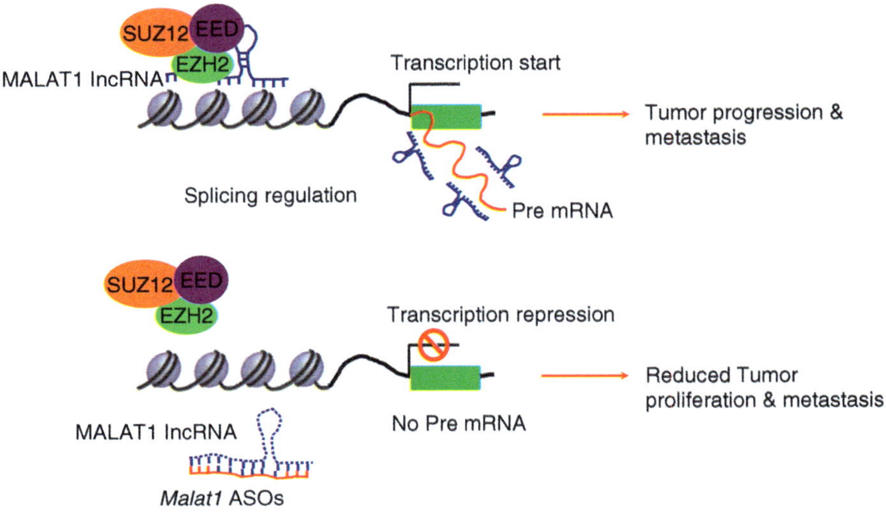

Fig. 7.3 Schematic representation of ASOs mediated knockdown of Malat1 in MMTV-PyMT breast cancer model

Though ASOs have yet to be proven as an anticancer therapy, MALAT-1 ASOs have shown efficacy in a preclinical mouse MMTV-PyMT breast cancer model where Malat1 drives tumor growth and metastasis. In this model, ASOs targeting Malat1 have demonstrated therapeutic efficacy *in vivo* by promoting cystic differentiation, increased cell adhesion, and decreased migration [53] (Fig. 7.3).

Activation of gene expression by the use of LNA targeted specific PRC2-lncRNA interactions resulted in upregulation of the target genes [121].

7.5.2 Aptamers

Many lncRNAs most likely form complex secondary structures, which could limit the access of the lncRNA to oligonucleotide targeting. The use of aptamers could provide a solution to this problem. Aptamers are structured oligonucleotides that are developed by *in vitro*- or *in vivo*-directed evolution and bind to protein or nucleic acid targets with high affinity and specificity. RNA aptamers can be generated against novel targets *in vitro* by using a combinatorial chemistry method termed SELEX (systematic evolution of ligands by exponential enrichment). That these aptamers can have therapeutic applications has been demonstrated by the expression of the short, stem-loop HIV-derived TARncRNA in CD4+ T cells, where it bound and repressed the viral protein Tat and thereby inhibited HIV replication [122]. SELEX allows the incorporation of modified nucleotides so that RNA aptamers with high nuclease resistance can be generated, which are therefore suitable for animal and clinical studies.

7.5.3 Exploiting lncRNA Expression Patterns

Another strategy takes advantage of the restricted expression of lncRNAs by using lncRNA regulatory elements. BC-819 is a plasmid containing the diphtheria toxin gene under H19 promoter control that has shown promising results as a cytoreduction agent in bladder, ovarian, and pancreatic tumors [123].

Cancer cells lacking the SWI/SNF component ATRX maintain persistent TERRA loci at telomeres as cells transition from S phase to G2. This results in persistent RPA occupancy on the single-stranded telomeric DNA preventing telomerase-dependent telomere lengthening. These cells therefore rely on the recombination-dependent alternative pathway of telomere lengthening which requires ATR, rendering ATRX-deficient cancer cells highly sensitive to ATR inhibitors [124].

7.6 Conclusions

Although a large amount of data is indicating the involvement of many lncRNA in cancer, this will not immediately translate for most of them in clinical implication for treatment. Indeed for most lncRNAs, we have yet to understand the functions better. Very informative functional studies rely on animal models. Modeling lncRNA function in mice is difficult. Indeed, lncRNAs are conserved at much lower rates than protein-coding genes; therefore many human lncRNAs have not been identified in mice. H19-, Malat-1-, and Neat-1-deficient mice show normal phenotype [125]. Deletion of Xist in hematopoietic progenitors causes hematologic cancer [54]. Differing results were observed in Hotair knockout mice. Deletion of Hoxc locus, which includes Hotair, in mouse was reported to have little effect *in vivo* [126]. However, mild but reproducible homeotic phenotypes were observed in two more targeted Hotair knockout mice [127, 128]. These variances demonstrate the difficulty and importance of designing "clean" lncRNA mouse models. These difficulties could be potentially addressed by developing novel transgenic mouse models, wherein larger human genome portions, comprising whole chromosomes, are added to or exchange portions of the mouse genome [129].

In order to prioritize which lncRNA might be the most relevant in a given cancer type, it has been suggested that using the TCGA lncRNome information as a clinical filter, one would be able to generate a concentrated and clinically relevant lncRNA list that could be used for a candidate-oriented functional screening.

There is still much work in perspective; nevertheless based on current knowledge it is possible to predict that a better understanding of lncRNA in cancer has the potential to open a new way of intervention, possibly at the level of so-called cancer progenitor genes [130].

References

1. Djebali S, Davis CA, Merkel A, Dobin A, Lassmann T, Mortazavi A, Tanzer A, Lagarde J, Lin W, Schlesinger F et al (2012) Landscape of transcription in human cells. Nature 489:101–108
2. Consortium EP, Birney E, Stamatoyannopoulos JA, Dutta A, Guigo R, Gingeras TR, Margulies EH, Weng Z, Snyder M, Dermitzakis ET et al (2007) Identification and analysis of functional elements in 1% of the human genome by the ENCODE pilot project. Nature 447:799–816
3. Han Li C, Chen Y (2015) Small and long non-coding RNAs: novel targets in perspective cancer therapy. Curr Genomics 16:319–326
4. Carninci P, Hayashizaki Y (2007) Noncoding RNA transcription beyond annotated genes. Curr Opin Genet Dev 17:139–144
5. Iyer MK, Niknafs YS, Malik R, Singhal U, Sahu A, Hosono Y, Barrette TR, Prensner JR, Evans JR, Zhao S et al (2015) The landscape of long noncoding RNAs in the human transcriptome. Nat Genet 47:199–208
6. Kapranov P, Cheng J, Dike S, Nix DA, Duttagupta R, Willingham AT, Stadler PF, Hertel J, Hackermuller J, Hofacker IL et al (2007) RNA maps reveal new RNA classes and a possible function for pervasive transcription. Science 316:1484–1488
7. Ulitsky I, Bartel DP (2013) lincRNAs: genomics, evolution, and mechanisms. Cell 154:26–46
8. Kung JT, Colognori D, Lee JT (2013) Long noncoding RNAs: past, present, and future. Genetics 193:651–669
9. Jia H, Osak M, Bogu GK, Stanton LW, Johnson R, Lipovich L (2010) Genome-wide computational identification and manual annotation of human long noncoding RNA genes. RNA 16:1478–1487
10. Pauli A, Valen E, Lin MF, Garber M, Vastenhouw NL, Levin JZ, Fan L, Sandelin A, Rinn JL, Regev A et al (2012) Systematic identification of long noncoding RNAs expressed during zebrafish embryogenesis. Genome Res 22:577–591
11. Sun L, Zhang Z, Bailey TL, Perkins AC, Tallack MR, Xu Z, Liu H (2012) Prediction of novel long non-coding RNAs based on RNA-Seq data of mouse Klf1 knockout study. BMC Bioinformatics 13:331
12. Huarte M, Rinn JL (2010) Large non-coding RNAs: missing links in cancer? Hum Mol Genet 19:R152–R161
13. Taft RJ, Pang KC, Mercer TR, Dinger M, Mattick JS (2010) Non-coding RNAs: regulators of disease. J Pathol 220:126–139
14. Kornienko AE, Guenzl PM, Barlow DP, Pauler FM (2013) Gene regulation by the act of long non-coding RNA transcription. BMC Biol 11:1–14
15. Rinn JL, Kertesz M, Wang JK, Squazzo SL, Xu X, Brugmann SA, Goodnough LH, Helms JA, Farnham PJ, Segal E et al (2007) Functional demarcation of active and silent chromatin domains in human HOX Loci by noncoding RNAs. Cell 129:1311–1323
16. Prensner JR, Iyer MK, Balbin OA, Dhanasekaran SM, Cao Q, Brenner JC, Laxman B, Asangani IA, Grasso CS, Kominsky HD et al (2011) Transcriptome sequencing across a prostate cancer cohort identifies PCAT-1, an unannotated lincRNA implicated in disease progression. Nat Biotechnol 29:742–749
17. Tsai MC, Spitale RC, Chang HY (2011) Long intergenic noncoding RNAs: new links in cancer progression. Cancer Res 71:3–7
18. Parasramka MA, Maji S, Matsuda A, Yan IK, Patel T (2016) Long non-coding RNAs as novel targets for therapy in hepatocellular carcinoma. Pharmacol Ther 161:67–78
19. Xie H, Ma H, Zhou D (2013) Plasma HULC as a promising novel biomarker for the detection of hepatocellular carcinoma. Biomed Res Int 2013:5
20. Zhang L, Yang F, J-h Y, S-x Y, W-p Z, Huo X-s XD, H-s B, Wang F, S-h S (2013) Epigenetic activation of the MiR-200 family contributes to H19-mediated metastasis suppression in hepatocellular carcinoma. Carcinogenesis 34:577–586

21. Mariner PD, Walters RD, Espinoza CA, Drullinger LF, Wagner SD, Kugel JF, Goodrich JA (2008) Human Alu RNA is a modular transacting repressor of mRNA transcription during heat shock. Mol Cell 29:499–509
22. Yakovchuk P, Goodrich JA, Kugel JF (2009) B2 RNA and Alu RNA repress transcription by disrupting contacts between RNA polymerase II and promoter DNA within assembled complexes. Proc Natl Acad Sci U S A 106:5569–5574
23. Schwartz YB, Pirrotta V (2007) Polycomb silencing mechanisms and the management of genomic programmes. Nat Rev Genet 8:9–22
24. Zhao J, Sun BK, Erwin JA, Song JJ, Lee JT (2008) Polycomb proteins targeted by a short repeat RNA to the mouse X chromosome. Science 322:750–756
25. Sone M, Hayashi T, Tarui H, Agata K, Takeichi M, Nakagawa S (2007) The mRNA-like noncoding RNA Gomafu constitutes a novel nuclear domain in a subset of neurons. J Cell Sci 120:2498–2506
26. Tsuiji H, Yoshimoto R, Hasegawa Y, Furuno M, Yoshida M, Nakagawa S (2011) Competition between a noncoding exon and introns: Gomafu contains tandem UACUAAC repeats and associates with splicing factor-1. Genes Cells 16:479–490
27. Gu R, Zhang Z, DeCerbo JN, Carmichael GG (2009) Gene regulation by sense-antisense overlap of polyadenylation signals. RNA 15:1154–1163
28. Matsui K, Nishizawa M, Ozaki T, Kimura T, Hashimoto I, Yamada M, Kaibori M, Kamiyama Y, Ito S, Okumura T (2008) Natural antisense transcript stabilizes inducible nitric oxide synthase messenger RNA in rat hepatocytes. Hepatology (Baltimore, Md) 47:686–697
29. Leucci E, Vendramin R, Spinazzi M, Laurette P, Fiers M, Wouters J, Radaelli E, Eyckerman S, Leonelli C, Vanderheyden K et al (2016) Melanoma addiction to the long non-coding RNA SAMMSON. Nature 531:518–522
30. Prensner JR, Chinnaiyan AM (2011) The emergence of lncRNAs in cancer biology. Cancer Discov 1:391–407
31. Barsyte-Lovejoy D, Lau SK, Boutros PC, Khosravi F, Jurisica I, Andrulis IL, Tsao MS, Penn LZ (2006) The c-Myc oncogene directly induces the H19 noncoding RNA by allele-specific binding to potentiate tumorigenesis. Cancer Res 66:5330–5337
32. Matouk IJ, DeGroot N, Mezan S, Ayesh S, Abu-lail R, Hochberg A, Galun E (2007) The H19 non-coding RNA is essential for human tumor growth. PLoS One 2:e845
33. Kim T, Cui R, Jeon YJ, Lee JH, Lee JH, Sim H, Park JK, Fadda P, Tili E, Nakanishi H et al (2014) Long-range interaction and correlation between MYC enhancer and oncogenic long noncoding RNA CARLo-5. Proc Natl Acad Sci U S A 111:4173–4178
34. Xiang JF, Yin QF, Chen T, Zhang Y, Zhang XO, Wu Z, Zhang S, Wang HB, Ge J, Lu X et al (2014) Human colorectal cancer-specific CCAT1-L lncRNA regulates long-range chromatin interactions at the MYC locus. Cell Res 24:513–531
35. Li Y, Wang Z, Wang Y, Zhao Z, Zhang J, Lu J, Xu J, Li X (2016) Identification and characterization of lncRNA mediated transcriptional dysregulation dictates lncRNA roles in glioblastoma. Oncotarget 7(29):45027–45041
36. Wang KC, Yang YW, Liu B, Sanyal A, Corces-Zimmerman R, Chen Y, Lajoie BR, Protacio A, Flynn RA, Gupta RA et al (2011) A long noncoding RNA maintains active chromatin to coordinate homeotic gene expression. Nature 472:120–124
37. Zhang X, Lian Z, Padden C, Gerstein MB, Rozowsky J, Snyder M, Gingeras TR, Kapranov P, Weissman SM, Newburger PE (2009) A myelopoiesis-associated regulatory intergenic noncoding RNA transcript within the human HOXA cluster. Blood 113:2526–2534
38. Cooper C, Guo J, Yan Y, Chooniedass-Kothari S, Hube F, Hamedani MK, Murphy LC, Myal Y, Leygue E (2009) Increasing the relative expression of endogenous non-coding steroid receptor RNA activator (SRA) in human breast cancer cells using modified oligonucleotides. Nucleic Acids Res 37:4518–4531
39. Hung T, Wang Y, Lin MF, Koegel AK, Kotake Y, Grant GD, Horlings HM, Shah N, Umbricht C, Wang P et al (2011) Extensive and coordinated transcription of noncoding RNAs within cell-cycle promoters. Nat Genet 43:621–629

40. Negishi M, Wongpalee SP, Sarkar S, Park J, Lee KY, Shibata Y, Reon BJ, Abounader R, Suzuki Y, Sugano S et al (2014) A new lncRNA, APTR, associates with and represses the CDKN1A/p21 promoter by recruiting polycomb proteins. PLoS One 9:e95216
41. Tripathi V, Shen Z, Chakraborty A, Giri S, Freier SM, Wu X, Zhang Y, Gorospe M, Prasanth SG, Lal A et al (2013) Long noncoding RNA MALAT1 controls cell cycle progression by regulating the expression of oncogenic transcription factor B-MYB. PLoS Genet 9:e1003368
42. Beltran M, Puig I, Pena C, Garcia JM, Alvarez AB, Pena R, Bonilla F, de Herreros AG (2008) A natural antisense transcript regulates Zeb2/Sip1 gene expression during Snail1-induced epithelial-mesenchymal transition. Genes Dev 22:756–769
43. Rossignol F, Vache C, Clottes E (2002) Natural antisense transcripts of hypoxia-inducible factor 1alpha are detected in different normal and tumour human tissues. Gene 299:135–140
44. Cayre A, Rossignol F, Clottes E, Penault-Llorca F (2003) aHIF but not HIF-1alpha transcript is a poor prognostic marker in human breast cancer. Breast Cancer Res 5:R223–R230
45. Kim YK, Furic L, Parisien M, Major F, DesGroseillers L, Maquat LE (2007) Staufen1 regulates diverse classes of mammalian transcripts. EMBO J 26:2670–2681
46. Tay Y, Rinn J, Pandolfi PP (2014) The multilayered complexity of ceRNA crosstalk and competition. Nature 505:344–352
47. Wang J, Liu X, Wu H, Ni P, Gu Z, Qiao Y, Chen N, Sun F, Fan Q (2010) CREB up-regulates long non-coding RNA, HULC expression through interaction with microRNA-372 in liver cancer. Nucleic Acids Res 38:5366–5383
48. Fan M, Li X, Jiang W, Huang Y, Li J, Wang Z (2013) A long non-coding RNA, PTCSC3, as a tumor suppressor and a target of miRNAs in thyroid cancer cells. Exp Ther Med 5:1143–1146
49. Nie W, Ge HJ, Yang XQ, Sun X, Huang H, Tao X, Chen WS, Li B (2016) LncRNA-UCA1 exerts oncogenic functions in non-small cell lung cancer by targeting miR-193a-3p. Cancer Lett 371:99–106
50. Peng W, Si S, Zhang Q, Li C, Zhao F, Wang F, Yu J, Ma R (2015) Long non-coding RNA MEG3 functions as a competing endogenous RNA to regulate gastric cancer progression. J Exp Clin Cancer Res 34:79
51. Liu B, Sun L, Liu Q, Gong C, Yao Y, Lv X, Lin L, Yao H, Su F, Li D et al (2015) A cytoplasmic NF-kappaB interacting long noncoding RNA blocks IkappaB phosphorylation and suppresses breast cancer metastasis. Cancer Cell 27:370–381
52. Lee S, Kopp F, Chang TC, Sataluri A, Chen B, Sivakumar S, Yu H, Xie Y, Mendell JT (2016) Noncoding RNA NORAD regulates genomic stability by sequestering PUMILIO proteins. Cell 164:69–80
53. Arun G, Diermeier S, Akerman M, Chang KC, Wilkinson JE, Hearn S, Kim Y, MacLeod AR, Krainer AR, Norton L et al (2016) Differentiation of mammary tumors and reduction in metastasis upon Malat1 lncRNA loss. Genes Dev 30:34–51
54. Yildirim E, Kirby JE, Brown DE, Mercier FE, Sadreyev RI, Scadden DT, Lee JT (2013) Xist RNA is a potent suppressor of hematologic cancer in mice. Cell 152:727–742
55. Ning S, Zhang J, Wang P, Zhi H, Wang J, Liu Y, Gao Y, Guo M, Yue M, Wang L et al (2016) Lnc2Cancer: a manually curated database of experimentally supported lncRNAs associated with various human cancers. Nucleic Acids Res 44:D980–D985
56. Hanahan D, Weinberg RA (2000) The hallmarks of cancer. Cell 100:57–70
57. Hanahan D, Weinberg RA (2011) Hallmarks of cancer: the next generation. Cell 144:646–674
58. Bartonicek N, Maag JL, Dinger ME (2016) Long noncoding RNAs in cancer: mechanisms of action and technological advancements. Mol Cancer 15:43
59. Gutschner T, Diederichs S (2012) The hallmarks of cancer: a long non-coding RNA point of view. RNA Biol 9:703–719
60. Schmitt AM, Chang HY (2016) Long noncoding RNAs in cancer pathways. Cancer Cell 29:452–463
61. Tseng YY, Moriarity BS, Gong W, Akiyama R, Tiwari A, Kawakami H, Ronning P, Reuland B, Guenther K, Beadnell TC et al (2014) PVT1 dependence in cancer with MYC copy-number increase. Nature 512:82–86
62. Hung CL, Wang LY, Yu YL, Chen HW, Srivastava S, Petrovics G, Kung HJ (2014) A long noncoding RNA connects c-Myc to tumor metabolism. Proc Natl Acad Sci U S A 111:18697–18702

63. Yang F, Xue X, Zheng L, Bi J, Zhou Y, Zhi K, Gu Y, Fang G (2014) Long non-coding RNA GHET1 promotes gastric carcinoma cell proliferation by increasing c-Myc mRNA stability. FEBS J 281:802–813
64. Li Z, Zhao X, Zhou Y, Liu Y, Zhou Q, Ye H, Wang Y, Zeng J, Song Y, Gao W et al (2015) The long non-coding RNA HOTTIP promotes progression and gemcitabine resistance by regulating HOXA13 in pancreatic cancer. J Transl Med 13:84
65. Trimarchi T, Bilal E, Ntziachristos P, Fabbri G, Dalla-Favera R, Tsirigos A, Aifantis I (2014) Genome-wide mapping and characterization of notch-regulated long noncoding RNAs in acute leukemia. Cell 158:593–606
66. Yan X, Hu Z, Feng Y, Hu X, Yuan J, Zhao SD, Zhang Y, Yang L, Shan W, He Q et al (2015) Comprehensive genomic characterization of long non-coding RNAs across human cancers. Cancer Cell 28:529–540
67. Panzitt K, Tschernatsch MM, Guelly C, Moustafa T, Stradner M, Strohmaier HM, Buck CR, Denk H, Schroeder R, Trauner M et al (2007) Characterization of HULC, a novel gene with striking up-regulation in hepatocellular carcinoma, as noncoding RNA. Gastroenterology 132:330–342
68. Du Y, Kong G, You X, Zhang S, Zhang T, Gao Y, Ye L, Zhang X (2012) Elevation of highly up-regulated in liver cancer (HULC) by hepatitis B virus X protein promotes hepatoma cell proliferation via down-regulating p18. J Biol Chem 287:26302–26311
69. Cui M, Xiao Z, Wang Y, Zheng M, Song T, Cai X, Sun B, Ye L, Zhang X (2015) Long noncoding RNA HULC modulates abnormal lipid metabolism in hepatoma cells through an miR-9-mediated RXRA signaling pathway. Cancer Res 75:846–857
70. Wu Y, Tan C, Weng WW, Deng Y, Zhang QY, Yang XQ, Gan HL, Wang T, Zhang PP, Xu MD et al (2016) Long non-coding RNA Linc00152 is a positive prognostic factor for and demonstrates malignant biological behavior in clear cell renal cell carcinoma. Am J Cancer Res 6:285–299
71. Wang F, Li X, Xie X, Zhao L, Chen W (2008) UCA1, a non-protein-coding RNA up-regulated in bladder carcinoma and embryo, influencing cell growth and promoting invasion. FEBS Lett 582:1919–1927
72. Bian Z, Jin L, Zhang J, Yin Y, Quan C, Hu Y, Feng Y, Liu H, Fei B, Mao Y et al (2016) LncRNA-UCA1 enhances cell proliferation and 5-fluorouracil resistance in colorectal cancer by inhibiting miR-204-5p. Sci Rep 6:23892
73. Tsang WP, Ng EK, Ng SS, Jin H, Yu J, Sung JJ, Kwok TT (2010) Oncofetal H19-derived miR-675 regulates tumor suppressor RB in human colorectal cancer. Carcinogenesis 31:350–358
74. Wang G, Lunardi A, Zhang J, Chen Z, Ala U, Webster KA, Tay Y, Gonzalez-Billalabeitia E, Egia A, Shaffer DR et al (2013) Zbtb7a suppresses prostate cancer through repression of a Sox9-dependent pathway for cellular senescence bypass and tumor invasion. Nat Genet 45:739–746
75. Luo M, Li Z, Wang W, Zeng Y, Liu Z, Qiu J (2013) Long non-coding RNA H19 increases bladder cancer metastasis by associating with EZH2 and inhibiting E-cadherin expression. Cancer Lett 333:213–221
76. Yu W, Gius D, Onyango P, Muldoon-Jacobs K, Karp J, Feinberg AP, Cui H (2008) Epigenetic silencing of tumour suppressor gene p15 by its antisense RNA. Nature 451:202–206
77. Dimitrova N, Zamudio JR, Jong RM, Soukup D, Resnick R, Sarma K, Ward AJ, Raj A, Lee JT, Sharp PA et al (2014) LincRNA-p21 activates p21 in cis to promote Polycomb target gene expression and to enforce the G1/S checkpoint. Mol Cell 54:777–790
78. Huarte M, Guttman M, Feldser D, Garber M, Koziol MJ, Kenzelmann-Broz D, Khalil AM, Zuk O, Amit I, Rabani M et al (2010) A large intergenic noncoding RNA induced by p53 mediates global gene repression in the p53 response. Cell 142:409–419
79. Morris KV, Santoso S, Turner AM, Pastori C, Hawkins PG (2008) Bidirectional transcription directs both transcriptional gene activation and suppression in human cells. PLoS Genet 4:e1000258
80. Zhou Y, Zhong Y, Wang Y, Zhang X, Batista DL, Gejman R, Ansell PJ, Zhao J, Weng C, Klibanski A (2007) Activation of p53 by MEG3 non-coding RNA. J Biol Chem 282:24731–24742

81. Poliseno L, Salmena L, Zhang J, Carver B, Haveman WJ, Pandolfi PP (2010) A coding-independent function of gene and pseudogene mRNAs regulates tumour biology. Nature 465:1033–1038
82. Salameh A, Lee AK, Cardo-Vila M, Nunes DN, Efstathiou E, Staquicini FI, Dobroff AS, Marchio S, Navone NM, Hosoya H et al (2015) PRUNE2 is a human prostate cancer suppressor regulated by the intronic long noncoding RNA PCA3. Proc Natl Acad Sci U S A 112:8403–8408
83. Hudson WH, Pickard MR, de Vera IM, Kuiper EG, Mourtada-Maarabouni M, Conn GL, Kojetin DJ, Williams GT, Ortlund EA (2014) Conserved sequence-specific lincRNA-steroid receptor interactions drive transcriptional repression and direct cell fate. Nat Commun 5:5395
84. Kino T, Hurt DE, Ichijo T, Nader N, Chrousos GP (2010) Noncoding RNA gas5 is a growth arrest- and starvation-associated repressor of the glucocorticoid receptor. Sci Signal 3:ra8
85. Renganathan A, Kresoja-Rakic J, Echeverry N, Ziltener G, Vrugt B, Opitz I, Stahel RA, Felley-Bosco E (2014) GAS5 long non-coding RNA in malignant pleural mesothelioma. Mol Cancer 13:119
86. Pickard MR, Williams GT (2015) Molecular and cellular mechanisms of action of tumour suppressor GAS5 LncRNA. Genes (Basel) 6:484–499
87. Khaitan D, Dinger ME, Mazar J, Crawford J, Smith MA, Mattick JS, Perera RJ (2011) The melanoma-upregulated long noncoding RNA SPRY4-IT1 modulates apoptosis and invasion. Cancer Res 71:3852–3862
88. Singh R, Gupta SC, Peng WX, Zhou N, Pochampally R, Atfi A, Watabe K, Lu Z, Mo YY (2016) Regulation of alternative splicing of Bcl-x by BC200 contributes to breast cancer pathogenesis. Cell Death Dis 7:e2262
89. Walsh KM, Codd V, Smirnov IV, Rice T, Decker PA, Hansen HM, Kollmeyer T, Kosel ML, Molinaro AM, LS MC et al (2014) Variants near TERT and TERC influencing telomere length are associated with high-grade glioma risk. Nat Genet 46:731–735
90. Dorji T, Monti V, Fellegara G, Gabba S, Grazioli V, Repetti E, Marcialis C, Peluso S, Di Ruzza D, Neri F et al (2015) Gain of hTERC: a genetic marker of malignancy in oral potentially malignant lesions. Hum Pathol 46:1275–1281
91. Rippe K, Luke B (2015) TERRA and the state of the telomere. Nat Struct Mol Biol 22:853–858
92. Flynn RL, Centore RC, O'Sullivan RJ, Rai R, Tse A, Songyang Z, Chang S, Karlseder J, Zou L (2011) TERRA and hnRNPA1 orchestrate an RPA-to-POT1 switch on telomeric single-stranded DNA. Nature 471:532–536
93. Prensner JR, Chen W, Iyer MK, Cao Q, Ma T, Han S, Sahu A, Malik R, Wilder-Romans K, Navone N et al (2014) PCAT-1, a long noncoding RNA, regulates BRCA2 and controls homologous recombination in cancer. Cancer Res 74:1651–1660
94. Sharma V, Khurana S, Kubben N, Abdelmohsen K, Oberdoerffer P, Gorospe M, Misteli T (2015) A BRCA1-interacting lncRNA regulates homologous recombination. EMBO Rep 16:1520–1534
95. Wan G, Hu X, Liu Y, Han C, Sood AK, Calin GA, Zhang X, Lu X (2013) A novel non-coding RNA lncRNA-JADE connects DNA damage signalling to histone H4 acetylation. EMBO J 32:2833–2847
96. Adriaens C, Standaert L, Barra J, Latil M, Verfaillie A, Kalev P, Boeckx B, Wijnhoven PW, Radaelli E, Vermi W et al (2016) p53 induces formation of NEAT1 lncRNA-containing paraspeckles that modulate replication stress response and chemosensitivity. Nat Med 22:861–868
97. Ji P, Diederichs S, Wang W, Boing S, Metzger R, Schneider PM, Tidow N, Brandt B, Buerger H, Bulk E et al (2003) MALAT-1, a novel noncoding RNA, and thymosin beta4 predict metastasis and survival in early-stage non-small cell lung cancer. Oncogene 22:8031–8041
98. Nakagawa S, Ip JY, Shioi G, Tripathi V, Zong X, Hirose T, Prasanth KV (2012) Malat1 is not an essential component of nuclear speckles in mice. RNA 18:1487–1499
99. Zhang B, Arun G, Mao YS, Lazar Z, Hung G, Bhattacharjee G, Xiao X, Booth CJ, Wu J, Zhang C et al (2012) The lncRNA Malat1 is dispensable for mouse development but its transcription plays a cis-regulatory role in the adult. Cell Rep 2:111–123

100. Gutschner T, Hammerle M, Eissmann M, Hsu J, Kim Y, Hung G, Revenko A, Arun G, Stentrup M, Gross M et al (2013) The noncoding RNA MALAT1 is a critical regulator of the metastasis phenotype of lung cancer cells. Cancer Res 73:1180–1189
101. Li L, Chen H, Gao Y, Wang YW, Zhang GQ, Pan SH, Ji L, Kong R, Wang G, Jia YH et al (2016) Long noncoding RNA MALAT1 promotes aggressive pancreatic cancer proliferation and metastasis via the stimulation of autophagy. Mol Cancer Ther 15(9):2232–2243
102. Gupta RA, Shah N, Wang KC, Kim J, Horlings HM, Wong DJ, Tsai MC, Hung T, Argani P, Rinn JL et al (2010) Long non-coding RNA HOTAIR reprograms chromatin state to promote cancer metastasis. Nature 464:1071–1076
103. Prensner JR, Zhao S, Erho N, Schipper M, Iyer MK, Dhanasekaran SM, Magi-Galluzzi C, Mehra R, Sahu A, Siddiqui J et al (2014) RNA biomarkers associated with metastatic progression in prostate cancer: a multi-institutional high-throughput analysis of SChLAP1. Lancet Oncol 15:1469–1480
104. Prensner JR, Iyer MK, Sahu A, Asangani IA, Cao Q, Patel L, Vergara IA, Davicioni E, Erho N, Ghadessi M et al (2013) The long noncoding RNA SChLAP1 promotes aggressive prostate cancer and antagonizes the SWI/SNF complex. Nat Genet 45:1392–1398
105. Yang F, Huo XS, Yuan SX, Zhang L, Zhou WP, Wang F, Sun SH (2013) Repression of the long noncoding RNA-LET by histone deacetylase 3 contributes to hypoxia-mediated metastasis. Mol Cell 49:1083–1096
106. Eades G, Wolfson B, Zhang Y, Li Q, Yao Y, Zhou Q (2015) lincRNA-RoR and miR-145 regulate invasion in triple-negative breast cancer via targeting ARF6. Mol Cancer Res 13:330–338
107. Yuan JH, Yang F, Wang F, Ma JZ, Guo YJ, Tao QF, Liu F, Pan W, Wang TT, Zhou CC et al (2014) A long noncoding RNA activated by TGF-beta promotes the invasion-metastasis cascade in hepatocellular carcinoma. Cancer Cell 25:666–681
108. Sas-Chen A, Aure MR, Leibovich L, Carvalho S, Enuka Y, Korner C, Polycarpou-Schwarz M, Lavi S, Nevo N, Kuznetsov Y et al (2016) LIMT is a novel metastasis inhibiting lncRNA suppressed by EGF and downregulated in aggressive breast cancer. EMBO Mol Med 8(9):1052–1064
109. Battistelli C, Cicchini C, Santangelo L, Tramontano A, Grassi L, Gonzalez FJ, de Nonno V, Grassi G, Amicone L, Tripodi M (2017) The Snail repressor recruits EZH2 to specific genomic sites through the enrollment of the lncRNA HOTAIR in epithelial-to-mesenchymal transition. Oncogene 36(7):942–955
110. Majidinia M, Yousefi B (2016) Long non-coding RNAs in cancer drug resistance development. DNA Repair (Amst) 45:25–33
111. Takahashi K, Yan IK, Wood J, Haga H, Patel T (2014) Involvement of extracellular vesicle long noncoding RNA (linc-VLDLR) in tumor cell responses to chemotherapy. Mol Cancer Res 12:1377–1387
112. Tsang WP, Kwok TT (2007) Riboregulator H19 induction of MDR1-associated drug resistance in human hepatocellular carcinoma cells. Oncogene 26:4877–4881
113. Zhang XW, Bu P, Liu L, Zhang XZ, Li J (2015) Overexpression of long non-coding RNA PVT1 in gastric cancer cells promotes the development of multidrug resistance. Biochem Biophys Res Commun 462:227–232
114. Modarresi F, Faghihi MA, Lopez-Toledano MA, Fatemi RP, Magistri M, Brothers SP, van der Brug MP, Wahlestedt C (2012) Inhibition of natural antisense transcripts in vivo results in gene-specific transcriptional upregulation. Nat Biotechnol 30:453–459
115. Lennox KA, Behlke MA (2016) Cellular localization of long non-coding RNAs affects silencing by RNAi more than by antisense oligonucleotides. Nucleic Acids Res 44:863–877
116. Choung S, Kim YJ, Kim S, Park HO, Choi YC (2006) Chemical modification of siRNAs to improve serum stability without loss of efficacy. Biochem Biophys Res Commun 342:919–927
117. Rettig GR, Behlke MA (2012) Progress toward in vivo use of siRNAs-II. Mol Ther 20:483–512
118. Dias N, Stein CA (2002) Antisense oligonucleotides: basic concepts and mechanisms. Mol Cancer Ther 1:347–355

119. Jepsen JS, Wengel J (2004) LNA-antisense rivals siRNA for gene silencing. Curr Opin Drug Discov Devel 7:188–194
120. Ivanova G, Reigadas S, Ittig D, Arzumanov A, Andreola ML, Leumann C, Toulme JJ, Gait MJ (2007) Tricyclo-DNA containing oligonucleotides as steric block inhibitors of human immunodeficiency virus type 1 tat-dependent trans-activation and HIV-1 infectivity. Oligonucleotides 17:54–65
121. Sarma K, Levasseur P, Aristarkhov A, Lee JT (2010) Locked nucleic acids (LNAs) reveal sequence requirements and kinetics of Xist RNA localization to the X chromosome. Proc Natl Acad Sci U S A 107:22196–22201
122. Sullenger BA, Gallardo HF, Ungers GE, Gilboa E (1990) Overexpression of TAR sequences renders cells resistant to human immunodeficiency virus replication. Cell 63:601–608
123. Smaldone MC, Davies BJ (2010) BC-819, a plasmid comprising the H19 gene regulatory sequences and diphtheria toxin A, for the potential targeted therapy of cancers. Curr Opin Mol Ther 12:607–616
124. Flynn RL, Cox KE, Jeitany M, Wakimoto H, Bryll AR, Ganem NJ, Bersani F, Pineda JR, Suva ML, Benes CH et al (2015) Alternative lengthening of telomeres renders cancer cells hypersensitive to ATR inhibitors. Science 347:273–277
125. Lee JT (2012) Epigenetic regulation by long noncoding RNAs. Science 338:1435–1439
126. Schorderet P, Duboule D (2011) Structural and functional differences in the long non-coding RNA hotair in mouse and human. PLoS Genet 7:e1002071
127. Lai KM, Gong G, Atanasio A, Rojas J, Quispe J, Posca J, White D, Huang M, Fedorova D, Grant C et al (2015) Diverse phenotypes and specific transcription patterns in twenty mouse lines with ablated LincRNAs. PLoS One 10:e0125522
128. Li L, Liu B, Wapinski OL, Tsai MC, Qu K, Zhang J, Carlson JC, Lin M, Fang F, Gupta RA et al (2013) Targeted disruption of Hotair leads to homeotic transformation and gene derepression. Cell Rep 5:3–12
129. Devoy A, Bunton-Stasyshyn RK, Tybulewicz VL, Smith AJ, Fisher EM (2012) Genomically humanized mice: technologies and promises. Nat Rev Genet 13:14–20
130. Feinberg AP, Koldobskiy MA, Gondor A (2016) Epigenetic modulators, modifiers and mediators in cancer aetiology and progression. Nat Rev Genet 17:284–299

Chapter 8
Long Noncoding RNAs in Pluripotency of Stem Cells and Cell Fate Specification

Debosree Pal and M.R.S. Rao

Abstract Since the annotation of the mouse genome (FANTOM project) [Kawai J et al (2001) Functional annotation of a full-length mouse cDNA collection. Nature 409(6821):685–690] or the human genome [An integrated encyclopedia of DNA elements in the human genome. (2012) Nature 489(7414):57–74; Harrow J et al (2012) GENCODE: the reference human genome annotation for the ENCODE project. Genome Res 22(9):1760–1774], the roles of long noncoding RNAs in coordinating specific signaling pathways have been established in a wide variety of model systems. They have emerged as crucial and key regulators of stem cell maintenance and/or their differentiation into different lineages. In this chapter we have discussed the recently discovered lncRNAs that have been shown to be necessary for the maintenance of pluripotency of both mouse and human ES cells. We have also highlighted the different lncRNAs which are involved in directed differentiation of stem cells into any of the three germ layers. In recent years stem cell therapies including bone marrow transplantation are becoming an integral part of modern medicinal practices. However, there are still several challenges in making stem cell therapy more reproducible so that the success rate reaches a high percentage in the clinic. It is hoped that understanding the molecular mechanisms pertaining to the role of these newly discovered lncRNAs in the differentiation process of stem cells to specific lineages should pave the way to make stem cell therapy and regenerative medicine as a normal clinical practice in the near future.

Keywords lncRNA • Stem cells • Induced pluripotent stem cells • Pluripotency • Differentiation • Enhancer RNAs • miRNA sponge

D. Pal • M.R.S. Rao (✉)
Molecular Biology and Genetics Unit, Jawaharlal Nehru Centre for Advanced Scientific Research, Jakkur, Bangalore 560064, India
e-mail: mrsrao@jncasr.ac.in

M.R.S. Rao (ed.), *Long Non Coding RNA Biology*, Advances in Experimental Medicine and Biology 1008, DOI 10.1007/978-981-10-5203-3_8

8.1 Introduction

The functional relevance of long noncoding RNAs, previously thought of as by-products of transcription, is no longer a debatable topic. Even as the repertoire of lncRNAs is constantly on the rise, we ought to note that with increasing complexity of the living organisms, the percentage of the noncoding genome has also considerably increased [1]. One may attribute this feature to a concomitant increase in the genome size and hence an explosion in the proportion of "junk sequences." But, increasing amount of evidence suggests that these noncoding transcripts play indispensible roles in the context of regulating developmental cues and signals, and their functional contribution becomes only more diverse when one moves up the evolutionary ladder. LncRNAs have been shown to participate in a wide variety of developmental processes like in regulating lineage commitment, specifying cellular identities and fates, in organogenesis, in imprinting of alleles during early development, and also in specification of the body pattern. A few of the first lncRNAs that were discovered through traditional gene mapping approaches are *Xist* [2] and *H19* [3], and interestingly enough they both play roles in regulating specific developmental processes, reiterating the aforesaid point that the evolution of the noncoding transcriptome in higher organisms has a functional significance and is not just an offshoot of the genomic size.

In the later part of the twentieth century, scientists were coalescing their efforts toward understanding how the genetic makeup of an individual regulates or predicts the development of various hereditary or familial diseases. While the field of genetics was resonating with breakthrough discoveries all over the world, cell biologists were not far behind in making discoveries that would ultimately form the basic model systems of study for the infinite complexities akin to the higher eukaryotes and mammals. In 1981, a report published by Martin Evans along with Matthew Kaufman [4] and another report published independently by Gail R. Martin [5] described the isolation of embryonic stem cells from the inner cell mass of blastocyst stage embryos and their subsequent maintenance under conditions of cell culture. These embryonic stem cells would, in the future, form the platform for carrying out research to understand the intricate signaling pathways and mechanisms governing mammalian development. They would further become the foundation for stem cell technology and stem cell therapy wherein damaged or defective tissues or organs would become replaceable due to the inherent properties of these cells (as will be discussed later). As a matter of fact, the groundwork for this technology was laid in the year 1995 by James Thomson and his colleagues at the Wisconsin Regional Primate Center (WPRC), University of Wisconsin-Madison, when they successfully isolated embryonic stem cells from the inner cell mass of rhesus monkeys making it the first report for the culture of nonhuman primate embryonic stem cells [6]. This led to the next achievement in 1998, whereby after an approval from bioethicists at the university, Thomson et al. derived human embryonic stem cells from leftover in vitro fertilized human embryos [7] that won him the *Science*'s "1999 Scientific Breakthrough of the Year" award. At the same time, the group led by John Gearhart obtained embryonic or primordial germ cells from the gonadal ridge of 5–9-week fetal tissue of electively aborted fetuses [8]. But ethical concerns over the use of human embryos for research purposes have paved the way for the generation of induced pluripotent stem cells (iPSCs), a groundbreaking discovery made independently by Thomson in his own lab [9] and Shinya Yamanaka [10] at the Kyoto University. Prof. Thomson reprogrammed adult human somatic cells into induced pluripotent stem cells by using a cocktail of four genes that were sufficient to impart "stemness" to the somatic cells. Research on the same lines carried by Yamanaka led to the identification of what is popularly known as the Yamanaka factors, namely, OCT3/4, SOX2, c-MYC, and KLF4, that could reprogram adult or embryonic fibroblasts into pluripotent stem cells. This discovery earned him the Nobel Prize for Physiology or Medicine in 2012. The implications of this discovery were immense because now theoretically, the cells from say, the skin of a person could be isolated and the clock turned backward to generate iPSCs which could be further differentiated to any cell type of the body and be used for the treatment of diseases like Parkinson's, spinal cord injury, Duchenne's muscular dystrophy, and so on, removing risks of transplants attacking their hosts.

In lieu of the importance of stem cell research, it becomes paramount to delve deeper into the mechanisms and key pathways that regulate the pluripotent nature of stem cells or guide them toward differentiation into various lineages. The term "pluripotency" has been derived from the Latin term *plurimus* meaning very many and *potens* meaning having power referring to the capability of stem cells to form various types of cells pertaining to any of the three germ layers of the body, namely, ectoderm, mesoderm, and endoderm. They also possess the power to divide and self-renew through continuous cell divisions, theoretically indefinitely (Fig. 8.1). Embryonic stem cells are those which are present in the embryo within the inner cell mass of the blastocysts, whereas adult stem cells reside in mature organs like the brain, skin, muscle, and bone marrow which act to regenerate parts of the tissues lost during processes of wear and tear or injury.

Soon after the establishment of stem cell cultures, widespread studies began on elucidating the molecular features of these cells. What factors maintain the "stemness" of these cells? What factors guide them into differentiation of either one or the other lineage? How can a bunch of similar cells give rise to an entire organism? While most of these questions have been addressed thoroughly by scientists around the world, nature never seems to exhaust us by posing new surprises and challenges. The discovery of noncoding RNAs revolutionized the understanding of the central dogma of biology and opened up a whole new avenue for exploration. Widespread studies that followed this discovery unraveled the ways in which these noncoding RNAs regulate crucial cellular pathways that govern the functioning of the individual cell and that ultimately manifests into functioning of the entire organism.

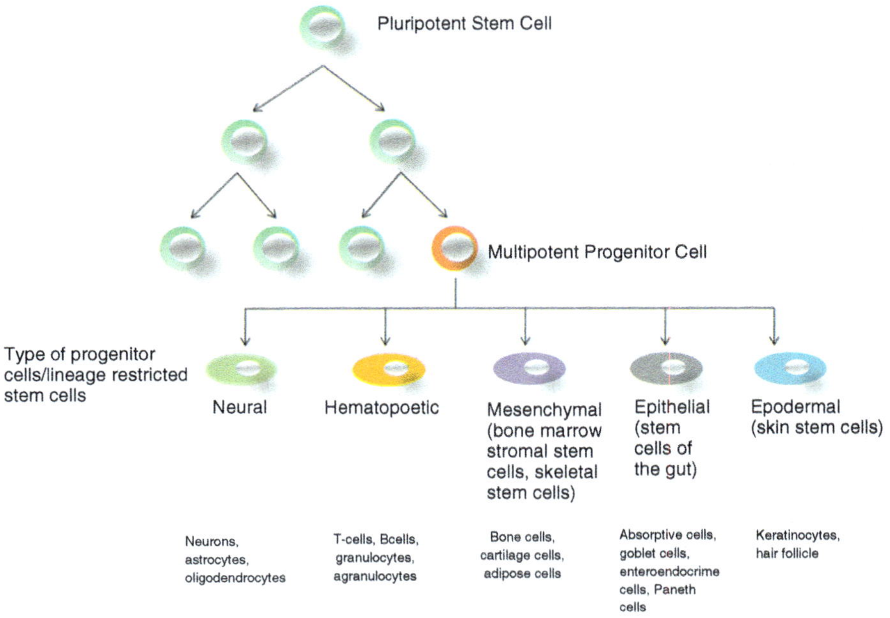

Fig. 8.1 A pluripotent stem cell upon asymmetric division gives rise to a multipotent progenitor cell which can be of various categories as illustrated. Each category of multipotent progenitor cell or, lineage-restricted stem cell, can again undergo limited rounds of cell division or differentiate into cells of the corresponding lineage

8.2 Long Noncoding RNAs in Pluripotent Embryonic Stem Cells

8.2.1 Long Noncoding RNAs in Mouse Embryonic Stem Cells

As has been discussed in the previous chapters, lncRNAs play a significant role in modulating gene expression in several of the model systems. In this context, studies were initiated at a genome-wide level to unravel the cohort of long noncoding RNAs involved in the regulation of stem cell pluripotency. In biology, in order to understand the functional relevance of a molecule, a common approach is to selectively deplete it from the cell and observe the downstream effects with the help of techniques like microarray or RNA sequencing that shed light about the perturbations in expression of transcripts at the genome level. Guttman et al. [11] adopted such a methodology to address the function of a select class of lncRNAs known as long intergenic noncoding RNAs (lincRNAs) which as their name suggests are expressed from regions of the genomic segment present between two protein-coding genes. In this report, 226 lincRNAs were knocked down or depleted from embryonic stem cells by using short hairpin RNAs, and microarray was performed to analyze the effect. An interesting outcome of this study was that most of the lincRNAs act in trans, at locations that are genomically farther away from their own site of transcription, adding a new dimension to the already known cis mechanism of action of lncRNAs. The more relevant outcome was, however, the discovery of 26 lincRNAs, knockdowns of which showed reduction in luciferase reporter activity, the expression of the luciferase gene being driven by the Nanog promoter. This observation established the fact that these lincRNAs contribute to the maintenance of pluripotency. Further experiments showed that ES cells depleted of these lincRNAs lead to loss of ES cell morphology characteristic to their pluripotent state along with a reduction in the expression of the core pluripotency factors. The fact that lincRNAs directly maintain the pluripotency of stem cells was subsequently corroborated by a more detailed analyses wherein knockdown of these lincRNAs resulted in the differentiation of stem cells toward one or the other lineage, recapitulating the phenomena that occurs when OCT4 or NANOG themselves are depleted from stem cells. It is interesting to note that at the molecular level, the lincRNAs are themselves directly regulated by the occupancy of one or more of the core pluripotency transcription factors at their promoters, establishing the importance of lncRNAs in coordinating mechanisms to maintain the pluripotent state of stem cells or repress their differentiation into various lineages.

While such holistic approaches as above have turned out to be crucial in discerning the function of the multitude of lncRNAs involved in ES cell circuitry, more direct studies with specific examples of lncRNAs have proved their indispensability for the proper functioning of ES cells. A study by Mohammed et al. [12] initiated at the genome level, to identify lncRNAs that are closely associated on the genomic loci serving as binding sites for OCT4 and NANOG, focused on two specific lncRNAs that play roles in fine-tuning the ES cell pluripotency/differentiation states. Directed knockdown of lncRNA AK028326, in essential a 3′ fragment of the annotated 9 kb long lncRNA GOMAFU/MIAT, results in downregulation of Oct4 and other pluripotency markers and upregulation of markers of the trophectodermal and mesodermal lineages. Similar results were observed with lncRNA AK141205 although in this case, it was only OCT4 whose expression was concomitantly downregulated but not of Nanog. In accordance with these observations, AK028326 depletion in ES cells also resulted in a loss of ES cell colony morphology, suggesting a loss of pluripotent state, hence proving the necessity of this lncRNA in maintaining stem cell character. But an intriguing fact lay in the overexpression studies, wherein ectopic expression of these lncRNAs resulted in ES cells differentiating toward the neuroectodermal or mesodermal/ectodermal lineages, respectively. This suggests the diversity and complexity of functions of lncRNAs in stem cell biology. Basal levels of these lncRNAs might be important in maintaining the pluripotency of stem cells, whereas their overexpression may alter separate pathways altogether and guide the cells toward differentiation. *Linc86023*, named as Tcl1 upstream neuron-associated lincRNA (TUNA or MEGAMIND), was similarly identified by Lin et al. [13] as a crucial molecule necessary for maintaining the pluripotent state of mouse embryonic stem cells. Being conserved remarkably across vertebrates, its loss of function resulted in altered cell morphology, reduced expression of pluripotency factors, and decreased cell proliferation, all of which are signatures of differentiation of otherwise self-renewing stem cells. TUNA was shown to form a multi-protein complex with RNA-binding proteins PTBP1, hnRNP-K, and NCL which occupy promoters of Nanog, Sox2, and Fgf4 to maintain the pluripotent nature of stem cells. Again, in this case too, it was observed that TUNA is essential for the formation of neural precursors from stem cells in monolayer-adherent cultures, and its knockdown abolished the capacity of the stem cells to progress toward the neural lineage, emphasizing the pleiotropic nature of regulation of stem cell pathways by lncRNAs.

In another study, Chakraborty et al. [14] employed esiRNAs to downregulate around 594 previously annotated lncRNAs in mouse embryonic stem cells. The same esiRNA sequences, transcribed either in the sense or the antisense direction, were used to understand the cellular localization of the lncRNAs by FISH (fluorescent in situ hybridization). ES cells expressing GFP under the Oct4 promoter were transfected with the esiRNAs against the lncRNAs and scored for loss of GFP expression. Loss of GFP expression in the presence of esiRNA against a particular lncRNA would imply the probable involvement of that lncRNA in the maintenance of pluripotency. By this method, three lncRNAs were short-listed and were named pluripotency associated noncoding transcripts 1–3 or PANCT 1–3. Among them, PANCT 1 was characterized specifically because it showed the strongest effect on the expression of GFP. It was observed that PANCT 1 levels decreased steadily when ES cells were subjected to differentiation, and this was further confirmed by PANCT 1 knockdown studies wherein the cells showed reduction in pluripotency markers, reduction in DNA synthesis (exit from the dividing pluripotent state), and upregulation of various lineage-specific markers, suggesting a role for PANCT 1 in ES cell pluripotency regulation.

8.2.2 Long Noncoding RNAs in Human Embryonic Stem Cells

Studies on similar lines were performed in human embryonic stem cells by Ng et al. [15] who identified three lncRNAs, lncRNA_ES1 (AK056826), lncRNA_ES2 (EF565083), and lncRNA_ES3 (BC026300) which had Oct4 or Nanog binding sites near their transcription start sites. OCT4 or NANOG RNAi experiments showed reduction in the expression of lncRNA ES1and lncRNA ES2 and ES3, respectively. Downregulation of any of these three lncRNAs also resulted in loss of OCT4 expression, decrease in expression of a panel of pluripotency markers, and upregulation of genes involved in the formation of neuroectodermal, endodermal, and mesodermal markers. In accordance with studies performed before, it was observed that the lncRNAs mentioned above interact directly with either the core pluripotency factors or components of chromatin remodelers like SUZ12 (of the PRC2 complex) to determine active or silenced states of genes required for the maintenance of pluripotency or lineage differentiation. Linc-RoR (to be discussed in the next section) is yet another lincRNA that is necessary for the maintenance of the undifferentiated state of human embryonic stem cells [16]. Linc-RoR presents forth a unique example of

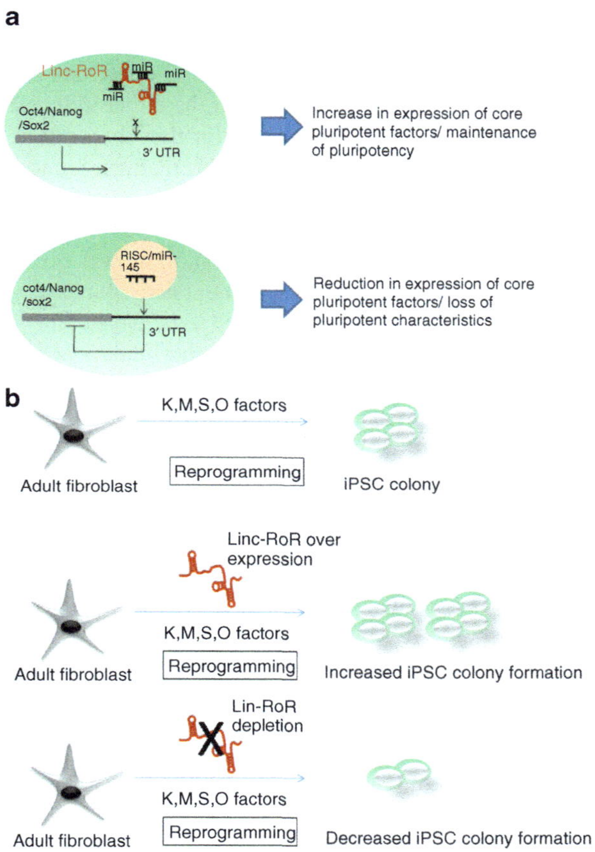

Fig. 8.2 Mechanisms of action of linc-RoR. (**a**) In human embryonic stem cells, linc-RoR acts as a competing endogenous miRNA sponge and titers away miR-145 from its targets which include the core pluripotency transcription factors, Oct4, Nanog, and Sox2; (**b**) During reprogramming, linc-RoR acts to regulate p53 DNA damage and cell apoptotic pathways to aid the formation of induced pluripotent stem cells (iPSCs)

the diverse mechanisms of action of lncRNAs. It possesses binding sites for several of the microRNAs that target and reduce the expression of the core pluripotency factors. By binding to and sequestering these miRNAs, linc-RoR acts as a "sponge" and prevents these miRNAs from degrading their target mRNAs that is required for the proper self-renewal of the human stem cells (Fig. 8.2a). Interestingly, linc-RoR transcription is itself regulated by the core transcription factors OCT4, NANOG, and SOX2, conforming to the well-known biological phenomenon of autofeedback regulatory loop.

8.3 LncRNAs in Induced Pluripotent Stem Cells

iPSCs (induced pluripotent stem cells) are being explored as a promising candidate for stem cell-based therapies, albeit scientists are still trying to understand the pathways and regulatory mechanisms governing the framework and functioning of these cells. In 2011, 5 years after the groundbreaking discovery of iPSCs, Loewer et al. [17] generated iPSCs from adult fibroblasts and analyzed gene expression changes on a microarray platform probing ~900 lincRNAs encoded in the human genome. About 207 lincRNAs were found to be either induced or repressed upon iPSC formation. One possible explanation for this observation is that reprogramming leads to changes in conformation of the chromatin genome wide, and opening up or compaction of protein-coding chromatin domains might directly affect the expression of the neighboring lincRNAs. However this possibility was ruled out because for each of the lincRNAs under consideration, there was no significant correlation between the neighboring protein-coding gene status. LincRNA-SFMBT2, lincRNA-VLDLR, and lincRNA-ST8SIA3 were found to be physically occupied at their promoters by Oct4, Sox2, and Nanog, indicating the functional intertwining of these lincRNAs and the core pluripotency factors in the formation of iPSCs. Furthermore it was observed that ES cells subjected to depletion of these lincRNAs by short hairpins showed a reduction in the formation of iPSC colonies in the case of lincST8SIA3, demonstrating the functional requirement of this lincRNA in iPSC formation. RACE (rapid amplification of cDNA ends) analysis recovered a transcript 2.6 kb long comprising four exons and no protein-coding activity. Overexpression of this lincRNA in fibroblasts followed by their reprogramming into iPSCs showed a two-fold increase in the formation of iPSC colonies (Fig. 8.2b). When a microarray analysis was performed upon knockdown of lincST8SIA3, it was found that genes of the p53 DNA damage response, and cell apoptotic pathways were upregulated, consistent with the phenotype observed when the lincRNA is depleted from the cells. p53 knockdown under the lincRNA knockdown conditions partially rescued the phenotype. This was one of the first reports to establish the role of a lincRNA in the formation and maintenance of iPSCs, opening up a whole new avenue of stem cell therapy and research. The lincRNA was aptly named linc-RoR or regulator of reprogramming (Table 8.1).

8.4 LncRNAs in Lineage-Restricted Stem Cells and Differentiation

While pluripotent stem cells can give rise to any of the cells specific to the three germ layers, multipotent cells are more specialized or committed in their differentiation capacity and can generate cells of a particular lineage, for example, only the neural lineage or the hematopoietic lineage. Since they possess the ability to self-renew and form a specific set of cell types, they are classified under stem cells. Multipotent stem cells exist both in the embryonic and the adult stages. In the embryonic stages, they act to generate nascent mature cells of the corresponding type, whereas adult stem cells are mainly responsible for the regeneration and repair of damaged adult tissues. In the following section, we discuss how multipotent stem cell networks are regulated by lncRNAs.

Table 8.1 List of lncRNAs involved in maintenance of stem cell properties in mouse/human/induced stem cells

Name	Identification	Phenotype	Methods	Mechanism of action	Reference
LincRNAs containing "K4-K36" domain	ChIP followed by parallel sequencing	Loss of ES cell morphology and pluripotent characteristics	Lentiviral shRNA-mediated knockdown	LincRNAs are bound at their promoters by one or more of the pluripotency factors; they also associate with chromatin remodelers	[11, 18]
LncRNA AK028326 (3′ fragment of GOMAFU/MIAT) LncRNA AK141205	Analysis of ChIP-PET results for Oct4 and Nanog	Loss of ES cell pluripotency; upregulation of markers of one or more of the three germ layers	siRNA-mediated silencing	LncRNAs are direct targets of Oct4 and Nanog, respectively; AK028326 participates in an autofeedback loop with Oct4 to regulate pluripotency	[12]
LincRNA TUNA/Megamind	Genome-scale shRNA library targeting lincRNAs	Reduced expression of pluripotent factors; decreased cell proliferation	shRNA-mediated knockdown	TUNA forms a RNP complex with PTBP1, hnRNP-K, and NCL to occupy promoters of Nanog, Sox2, and Fgf4 in order to regulate pluripotency	[13]
PANCT 1	esiRNA library targeting 594 lncRNAs (RNAi screening)	Reduction in pluripotency markers; reduction on DNA synthesis; upregulation of lineage-specific markers	esiRNA-mediated silencing	Still to be addressed	[14]

(continued)

Table 8.1 (continued)

Name	Identification	Phenotype	Methods	Mechanism of action	Reference
LncRNA_ES1 (AK056826), LncRNA_ES2 (EF565083), and LncRNA_ES3 (BC026300)	Custom-designed microarray platform for lncRNAs	Loss of Oct4 expression; reduction in pluripotency markers; upregulation of lineage-specific markers	siRNA-mediated silencing	LncRNA ES1: expression decreased upon Oct4 RNAi LncRNA ES2 and ES3: similar effect upon Nanog RNAi; ES1 and ES2 associate with Suz12 (PRC2 component)	[15]
Linc-RoR	Previously identified	Increase in expression of core pluripotency factors or their reduction in expression	Overexpression of Linc-RoR or siRNA-mediated silencing	Functions as a competing endogenous miRNA sponge for miR-145 and prevents miR-145-mediated degradation for Oct4, Nanog, and Sox2	[16]
Linc-RoR (Linc ST8SIA3)	Microarray platform screening for ~900 lincRNAs	Increase in the formation of iPSC colonies upon overexpression and a concomitant decrease upon downregulation	Viral-mediated overexpression or shRNA-mediated knock down	Contributes to iPSC formation by regulating p53-mediated apoptotic pathways	[17]

8.4.1 Long Noncoding RNAs in Neural Stem Cells and Differentiation

One of the most evolutionarily susceptible and complex organs, the brain, consists of neurons that impart the sensory and motor functions and glia that act more as a support system for the cells of the brain itself. In the mammalian embryo, the forebrain harbors the stem cells or the radial glia cells that divide and specialize to form both neurons and glia, i.e., astrocytes and oligodendrocytes. In the neonatal and subsequently in the adult stages, the quiescent neural stem cells are present in specific areas known as neurogenic niches which include the ventricular and subventricular zones and the subgranular zone of the dentate gyrus in the hippocampus [19]. In one of the genome-wide studies by Ng et al. [15], 35 lncRNAs were found which were highly expressed in mature neurons when compared to human embryonic stem cells or neural progenitors, among which knockdown of RMST (rhabdomyosarcoma 2-associated transcript), lncRNA_N1, lncRNA_N2, and lncRNA_N3 led to lack of neuron generation in vitro. Overexpression studies showed the generation of an increased percentage of neurons, underlining the importance of lncRNA RMST in neuronal differentiation of human embryonic stem cells. RNA pulldown experiments revealed that RMST physically interacts with SOX2. Subsequently an overlap of the microarray datasets for siRMST and siSOX2 cells showed that they both co-regulate a specific subset of genes which are important for neurogenesis [20]. In fact, in cells where RMST was depleted by siRNA, it was observed that SOX2 binding to the target genes was ablated, underlining the importance of this lncRNA in acting as a co-regulator of SOX2-mediated neurogenesis.

Pax6 upstream antisense RNA (PAUPAR) is a lncRNA [21] situated 8.5 kb upstream of the Pax6 gene which codes for Pax6, a crucial transcription factor involved in neural progenitor cell proliferation, subtype specification, and spatial patterning in the brain. Downregulation of PAUPAR in neuroblastoma cells revealed that this lncRNA acts to maintain self-renewal of neural progenitor cells since its depletion led to increased neurite growth and increased appearance of neuronal differentiation markers in the cells. At the genic level, PAUPAR was found to be a large-scale regulator of gene expression in neural progenitor cells, affecting the expression of around 942 genes most of which belonged to synaptic regulation and cell cycle control. Interestingly, it was observed that Pax6 and PAUPAR not only co-occupy a common and distinct set of genes but also co-regulate several of them. Depletion of PAUPAR, however, does not affect the Pax6 occupancy at those genes, indicating that PAUPAR might act to recruit transcriptional coactivators at these sites of the genome and regulate their expression.

Much of the studies reported in the literature have focused on the functional significance of noncoding transcripts emanating from regions neighboring to protein-coding genes important for a specific developmental regime. LncRNA DALI [22], situated downstream from *Pou3f3* locus, exhibits concomitant expression pattern in the embryonic brain and in retinoic acid-treated ES cells with respect to *Pou3f3*, a protein known to have a role in the development of the nervous system.

In neuroblastoma cells, depletion of DALI leads to reduction in neurite growth, indicating DALI is required for proper differentiation of these cells. Genome-wide studies showed that DALI regulates genes like *E2f2*, *Fam5b*, *Sparc*, and *Dkk1* which are known to be pro-differentiation factors and negatively regulates genes that prevent the formation of neurites. An intriguing feature of this lncRNA is that it acts in cis on the neighboring *Pou3f3* gene where it physically contacts the gene at several locations as shown by 3C (chromosome conformation capture) technique. Simultaneously, it also acts in trans on genes involved in neuronal differentiation, cell cycle, neuronal projection formation, and intracellular signaling as shown by CHART-Seq (capture hybridization analysis of RNA targets). Furthermore, it also interacts with DNMT1, a DNA methyltransferase, and regulates DNA methylation at specific gene loci. DALI knockdown was shown to increase methylation at the CpG islands of *Dlgap5*, *Hmgb2*, and *Nos1* promoters, revealing an intricate network of neuronal gene regulation by lncRNA DALI.

A more recent study characterized PINKY (PNKY) lncRNA [23], a nuclear restricted neural-specific noncoding transcript, that maintains the neural stem cells of the ventricular zone in embryonic brains or ventricular-subventricular zones in adult brains. PNKY is expressed in neural stem cells but upon differentiation gets restricted specifically to the $GFAP^+$ astrocyte lineage. Knockdown of PNKY in monolayer cultures resulted in the generation of increased numbers of $Tuj1^+$ neuronal cells. When the shRNA construct of PNKY was electroporated into the embryonic brain and compared against the control brain, it was observed that the proportion of $Sox2^+$ stem cells were reduced but that of $TBR2^+$ transit-amplifying cells (an intermediate stage between stem cells and neurons) was not affected albeit there was an increase in $Satb2^+$ young neurons, indicating that PNKY maintains neural stem cells in the embryonic brain. Further exploration into its mechanism revealed that PNKY interacted with PTBP1, a repressor of neuronal differentiation. PTBP1 is known to regulate alternative splicing. Independently knocked down cells of PNKY and PTBP1 when subjected to RNA sequencing revealed that they regulate a common set of differentially perturbed genes and a common set of splice variants, suggesting a close coordination between these two molecules to maintain the neural stem cells in the brain.

8.4.2 Long Noncoding RNAs in Hematopoietic Stem Cells and Differentiation

The hematopoietic system of our body comprises of blood cells and the cells of the immune system both of which are critical for maintaining the body homeostasis. While red blood cells are the central pivots of oxygen transportation in the body and platelets of blood coagulation, white blood cells act to protect the body from the millions of pathogens it gets exposed to everyday, thereby forming the pillars of the immune system. Till and McCulloch, back in the early 1960s, [24] probed into the components of blood that leads to its regeneration which led to the discovery of hematopoietic stem cells (HSCs). Like any other multipotent stem cells, they too can self-renew and give rise to all cell types of the blood. A mouse that has received an irradiation dose to kill its own blood-producing cells can survive if injected with these stem cells. However, HSCs can be either long-term stem cells that can constantly self-renew and support the blood system of an irradiated mouse (irradiation-depleted blood-producing cells) over several divisions or short-term progenitor or precursor cells that are restricted by the number of divisions that they can undergo. Since there are many types of blood cells, the differentiation of the HSCs has been characterized in the following manner: each stem cell can give rise to a myeloid progenitor cell and a lymphoid progenitor cell. Myeloid progenitor cells form the red blood cells, platelets, and the white blood cells which can again be divided into granulocytes (eosinophils, neutrophils, basophils) or agranulocytes (lymphocytes/macrophages). On the other hand, lymphoid progenitor cells give rise to T-lymphocytes, B-lymphocytes, and natural killer cells. HSCs have found widespread applications in the clinic. They are used for the treatment of leukemia and lymphoma wherein the patient's own blood cells are destroyed by radiation and replaced with a bone marrow transplant from a matched donor. Bone marrow transplants are also used for the treatment of genetic disorders of the blood like anemia and thalassemia.

One of the first ever lncRNAs reported to be involved in the maintenance of the hematopoiesis, specifically erythropoiesis, is lincRNA-EPS. Hu et al. [25] isolated cells from embryonic liver, a site for active erythropoiesis with cells of the erythroid lineage forming >90% of the liver and performed RNA-Seq analysis to identify the repertoire of lncRNAs which might be involved in the erythroid lineage. They concentrated their efforts on three types of cells, burst-forming erythroids, colony-forming erythroids, and Ter 119^{+} cells that represent the three key stages of erythropoietic development and found that greater than 400 lncRNAs are perturbed during erythropoiesis. Out of these, 163 putative lncRNAs are upregulated and 42 are downregulated. They focused on those that show an increase in expression between colony-forming erythroids (progenitors) and Ter 119^{+}-differentiated erythroblasts with an aim to understand the regulation of erythroid differentiation by lncRNAs. A probe into the functional aspects of lincRNA-EPS revealed that its depletion in erythroid progenitors led to increased apoptosis and reduction in proliferation of the progenitors in the presence of erythropoietin (erythropoietin promotes proliferation and subsequent differentiation of progenitors). This resulted in the reduced conversion of progenitors into terminally differentiated cells. On the other hand, under erythropoietin-starved conditions, progenitors that overexpressed lincRNA-EPS did not undergo apoptosis implying that lincRNA-EPS conferred anti-apoptotic phenotype to these progenitor cells. Microarray analyses in lincRNA-EPS overexpressing progenitors revealed the repression of a proapoptotic gene *Pycard*, which under normal circumstances activates caspase in apoptosis. Thus, lincRNA-EPS acts as an anti-apoptotic regulator during erythroid differentiation and development.

In a parallel study, Paralkar et al. [26] were interested in identifying the cohort of lncRNAs that are expressed in megakaryocyte-erythroid precursors from the bone marrow, megakaryocytes from cultured fetal liver progenitors, and fetal liver erythroblasts in mouse as well as in human cord blood erythroblasts. This comparative analysis identified approximately 1100 lncRNAs expressed during murine erythro-megakaryopoiesis, out of which about 85% are present both in fetal and adult erythroblasts, suggesting the involvement of these lncRNAs in erythropoiesis. Interestingly, ~75% of the identified lncRNAs are expressed from promoter regions of genes, whereas ~25% are expressed from enhancer regions as evident from CHIP-Seq studies with transcription activation histone modification mark (H3K4me3) or enhancer modification mark (H3K4 me1). Further CHIP-Seq studies with key erythropoietic transcription factors GATA1 and TAL1 in erythroblasts and GATA1, GATA2, TAL1, and FLI1 in megakaryocytes showed occupancy of most of the lncRNA loci with these transcription factors. Knockdown studies with shRNA constructs against several of these lncRNAs inhibited enucleation and maturation of erythroblasts into reticulocytes when the erythroblasts were subjected to differentiation in erythropoietin-containing medium. Lnc051, annotated previously as LINCRED1 along with ERYTHRA and SCARLETLTR, were a few of the candidate lncRNAs with potential roles in erythroid terminal maturation.

Eosinophils are another cell type that arise from the common myeloid progenitor and have a role to play in parasitic immunity and allergic diseases. $CD34^+$ human hematopoietic stem cells supplemented with IL-5, an eosinophil-specific cytokine for 24 h, were subjected to gene expression profiling by microarray upon which a novel transcript encoded within an intron on the opposite strand of the inositol triphosphate receptor type 1 (*Itpr1*) gene was discovered [27]. It was named as EGO for eosinophil granule ontogeny lncRNA. The EGO transcript has two splice variant transcripts, EGO-A and EGO-B, and both of them are highly overexpressed upon stimulation of umbilical cord blood cells or bone marrow cells ($CD34^+$) with IL-5 and only slightly induced in the presence of other cytokines like epoetin-α, SCF, GM-CSF, etc. RNA silencing experiments were performed in erythroleukemic cells to understand the functional significance of EGO lncRNA. Interestingly, it was found that levels of the eosinophil proteins MBP (major basic protein) and EDN (eosinophil-derived neurotoxin) were concomitantly reduced. CD34+ umbilical cord blood cells expressing shRNA against EGO show incomplete development and die within 5 days of growth in IL-5 medium with respect to the control cells. Also, MBP and EDN levels were reduced considerably, suggesting that EGO lncRNA is necessary for the expression of these eosinophil proteins and hence normal eosinophilosis although the exact mechanism of action remains to be elucidated.

In another study, transcriptome profiling by microarray was performed on human peripheral blood neutrophils and on NB4 and HL-60 cells treated with all-trans-retinoic acid (ATRA) (cells directed toward granulocytic differentiation). This led to the identification of transcriptionally active regions between *HoxA1* and *HoxA2* genes [28]. The transcript was identified as a 483 nt RNA-spliced product from a primary transcript consisting of two exons and was subsequently named as HOTAIRM1 (HOX antisense intergenic myeloid 1). The expression of HOTAIRM1 was significantly induced when NB4 cells were treated with retinoic acid, but this phenomenon was not observed in the ATRA-resistant NB4r2 cell line. In fact, the expression of HOTAIRM1 was highly specific to the myeloid lineage as was evident by its specific upregulation in ATRA-treated NB4 or ATRA-treated K562 cells as compared to its baseline expression levels in the promyelocytic stages of NB4 cells. It was also found to exhibit low expression in hematopoietic stem or progenitor cells and was seen to be almost lacking expression in other organs like the brain, heart, pancreas, or skeletal muscle. In cells treated with shRNA against HOTAIRM1, induction of expression of *HoxA1*, *HoxA4*, and to some extent *HoxA5* was significantly attenuated in comparison to control cells, both the cell types being subjected to granulocytic differentiation by ATRA. Induction of beta2 integrin molecules, CD11B and CD18 (hallmarks of granulocyte maturation), was also abrogated, implying important roles for HOTAIRM1 in myelopoiesis. Studies by Wei et al. [29] provided insights into the mechanistic aspects whereby they observed that the transcription factor PU.1 binds to and regulates the levels of HOTAIRM1. PU.1 itself is an important transcription factor involved during myeloid differentiation, reaching highest levels in mature granulocytes and monocytes. Indeed in acute promyelocytic leukemic cells, dysregulation of HOTAIRM1 is due to the binding of PML-RARα to PU.1 and subsequent prevention of PU.1-mediated transactivation of various myeloid differentiation genes.

An extensive study carried out by Hu et al. [30] was aimed at cataloging the long intergenic ncRNAs involved in T-cell maturation and differentiation. They obtained 42 subsets of T-cells which included CD4-CD8 double negative (DN), double positive (DP), single positive (SP) thymic T-cells, T-regulatory (Treg) cells from the lymph nodes of mice, and T_H1, T_H2, T_H17 (T-helper cells), and induced Treg (iTreg) cells from in vitro cultures derived from naïve $CD4^+$ T-cells. Across all of the T-cell types, they identified 1542 genomic regions that were expressing lincRNAs individually or in clusters (more than one lincRNA expressed from the same locus). Quite intriguingly, when the data was classified based on the expression status of lincRNAs or protein-coding genes in specific subsets like only DN cells, DP^+SP^+Treg cells, and naïve $CD4^+$ T_H cells, it was observed that 48–57% of the expressed lincRNAs were lineage specific as compared to 6–8% of mRNAs, and only 13–16% of lincRNAs were shared between subsets of T-cells in contrast to 70–80% of protein-coding transcripts. When followed over a time scale of differentiation, many of the lincRNAs were downregulated at 4 h of T-cell differentiation from naïve $CD4^+$ T-cells only to again regain the expression at 48–72 h implying their role in T-cell activation. Many of them, like LincR-Chd2-5′-74 K, remained mostly silenced after differentiation, while many others, like LincR-Sla-5′AS, were induced at 4 h of differentiation with a gradual subsidence of expression at later stages. CHIP-Seq and knockdown studies of two important transcription factors STAT4 and STAT6 revealed that STAT4 preferentially binds to and potentially regulates lincRNAs specific to T_H1 cells and STAT6 for T_H2 cells. Linc-Ccr2-5′-AS was further studied whereby it was found that depletion of this lncRNA resulted in reduction of expression of *CCr 1*, *2*, *3*, and *5* genes (chemokine receptors), all of which are located neighboring to the lincRNA genomic locus. Moreover, in vivo depletion of this lincRNA led to decreased migration of T_H2 cells to the lung, a process which is dependent on chemokine signaling. This study along with a study conducted by Ranzani et al. [31] gives a comprehensive insight into the lincRNAs with potential regulatory functions during lymphocyte differentiation, maturation, activation, and functioning. On similar lines, Casero et al. [32] studied the lncRNA profile of ten cell types of the lymphoid lineage: (1) $CD34^+$ $CD38^-$ Lin^- cells enriched in hematopoietic stem cells and obtained from the bone marrow; (2) three lymphoid progenitor populations such as common lymphoid progenitors, lymphoid-primed multipotent progenitors, and B-cell-committed progenitors from the bone marrow as well; (3) $CD34^+$ but CD4 CD8 double negative populations (Thy1, Thy2, Thy3) from the thymus; and (4) T-cell-committed populations from the thymus again. A set of 9444 lncRNA genes were identified among which 3348 are known. Yet again, most of these lncRNAs showed a highly stage-specific manner of expression, being restricted to one or the other lineage in comparison to their protein-coding counterparts. They were also positively correlated in expression with several of the protein-coding genes located either in trans or in cis to them, reinforcing the role of lncRNAs in the maintenance and/or differentiation of progenitors in the bone marrow and the thymus.

8.4.3 Long Noncoding RNAs in Muscle Stem Cells and Differentiation

Skeletal muscle, a striated muscle tissue comprising about ~40% of the body weight, is composed of multinucleated contractile muscle cells known as myofibers which in turn are generated by the fusion of progenitor cells or myoblasts [33]. Myofibers remain constant in number in the neonatal stages, but postnatally they grow in size by the fusion of a group of stem cells known as satellite cells. Satellite cells are the stem cell population of the adult muscle tissue, being quiescent under normal physiological conditions but quickly reenter active cell division in case of muscle injury to regenerate damaged or wounded tissue. Although the regenerative capacity of muscle tissue was observed as early as the nineteenth century, it was only in 1961 that two independent studies by Alexander Mauro and Bernard Katz actually proved their presence by electron microscopy in the sublaminar region of myofibers [34]. At the molecular level, quiescent satellite cells express Pax7, and only upon activation of mitosis, they start expressing myogenic transcription factors like MYOD, MYOGENIN, MYF5, and DESMIN [34]. About 24 kb upstream of the gene-encoding transcription factor MYOD1, two regulatory regions are present for the gene itself, referred to as CE (core enhancer) and DRR (distal regulatory region). Through a series of RNA-Seq experiments, it was observed that these enhancer regions, characterized by the presence of histone modifications H3K4me1 and H3K27ac along with p300/CBP/RNAP II occupancy, are actually transcriptionally active, giving rise to enhancer RNAs or eRNAs [35].

In an approach to dissect out the role of these eRNAs, a screening was done for ten siRNAs designed against various regulatory regions upstream of *MyoD*, and interestingly enough it was observed that the levels of MyoD diminished drastically only in the case of siRNA targeting the CE region. It was further observed that ^{CE}RNA acts in cis to regulate the transcription of *MyoD1* by enhancing the occupancy of RNAPol II at *MyoD1* proximal regions. On a similar note (yet with a twist in the tale), it was discovered that ^{DRR}RNA acts in trans to enhance the expression of *MyoG* and *Myh*, thereby acting to promote myogenic differentiation. The role of eRNAs, a class of lncRNAs, was established in this study, and their mechanisms of function which mainly includes modification of chromatin organization by either causing nucleosome repositioning or by effecting recruitment of various chromatin modifiers were elucidated. Parallel studies by Mueller et al. [36] on the *MyoD* upstream locus led to further characterization of a lncRNA MUNC (MyoD upstream noncoding) which initiates transcription in the ^{DRR}RNA locus. Downregulation and overexpression of MUNC in undifferentiated muscle cells in culture caused a respective decrease or increase in the levels of key myogenic transcription factors like MYOGENIN, MYH3, and MYOD itself to some extent. In vivo, when siRNA against MUNC was injected into the tibia anterior (TA) muscles of mice followed by muscle injury with cardiotoxin, it was observed that over a period of 2 weeks of muscle regeneration, the levels of MYOGENIN, MYH3, and MYOD were significantly lower in the siMUNC tissues. This was accompanied with a decrease in myofiber diameter and increase in inflammatory infiltrates in the regenerated tissue, reestablishing the importance of lncRNAs in myogenesis.

Analysis of the transcriptional start sites and promoter elements of the muscle-specific miRNA loci, pre-miRNA-133, and pre-miRNA-206 revealed the presence of lincRNA linc-MD1 [37], which indeed was the first identified muscle-specific lincRNA. Linc-MD1 is specifically activated when myoblasts, satellite cells, or MYOD-trans-differentiated fibroblasts (muscle cells derived from myoblasts) were subjected to differentiation. This lncRNA was found to be expressed in newly regenerating muscle fibers. Mechanistically, it acts as a competing endogenous RNA or ceRNA whereby it acts as a sponge or decoy to sequester miRNAs such as miR-133 and miR-135 which otherwise bind to their targets MEF2C and MAML1, both of which are important transcription factors required for myogenesis. In an independent study conducted by Legnini et al. [38], it was shown that another myogenically important RNA-binding protein, HuR, is involved in the cross talk between Linc-MD1 and miR-133. RNA interference experiments for HuR revealed a consistent decrease in the cytoplasmic accumulation of linc-MD1 and increase in the pools of miR-133a/miR-133b. A series of experiments thereafter confirmed that it is the binding of HuR to linc-MD1 that increases its presence in the cytoplasm, aiding its miRNA sponging activity at the expense of miR-133 biogenesis (miR-133 being a result of processing of linc-MD1 by Drosha). In a positive feed-forward loop, linc-MD1 and HuR regulate the differentiation of muscle progenitors and hence myogenesis.

One of the first lncRNAs to be discovered with respect to muscle differentiation was SRA (steroid receptor RNA activator). MYOD co-immunoprecipitates with p68/p72 DEAD box RNA helicases, and both of them were shown to interact with SRA in skeletal muscle cells through immunoprecipitation experiments followed by PCR to score for the associated RNA [39]. Luciferase reporter assay experiments were performed wherein the muscle-specific creatinine kinase enhancer was fused upstream of the luciferase gene and transfected into fibroblast cells along with p68, p72, or SRA expression vectors, individually or in combination. No effect was observed on the luciferase gene expression in any of the above cases. However, expression of MYOD either alone or in conjunction with either of the protein (p68/p72) or RNA (SRA) interactors enhanced the luciferase reporter activity. The highest enhancement was observed when all the three (p68/p72, SRA, and MYOD) were co-expressed, thereby establishing that p68/p72 and SRA act as transcriptional coactivators of MYOD. In fact RNA silencing experiments further proved that these three coactivators of MYOD are essential for the differentiation of muscle cells into myotubes. In another interesting study, it was shown that the SRA transcript is actually alternatively spliced to give rise to a protein counterpart SRAP [40]. In undifferentiated myoblasts versus differentiated myotubes, the ratio between the noncoding SRA and the coding SRAP is largely in favor of the noncoding counterpart. In primary human satellite cells subjected toward differentiation, a similar observation was made, SRA levels being observed to be higher than SRAP. Through a series of luciferase and chromatin immunoprecipitation experiments, SRAP was found to physically bind to SRA and prevent it from acting as the coactivator of MyoD, thus unraveling a network of proteins and RNA, fine-tuning the regulation of myogenic differentiation.

A large imprinted locus known as the Dlk1-Gtl2 (delta-like 1 homolog-gene trap locus 2) contains many protein-coding, noncoding, and paternally/maternally imprinted genes, GTL2 being one of the noncoding RNAs [41]. It is also known as MEG3 in humans. A knockout mouse was generated, the knockout locus encompassing the promoter region and exons 1–5 of the *Gtl2* gene. It was observed that while the mice carrying the deletion at the paternal locus survived and were healthy, the mice carrying the same at the maternal locus did not survive. Intriguingly enough, while the Glt2 knockout embryos showed no abnormalities in organs like the brain, heart, liver, kidney, lung, or spleen, their skeletal muscles showed severe defects of formation. The myofibers of the paraspinal muscles were not only small and rounded with peripherally placed nuclei; they were also lower in number. It was one of the first evidences of a lncRNA being necessary in vivo for the proper development of muscles.

Genome-wide binding studies for a transcription factor Yin yang 1 (YY1), a repressor of muscle differentiation genes in proliferating myoblasts, showed that it actually binds to many intergenic loci in the genome along with previously known or unknown protein-coding loci [42]. The potential linc RNA loci were 63 in number and were named as YAM (YY1-associated muscle lincRNA). One such loci, *Yam-1*, located on chromosome 17, was found to be positively regulated by YY1 in proliferating myoblasts. It was observed that YAM-1 was present in abundance in proliferating myoblasts or in the limb muscles of young mice displaying active myogenesis, whereas it was downregulated during myogenic differentiation of myoblasts in vitro or in vivo in older mice with reduced perinatal myogenesis. These observations were further confirmed by RNA silencing experiments. A probe into the mechanisms revealed that YAM-1 positively regulates the expression of its downstream effector miR-715 which in turn negatively regulates Wnt7b. Wnt7b is known to promote muscle differentiation. YAM-1 knockdown led to the upregulation of Wnt-7b, putting forth a mechanism whereby the anti-myogenic differentiation capacity of YAM-1 might be mediated through miR-715-mediated repression of Wnt7b. A study of the other YAMs showed that while YAM-2 and YAM-4 are pro-myogenic factors during the early stages of muscle differentiation, YAM-3 is again anti-myogenic, providing ample evidence of the tight regulation of muscle differentiation by lncRNAs.

Klattenhoff et al. [43] analyzed RNA-Seq data for the expression of lncRNAs in mouse embryonic stem cells as well as in differentiated tissues and focused on one such lncRNA AK143260. They observed that this lncRNA exhibited higher expression in the heart and hence termed it as *Braveheart (Bvht)*. BVHT was depleted from mouse ESCs by shRNA, and the cells were subjected to in vitro cardiomyocyte differentiation by the embryoid body method. Cardiomyocytes are the muscle cells of the heart. It was observed that in the control cells, ~25% of the embryoid bodies displayed spontaneous rhythmic beating as compared to only ~5% of the knockdown cells. Global gene expression analyses by RNA-Seq in BVHT-depleted cells revealed that a multitude of transcription factors coding genes like *Mesp1*, *Hand1*, *Hand2*, *Nkx2.5*, and *Tbx20* were not activated when the cells were differentiated into the cardiac lineage, establishing the importance of BVHT in cardiac lineage specification. An ES cell line harboring a doxycycline-inducible MESP1 overexpression plasmid, when subjected to cardiac differentiation along with MESP1 induction, was able to rescue the BVHT depletion phenotype. This proved that BVHT acts upstream of MESP1 during cardiac differentiation of ES cells. Studies by Xue et al. [44] were aimed at unraveling the secondary structure of BVHT. It was shown that BVHT possesses a AGIL motif in its 5′ domain. With the help of CRISPR/Cas9 system, they generated a 11 nt deletion in this motif ($bvht^{dagil}$). Interestingly, $bvht^{dagil}$ ES cells showed significantly reduced beating during the cardiac differentiation as compared to the wild-type cells. As observed earlier with BVHT knockdown cells, $bvht^{dagil}$ cells showed a lack of activation of major cardiac transcription factors like *Nkx2.5*, *Hand2*, *Gata4*, and *Gata6*. A protein microarray was employed to understand the interaction partners of $bvht^{dagil}$ wherein CNBP or ZNF9, a zinc finger transcription factor, was found to be an interesting interacting candidate for $bvht^{dagil}$ lncRNA. These studies suggested that the lncRNA protein interaction networks are crucial components of cell fate decisions and lineage commitment.

A brief representation of the various lncRNAs involved in the maintenance and/or differentiation of stem cells for the neural, hematopoietic, and muscle linage has been depicted in Fig. 8.3.

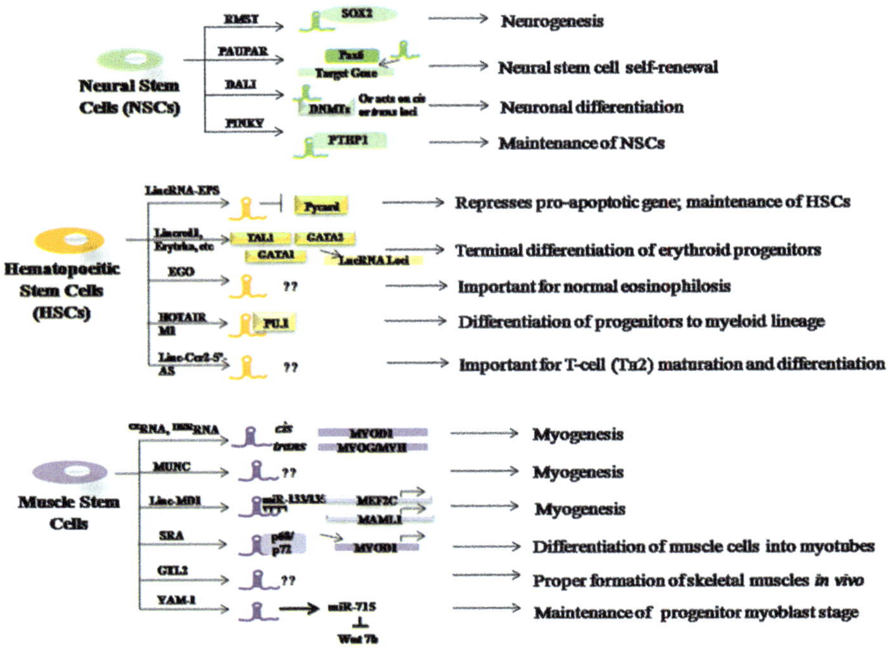

Fig. 8.3 Representative examples of lncRNAs that either maintain the stem cell state of somatic stem cells or promote their differentiation/terminal maturation. The mechanisms can either be through interaction with protein partners, regulating gene loci in cis or trans, or acting as competing endogenous RNAs

8.4.4 Long Noncoding RNAs in Epidermal Stem Cells and Differentiation

The skin is one of the most sturdy and versatile organs of the body in that it not only acts as a protective barrier, providing protection to the body against microbes and dehydration, but also constantly participates in maintaining homeostasis through withstanding temperature changes and providing tactile sense to the body. The stem cell niche of the skin is involved in constantly regenerating the epidermal hair and also in regenerating epidermal tissue after an injury or a wound. In the embryo, post-gastrulation, it is the neuroectoderm that gives rise to the epidermis that essentially starts as a single layer of uncommitted progenitor cells but finally forms a stratified structure, hair follicles, and the sebaceous glands or the apocrine (sweat) glands. In adults, the skin epithelium is made up of blocks, each block being made up of a pilosebaceous unit consisting of hair follicle (HF) and sebaceous gland along with the surrounding interfollicular epidermis (IFE). The HF contains multipotent stem cells that regenerate the hair as well as supply cells for replenishing damaged ones post injury for both the hair follicle and the epidermis. The IFE contains progenitor cells too that maintain tissue integrity and self-renewal under normal circumstances. Various types of signaling pathways including Wnt/β-catenin, BMP, Notch, and Shh have been implicated in the self-renewal and/or differentiation of the epidermal stem cells [45].

To understand the role of lncRNAs in keratinocyte differentiation from epidermal stem cells, Kretz et al. [46] performed high-throughput sequencing of human primary keratinocytes at various days of calcium-induced differentiation and uncovered 295 annotated and 835 unannotated putative lncRNAs. Keratinocytes are the major cell type of the epidermis. At 3 and 6 days of differentiation, the lncRNA reads obtained were compared with that of 0 day (progenitor population), and it was observed that there were significant perturbations at each of the stages of differentiation studied. To have a broader picture of previously unknown lncRNAs that may have a role to play in suppressing differentiation of various types of progenitors, RNA was obtained from keratinocytes, adipocytes, and osteoblasts in the progenitor and differentiated states and hybridized to tiling arrays. One interesting hit came in the form of the lncRNA NR_024031, termed hitherto as ANCR (antidifferentiation noncoding RNA) which was repressed in each of the model systems studied. ANCR, located in human chromosome 4, consists of three exons, miRNA4449-encoding sequence and a snoRNA-generating sequence in the introns 1 and 2, respectively. It codes for a 855-bp-long transcript that was found to be significantly downregulated at days 3 and 6 of keratinocyte differentiation. Interestingly, the ANCR lncRNA is expressed in multiple human tissues and is concomitantly repressed in many differentiated cell types, indicating its functional relevance in the transition from progenitor to differentiated states. RNAi against ANCR in progenitor keratinocytes induced the expression of many differentiation-related genes like filaggrin, loricrin, keratin 1, small proline-rich proteins 3 and 4, involucrin, S100 calcium-binding proteins A8 and A9, and ABCA12. Microarray analyses under such conditions revealed the perturbation of 388 genes including genes responsible for epidermal differentiation, keratinization, and cornification. Furthermore ANCR was depleted in regenerated, organotypic epidermal tissue, a system recapitulating most aspects of the human epidermis. Interestingly similar results were observed, with even the epidermal basal layer expressing differentiation genes which otherwise is not known to express such genes. Thus ANCR seems to be necessary to keep differentiation-related genes from expressing in the progenitor cell niche of the epidermis and hence in maintaining the identity of keratinocyte progenitors.

This group also identified TINCR (terminal differentiation-induced ncRNA) on chromosome 19 of the human genome encoding a 3.7 kb transcript, highly expressed, by greater than 150-fold, during epidermal differentiation [47]. It was shown to be enriched in the differentiated layers of human epidermal tissue, indicating its role in the differentiation of keratinocytes. When TINCR was downregulated by RNAi in organotypic culture system, expression of key differentiation genes was perturbed in expression although the epidermis stratified normally. Transcript profiling revealed 394 genes to be affected in expression, including those involved in the formation of the epidermal barrier. Specifically, caspase-14 required for proteolysis during the formation of the barrier was reduced drastically, and protein-rich keratohyalin granules and lipid-rich lamellar bodies were ill-formed in the epidermis. To elucidate the mechanism of action of TINCR, an interactome analysis was done using a protein microarray consisting of approximately 9400 recombinant proteins. STAU1 protein showed the highest affinity of binding with TINCR. Although STAU1 has not been previously implicated in epidermal differentiation, it was found that STAU1 depletion recapitulated effects of TINCR depletion, and there was a significant overlap of regulated genes between siSTAU1 and siTINCR cells with a predominance of genes involved in keratinocyte differentiation. Together, TINCR and STAU1 were shown to bind to and functionally stabilize mRNAs encoding key structural and regulatory proteins necessary for keratinocyte differentiation.

8.4.5 Long Noncoding RNAs in Spermatogonial Stem Cells and Differentiation

Spermatogenesis is a physiological process which defines the formation of the spermatozoa through a series of differentiations undergone by progenitor cells referred to as spermatogonial stem cells (SSCs). In the embryonic stages, primordial germ cells (PGCs) represent a population of cells that arise in the epiblast at 7–7.5 dpc of development and migrate to the gonadal ridges at around 12.5 dpc. Once they reach the gonadal ridge, the erstwhile proliferating PGCs enter into a mitotic arrest and reenter the cell cycle only after birth. They populate the basement membrane of seminiferous tubules generating a niche comprising the Sertoli cells, Leydig cells, and surrounding interstitial cells. They undergo constant self-renewal to generate millions of spermatozoa daily. Three types of spermatogonia were initially identified based on the nuclear architecture [48]: type A consisting of a more decompacted chromatin structure, type B spermatogonia consisting of a more heterochromatic chromatin, and an intermediate type between the both. Type A spermatogonia are the undifferentiated cells further classified into three types: A_{single} (A_s), A_{paired}(A_{pr}), and $A_{aligned}$(A_{al}) depending on the arrangement on the basement membrane of the seminiferous tubule. A single division of A_s leads to the formation of either (1) a A_{pr} that generates two A_s post-cytokinesis or (2) the two resulting cells remain connected by a cytoplasmic bridge that generates a chain of four A_{al} in the next round of division. The four A_{al} spermatogonia undergo mitotic divisions to generate 32 A_{al} spermatogonia, and 4–16 such chains are finally committed to differentiation. The A_{al} spermatogonia give rise to the type B spermatogonia which generate primary spermatocytes that undergo meiosis. Two rounds of meiosis give rise to secondary spermatocytes and haploid spermatids. The haploid spermatids then undergo morphological changes through 16 steps (in mouse) finally forming the mature spermatozoa.

One of the first identified lncRNAs in our laboratory which was shown to have a functional role in spermatogonial physiology is MRHL (mouse recombination hotspot locus) RNA [49]. It is a 2.4 kb transcript, expressed in the adult mouse testis and processed in vitro by the Drosha machinery to a 80 nt processed transcript [50]. To gain an understanding of its function in the mammalian testis [51], the RNA was downregulated in the mouse spermatogonial cell line (Gc1-Spg). Subsequent microarray analyses revealed a host of signaling pathways being affected, a prominent and noteworthy one being the Wnt signaling. Mass spectrometry identified p68/DDX5 helicase as one of the interacting proteins of MRHL following which it was shown that in mrhl RNA-depleted conditions, p68 translocates from the nucleus to the cytoplasm and aids the shuttling of Wnt signaling effector protein β-catenin into the nucleus resulting in subsequent activation of Wnt signaling. Thus, in mouse spermatogonial cells, mrhl RNA negatively regulates Wnt signaling through interaction with p68. Genome-wide occupancy studies of MRHL on the chromatin were performed through ChOP-Seq (chromatin oligoaffinity purification followed by sequencing) [52]. This study revealed that MRHL physically occupies 1400 loci among which 37 loci are regulated by this lncRNA. These loci are termed as the GRPAM loci (genes regulated by physical association of MRHL) which include genes involved in Wnt signaling, spermatogenesis, and differentiation. ChIP- and shRNA-mediated downregulation studies showed that Wnt signaling acts to downregulate MRHL RNA when spermatogonial cells are exposed to Wnt3a ligand. A detailed investigation into the mechanism of Wnt-mediated MRHL RNA downregulation revealed CTBP1 as the corepressor that increasingly occupies the promoter of *Mrhl* and establishes repressive histone modifications like H3K9me3 on the promoter leading to repression of transcription of the RNA [53]. Interestingly, it was also observed that upon Wnt treatment of spermatogonial cells, various premeiotic (*c-kit*, *Dmc1*, *Stra8*, *Lhx8*) as well as meiotic markers (*Zfp42*, *Hspa2*, *Mtl5*, and *Ccna1*) were significantly upregulated. Rescue of MRHL in trans did not abrogate these changes indicating that additional factors are necessary for the upregulation of these meiotic markers which are activated only under Wnt conditions. These studies thus proved that mrhl RNA acts at the chromatin level to regulate key aspects of spermatogonial differentiation initiated by Wnt signaling (Fig. 8.4).

A comprehensive genome-wide study was recently carried out by Sun et al. [54] wherein they performed lncRNA microarray analysis from 6-day-old (neonatal) and 8-week-old (adult) testis. They found that out of the ~14,000 lncRNA genes represented on the microarray, ~8000 (56%) exhibited expression above background, and 37% of these (~3000 lncRNAs) showed differential expression between the two stages studied. They classified all lncRNAs perturbed into specific groups such as exonic sense or antisense, intronic sense or antisense, and bidirectional or intergenic based on their locations and directions of transcription and found interesting correlations between the expression of theses lncRNAs and their neighboring protein-coding counterparts. For example, *Ccnd2*-coding gene expression occurs primarily in spermatogonia and is important for their self-renewal. Both *Ccnd2* and its associated sense lncRNA AK011429 were found to be downregulated in the adult testis tissue. Similarly, AK077193, expressed antisense to *Sycp2* (synaptonemal complex protein 2), was upregulated in the adult testis, and the expression was positively correlated with that of *Sycp2* itself, a gene required during meiosis in spermatocytes. LncRNA AK00574 was found to be specifically upregulated and highly expressed along with the protein-coding gene *Spata17* from whose intron it is transcribed in an antisense direction. *Spata17* is involved in male germ cell apoptosis

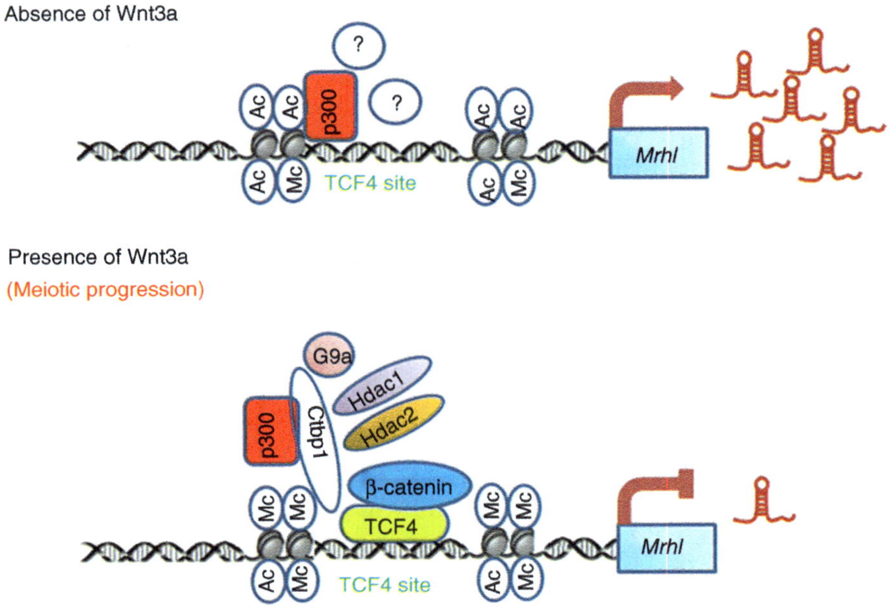

Fig. 8.4 Model summarizing the changes occurring at the proximal promoter region of mrhl RNA at the TCF4 binding site upon Wnt signaling activation with respect to the binding of different proteins like β-catenin-, TCF4-, Ctbp1-, and Ctbp1-associated proteins (p300, G9a, Hdac1, and Hdac2). p300 binds at the TCF4 binding site even in the absence of Wnt3a, and other proteins (?) could be associated with p300 for regulation of mrhlRNA expression. The changes in histone modifications are also shown. In the presence of Wnt3a, mrhlRNA downregulation possibly leads to meiotic commitment and differentiation

in the adult testis. Although the specific functions of these lncRNAs need to be elucidated, this study has listed a cohort of lncRNAs with possible functions in male germ cell differentiation and testes development.

Similar high-throughput transcriptome analysis was performed by Li et al. [55] on primary Thy1$^+$ spermatogonial stem cell cultures in various conditions such as (1) in the presence of the growth factor GDNF, (2) 18 h post-depletion of GDNF, and (3) post 8 h reexposure to GDNF in the depleted cultures. Interestingly, normal cultures growing in the presence of GDNF showed expression of twice the number of lncRNA transcripts as compared to protein-coding mRNAs, whereas in the depleted and replenished cultures, an equal proportion of both types of transcripts was perturbed. LncRNA 033862 was found to have the most significant expression changes upon GDNF withdrawal in SSC cultures. Its expression decreased upon GDNF withdrawal for 18 h, reappeared post 8 h of GDNF reexposure, and underwent almost 97% reduction upon 30 h of GDNF removal from cultures. Tissue-specific expression analysis revealed that this RNA is highly expressed in mouse testis and brain. In the mouse testis specifically, it was expressed during the immediate postnatal stages (P1–P3) with subsequent reduction in levels at P7 and P10, indicating its role in gene regulation in the spermatogonial progenitor cells of the testis. Indeed, in situ hybridization showed expression of this lncRNA in the spermatogonial cells located in the basement membrane of seminiferous tubules of testis. Chromatin isolation by RNA purification (ChIRP) experiments revealed that lncRNA 033862 bound physically to the *Gfra1* locus on mouse chromosome 19. LncRNA 033862 is transcribed in an antisense direction from exon 9 of *Gfra1* (GDNF family receptor). Knockdown experiments using lentiviral shRNA in SSC cultures led to increased apoptosis, significant changes in morphology with reduction in colony size and downregulation of SSC-associated self-renewal genes like *Bcl6b*, *Ccnd2*, and *Pou5f1*, and reduction in expression of *Gfra1* itself. Differentiation genes like *Stra8*, *Sycp1*, and *c-kit* were however not affected, thereby establishing that lncRNA 033862 is necessary for SSC self-renewal and maintenance. Furthermore, in vivo transplantation of the lncRNA knocked down cells into testis showed lower colonization of testis from donor cells as compared to controls. *Gfra1* encodes the co-receptor for GDNF in SSCs. The above studies proved the necessity of lncRNA 033862 in SSC maintenance and indicated that absence of GDNF signaling which led to reduction in expression of lncRNA 033862 might be the cause for transcriptional silencing of *Gfra1*, revealing an intricate role of this lncRNA in spermatogonial stem cell gene regulation.

TSX (testis-specific X-linked) is a lncRNA that is expressed from the highly characterized X-inactivation center in mammals being encoded upstream of the lncRNA locus *Xite* [56]. An expression pattern analysis revealed that while in female mice, TSX is expressed at higher levels in the brain than in the gonadal tissue; it is the reverse in males. Male gonadal tissue showed 10–100 times higher expression as compared to the brain. Isolation of male germ cells and further analyses showed that while in type A and B spermatogonia, TSX levels are comparatively lower; it is upregulated by 40-fold in the pachytene stage spermatocytes during meiosis with levels again decreasing thereafter, albeit maintaining steady-state levels in the postmeiotic stages. Generation of *Tsx* knockout mice did not affect viability of the offsprings or their Mendelian ratio although homozygous knockout female mice exhibited reduced fertility and preferred the birth of female offsprings. Closer inspection of 6-month-old testes of −/Y males showed smaller size in comparison to the wild-type ones. TUNEL experiments revealed increased apoptosis of germ cells, peaking at 14 days of development, coinciding with the first phase of pachytene stage. Further staining with SCP1 (synaptonemal complex protein 1) confirmed that it was indeed the pachytene spermatocytes that were undergoing apoptosis, thereby suggesting that lncRNA TSX might be required for germ cells to enter the meiotic phase of differentiation although its function might be redundant in the maturation of haploid spermatids during spermiogenesis.

8.5 Conclusions

Stem cells are an integral part of animal development. During the last two decades, we have seen an explosion in our basic understanding of stem cell biology. Stem cells are also being explored as an effective mode of human disease management and treatment. The first stem cell therapy ever to be performed was in 1968 when clinicians successfully carried out bone marrow transplantation. Bone marrow contains multipotent stem cells that can give rise to all the types of blood cells. Since then bone marrow transplantation has formed one of the major stem cell therapies, helping millions of patients suffering from cancers like leukemia. Not very far behind was the concept of using skin stem cells to replace burnt tissue in the form of skin grafts. Limbal stem cells in the eye have also huge potential in replacing lost corneal tissue by virtue of their stem cell properties. These are some of the successful stories of stem cell therapies. There are still a number of human diseases and disorders that need to be addressed via stem cell therapies. For example, Duchenne muscular dystrophy (DMD) is a genetic disease in which skeletal muscles and often heart muscles weaken over time due to prevention of formation of dystrophin protein. As we know, muscle harbors stem cells known as satellite cells which serve as great contenders for curing such genetic diseases. On the other hand, iPSCs also possess immense potential because adult somatic cells can be reprogrammed into iPSCs which can then theoretically be directed into the generation of any type of cell such as neurons for replacement in neurodegenerative diseases like Parkinson's and Alzheimer's diseases. One of the major challenges of stem cell therapies is the generation of a pure population of cells which can be transplanted into the human body without complications of tissue rejection and immune responses. In this direction, it is very important to understand the fine details of the molecular mechanisms of differentiation processes so that we can take care of every small detail that leads to the generation of the right type of cell with the expected phenotype. In this context, the emerging lncRNAs as key regulators of lineage-specific differentiation might serve as an important tool to fine-tune the differentiation pathway. This field although very nascent provides us with potential hope in making regenerative medicine a highly successful strategy in clinical practice in the near future.

Acknowledgments M.R.S. Rao thanks the Department of Science and Technology for J.C. Bose and SERB Distinguished fellowships. Funding has been granted by the Department of Biotechnology (BT/01/COE/07/09).

Glossary

Microarray It employs an array comprising of probes which can be DNA, cDNA, or oligonucleotides representing the sequences in a particular genome. Hybridization of query sequences to these probes can allow for the parallel analysis of gene expression for thousands of genes or for the identification of new genes.

ChIP-Seq Chromatin immunoprecipitation is a technique in which chromatin is isolated from cells or tissues, fragmented by sonication, chromatin associated with a particular protein is pulled down with the help of an antibody specific against the protein of interest, and the DNA is subsequently recovered. This is followed by sequencing of the DNA to decipher genomic binding loci of the concerned protein.

RNA-Seq RNA Sequencing uses a population of RNA (such as polyA+) to be converted to a library of cDNAs using adapters at one or both the ends. The library is then subjected to high-throughput sequencing where each molecule is sequenced to obtain reads that are typically 30–400 bp long. The reads are then aligned to a reference genome or reference transcriptome or assembled to generate a transcriptome for the particular system used for the RNA-Seq. This accurately depicts not only the transcriptome but the expression level of each gene for that system [57].

CHART In capture hybridization analysis of RNA targets, the RNA is cross-linked to its genomic binding sites on the chromatin, and the genome is isolated and fragmented. The RN-bound fragments are then enriched with the help of complementary locked or O2'-methylated oligonucleotides which are immobilized by beads. The corresponding DNA or protein fractions are then eluted to analyze either loci of binding or interacting partners for the RNA of interest.

ChIRP/ChOP Chromatin isolation by RNA purification or chromatin oligoaffinity purification. In this case, the complementary oligonucleotides are biotinylated, and the RNA-bound chromatin fragments are enriched by magnetic streptavidin beads. The DNA associated with the RNA or the interacting proteins can then be eluted for further analysis by sequencing or mass spectrometry, respectively.

siRNA/shRNA Mediated Knockdown Short-interfering RNAs are double-stranded RNA molecules consisting of a 3′ 2 nt overhang that activates the RNAi machinery inside the cytoplasm of cells upon delivery. After processing, one of the strands of the siRNA binds to its complementary sequence on the target mRNA leading to degradation by the RISC (RNA induced silencing complex). Short hairpin RNAs are transcribed from a plasmid in the form of a stem loop primary RNA which is processed by the Drosha machinery in the nucleus to generate siRNA.

CRISPR/Cas9 The clustered regularly interspaced short palindromic repeats is a bacterial immune system that is used to cleave invading foreign DNA. This technique is now used for genome engineering. The CRISPR system consists of a guide RNA and a nonspecific endonuclease, Cas9. The guide RNA "guides" the Cas9 endonuclease to the target region in the genome wherein Cas9 creates double-stranded breaks. The DNA sequence is then repaired with the help of either NHEJ- or HDR-mediated repair generating indels or desired knockouts/knockins.

Fluorescence In Situ Hybridization FISH technique is used to label or localize regions of interest in the genome or transcriptome with the help of short sequences known as probes. These probes are most often labeled with a fluorescent tag. The probes bind to the target regions of interest by complementary hybridization, and signals can be detected by fluorescent microscopy to understand the localization/copy number of the targets.

References

1. Huttenhofer A, Schattner P, Polacek N (2005) Non-coding RNAs: hope or hype? Trends Genet 21(5):289–297
2. Brown CJ et al (1992) The human XIST gene: analysis of a 17 kb inactive X-specific RNA that contains conserved repeats and is highly localized within the nucleus. Cell 71(3):527–542
3. Bartolomei MS, Zemel S, Tilghman SM (1991) Parental imprinting of the mouse H19 gene. Nature 351(6322):153–155
4. Evans MJ, Kaufman MH (1981) Establishment in culture of pluripotential cells from mouse embryos. Nature 292(5819):154–156
5. Martin GR (1981) Isolation of a pluripotent cell line from early mouse embryos cultured in medium conditioned by teratocarcinoma stem cells. Proc Natl Acad Sci U S A 78(12):7634–7638
6. Thomson JA et al (1995) Isolation of a primate embryonic stem cell line. Proc Natl Acad Sci U S A 92(17):7844–7848
7. Thomson JA et al (1998) Embryonic stem cell lines derived from human blastocysts. Science 282(5391):1145–1147
8. Shamblott MJ et al (1998) Derivation of pluripotent stem cells from cultured human primordial germ cells. Proc Natl Acad Sci U S A 95(23):13726–13731
9. Yu J et al (2007) Induced pluripotent stem cell lines derived from human somatic cells. Science 318(5858):1917–1920
10. Takahashi K, Yamanaka S (2006) Induction of pluripotent stem cells from mouse embryonic and adult fibroblast cultures by defined factors. Cell 126(4):663–676
11. Guttman M et al (2011) lincRNAs act in the circuitry controlling pluripotency and differentiation. Nature 477(7364):295–300
12. Sheik Mohamed J et al (2010) Conserved long noncoding RNAs transcriptionally regulated by Oct4 and Nanog modulate pluripotency in mouse embryonic stem cells. RNA 16(2):324–337
13. Lin N et al (2014) An evolutionarily conserved long noncoding RNA TUNA controls pluripotency and neural lineage commitment. Mol Cell 53(6):1005–1019
14. Chakraborty D et al (2012) Combined RNAi and localization for functionally dissecting long noncoding RNAs. Nat Methods 9(4):360–362
15. Ng SY, Johnson R, Stanton LW (2012) Human long non-coding RNAs promote pluripotency and neuronal differentiation by association with chromatin modifiers and transcription factors. EMBO J 31(3):522–533
16. Wang Y et al (2013) Endogenous miRNA sponge lincRNA-RoR regulates Oct4, Nanog, and Sox2 in human embryonic stem cell self-renewal. Dev Cell 25(1):69–80
17. Loewer S et al (2010) Large intergenic non-coding RNA-RoR modulates reprogramming of human induced pluripotent stem cells. Nat Genet 42(12):1113–1117
18. Guttman M et al (2009) Chromatin signature reveals over a thousand highly conserved large non-coding RNAs in mammals. Nature 458(7235):223–227
19. Urban N, Guillemot F (2014) Neurogenesis in the embryonic and adult brain: same regulators, different roles. Front Cell Neurosci 8:396
20. Ng SY et al (2013) The long noncoding RNA RMST interacts with SOX2 to regulate neurogenesis. Mol Cell 51(3):349–359
21. Vance KW et al (2014) The long non-coding RNA Paupar regulates the expression of both local and distal genes. EMBO J 33(4):296–311
22. Chalei V et al (2014) The long non-coding RNA Dali is an epigenetic regulator of neural differentiation. elife 3:e04530
23. Ramos AD et al (2015) The long noncoding RNA Pnky regulates neuronal differentiation of embryonic and postnatal neural stem cells. Cell Stem Cell 16(4):439–447
24. Till JE, Mc CE (1961) A direct measurement of the radiation sensitivity of normal mouse bone marrow cells. Radiat Res 14:213–222
25. Hu W et al (2011) Long noncoding RNA-mediated anti-apoptotic activity in murine erythroid terminal differentiation. Genes Dev 25(24):2573–2578

26. Paralkar VR, Weiss MJ (2011) A new 'Linc' between noncoding RNAs and blood development. Genes Dev 25(24):2555–2558
27. Wagner LA et al (2007) EGO, a novel, noncoding RNA gene, regulates eosinophil granule protein transcript expression. Blood 109(12):5191–5198
28. Zhang X et al (2009) A myelopoiesis-associated regulatory intergenic noncoding RNA transcript within the human HOXA cluster. Blood 113(11):2526–2534
29. Wei S et al (2016) PU.1 controls the expression of long noncoding RNA HOTAIRM1 during granulocytic differentiation. J Hematol Oncol 9(1):44
30. Hu G et al (2013) Expression and regulation of intergenic long noncoding RNAs during T cell development and differentiation. Nat Immunol 14(11):1190–1198
31. Ranzani V et al (2015) The long intergenic noncoding RNA landscape of human lymphocytes highlights the regulation of T cell differentiation by linc-MAF-4. Nat Immunol 16(3):318–325
32. Casero D et al (2015) Long non-coding RNA profiling of human lymphoid progenitor cells reveals transcriptional divergence of B cell and T cell lineages. Nat Immunol 16(12):1282–1291
33. Yin H, Price F, Rudnicki MA (2013) Satellite cells and the muscle stem cell niche. Physiol Rev 93(1):23–67
34. Yablonka-Reuveni Z (2011) The skeletal muscle satellite cell: still young and fascinating at 50. J Histochem Cytochem 59(12):1041–1059
35. Mousavi K et al (2013) eRNAs promote transcription by establishing chromatin accessibility at defined genomic loci. Mol Cell 51(5):606–617
36. Mueller AC et al (2015) MUNC, a long noncoding RNA that facilitates the function of MyoD in skeletal myogenesis. Mol Cell Biol 35(3):498–513
37. Cesana M et al (2011) A long noncoding RNA controls muscle differentiation by functioning as a competing endogenous RNA. Cell 147(2):358–369
38. Legnini I et al (2014) A feedforward regulatory loop between HuR and the long noncoding RNA linc-MD1 controls early phases of myogenesis. Mol Cell 53(3):506–514
39. Caretti G et al (2006) The RNA helicases p68/p72 and the noncoding RNA SRA are coregulators of MyoD and skeletal muscle differentiation. Dev Cell 11(4):547–560
40. Hube F et al (2011) Steroid receptor RNA activator protein binds to and counteracts SRA RNA-mediated activation of MyoD and muscle differentiation. Nucleic Acids Res 39(2):513–525
41. Zhou Y et al (2010) Activation of paternally expressed genes and perinatal death caused by deletion of the Gtl2 gene. Development 137(16):2643–2652
42. Lu L et al (2013) Genome-wide survey by ChIP-seq reveals YY1 regulation of lincRNAs in skeletal myogenesis. EMBO J 32(19):2575–2588
43. Klattenhoff CA et al (2013) Braveheart, a long noncoding RNA required for cardiovascular lineage commitment. Cell 152(3):570–583
44. Xue Z et al (2016) A G-rich motif in the lncRNA Braveheart interacts with a zinc-finger transcription factor to specify the cardiovascular lineage. Mol Cell 64(1):37–50
45. Blanpain C, Fuchs E (2006) Epidermal stem cells of the skin. Annu Rev Cell Dev Biol 22:339–373
46. Kretz M et al (2012) Suppression of progenitor differentiation requires the long noncoding RNA ANCR. Genes Dev 26(4):338–343
47. Kretz M et al (2013) Control of somatic tissue differentiation by the long non-coding RNA TINCR. Nature 493(7431):231–235
48. Luk AC et al (2014) Long noncoding RNAs in spermatogenesis: insights from recent high-throughput transcriptome studies. Reproduction 147(5):R131–R141
49. Nishant KT, Ravishankar H, Rao MR (2004) Characterization of a mouse recombination hot spot locus encoding a novel non-protein-coding RNA. Mol Cell Biol 24(12):5620–5634
50. Ganesan G, Rao SM (2008) A novel noncoding RNA processed by Drosha is restricted to nucleus in mouse. RNA 14(7):1399–1410
51. Arun G et al (2012) mrhl RNA, a long noncoding RNA, negatively regulates Wnt signaling through its protein partner Ddx5/p68 in mouse spermatogonial cells. Mol Cell Biol 32(15):3140–3152
52. Akhade VS et al (2014) Genome wide chromatin occupancy of mrhl RNA and its role in gene regulation in mouse spermatogonial cells. RNA Biol 11(10):1262–1279

53. Akhade VS et al (2016) Mechanism of Wnt signaling induced down regulation of mrhl long non-coding RNA in mouse spermatogonial cells. Nucleic Acids Res 44(1):387–401
54. Sun J, Lin Y, Wu J (2013) Long non-coding RNA expression profiling of mouse testis during postnatal development. PLoS One 8(10):e75750
55. Adriaens C et al (2016) p53 induces formation of NEAT1 lncRNA-containing paraspeckles that modulate replication stress response and chemosensitivity. Nat Med 22(8):861–868
56. Anguera MC et al (2011) Tsx produces a long noncoding RNA and has general functions in the germline, stem cells, and brain. PLoS Genet 7(9):e1002248
57. Wang Z, Gerstein M, Snyder M (2009) RNA-Seq: a revolutionary tool for transcriptomics. Nat Rev Genet 10(1):57–63

Chapter 9
Understanding the Role of lncRNAs in Nervous System Development

Brian S. Clark and Seth Blackshaw

Abstract The diversity of lncRNAs has expanded within mammals in tandem with the evolution of increased brain complexity, suggesting that lncRNAs play an integral role in this process. In this chapter, we will highlight the identification and characterization of lncRNAs in nervous system development. We discuss the potential role of lncRNAs in nervous system and brain evolution, along with efforts to create comprehensive catalogues that analyze spatial and temporal changes in lncRNA expression during nervous system development. Additionally, we focus on recent endeavors that attempt to assign function to lncRNAs during nervous system development. We highlight discrepancies that have been observed between *in vitro* and *in vivo* studies of lncRNA function and the challenges facing researchers in conducting mechanistic analyses of lncRNAs in the developing nervous system. Altogether, this chapter highlights the emerging role of lncRNAs in the developing brain and sheds light on novel, RNA-mediated mechanisms by which nervous system development is controlled.

Keywords Development • Neuron • Brain • Transcription • Evolution

9.1 Evolution of the Brain and Emergence of lncRNAs

The emergence of a true nervous system can be traced back to the evolution of the Bilateria, organisms that displayed two sides that are virtual mirror images, a hollow gut tube and a clustering of nerve cells into a nerve cord. The brain evolved from the clustering of the nerve cells at the anterior pole of the organism, connecting to other clusters of nerve cells, or ganglia, distributed along the central nerve cord. With the evolution of the vertebrates, this ventrally located nerve cord evolved into the dorsally located spinal cord. Likewise, throughout evolution, nervous system development is controlled by a largely conserved set of transcription

B.S. Clark, Ph.D. • S. Blackshaw, Ph.D. (✉)
Johns Hopkins University School of Medicine, Baltimore, MD, USA
e-mail: bclark28@jhmi.edu; sblack@jhmi.edu

M.R.S. Rao (ed.), *Long Non Coding RNA Biology*, Advances in Experimental Medicine and Biology 1008, DOI 10.1007/978-981-10-5203-3_9

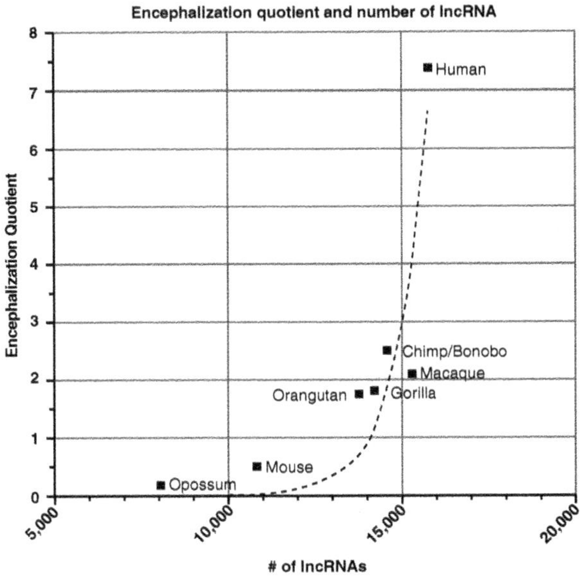

Fig. 9.1 Graph depicting the number of lncRNAs identified in given species relative to the encephalization quotient (EQ). EQ is indicative of the deviation from the expected brain size based on body mass, with larger numbers indicative of larger than expected brain size (normalized to the cat brain/body ratio equivalent) [17, 18]. Number of lncRNAs for a given species is taken from Gencode (*Homo sapiens*, Version 25 (March 2016, GRCh38)—Ensembl 85) [19, 20] or from published reports [21]. Trendline represents nonlinear fit with $R^2 = 0.8343$

factors and signaling molecules [1, 2]. Many of the same gene sets are even present and function orthologously in the more evolutionary primitive "nerve nets" of Cnidaria [3, 4]. These findings raise the question: How do the same gene sets function to control the great diversity of structures and cell types that are found in the nervous system across evolution? While some of this diversity is undoubtedly generated by gene duplications and repurposing of orthologous gene functions [5, 6], other mechanisms are certainly at work. These questions are most pertinent for the development of highly complex mammalian brains, particularly those of higher primates. In this section, we explore the potential central role of lncRNAs in the evolution of the mammalian nervous system.

During vertebrate and more specifically hominid evolution, the brain has undergone an evolutionarily rapid expansion in size. In mammals with larger, more convoluted cortices, the expansion in size correlates with an expanded diversity of progenitor cells [7] that possess an increased proliferative capacity [8–14]. In primates, the greatly increased size of the cerebral cortex appears to result from a dramatic expansion of progenitors in the outer subventricular zone [15, 16]. Interestingly, when brain size is normalized to body size using the encephalization quotient (EQ), the expansion in size of the brain across mammalian evolution shows a strong, nonlinear correlation with the expansion in the numbers of individual lncRNAs (Fig. 9.1).

Consistent with the notion that lncRNAs have co-evolved with the expanded repertoire of brain functions, a majority of lncRNAs examined to date display specific expression within neuroanatomical regions or neuronal cell types in mouse [22, 23]. Many of these brain-enriched lncRNAs are co-expressed with, and display genomic localizations in close proximity to, known neurodevelopmental regulators

[24] and likely regulate similar processes during neurodevelopment. Together, this has led to the general hypothesis that the expanded diversity in lncRNAs is pivotal to the expansion in higher-order cognitive ability of humans and primates and the diversity of neuronal cell types and function. Accordingly, roughly one-third of ~13,800 lncRNAs examined are specific to the primate lineage, with ~40% of the lncRNAs displaying nervous system-specific expression [19, 25–27]. The emergence of brain-specific lncRNAs during primate/human evolution likely occurred through gene duplication, since brain-specific lncRNAs are more likely to originate from genomic regions that have undergone recent duplication than are more ubiquitously expressed or non-brain-enriched lncRNAs [25, 28, 29]. With the continued annotation of genomes and transcriptomes of various species across evolution, we are better able to assess the evolutionary conservation of lncRNAs and their role in the emergence of human-specific traits [30].

Interestingly, although lncRNAs display poor overall primary sequence conservation when compared to protein-coding genes [29, 31], brain-specific lncRNAs display two interesting evolutionary attributes: (1) brain-specific lncRNAs display higher sequence conservation than lncRNAs expressed in other tissues [24, 32] and (2) the spatiotemporal expression patterns of orthologous brain-enriched lncRNAs are maintained across multiple species [32]. This suggests that the expansion in the number of lncRNAs has played a critical role in the development of brain structures throughout the mammalian lineage. To further support the hypothesis that lncRNAs are vital to the evolutionary expansion of relative brain size and cognitive ability, researchers have identified genomic loci that display high conservation throughout vertebrate evolution but have undergone rapid evolution in humans [33–36]. These sequences are postulated to, therefore, play a role in human-specific brain functions. ~2700 "human accelerated regions (HARs)," which had selectively undergone rapid evolution following the divergence of the ancestors of humans and chimpanzees, were identified in these studies. Most HARs mapped to noncoding regions throughout the human genome. Of these, an estimated 30% of HARs mapped to identified brain-specific enhancers. A total of 15 HARs mapped to sequences annotated as long intergenic noncoding RNAs [37]; however, the extent to which the majority of HARs overlap unannotated, intronic, or antisense lncRNAs remains to be analyzed.

Most notably, researchers identified one specific HAR, HAR1, which overlaps the *HAR1F* brain-expressed lncRNA [35]. HAR1 showed the most accelerated substitution rate of any genomic region examined (18 bp of substitution in 118 bp since the last common ancestor of humans and chimpanzees). The *HAR1F* lncRNA was further examined and is expressed developmentally in the Cajal-Retzius neurons of the cortex, the upper cortical plate, the hippocampal primordium, dentate gyrus, cerebellar cortex, and a handful of hindbrain nuclei. *HAR1F* expression in both cortical and extra-cortical regions overlapped with expression of reelin, a known regulator of neurodevelopment. Of particular interest are the Cajal-Retzius neurons. The Cajal-Retzius cells populate the subpial granular layer, a region of the brain that is enlarged in humans [38–41]. To date, the function of *HAR1F*, and many other HARs, remains unknown. It will be of great interest to determine the role of *HAR1F* in Cajal-Retzius cell development.

Aside from HARs that overlap annotated lncRNA sequences, the degree of lncRNA contribution to the evolution of the brain is difficult to assess, given the poor degree of primary sequence conservation that is seen for most lncRNAs [42]. To further address the evolutionary emergence and contributions of lncRNAs in organisms with more complex brains, researchers examined "micro-synteny" of human genomic regions that contained lncRNAs across both large- and small-brained species [43]. These efforts first identified 187 human lncRNAs that are differentially expressed in progenitors or mature neurons of the developing embryonic human brain. When comparing the degree of conservation in genomic architecture surrounding the lncRNA across 30 species (29 mammals plus the chicken), species with large brains (high gyrencephalic indices (GI) >1.5; corresponding on average to approximately one billion cortical neurons [11]) displayed higher than expected conservation of the syntenic genomic landscape surrounding the lncRNA [43]. Conversely, smaller-brained species, with fewer sulci and gyri, displayed lower than expected lncRNA gene-neighborhood conservation. There were, however, two key exceptions: (1) the marmoset, a low-GI primate thought to have recently evolved from a high-GI ancestor [44], and (2) the manatee, a large but lissencephalic species [45]. Both of these species had higher than anticipated degrees of micro-synteny conservation. Importantly, when examining lncRNAs that are expressed in non-neuronal cells across all species, there was no similar correlation between the degree of micro-synteny and brain size [43]. This data suggests an evolutionary pressure to maintain the genomic architecture of regions that include lncRNAs, which in turn is likely to be important for regulation of neurogenesis and brain size. Further supporting this hypothesis, the researchers observed that the degree of micro-synteny conservation of lncRNAs was highest when the lncRNAs were positioned in close proximity to transcription factors that control neuronal development [43].

Additional research both identifying and characterizing nervous system-expressed lncRNAs will continue to aid in our understanding of the evolutionary changes that have enabled the development of the human brain. Our current understanding of the identity (Sect. 9.2) and function (Sect. 9.3) of lncRNAs involved in nervous system development comprises the remainder of the chapter.

9.2 Building a Catalogue of lncRNAs Expressed in the Developing Nervous System

While functional studies on lncRNAs in nervous system development are still lagging considerably (see Sect. 9.3), transcriptomic analyses have identified thousands of lncRNAs. In fact, with much greater coverage of the developing and mature brain by RNA-Seq analysis, recent studies have identified numbers of lncRNAs within given species that approach, or even exceed, the number of protein-coding genes [46]. For example, the NONCODEv4 collection estimates ~56,000 or ~46,000 independent lncRNAs for human and mouse, respectively [47].

While these numbers likely overestimate the true number of lncRNA transcripts, we expect the number of validated lncRNAs to increase beyond the annotated numbers from GENCODE (~15,000 for human, ~9000 for mouse [19]) for two main reasons. First, recent analysis has revealed that lncRNAs show a higher degree of tissue- and cell-type specificity than protein-coding genes [22, 23, 48–52]. It is thus almost certain that large numbers of lncRNAs may have escaped detection in previous RNA-Seq experiments—in particular, many have likely been lost in libraries prepared from bulk tissue, due to highly specific expression in rare cell types and/or low levels of overall expression. This is especially relevant in the nervous system, where the numerous brain structures and nuclei are comprised of highly diverse neuronal subtypes. Advances in single-cell RNA sequencing (scRNA-Seq) along with systematic characterization of individual cell types through efforts such as the BRAIN Initiative [53] will overcome these technical limitations. The power of scRNA-Seq analysis was underlined in a recent study that identified >5500 novel lncRNAs from single cells of the mouse cleavage stage embryo [54]. In addition, experimental design may also be limiting our detection of lncRNAs. Many RNA-Seq experiments are designed to capture only polyadenylated transcripts, in order to deplete the fraction of regulatory RNAs from the sequencing runs. While many lncRNAs are polyadenylated, significant fractions of lncRNAs persist as non-polyadenylated transcripts [19]. Moreover, detection and identification of antisense lncRNAs remains difficult unless strand-specific RNA-Seq libraries are generated, a technique currently not employed for many publically available datasets.

The second reason is our rapidly advancing understanding of the immense complexity of the mammalian transcriptome. Our current knowledge of the transcriptome is substantially limited by the sequencing technologies we employ. Sequencing reads that do not map in a linear fashion to the reference genome are frequently discarded as aberrant or false sequences. The presence of circular RNAs is an example of novel lncRNA species that have only recently been detected in large numbers through more rigorous analyses of RNA-Seq datasets [55–58]. Additional complexity of the transcriptome is being uncovered through the use of targeted RNA-Seq or Capture-Seq [59]. This has identified intragenic splicing events and enabled reliable identification of novel lncRNAs, including lowly expressed or rare lncRNA variants [60–62].

Many studies have begun to examine the expression of lncRNAs in embryonic stem (ES) that have undergone controlled differentiation both in vitro and within progenitor and neural precursor cells within the native developing nervous system in vivo. Here, we summarize these results.

9.2.1 Identification of lncRNAs from In Vitro Studies

Many exploratory studies characterizing lncRNA expression during neuronal development, particularly those assessing human development, focus on analysis of neural progenitors generated through in vitro controlled differentiation from pluripotent stem cells. These studies allow researchers to both easily obtain large quantities of relatively pure cells for in-depth analysis of transcript profiles and to control the precise developmental environment to analyze temporal changes in gene expression. The high degree of cell purity that can often be obtained using these approaches also aids in the detection of transcripts expressed at low levels, a category that includes many lncRNAs. A series of recent studies have shed considerable light on the identity of lncRNAs expressed in pluripotent stem cells and the highly dynamic patterns of lncRNA expression seen during directed differentiations toward specific neuronal cell type fates.

Initial studies of pluripotent stem cells reported that twice as many lncRNAs were selectively expressed in undifferentiated ES cells relative to more differentiated stages [63]. This is consistent with previous observations that analyzed the number of protein-coding genes expressed during pluripotent stages [64–66] and is also consistent with the high overall fraction of the genome that is present as euchromatin in ES cells [67–69]. All told, studies have identified over 250 lncRNAs that are selectively expressed in pluripotent stem cells [63, 70, 71].

Other studies have aimed to identify lncRNAs that are candidates for controlling neuronal identity based on differential expression of lncRNAs during progressive differentiation of stem cells toward neuronal lineages. In one study, researchers examined the expression profiles of lncRNAs during the differentiation of mouse embryonic forebrain-derived neural stem cells using microarrays [72]. These studies examined the bipotent sonic hedgehog-responsive, *Nkx2.1*-positive stem cells that generate both cortical GABAergic interneurons and oligodendrocytes. Comparing the bipotent progenitor cells, GABAergic interneurons, oligodendrocytes, and oligodendrocyte progenitor cells to neural stem cells, the researchers identified 169 lncRNAs (out of 3659 probes on the arrays) with differential expression during neural progenitor cell differentiation. Of particular interest, four lncRNAs were selectively activated upon GABAergic neuronal commitment. One of these, *Ak044422*, appears to function as a pre-miR for *miR-124a*, which accounts for nearly half of all brain-expressed miRNAs [73]. *miR-124a* is known to promote neuronal differentiation at least in part through repression of *Ptbp1* [74]. Consistent with this, *Ak044422* shows complementary expression to *Ptbp1* during neuronal differentiation [72]. However, the researchers suggest that expression of *Ak044422* transcript in the mature nervous system, and posttranscriptional modifications that include alternative splicing and polyadenylation, imply that the *Ak044422* transcript may have additional functions in nervous system development independent of *miR-124a*. This same study identified 100 additional lncRNAs that displayed differential expression upon oligodendrocyte progenitor specification [72].

Similar experiments profiling lncRNA expression were performed on the directed differentiations of human ES cells to dopamine neurons [75]. Microarray profiling of lncRNA expression in ES cells, neurogenic progenitors, and mature dopamine neurons were used to identify lncRNAs that were candidates for regulating maintenance of pluripotency or neurogenic commitment. Over 900 lncRNAs were identified as differentially expressed during neurogenic commitment, with three lncRNAs identified as exclusively expressed in undifferentiated ES cells and 35 lncRNAs highly enriched in neural progenitor cells.

Together, these studies have provided a foundation for examining the considerable diversity of lncRNA expression during neurogenic differentiation and neuronal cell fate commitment. However, in most cases, the extent to which the in vitro expression of individual lncRNAs correlate with their in vivo expression patterns remains undetermined.

9.2.2 LncRNAs Identified Through In Vivo Studies of Neuronal Differentiation

While in vitro studies of cultured cells have provided a wealth of data identifying lncRNA expression with respect to neuronal differentiation from pluripotent stem cells, the extent of lncRNA expression in the nervous system has been further advanced through the profiling of primary tissue samples. Techniques such as customized microarrays for lncRNAs, serial analysis of gene expression (SAGE), and RNA-Seq analysis of primary nervous system tissue have identified thousands of lncRNAs expressed in the nervous system of multiple species [19, 20, 22, 23, 30, 46–50, 61, 76–78]. Additionally, highly cell and tissue-specific lncRNAs can be identified by expression profiling of micro-dissected or sorted tissue and/or cell populations.

Early studies examined the global transcript expression across a time series of retinal development using SAGE [49], identifying multiple noncoding transcripts that showed both temporally dynamic and spatially restricted expression patterns. These analyses identified and examined the retinal expression patterns of lncRNAs including *Six3os* (*Rncr1*), *Neat1* or *Gomafu* (*Rncr2*), and *RncrR3* (the previously mentioned *Ak044422*). Importantly, these studies indicated that some lncRNAs display exceptionally high levels of expression during retinal development, including *Rncr2* which comprised ~0.2% of all polyadenylated RNA transcripts in the neonatal retina [49, 79].

More recent studies have begun to examine the complexity of the transcriptome within defined progenitors and neuronal cell types. In one such study, researchers examined the diversity of transcript expression within three defined subtypes of cortical pyramidal neurons including the sub-cerebral projection neurons, callosal projection neurons, and corticothalamic projection neurons [80]. RNA-Seq analysis of FACS-sorted cells across neurodevelopment identified 806 lncRNAs with significant differential expression between cell types and developmental stages. Four hundred forty-nine of these lncRNAs were selectively expressed in one of these pyramidal cell subtypes, supporting the high degree of cell-type specificity of lncRNAs [23].

LncRNA expression in adult neural stem cells of the subventricular zone (SVZ) of the lateral ventricles and the subgranular zone (SGZ) has been profiled extensively. The first study to address this question was a large-scale in situ hybridization analysis conducted by the Allen Brain Atlas. This effort identified 849 brain-expressed lncRNAs, a number of which were selectively expressed in adult neural stem cells [22]. Later studies have examined the expression of lncRNAs within neurogenic progenitor niches of the adult mouse subventricular zone [77]. Transcript profiles were compared to neurons of the mature olfactory bulb, to which the differentiating cells of the SVZ migrate, neural stem cells of the SGZ, ES cells, and ES cell-derived neurogenic progenitors. This study identified 6876, 5044, or 3680 novel lncRNA transcripts beyond those annotated in the RefSeq, UCSC, or Ensembl reference genome builds, respectively [77]. Consistent with previous reports, lncRNAs displayed more highly spatially and temporally restricted expression than protein-coding genes. To further profile these cell types, RNA Capture-Seq was used and identified an additional 3500 lncRNA transcripts within the SVZ, olfactory bulb, and dentate gyrus [77].

As previously mentioned, additional lncRNAs continue to be identified as sequencing technology advances, particularly with the recent optimization of single-cell RNA-Seq [54]. Recently, researchers profiled the transcript profiles of both bulk tissue samples and individual cells from micro-dissected human neocortices [78]. In bulk-sequencing experiments of tissue across human neocortical development, over 8000 novel lncRNAs were identified. When examining lncRNA expression across 276 individual cells, over 1400 lncRNAs were detected. Interestingly, when the expression levels of individual lncRNAs were analyzed in individual cells, it was found that lncRNAs displayed similar expression levels to protein-coding genes. However, when analyzing expression of the same lncRNAs in bulk tissue samples or within the pooled reads of the 276 individual cells, lncRNA expression levels were detected at much lower levels. This further supports the hypothesis that while lncRNAs are expressed at similar levels to protein-coding genes within individual cells, they display a much higher level of cell type-specific expression.

Numerous studies have now indicated the extensive expression of lncRNAs during all stages of brain development. Global sequencing/profiling experiments have identified thousands of brain-specific lncRNAs. Large-scale efforts, including the Allen Brain Atlas, have begun to examine both the spatial and temporal expression of individual lncRNAs [22, 81]. Additional studies have complemented these large-scale efforts, focusing on more discrete cell populations, including the primary auditory cortex and medial geniculate body [76] or restricted numbers of lncRNAs including, but not limited to, *linc-RBE* [82], *linc-00320* [83], *Dio3os* [84], *Evf1* [85, 86], and *Evf2* [87]. Other studies have observed changes in brain lncRNA expression that are associated with genetic mutants in neurological and psychiatric diseases [88–94] and pharmacological treatments [95, 96] in the brain. Yet despite the large number of lncRNAs that show highly dynamic expression patterns during brain development, many researchers still remain skeptical of their functional importance. A key challenge moving forward is the need to carefully design studies to both address the function of lncRNAs in nervous system development and to identify the mechanisms by which they act.

9.3 LncRNA Function in Nervous System Development

We have previously conducted an extensive review of the role of lncRNAs in regulation of neural development [97]. Here we will briefly discuss the major findings previously reviewed and highlight more recent studies that further demonstrate the importance of lncRNAs in nervous system development.

9.3.1 Lessons Learned from In Vitro Studies

Most large-scale studies of lncRNA function have focused on identifying the regulation of pluripotency states and neural induction. As previously mentioned, numerous lncRNAs display dynamic expression during neural differentiation of ES cells. To identify regulatory role of these lncRNAs, researchers have performed loss of function studies using short hairpin RNA (shRNA)-mediated knockdown and analyzed effects on ES cell differentiation. In particular, inhibition of five lncRNAs resulted in a propensity of the stem cells to adopt a neuroectoderm lineage [98], suggesting a role of these lncRNAs in repressing neural commitment. Additionally, 30% of lncRNAs with selective expression in ES cells interacted with chromatin-modifying proteins, leading the authors to suggest that these lncRNAs function to promote pluripotency through regulation of chromatin architecture [98]. Similarly, further analysis of *lincRNA1230* (*linc1230*) identified that this lncRNA is both necessary and sufficient to repress neural commitment of mouse ES cells [99]. *Linc1230* modulates H3K4me3 accumulation on the promoters of the transcription factors *Pax6* and *Sox1* by interacting with the Trithorax complex component WDR5 [99]. Overexpression of *linc1230* results in reduced WDR5 occupancy and H3K4me3 histone marks at promoters of genes that promote neuronal differentiation, suggesting that *linc1230* inhibits neural induction by sequestering WDR5.

The lncRNA *Tuna* (also known as *megamind* [29]) and 19 additional lncRNAs were identified as regulators of pluripotency through a large-scale RNA-interference screen in mouse ES cells [100]. Interestingly, *Tuna* shows a high degree of sequence homology across vertebrates and is selectively expressed in the nervous system in zebrafish, mouse, and humans. In ES cells, knockdown of *Tuna* results in reduced proliferation, while overexpression of *Tuna* in ES cells resulted in the opposite phenotype, leading to increased proliferation. In differentiating neuronal cultures, loss of *Tuna* expression resulted in reduced expression of neural progenitor markers and genes involved in neural lineage commitment. Consistent with this, knockdown of *megamind* in zebrafish resulted in embryos with small brains and eyes, a phenotype that was rescued by expression of the orthologous zebrafish, human, or mouse isoforms [29]. Further analysis showed that *Tuna* interacts with three RNA-binding proteins (RBPs) [100]. Knockdown of each of these RBPs mimicked the loss of *Tuna* expression within ES cells. Further analyses indicated that *Tuna* expression is required to recruit the RBPs to the promoters of pluripotency factors including *Nanog*, *Sox2*, and *Fgf4*. Chromatin isolation by RNA purification (ChIRP) was used to identify *Tuna*/DNA interactions [101]. ChiRP experiments revealed that *Tuna* was associated with the promoters of pluripotency factors in ES cells [100]. Similarly, the lncRNA *LOC646329* was found to be expressed in radial glia, which functions as neural stem cells during brain development, and in both primary glioblastoma multiforme tumors and glioblastoma-derived cell lines [78]. Inhibition of *LOC646329* expression reduced the propagation of the tumor cell line, identifying an additional lncRNA that regulates the proliferative capacity of stem cells [78].

Other studies have focused on the function of lncRNAs selectively expressed upon neuronal induction or in regulation of cell fate specification in subventricular zone neural stem cells (SVZ NSCs) [77]. *Six3os* expression is enriched in the stem cells of the SVZ relative to SVZ-derived neural precursors, and its knockdown resulted in fewer neurons and oligodendrocytes and an increase in astrocytes [77]. In contrast, *Dlx1as* also displays robust expression in the SVZ, and inhibition of *Dlx1as* expression inhibited neurogenesis and decreased astrocyte formation, but had no effect on oligodendrocyte production [77]. This phenotype, seen following *Dlx1as* knockdown, may result from altered expression of nearby protein coding genes, as this study observed a decrease in transcript levels of both *Dlx1* and *Dlx2*. Short interfering RNA (siRNA) loss of function was used to analyze the function of four lncRNAs (*Rmst*, *Ak124684*, *Ak091713*, and *Ak055040*) that displayed enriched expression upon neuronal induction of human ES cells. In each case, knockdown of the lncRNA resulted a roughly fivefold decrease in the number of neurons generated, instead promoting oligodendrocyte production [75]. Further analysis of these lncRNAs suggests that they control neuronal fate specification through a variety of different mechanisms. These include regulation of chromatin structure through interactions with SUZ12 (*Ak055040*), regulating expression of the neurogenic miRNAs *miR-125b* and *let-7* (*Ak091713*), interaction with the REST/coREST complex (*Ak124684*), and by functioning as a transcriptional co-regulator, recruiting SOX2 to its transcriptional targets (*RMST*) [75, 102]. Further characterization of the lncRNA *Rmst* has discovered that the miRNA *miR-135a2* is encoded in the last intron of *Rmst* [103]. Recent studies have identified a feedback loop where Lmx1b, in response to Wnt/beta-catenin pathway activation, increases expression of *Rmst/miR-135a2* and *miR-135a2*, which in turn decreases *Wnt1* expression levels. This regulatory circuit thus controls the size of dopaminergic progenitor pool of the midbrain [103–105]. In light of this, it will be interesting to determine the extent of which *Rmst* regulates neural induction independent of *miR-135a2* expression.

LncRNAs have also been identified as regulators of oligodendrocyte specification. While many lncRNAs have been identified as selectively expressed in intermediate neural progenitors prior to oligodendrocyte specification [72], relatively few have been directly identified in oligodendrocyte precursors. In one study, the regulatory function of the antisense transcript *Nkx2.2as* was identified to be a positive regulator of oligodendrocyte specification. *Nkx2.2as* overexpression resulted in an increased number of Nestin + stem cells and a bias toward oligodendrocyte lineage during differentiation of neural stem cells [106]. As *Nkx2.2* is required for oligodendrocyte specification [107], the result of overexpression of *Nkx2.2as* on *Nkx2.2* transcript abundance was examined. It was determined that the sense-antisense pairing of *Nkx2.2as* and *Nkx2.2* stabilized *Nkx2.2* mRNA [106]. However, the effect on Nkx2.2 protein levels remains undetermined. In other studies, the expression of lncRNAs during the controlled differentiation of oligodendrocyte precursor cells (OPC) from neural stem cells was examined. These studies identified that *lnc-OPC* (long noncoding RNA-oligodendrocyte precursor cell) shows highly specific expression in OPCs [108]. Olig2, a transcription factor that is necessary and in some contexts sufficient, for OPC specification from neural progenitors [109–111], was found to bind the proximal promoter of *lnc-OPC* and induce its expression upon OPC specification [108]. These data implicate lncRNAs in the regulation of both neuronal and glial differentiation.

As many lncRNAs are primate or even human specific [52], studies of such lncRNAs remain limited to cultured cells. Recently, researchers identified the lncRNA *LncND* in a screen for lncRNA transcripts that may function as miRNA sponges during human brain development [112]. LncRNAs, along with transcribed pseudogenes and circular RNAs [58, 113–120], can fine-tune miRNA concentration by sequestering and stabilizing miRNAs within Argonaute protein complexes [121], thereby controlling translation during development [113, 116, 119, 122]. Interestingly, *LncND* is expressed from a genomic locus that is deleted in individuals with certain neurodevelopmental disorders [123–126]. Expression of *LncND* increases during neurogenesis and rapidly drops upon neuronal differentiation [112]. Similarly, *LncND* is expressed at high levels within the ventricular zone of the developing cerebral cortex [112]. In silico analysis predicted 16 putative miR-143-3p seed sites within *LncND*, which were confirmed using luciferase assays [112]. Interaction of *LncND* with AGO2, a component of the RISC complex, further supported the hypothesis that *LncND* functions as a miRNA sponge [112]. Analysis of mRNA transcripts for miR-143-3p binding sites identified putative binding sites of miR-143-3p in the 3′ UTRs of both *Notch1* and *Notch2* [112]. Consistent with a role in regulating the Notch signaling pathway, knockdown of *LncND* resulted in decreased Notch pathway activation and a corresponding increase in neurogenesis [112]. Overexpression of *LncND* in cerebral organoids resulted in expansion of the radial glial cell population [112], which phenocopies the effects of increased Notch pathway activation [127–129]. These results suggest that *LncND* functions to sequester and stabilize miR-143-3p within neural progenitors, in order to maintain Notch signaling and prevent premature neuronal differentiation [112].

Many lncRNAs, such as *Meg3* and *Dio3os*, are expressed in the brain and other tissues from imprinted loci [84, 130]. The lncRNA *Meg3* acts as a tumor suppressor, likely by regulating apoptosis and angiogenesis [130]. *Dio3os* is also expressed in the brain from an imprinted locus. In contrast to the usual pattern seen with imprinted lncRNAs and associated protein-coding genes, *Dio3os* and *Dio3* are both expressed from the same chromosome [84, 131]. It will be interesting to determine if *Dio3os* facilitates imprinting of its locus through silencing of the opposite chromosome.

9.3.2 Lessons Learned from In Vivo Analysis of lncRNAs

9.3.2.1 lncRNAs in Retinal Development

One neuronal tissue that has provided a wealth of information regarding lncRNA regulation of nervous system development is the developing retina. The retina serves as a simplified neural tissue that arises from a multipotent pool of progenitor cells capable of generating each of the seven major classes of retinal cell types (six neural—retinal ganglion cells (RGCs), amacrine cells (ACs), bipolar cells (BCs), rod photoreceptors, and cone photoreceptors; one glial cell—Müller glial cells (MG)). SAGE analysis, qRT-PCR, and in situ hybridization experiments on retinal tissue across mouse retinal development indicated the presence and abundance of lncRNAs and identified cell-type specific expression of many lncRNAs within discrete retinal cell types [49, 132]. Characterization of the function of these lncRNAs in retinal development has subsequently been performed through in vivo gain and loss of function studies [49, 133–136].

The lncRNA *Tug1* was identified in a screen examining genes upregulated after exposure of RPCs to taurine, which induces rod photoreceptor differentiation [133]. Knockdown of *Tug*1 resulted in abnormal inner and outer segments of photoreceptors, increased cell death, and an increase in the cone photoreceptor marker PNA, consistent with a role of *Tug*1 in promoting rod genesis and inhibiting production of cones [133]. Additional studies of *Tug1* first indicated that *Tug1* regulates cell fate decisions through its interaction with the polycomb repressive complex 2 (PRC2) and through regulation chromatin structure [137]. Interestingly, *Tug1* expression is induced by p53, and loss of *Tug1* expression resulted in an increase of cell-cycle regulator transcript expression, implying that *Tug1* inhibits cell proliferation during cellular damage/stress [133].

The lncRNA *Gomafu*, also known as *RNCR2* or *Miat*, is the most abundantly expressed lncRNA in the developing retina, comprising 0.2% of all polyadenylated transcripts in neonatal mouse retina [49, 135, 138, 139]. Functional studies indicate that *Gomafu* negatively regulates both AC and MG cell differentiation, with loss of function resulting in increased production of ACs and MG [135]. More recently, it was shown that *Gomafu* functions by regulating alternative splicing through interaction with the splicing regulators QKI and SF1 [140]. It will be interesting to determine if *Gomafu*'s role in controlling retinal cell fate specification is mediated by regulation of alternative splicing.

Six3os is a lncRNA that is both divergently transcribed and co-expressed with the homeodomain transcription factor *Six3*. *Six3os* is shown to promote BC specification and inhibit MG development [136]. *Six3os* regulates SIX3 transcriptional activity by acting as a transcriptional scaffold, stabilizing a complex including SIX3, EYA1, and EZH2 and directly regulating expression of SIX3 target genes [136, 141]. In other studies, researchers characterized the expression and function of the natural antisense transcript *Vax2os1*. Overexpression of *Vax2os1* indicates that it functions to maintain the proliferative potential of retinal progenitor cells and prevents premature differentiation of rods [142].

Finally, recent work in the retina has examined functional role of *RNCR4*. *RNCR4* is divergently transcribed from a locus that contains the pre-miRNA cluster miR-183/96/182. The mature miR-183/96/182 and *RNCR4* both display robust expression in photoreceptor cells, beginning at P5 and increasing into adulthood [134]. It was shown that *RNCR4* expression results in increased processing of the pre-mIR-183/96/182 to the mature miRNAs by acting as a repressor to the pri-miR-183/96/182 processing inhibitor *Ddx3x* [134]. Increases in the mature miR-183/96/182 expression levels result in aberrant cellular organization of multiple retinal cell types and the appearance of whorls and rosettes in the outer retina as a consequence of premature miR-183/96/182 expression, which in turn disrupts outer limiting membrane formation by altering Crb1 expression [134]. This suggests that *RNCR4* expression controls the timing of pri-miR-183/96/182 processing to guide retinal histogenesis and outer limiting membrane formation.

9.3.2.2 lncRNAs in Brain Development

As our understanding of genome complexity expands, and tools to manipulate the genome to assess gene function improve, researchers are beginning to assess the requirement of lncRNAs in vivo during brain and nervous system development. In particular, genetic knockout or knockdown experiments are being used to assess the necessity of individual lncRNAs in control of nervous system development. Surprisingly, in many cases in vivo gain and loss of function analysis of individual lncRNAs gives discordant results when compared to in vitro manipulations. Additionally, phenotypes observed following targeted deletions of large regions of DNA that include lncRNA transcript sequence have often been viewed skeptically, due to the potential loss of important cis-regulatory elements of neighboring genes. In this section, we will review our current understanding of the regulation of neural development by lncRNAs in vivo (Sect. 9.3.2.2.1) and highlight the functions attributed to lncRNAs resulting from genetic manipulations of lncRNA expression on nervous system development (Sect. 9.3.2.2.2).

Regulation of Nervous System Development, by lncRNAs In Vivo

Recent studies have identified the neural-specific lncRNA *Pinky* (*Pnky*) as a regulator of neurogenesis in the embryonic and postnatal brain [143]. *Pnky* is expressed at high levels in the ventricular-subventricular zone of the adult brain, a neurogenic niche that is maintained into adulthood [143, 144]. However, *Pnky* expression is downregulated upon activation of the differentiation program of the neural stem cells [143]. Interestingly, when *Pnky* expression is decreased using shRNAs either in vitro or in vivo, the researchers observed an increase in neurogenesis, with a concomitant decrease in the fraction of cells expressing markers of neural stem cells. This likely resulted from an increase in proliferation of transient amplifying cells and decrease in cell death [143]. To further investigate the mechanisms by which

Pnky functions, the researchers analyzed its protein partners using RNA pulldown followed by mass spectrometry, demonstrating that *Pnky* interacted with PTBP1 [143]. PTBP1 has previously been shown to function as a repressor of neuronal differentiation by both regulating pre-mRNA splicing and by inhibiting expression of the neurogenesis-promoting gene *Ptbp2* [145–150]. Like *Pnky* knockdown, *Ptbp1* knockdown also resulted in expanded production of neurons, regulating a highly overlapping gene set as *Pnky* [143]. The physical interaction between *Pnky* and PTBP1 interaction, the identical phenotype seen following loss of expression, the highly overlapping set of regulated genes, and additional epistasis experiments allowed the researchers to conclude that *Pnky* regulates neurogenesis of ventricular-subventricular zone stem cells through regulation of PTBP1 function [143].

Genetic Loss of Function Studies of lncRNAs in Brain Development

While much of our knowledge about the mechanisms by which lncRNAs regulate neural development stems from in vitro studies, recent efforts have begun to increase the number of genetic models of lncRNA loss of function. Here we will highlight the importance of carefully designed loss of function studies (Fig. 9.2) and compare the results from in vivo genetic manipulations to those seen in cell line-based in vitro studies.

One of the more thoroughly analyzed lncRNAs that regulates brain development is *Evf2* (*Dlx6as1*) [87, 151–153]. *Evf2* is expressed in the Shh-responsive cells of the ventral telencephalon during embryonic development and is transcribed from a region that partially overlaps ei, one of the two ultra-conserved enhancers (ei + eii) for the neighboring genes *Dlx5* and *Dlx6* [87]. Transcriptional initiation of *Evf2* occurs just 3′ to the eii enhancer [87]. Initial in vitro experiments suggested that *Evf2* supplied in trans was required for DLX2 recruitment and activation of the *Dlx5*/*Dlx6* enhancer sequence, which in turn activated *Dlx5* and *Dlx6* transcription [87]. However, when *Evf2* expression was inhibited by targeted insertion of a premature polyadenylation signal, *Dlx6* expression was actually increased, suggesting that *Evf2* transcription can also function to inhibit *Dlx6* expression [151]. Interestingly, the regulation of *Dlx6* expression by *Evf2* seems to occur in cis, as low levels of ectopic *Evf2* expression fail to rescue *Dlx6* expression in *Evf2*-mutant mice [151]. However, when high levels of *Evf2* are ectopically expressed in *Evf2* mutants, both *Dlx5* and *Dlx6* transcript levels increase, suggesting a trans-acting effect of *Evf2* similar to those that are observed in vitro [87, 151]. Genetic disruption of *Evf2* expression initially resulted in a decrease in the number of hippocampal interneurons, which resolved as the mice matured [151]. However, although the number of interneurons in the adult mice remained similar to wild-type controls, the researchers observed reduced inhibition of the CA1 pyramidal neuron activity [151], the postsynaptic targets of the hippocampal interneurons, implying the presence of a persistent functional defect in these cells.

Other experiments investigating *Evf2* function suggested that loss of *Evf2* expression resulted in a failure to recruit both DLX proteins and the transcriptional

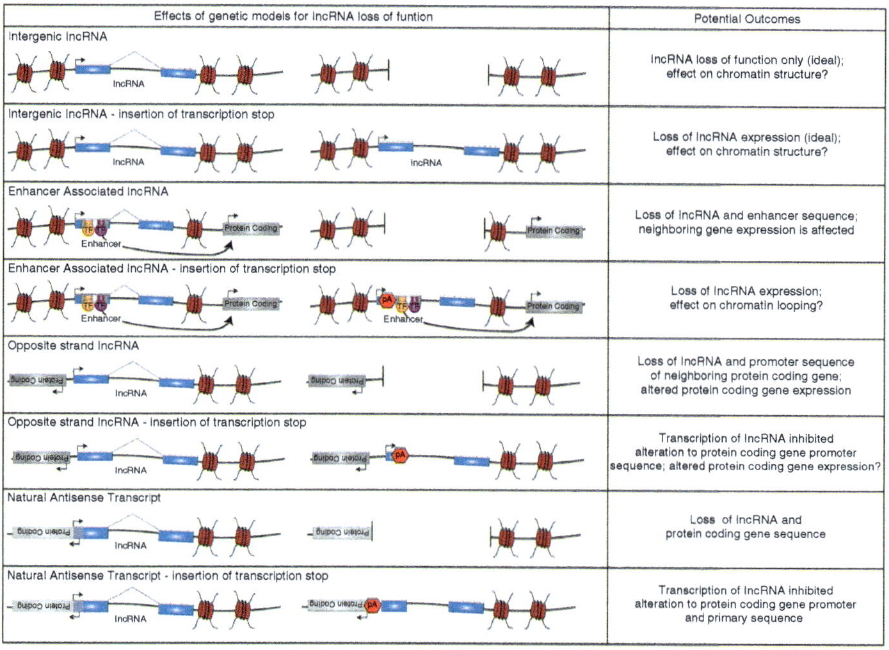

Fig. 9.2 Examples of the effects of lncRNA deletion or insertion of a strong transcriptional stop (pA) on genomic architecture and neighboring gene expression. Careful design of genetic strategies targeting lncRNA loss of function must be implored to ensure that resulting outcomes are the result of lncRNA function and not a consequence of changes to the genome architecture that result in unintended outcomes

repressor MECP2 to the *Dlx5/Dlx6* enhancers [151]. *Evf2* was found to prevent DNA methylation of enhancer ei, suggesting that regulation of ei methylation modulates the binding affinities of DLX1/DLX2 and MECP2 to the enhancer, which in turn regulates *Dlx5/Dlx6* expression [152]. Since *Evf2* is required for recruitment of DLX1/DLX2 to the *Dlx5/Dlx6* enhancer, but does not bind DLX1/DLX2 directly [87, 151], the researchers employed immuno-affinity purification followed by mass spectrometry to identify additional proteins that are part of the *Evf2*-DLX1/DLX2 complex, and which potentially contribute to the *Evf2*-mutant phenotype [153]. These experiments indicated that DLX1 interacts directly with the chromatin remodeling proteins BRG1 and BAF170 in the developing mouse forebrain [153]. DLX1-BRG1 complexes were found to associate with the *Dlx5/Dlx6* enhancers and were enriched in the presence of *Evf2*, suggesting a functional role of the DLX1-BRG1 complex formation in regulation of *Dlx5/Dlx6* expression [153]. Furthermore, BRG1 was found to bind *Evf2* through its RNA-binding domain [153]. Absence of *Evf2* expression decreases DLX1-BRG1 complex formation at the enhancers that control *Dlx5/Dlx6* transcription and also leads to a corresponding decrease in both H3AcK9 and H3AcK18 histone modifications locally [153]. Interestingly, *Evf2* inhibits the ATPase domain of BRG1, suggesting that *Evf2* directs the BRG1-DLX1/

DLX2 complex to the *Dlx5/Dlx6* enhancer but that high levels of *Evf2* transcripts within the complex also inhibit the chromatin remodeling activity of BRG1, thus inhibiting *Dlx5/Dlx6* enhancer activity [153].

Much like the approach used to generate *Evf2* knockout animals, *Dlx1as* expression was genetically disrupted through targeted insertion of a premature polyadenylation sequence [154]. Contrary to in vitro reports, where loss of *Dlx1as* resulted in compensatory decrease in *Dlx1* and *Dlx2* expression [77], loss of *Dlx1as* in vivo results in a modest increase *Dlx1* and *Mash1* transcript expression [154]. Since *Dlx1*-mutant mice display profound defects in hippocampal GABAergic interneuron specification [155–157], the researchers next examined expression of *Gad67*, a marker of GABAergic cells. However, the increase in *Dlx1* expression induced as a consequence of *Dlx1as* loss had no effect on GABAergic interneuron number [154], consistent with previous reports where *Dlx1* overexpression did not induce GABAergic interneuron specification, in sharp contrast to its paralogues *Dlx2* and *Dlx5* [158, 159]. While expression of genes that control GABAergic neuronal specification such as *Dlx1, Mash1,* and *Lhx6* is altered within *Dlx1as* brains, no other neurological phenotypes were observed [154]. Instead, slight defects in the alicochlear commissure were observed in *Dlx1as* mutants [154]. Further investigations will be required to determine if any behavioral phenotypes are observed as a consequence of *Dlx1as* loss of function, akin to those observed for *Evf2*-mutant mice, despite the absence of altered GABAergic interneuron cell number [151].

Recently, a consortium of researchers generated a cohort of targeted deletions of individual mouse lncRNAs to explore their developmental function in vivo [160]. As many lncRNAs overlap protein-coding sequences, the researchers focused on intergenic lncRNAs, so as to investigate phenotypes not attributable to loss of protein-coding gene sequence. Using a combination of cell-based functional assays, RNA-sequencing and computational analyses, the group selected 18 lncRNAs, many with shared homology to human transcript sequences, for targeted deletion in mouse [160]. Of relevance to this review, 12 of the 18 lncRNAs display expression within adult mouse brain or ES cell-derived neural stem cells. Seven of the 12 brain-expressed lncRNAs have human orthologues that display differential expression during human neuronal stem cell differentiation [160].

In order to generate the lncRNA knockouts, the entire lncRNA transcript sequence was replaced with a *lacZ* reporter cassette, allowing a highly sensitive assessment of lncRNA expression patterns using *lacZ/beta-Gal* staining [160]. Targeted inactivation of the lncRNA *Peril*, which is expressed at high levels in mouse ES cells, and shows both temporally and spatially restricted expression within the brain and spinal cord of developing mouse embryos, resulted in reduced viability relative to wild-type or heterozygous littermates [160]. RNA-seq and gene set enrichment analysis (GSEA) of *Peril*$^{-/-}$ brains revealed that genes involved in multiple essential processes including cell cycle regulation and energy metabolism were downregulated [160]. However, the mechanism by which *Peril* loss contributes to lethality remains unknown. The knockout of an additional lncRNA, *Fendrr*, which is expressed at low levels in the developing brain, likewise results in increased lethality. In two separate knockout models, either by gene replacement [160] or insertion of premature polyadenylation signals [161, 162], lethality was observed, as a result of either cardiac [162] or respiratory defects [160]. The role of this lncRNA in brain development, however, remains unexplored.

Interestingly, the locus that encodes the transcription factor *Brn1* (Pou3f3), a well-studied regulator of cortical development [163–165], also contains two lncRNAs (*Pantr1*, also known as *linc-Brn1a*, and *Pantr2*, also known as *linc-Brn1b*) that were both deleted as part of this knockout project [160]. *Pantr1* and *Pantr2* both have conserved human orthologues and were identified as differentially expressed during neural stem cell differentiations [160]. *Pantr1* is transcribed from a region immediately upstream of the transcriptional start site of *Brn1*, and on the opposite strand. Deletion of *Pantr1*, therefore, may also delete portions of the proximal promoter of *Brn1* and is likely to lead to a decrease in *Brn1* transcription. *Pantr2*, however, is transcribed from a region ~10 kb downstream of the *Brn1* locus, again on the opposite DNA strand. Using the *lacZ* knock-in, it was shown that *Pantr2* is expressed within neural progenitors of the mouse dorsal and ventral telencephalon at E13.5. Expression is maintained within both the subventricular and ventricular zones at E15.5, but more restricted expression is observed in the superior cortical layers by E18.5. Upon deletion of *Pantr2*, *Brn1* transcript and BRN1 protein levels were both decreased by ~50%. *Pantr1* expression, however, was increased [160]. Deletion of *Pantr2* resulted in a reduction in the thickness of all cortical layers, likely the result of reduced proliferation of the intermediate progenitors of the subventricular zone that subsequently give rise to cortical neurons. Examination of the cortex of *Pantr2*$^{-/-}$ mice revealed that a subset of upper-layer cortical neurons were converted to deep layer neurons [160]. This cortical thinning phenotype is similar to what is observed for *Brn1/Brn2* (*Pou3f3/Pou3f2*) double, but not single, mutants [163–165], suggesting that *Pantr2* functions in the specification of upper cortical neuron identity independent of its role in regulating *Brn1* expression [160].

To expand on the preliminary studies of each of the knockout lines, RNA-Seq was conducted at E14.5 and adult stages on brains from knockout and wild-type littermates for the 13 lncRNAs that displayed any brain expression [166]. Interestingly, loss of *Pantr1* or *Pantr2* did not affect *Brn1* expression in whole brains at either time point. Conversely, *Brn1* (*Pou3f3*), *Brn2* (*Pou3f2*), and *Brn4* (*Pou3f4*) all displayed increased expression in *Pantr1* knockout brains at E14.5 [166]. Additionally, of the 13 lncRNA knockout lines studied in these experiments, only five showed significant differences in the expression levels of neighboring gene expression at either the time point [166]. Together, these data further suggest that brain-expressed lncRNAs may function to regulate gene expression both locally in cis but also in trans [166]. Further characterization of each knockout line will be required to determine the specific function of each individual lncRNA.

Some lncRNAs expressed during nervous system development, however, show no or very mild phenotypes following targeted inactivation. Two examples of brain-expressed lncRNAs in which knockout models that fail to produce obvious phenotypes are *Malat1* and *Visc2* [167–170]. Other lncRNA knockouts display only minor phenotypes. The lncRNA *Neat1*, which displays enriched expression in neurons compared to their neural precursors [72], is expressed in a nuclear subdomain known as paraspeckles [171–173]. Paraspeckles are nuclear bodies composed of more than 40 RNA-binding proteins [174]. *Neat1* is required for paraspeckle formation in both in vitro and in vivo studies [171–173, 175, 176]. However, the physiological function of paraspeckle formation is unclear, as mice lacking *Neat1* expression and paraspeckle formation fail to display any clear developmental phenotype [176], with one notable exception. It was recently determined that *Neat1* knockout mice display a stochastic infertility resulting from corpus luteum dysfunction [177, 178]. However, the contribution of *Neat1* and paraspeckle formation to brain development or nervous system diseases such as amyotrophic lateral sclerosis (ALS), where elevated *Neat1* expression is observed during early stages of the disease [179], remains to be determined.

The lncRNA *Gomafu* (also known as *RNCR2* or *Miat*) was previously identified to function in retinal cell fate specification [135], and decrease in *Gomafu* expression is associated with mRNA splicing defects in schizophrenia [140]. Knockout mice, however, display no gross developmental defects. Instead, behavioral tests performed on *Gomafu* knockouts suggest that these mice display a mild hyperactivity phenotype and enhanced hyperactivity to repeated psychostimulant exposure [180]. Analysis of extracellular dopamine within the nucleus accumbens revealed increased dopamine levels compared to wild-type controls, consistent with the observed hyperactivity [180]. Likewise, schizophrenia patients also exhibit hyperactivity, and *Gomafu* expression is downregulated in postmortem brain samples of schizophrenia patients [140]. However, RNA-Seq analysis of hippocampal cultures from wild-type and *Gomafu* knockout mice revealed only 18 transcripts that displayed differential expression [180]. As *Gomafu* is previously predicted to regulate mRNA splicing through its interactions with the splicing regulators QKI, SF1, and CELF3 [140, 181, 182], alternative splicing of a handful of transcripts was assessed in hippocampal cells of *Gomafu* knockout mice [180]. However, unlike the changes observed in postmortem brain samples from schizophrenia patients, where *Gomafu* showed decreased expression [140], *Gomafu*$^{-/-}$ mice displayed little change in alternative splicing [180].

Despite these studies, our knowledge of the in vivo contributions of lncRNAs to nervous system development remains clouded by emerging discrepancies between in vitro and in vivo results. To further complicate matters, the design of knockout targeting strategies for in vivo loss of function studies can significantly affect interpretations of any phenotypes obtained (Fig. 9.2). For example, three different targeting strategies were used to generate *Malat1* knockout mice. Importantly, while no gross phenotypes were observed in any of the studies [167–169], the effect of *Malat1* loss of function on *Neat1* expression depended on the mechanism by which *Malat1* expression was inhibited. Deletion of either the promoter and proximal 5′ transcript sequence or the entire gene body of *Malat1* resulted in an increase in *Neat1* expression [167, 169]. However, insertion of *lacZ* and two premature polyadenylation sequences into the *Malat1* locus resulted in decreased *Neat1* expression [168]. Together, these data indicate the context-dependent sequence requirement for genome architecture. It remains to be determined, however, if *Malat1*-dependent regulation of *Neat1* expression occurs solely by controlling the activity of cis-regulatory elements or whether *Malat1* also regulates *Neat1* transcription in trans.

9.4 Conclusions

With the emergence of vastly improved sequencing technologies, we are beginning to understand the full complexity of the transcriptome. These analyses have revealed that the numbers of lncRNAs have expanded in parallel with the evolutionary increase in brain complexity. Emerging experiments profiling the transcriptomes of nervous system tissue continue to identify many novel lncRNAs. As we continue to identify and characterize the diverse cell types of the brain across development through single-cell RNA-Seq, and continue to explore the complexity of alternative splicing through Capture-seq profiling, we expect that the number of validated lncRNAs will expand dramatically. While considerable effort is now going into investigating the function of these lncRNAs during nervous system development, it is important to keep in mind exactly how these studies are performed. In vitro studies expand the repertoire of mechanistic analyses that we can perform, but results from such studies require in vivo validation, as the lncRNAs are likely functioning in a cell type- and context-specific manner that is often only imperfectly recapitulated in cultured cells. Furthermore, in vivo studies of lncRNA function need to be carefully designed to directly examine the function of the lncRNA transcripts themselves. This is especially important for genetic loss of function studies, where changing the genomic locus that encodes the lncRNA in question may disrupt the activity of important cis-regulatory elements. Additional challenges remain in the design of efforts to address the function of natural-antisense or opposite strand transcripts, due to their genomic proximity to protein-coding genes.

Given the abundance of functions being attributed to lncRNAs, it is especially important to understand their mechanisms of action. Since lncRNAs in many cases function as molecular scaffolds—that bind DNA, RNA, protein, or combinations of these biomolecules—understanding the precise composition of these complexes will be pivotal. However, given the low cellular expression levels and/or scarcity of cell types in which the lncRNAs are expressed, traditional pulldown/mass spectrometry experiments will prove challenging. In any case, recent years have clearly shown that lncRNAs are central to regulation of neuronal differentiation, and our appreciation of their importance will likely grow substantially in the years ahead.

Acknowledgments B.S.C. was supported by F32EY024201 from the NIH.

References

1. Arendt D, Denes AS, Jekely G, Tessmar-Raible K (2008) The evolution of nervous system centralization. Philos Trans R Soc Lond Ser B Biol Sci 363(1496):1523–1528
2. Hirth F (2010) On the origin and evolution of the tripartite brain. Brain Behav Evol 76(1):3–10
3. Galliot B, Quiquand M, Ghila L, de Rosa R, Miljkovic-Licina M, Chera S (2009) Origins of neurogenesis, a cnidarian view. Dev Biol 332(1):2–24

4. Nakanishi N, Renfer E, Technau U, Rentzsch F (2012) Nervous systems of the sea anemone Nematostella vectensis are generated by ectoderm and endoderm and shaped by distinct mechanisms. Development 139(2):347–357
5. Ohno S (1970) Evolution by gene duplication. Springer, Berlin
6. Zhang J (2003) Evolution by gene duplication: an update. TRENDS Ecol Evol 18(6):292–298
7. Betizeau M, Cortay V, Patti D, Pfister S, Gautier E, Bellemin-Menard A, Afanassieff M, Huissoud C, Douglas RJ, Kennedy H, Dehay C (2013) Precursor diversity and complexity of lineage relationships in the outer subventricular zone of the primate. Neuron 80(2):442–457
8. Borrell V, Calegari F (2014) Mechanisms of brain evolution: regulation of neural progenitor cell diversity and cell cycle length. Neurosci Res 86:14–24
9. Florio M, Huttner WB (2014) Neural progenitors, neurogenesis and the evolution of the neocortex. Development 141(11):2182–2194
10. Lewitus E, Kelava I, Huttner WB (2013) Conical expansion of the outer subventricular zone and the role of neocortical folding in evolution and development. Front Hum Neurosci 7:424
11. Lewitus E, Kelava I, Kalinka AT, Tomancak P, Huttner WB (2014) An adaptive threshold in mammalian neocortical evolution. PLoS Biol 12(11):e1002000
12. Lui JH, Hansen DV, Kriegstein AR (2011) Development and evolution of the human neocortex. Cell 146(1):18–36
13. Stahl R, Walcher T, De Juan Romero C, Pilz GA, Cappello S, Irmler M, Sanz-Aquela JM, Beckers J, Blum R, Borrell V, Gotz M (2013) Trnp1 regulates expansion and folding of the mammalian cerebral cortex by control of radial glial fate. Cell 153(3):535–549
14. Sun T, Hevner RF (2014) Growth and folding of the mammalian cerebral cortex: from molecules to malformations. Nat Rev Neurosci 15(4):217–232
15. Kaskan PM, Finlay BL (2001) Bigger is better: primate brain size in relationship to cognition. In: Falk D, Gibson KR (eds) Evolutionary anatomy of the primate cerebral cortex. Cambridge University Press, Cambridge, pp 79–97
16. Finlay BL, Darlington RB (1995) Linked regularities in the development and evolution of mammalian brains. Science (New York, NY) 268(5217):1578–1584
17. Roth G, Dicke U (2005) Evolution of the brain and intelligence. Trends Cogn Sci 9(5):250–257
18. Jerison HJ (1973) Evolution of the brain and intelligence, XIV edn. Academic, New York
19. Derrien T, Johnson R, Bussotti G, Tanzer A, Djebali S, Tilgner H, Guernec G, Martin D, Merkel A, Knowles DG, Lagarde J, Veeravalli L, Ruan X, Ruan Y, Lassmann T, Carninci P, Brown JB, Lipovich L, Gonzalez JM, Thomas M, Davis CA, Shiekhattar R, Gingeras TR, Hubbard TJ, Notredame C, Harrow J, Guigo R (2012) The GENCODE v7 catalog of human long noncoding RNAs: analysis of their gene structure, evolution, and expression. Genome Res 22(9):1775–1789
20. Harrow J, Frankish A, Gonzalez JM, Tapanari E, Diekhans M, Kokocinski F, Aken BL, Barrell D, Zadissa A, Searle S, Barnes I, Bignell A, Boychenko V, Hunt T, Kay M, Mukherjee G, Rajan J, Despacio-Reyes G, Saunders G, Steward C, Harte R, Lin M, Howald C, Tanzer A, Derrien T, Chrast J, Walters N, Balasubramanian S, Pei B, Tress M, Rodriguez JM, Ezkurdia I, van Baren J, Brent M, Haussler D, Kellis M, Valencia A, Reymond A, Gerstein M, Guigo R, Hubbard TJ (2012) GENCODE: the reference human genome annotation for the ENCODE project. Genome Res 22(9):1760–1774
21. Necsulea A, Soumillon M, Warnefors M, Liechti A, Daish T, Zeller U, Baker JC, Grutzner F, Kaessmann H (2014) The evolution of lncRNA repertoires and expression patterns in tetrapods. Nature 505(7485):635–640
22. Mercer TR, Dinger ME, Sunkin SM, Mehler MF, Mattick JS (2008) Specific expression of long noncoding RNAs in the mouse brain. Proc Natl Acad Sci U S A 105(2):716–721
23. Luo H, Sun S, Li P, Bu D, Cao H, Zhao Y (2013) Comprehensive characterization of 10,571 mouse large intergenic noncoding RNAs from whole transcriptome sequencing. PLoS One 8(8):e70835
24. Ponjavic J, Oliver PL, Lunter G, Ponting CP (2009) Genomic and transcriptional co-localization of protein-coding and long non-coding RNA pairs in the developing brain. PLoS Genet 5(8):e1000617
25. Francescatto M, Vitezic M, Heutink P, Saxena A (2014) Brain-specific noncoding RNAs are likely to originate in repeats and may play a role in up-regulating genes in cis. Int J Biochem Cell Biol 54:331–337

26. Kaushik K, Leonard VE, Kv S, Lalwani MK, Jalali S, Patowary A, Joshi A, Scaria V, Sivasubbu S (2013) Dynamic expression of long non-coding RNAs (lncRNAs) in adult zebrafish. PLoS One 8(12):e83616
27. Washietl S, Kellis M, Garber M (2014) Evolutionary dynamics and tissue specificity of human long noncoding RNAs in six mammals. Genome Res 24(4):616–628
28. Kelley D, Rinn J (2012) Transposable elements reveal a stem cell-specific class of long non-coding RNAs. Genome Biol 13(11):R107. doi:10.1186/gb-2012-13-11-r107
29. Ulitsky I, Shkumatava A, Jan CH, Sive H, Bartel DP (2011) Conserved function of lincRNAs in vertebrate embryonic development despite rapid sequence evolution. Cell 147(7):1537–1550
30. Lipovich L, Hou ZC, Jia H, Sinkler C, McGowen M, Sterner KN, Weckle A, Sugalski AB, Pipes L, Gatti DL, Mason CE, Sherwood CC, Hof PR, Kuzawa CW, Grossman LI, Goodman M, Wildman DE (2016) High-throughput RNA sequencing reveals structural differences of orthologous brain-expressed genes between western lowland gorillas and humans. J Comp Neurol 524(2):288–308
31. Basu S, Muller F, Sanges R (2013) Examples of sequence conservation analyses capture a subset of mouse long non-coding RNAs sharing homology with fish conserved genomic elements. BMC Bioinformatics 14(Suppl 7):S14. doi:10.1186/1471-2105-14-S7-S14. Epub 2013 Apr 22.
32. He Z, Bammann H, Han D, Xie G, Khaitovich P (2014) Conserved expression of lincRNA during human and macaque prefrontal cortex development and maturation. RNA (New York, NY) 20(7):1103–1111
33. Bird CP, Stranger BE, Liu M, Thomas DJ, Ingle CE, Beazley C, Miller W, Hurles ME, Dermitzakis ET (2007) Fast-evolving noncoding sequences in the human genome. Genome Biol 8(6):R118
34. Bush EC, Lahn BT (2008) A genome-wide screen for noncoding elements important in primate evolution. BMC Evol Biol 8:17. doi:10.1186/1471-2148-8-17
35. Pollard KS, Salama SR, Lambert N, Lambot MA, Coppens S, Pedersen JS, Katzman S, King B, Onodera C, Siepel A, Kern AD, Dehay C, Igel H, Ares M Jr, Vanderhaeghen P, Haussler D (2006) An RNA gene expressed during cortical development evolved rapidly in humans. Nature 443(7108):167–172
36. Prabhakar S, Visel A, Akiyama JA, Shoukry M, Lewis KD, Holt A, Plajzer-Frick I, Morrison H, Fitzpatrick DR, Afzal V, Pennacchio LA, Rubin EM, Noonan JP (2008) Human-specific gain of function in a developmental enhancer. Science (New York, NY) 321(5894):1346–1350
37. Capra JA, Erwin GD, McKinsey G, Rubenstein JL, Pollard KS (2013) Many human accelerated regions are developmental enhancers. Philos Trans R Soc Lond Ser B Biol Sci 368(1632):20130025
38. Meyer G, Soria JM, Martinez-Galan JR, Martin-Clemente B, Fairen A (1998) Different origins and developmental histories of transient neurons in the marginal zone of the fetal and neonatal rat cortex. J Comp Neurol 397(4):493–518
39. Rakic S, Zecevic N (2003) Emerging complexity of layer I in human cerebral cortex. Cereb Cortex (New York, NY: 1991) 13(10):1072–1083
40. Meyer G, Wahle P (1999) The paleocortical ventricle is the origin of reelin-expressing neurons in the marginal zone of the foetal human neocortex. Eur J Neurosci 11(11):3937–3944
41. Zecevic N, Rakic P (2001) Development of layer I neurons in the primate cerebral cortex. J Neurosci 21(15):5607–5619
42. Pang KC, Frith MC, Mattick JS (2006) Rapid evolution of noncoding RNAs: lack of conservation does not mean lack of function. Trends Genet 22(1):1–5
43. Lewitus E, Huttner WB (2015) Neurodevelopmental LincRNA Microsyteny conservation and mammalian brain size evolution. PLoS One 10(7):e0131818
44. Kelava I, Reillo I, Murayama AY, Kalinka AT, Stenzel D, Tomancak P, Matsuzaki F, Lebrand C, Sasaki E, Schwamborn JC, Okano H, Huttner WB, Borrell V (2012) Abundant occurrence of basal radial glia in the subventricular zone of embryonic neocortex of a lissencephalic primate, the common marmoset Callithrix jacchus. Cereb Cortex (New York, NY: 1991) 22(2):469–481
45. Kelava I, Lewitus E, Huttner WB (2013) The secondary loss of gyrencephaly as an example of evolutionary phenotypical reversal. Front Neuroanat 7:16

46. Carninci P, Kasukawa T, Katayama S, Gough J, Frith MC, Maeda N, Oyama R, Ravasi T, Lenhard B, Wells C, Kodzius R, Shimokawa K, Bajic VB, Brenner SE, Batalov S, Forrest AR, Zavolan M, Davis MJ, Wilming LG, Aidinis V, Allen JE, Ambesi-Impiombato A, Apweiler R, Aturaliya RN, Bailey TL, Bansal M, Baxter L, Beisel KW, Bersano T, Bono H, Chalk AM, Chiu KP, Choudhary V, Christoffels A, Clutterbuck DR, Crowe ML, Dalla E, Dalrymple BP, de Bono B, Della Gatta G, di Bernardo D, Down T, Engstrom P, Fagiolini M, Faulkner G, Fletcher CF, Fukushima T, Furuno M, Futaki S, Gariboldi M, Georgii-Hemming P, Gingeras TR, Gojobori T, Green RE, Gustincich S, Harbers M, Hayashi Y, Hensch TK, Hirokawa N, Hill D, Huminiecki L, Iacono M, Ikeo K, Iwama A, Ishikawa T, Jakt M, Kanapin A, Katoh M, Kawasawa Y, Kelso J, Kitamura H, Kitano H, Kollias G, Krishnan SP, Kruger A, Kummerfeld SK, Kurochkin IV, Lareau LF, Lazarevic D, Lipovich L, Liu J, Liuni S, McWilliam S, Madan Babu M, Madera M, Marchionni L, Matsuda H, Matsuzawa S, Miki H, Mignone F, Miyake S, Morris K, Mottagui-Tabar S, Mulder N, Nakano N, Nakauchi H, Ng P, Nilsson R, Nishiguchi S, Nishikawa S, Nori F, Ohara O, Okazaki Y, Orlando V, Pang KC, Pavan WJ, Pavesi G, Pesole G, Petrovsky N, Piazza S, Reed J, Reid JF, Ring BZ, Ringwald M, Rost B, Ruan Y, Salzberg SL, Sandelin A, Schneider C, Schonbach C, Sekiguchi K, Semple CA, Seno S, Sessa L, Sheng Y, Shibata Y, Shimada H, Shimada K, Silva D, Sinclair B, Sperling S, Stupka E, Sugiura K, Sultana R, Takenaka Y, Taki K, Tammoja K, Tan SL, Tang S, Taylor MS, Tegner J, Teichmann SA, Ueda HR, van Nimwegen E, Verardo R, Wei CL, Yagi K, Yamanishi H, Zabarovsky E, Zhu S, Zimmer A, Hide W, Bult C, Grimmond SM, Teasdale RD, Liu ET, Brusic V, Quackenbush J, Wahlestedt C, Mattick JS, Hume DA, Kai C, Sasaki D, Tomaru Y, Fukuda S, Kanamori-Katayama M, Suzuki M, Aoki J, Arakawa T, Iida J, Imamura K, Itoh M, Kato T, Kawaji H, Kawagashira N, Kawashima T, Kojima M, Kondo S, Konno H, Nakano K, Ninomiya N, Nishio T, Okada M, Plessy C, Shibata K, Shiraki T, Suzuki S, Tagami M, Waki K, Watahiki A, Okamura-Oho Y, Suzuki H, Kawai J, Hayashizaki Y, FANTOM Consortium & RIKEN Genome Exploration Research Group and Genome Science Group (Genome Network Project Core Group) (2005) The transcriptional landscape of the mammalian genome. Science (New York, NY) 309(5740):1559–1563
47. Xie C, Yuan J, Li H, Li M, Zhao G, Bu D, Zhu W, Wu W, Chen R, Zhao Y (2014) NONCODEv4: exploring the world of long non-coding RNA genes. Nucleic Acids Res 42:D98–103
48. Lv J, Liu H, Huang Z, Su J, He H, Xiu Y, Zhang Y, Wu Q (2013) Long non-coding RNA identification over mouse brain development by integrative modeling of chromatin and genomic features. Nucleic Acids Res 41(22):10044–10061
49. Blackshaw S, Harpavat S, Trimarchi J, Cai L, Huang H, Kuo WP, Weber G, Lee K, Fraioli RE, Cho SH, Yung R, Asch E, Ohno-Machado L, Wong WH, Cepko CL (2004) Genomic analysis of mouse retinal development. PLoS Biol 2(9):E247
50. Aprea J, Prenninger S, Dori M, Ghosh T, Monasor LS, Wessendorf E, Zocher S, Massalini S, Alexopoulou D, Lesche M, Dahl A, Groszer M, Hiller M, Calegari F (2013) Transcriptome sequencing during mouse brain development identifies long non-coding RNAs functionally involved in neurogenic commitment. EMBO J 32(24):3145–3160
51. Mehler MF, Mattick JS (2007) Noncoding RNAs and RNA editing in brain development, functional diversification, and neurological disease. Physiol Rev 87(3):799–823
52. Cabili MN, Trapnell C, Goff L, Koziol M, Tazon-Vega B, Regev A, Rinn JL (2011) Integrative annotation of human large intergenic noncoding RNAs reveals global properties and specific subclasses. Genes Dev 25(18):1915–1927
53. Jorgenson LA, Newsome WT, Anderson DJ, Bargmann CI, Brown EN, Deisseroth K, Donoghue JP, Hudson KL, Ling GS, MacLeish PR, Marder E, Normann RA, Sanes JR, Schnitzer MJ, Sejnowski TJ, Tank DW, Tsien RY, Ugurbil K, Wingfield JC (2015) The BRAIN initiative: developing technology to catalyse neuroscience discovery. Philos Trans R Soc Lond Ser B Biol Sci 370(1668). doi:10.1098/rstb.2014.0164
54. Zhang K, Huang K, Luo Y, Li S (2014) Identification and functional analysis of long non-coding RNAs in mouse cleavage stage embryonic development based on single cell transcriptome data. BMC Genomics 15:845. doi:10.1186/1471-2164-15-845

55. Jeck WR, Sorrentino JA, Wang K, Slevin MK, Burd CE, Liu J, Marzluff WF, Sharpless NE (2013) Circular RNAs are abundant, conserved, and associated with ALU repeats. RNA (New York, NY) 19(2):141–157
56. Salzman J, Gawad C, Wang PL, Lacayo N, Brown PO (2012) Circular RNAs are the predominant transcript isoform from hundreds of human genes in diverse cell types. PLoS One 7(2):e30733
57. Barrett SP, Salzman J (2016) Circular RNAs: analysis, expression and potential functions. Development 143(11):1838–1847
58. Memczak S, Jens M, Elefsinioti A, Torti F, Krueger J, Rybak A, Maier L, Mackowiak SD, Gregersen LH, Munschauer M, Loewer A, Ziebold U, Landthaler M, Kocks C, le Noble F, Rajewsky N (2013) Circular RNAs are a large class of animal RNAs with regulatory potency. Nature 495(7441):333–338
59. Bussotti G, Leonardi T, Clark MB, Mercer TR, Crawford J, Malquori L, Notredame C, Dinger ME, Mattick JS, Enright AJ (2016) Improved definition of the mouse transcriptome via targeted RNA sequencing. Genome Res 26(5):705–716
60. Mercer TR, Gerhardt DJ, Dinger ME, Crawford J, Trapnell C, Jeddeloh JA, Mattick JS, Rinn JL (2011) Targeted RNA sequencing reveals the deep complexity of the human transcriptome. Nat Biotechnol 30(1):99–104
61. Clark MB, Mercer TR, Bussotti G, Leonardi T, Haynes KR, Crawford J, Brunck ME, Cao KA, Thomas GP, Chen WY, Taft RJ, Nielsen LK, Enright AJ, Mattick JS, Dinger ME (2015) Quantitative gene profiling of long noncoding RNAs with targeted RNA sequencing. Nat Methods 12(4):339–342
62. Mercer TR, Clark MB, Crawford J, Brunck ME, Gerhardt DJ, Taft RJ, Nielsen LK, Dinger ME, Mattick JS (2014) Targeted sequencing for gene discovery and quantification using RNA Capture-Seq. Nat Protoc 9(5):989–1009
63. Dinger ME, Amaral PP, Mercer TR, Pang KC, Bruce SJ, Gardiner BB, Askarian-Amiri ME, Ru K, Solda G, Simons C, Sunkin SM, Crowe ML, Grimmond SM, Perkins AC, Mattick JS (2008) Long noncoding RNAs in mouse embryonic stem cell pluripotency and differentiation. Genome Res 18(9):1433–1445
64. Bruce SJ, Gardiner BB, Burke LJ, Gongora MM, Grimmond SM, Perkins AC (2007) Dynamic transcription programs during ES cell differentiation towards mesoderm in serum versus serum-freeBMP4 culture. BMC Genomics 8:365
65. Ivanova NB, Dimos JT, Schaniel C, Hackney JA, Moore KA, Lemischka IR (2002) A stem cell molecular signature. Science (New York, NY) 298(5593):601–604
66. Ramalho-Santos M, Yoon S, Matsuzaki Y, Mulligan RC, Melton DA (2002) "Stemness": transcriptional profiling of embryonic and adult stem cells. Science (New York, NY) 298(5593):597–600
67. Meshorer E, Yellajoshula D, George E, Scambler PJ, Brown DT, Misteli T (2006) Hyperdynamic plasticity of chromatin proteins in pluripotent embryonic stem cells. Dev Cell 10(1):105–116
68. Meshorer E, Misteli T (2006) Chromatin in pluripotent embryonic stem cells and differentiation. Nat Rev Mol Cell Biol 7(7):540–546
69. Gaspar-Maia A, Alajem A, Polesso F, Sridharan R, Mason MJ, Heidersbach A, Ramalho-Santos J, McManus MT, Plath K, Meshorer E, Ramalho-Santos M (2009) Chd1 regulates open chromatin and pluripotency of embryonic stem cells. Nature 460(7257):863–868
70. Guttman M, Amit I, Garber M, French C, Lin MF, Feldser D, Huarte M, Zuk O, Carey BW, Cassady JP, Cabili MN, Jaenisch R, Mikkelsen TS, Jacks T, Hacohen N, Bernstein BE, Kellis M, Regev A, Rinn JL, Lander ES (2009) Chromatin signature reveals over a thousand highly conserved large non-coding RNAs in mammals. Nature 458(7235):223–227
71. Guttman M, Garber M, Levin JZ, Donaghey J, Robinson J, Adiconis X, Fan L, Koziol MJ, Gnirke A, Nusbaum C, Rinn JL, Lander ES, Regev A (2010) Ab initio reconstruction of cell type-specific transcriptomes in mouse reveals the conserved multi-exonic structure of lincRNAs. Nat Biotechnol 28(5):503–510

72. Mercer TR, Qureshi IA, Gokhan S, Dinger ME, Li G, Mattick JS, Mehler MF (2010) Long noncoding RNAs in neuronal-glial fate specification and oligodendrocyte lineage maturation. BMC Neurosci 11:14. doi:10.1186/1471-2202-11-14
73. Lagos-Quintana M, Rauhut R, Yalcin A, Meyer J, Lendeckel W, Tuschl T (2002) Identification of tissue-specific microRNAs from mouse. Curr Biol 12(9):735–739
74. Makeyev EV, Zhang J, Carrasco MA, Maniatis T (2007) The MicroRNA miR-124 promotes neuronal differentiation by triggering brain-specific alternative pre-mRNA splicing. Mol Cell 27(3):435–448
75. Ng SY, Johnson R, Stanton LW (2012) Human long non-coding RNAs promote pluripotency and neuronal differentiation by association with chromatin modifiers and transcription factors. EMBO J 31(3):522–533
76. Guo Y, Zhang P, Sheng Q, Zhao S, Hackett TA (2016) lncRNA expression in the auditory forebrain during postnatal development. Gene 593(1):201–216
77. Ramos AD, Diaz A, Nellore A, Delgado RN, Park KY, Gonzales-Roybal G, Oldham MC, Song JS, Lim DA (2013) Integration of genome-wide approaches identifies lncRNAs of adult neural stem cells and their progeny in vivo. Cell Stem Cell 12(5):616–628
78. Liu SJ, Nowakowski TJ, Pollen AA, Lui JH, Horlbeck MA, Attenello FJ, He D, Weissman JS, Kriegstein AR, Diaz AA, Lim DA (2016) Single-cell analysis of long non-coding RNAs in the developing human neocortex. Genome Biol 17:67. doi:10.1186/s13059-016-0932-1
79. Rapicavoli NA, Blackshaw S (2009) New meaning in the message: noncoding RNAs and their role in retinal development. Dev Dyn 238(9):2103–2114
80. Molyneaux BJ, Goff LA, Brettler AC, Chen HH, Brown JR, Hrvatin S, Rinn JL, Arlotta P (2015) DeCoN: genome-wide analysis of in vivo transcriptional dynamics during pyramidal neuron fate selection in neocortex. Neuron 85(2):275–288
81. Lein ES, Hawrylycz MJ, Ao N, Ayres M, Bensinger A, Bernard A, Boe AF, Boguski MS, Brockway KS, Byrnes EJ, Chen L, Chen L, Chen TM, Chin MC, Chong J, Crook BE, Czaplinska A, Dang CN, Datta S, Dee NR, Desaki AL, Desta T, Diep E, Dolbeare TA, Donelan MJ, Dong HW, Dougherty JG, Duncan BJ, Ebbert AJ, Eichele G, Estin LK, Faber C, Facer BA, Fields R, Fischer SR, Fliss TP, Frensley C, Gates SN, Glattfelder KJ, Halverson KR, Hart MR, Hohmann JG, Howell MP, Jeung DP, Johnson RA, Karr PT, Kawal R, Kidney JM, Knapik RH, Kuan CL, Lake JH, Laramee AR, Larsen KD, Lau C, Lemon TA, Liang AJ, Liu Y, Luong LT, Michaels J, Morgan JJ, Morgan RJ, Mortrud MT, Mosqueda NF, Ng LL, Ng R, Orta GJ, Overly CC, Pak TH, Parry SE, Pathak SD, Pearson OC, Puchalski RB, Riley ZL, Rockett HR, Rowland SA, Royall JJ, Ruiz MJ, Sarno NR, Schaffnit K, Shapovalova NV, Sivisay T, Slaughterbeck CR, Smith SC, Smith KA, Smith BI, Sodt AJ, Stewart NN, Stumpf KR, Sunkin SM, Sutram M, Tam A, Teemer CD, Thaller C, Thompson CL, Varnam LR, Visel A, Whitlock RM, Wohnoutka PE, Wolkey CK, Wong VY, Wood M, Yaylaoglu MB, Young RC, Youngstrom BL, Yuan XF, Zhang B, Zwingman TA, Jones AR (2007) Genome-wide atlas of gene expression in the adult mouse brain. Nature 445(7124):168–176
82. Kour S, Rath PC (2015) Age-dependent differential expression profile of a novel intergenic long noncoding RNA in rat brain. Int J Dev Neurosci 47(Pt B):286–297
83. Mills JD, Chen J, Kim WS, Waters PD, Prabowo AS, Aronica E, Halliday GM, Janitz M (2015) Long intervening non-coding RNA 00320 is human brain-specific and highly expressed in the cortical white matter. Neurogenetics 16(3):201–213
84. Dietz WH, Masterson K, Sittig LJ, Redei EE, Herzing LB (2012) Imprinting and expression of Dio3os mirrors Dio3 in rat. Front Genet 3:279
85. Kohtz JD, Baker DP, Corte G, Fishell G (1998) Regionalization within the mammalian telencephalon is mediated by changes in responsiveness to sonic hedgehog. Development 125(24):5079–5089
86. Kohtz JD, Fishell G (2004) Developmental regulation of EVF-1, a novel non-coding RNA transcribed upstream of the mouse Dlx6 gene. Gene Expr Patterns 4(4):407–412
87. Feng J, Bi C, Clark BS, Mady R, Shah P, Kohtz JD (2006) The Evf-2 noncoding RNA is transcribed from the Dlx-5/6 ultraconserved region and functions as a Dlx-2 transcriptional coactivator. Genes Dev 20(11):1470–1484

88. Davis CJ, Taishi P, Honn KA, Koberstein JN, Krueger JM (2016) P2X7 receptors in body temperature, locomotor activity, and brain mRNA and lncRNA responses to sleep deprivation. Am J Physiol Regul Integr Comp Physiol 311(6):R1004–R1012. doi:10.1152/ajpregu.00167.2016
89. Zhong J, Jiang L, Cheng C, Huang Z, Zhang H, Liu H, He J, Cao F, Peng J, Jiang Y, Sun X (2016) Altered expression of long non-coding RNA and mRNA in mouse cortex after traumatic brain injury. Brain Res 1646:589–600
90. Jiang H, Good DJ (2016) A molecular conundrum involving hypothalamic responses to and roles of long non-coding RNAs following food deprivation. Mol Cell Endocrinol 438:52–60
91. D'haene E, Jacobs EZ, Volders PJ, De Meyer T, Menten B, Vergult S (2016) Identification of long non-coding RNAs involved in neuronal development and intellectual disability. Sci Rep 6:28396
92. Chen R, Liu L, Xiao M, Wang F, Lin X (2016) Microarray expression profile analysis of long noncoding RNAs in premature brain injury: a novel point of view. Neuroscience 319:123–133
93. Kraus TF, Haider M, Spanner J, Steinmaurer M, Dietinger V, Kretzschmar HA (2016) Altered long noncoding RNA expression precedes the course of Parkinson's disease-a preliminary report. Mol Neurobiol 54(4):2869–2877
94. Faedo A, Quinn JC, Stoney P, Long JE, Dye C, Zollo M, Rubenstein JL, Price DJ, Bulfone A (2004) Identification and characterization of a novel transcript down-regulated in Dlx1/Dlx2 and up-regulated in Pax6 mutant telencephalon. Dev Dyn 231(3):614–620
95. Kour S, Rath PC (2016) All-trans retinoic acid induces expression of a novel Intergenic long noncoding RNA in adult rat primary hippocampal neurons. J Mol Neurosci 58(2):266–276
96. Chen X, Zhou X, Lu D, Yang X, Zhou Z, Chen X, Chen Y, He W, Feng X (2016) Aberrantly expressed long noncoding RNAs are involved in sevoflurane-induced developing hippocampal neuronal apoptosis: a microarray related study. Metab Brain Dis 31(5):1031–1040
97. Clark BS, Blackshaw S (2014) Long non-coding RNA-dependent transcriptional regulation in neuronal development and disease. Front Genet 5:164
98. Guttman M, Donaghey J, Carey BW, Garber M, Grenier JK, Munson G, Young G, Lucas AB, Ach R, Bruhn L, Yang X, Amit I, Meissner A, Regev A, Rinn JL, Root DE, Lander ES (2011) lincRNAs act in the circuitry controlling pluripotency and differentiation. Nature 477(7364):295–300
99. Wang C, Li G, Wu Y, Xi J, Kang J (2016) LincRNA1230 inhibits the differentiation of mouse ES cells towards neural progenitors. Sci China Life Sci 59(5):443–454
100. Lin N, Chang KY, Li Z, Gates K, Rana ZA, Dang J, Zhang D, Han T, Yang CS, Cunningham TJ, Head SR, Duester G, Dong PD, Rana TM (2014) An evolutionarily conserved long noncoding RNA TUNA controls pluripotency and neural lineage commitment. Mol Cell 53(6):1005–1019
101. Chu C, Qu K, Zhong FL, Artandi SE, Chang HY (2011) Genomic maps of long noncoding RNA occupancy reveal principles of RNA-chromatin interactions. Mol Cell 44(4):667–678
102. Ng SY, Bogu GK, Soh BS, Stanton LW (2013) The long noncoding RNA RMST interacts with SOX2 to regulate neurogenesis. Mol Cell 51(3):349–359
103. Anderegg A, Lin HP, Chen JA, Caronia-Brown G, Cherepanova N, Yun B, Joksimovic M, Rock J, Harfe BD, Johnson R, Awatramani R (2013) An Lmx1b-miR135a2 regulatory circuit modulates Wnt1/Wnt signaling and determines the size of the midbrain dopaminergic progenitor pool. PLoS Genet 9(12):e1003973
104. Anderegg A, Awatramani R (2015) k the size of the midbrain and the dopaminergic progenitor pool. Neurogenesis (Austin, TX) 2(1):e998101
105. Caronia-Brown G, Anderegg A, Awatramani R (2016) Expression and functional analysis of the Wnt/beta-catenin induced mir-135a-2 locus in embryonic forebrain development. Neural Dev 11:9. doi:10.1186/s13064-016-0065-y
106. Tochitani S, Hayashizaki Y (2008) Nkx2.2 antisense RNA overexpression enhanced oligodendrocytic differentiation. Biochem Biophys Res Commun 372(4):691–696
107. Qi Y, Cai J, Wu Y, Wu R, Lee J, Fu H, Rao M, Sussel L, Rubenstein J, Qiu M (2001) Control of oligodendrocyte differentiation by the Nkx2.2 homeodomain transcription factor. Development 128(14):2723–2733

108. Dong X, Chen K, Cuevas-Diaz Duran R, You Y, Sloan SA, Zhang Y, Zong S, Cao Q, Barres BA, Wu JQ (2015) Comprehensive identification of long non-coding RNAs in purified cell types from the brain reveals functional LncRNA in OPC fate determination. PLoS Genet 11(12):e1005669
109. Yang N, Zuchero JB, Ahlenius H, Marro S, Ng YH, Vierbuchen T, Hawkins JS, Geissler R, Barres BA, Wernig M (2013) Generation of oligodendroglial cells by direct lineage conversion. Nat Biotechnol 31(5):434–439
110. Lu QR, Sun T, Zhu Z, Ma N, Garcia M, Stiles CD, Rowitch DH (2002) Common developmental requirement for Olig function indicates a motor neuron/oligodendrocyte connection. Cell 109(1):75–86
111. Zhou Q, Choi G, Anderson DJ (2001) The bHLH transcription factor Olig2 promotes oligodendrocyte differentiation in collaboration with Nkx2.2. Neuron 31(5):791–807
112. Rani N, Nowakowski TJ, Zhou H, Godshalk SE, Lisi V, Kriegstein AR, Kosik KS (2016) A primate lncRNA mediates notch signaling during neuronal development by sequestering miRNA. Neuron 90(6):1174–1188
113. Cesana M, Cacchiarelli D, Legnini I, Santini T, Sthandier O, Chinappi M, Tramontano A, Bozzoni I (2011) A long noncoding RNA controls muscle differentiation by functioning as a competing endogenous RNA. Cell 147(2):358–369
114. Ebert MS, Sharp PA (2010) Emerging roles for natural microRNA sponges. Curr Biol 20(19):R858–R861
115. Hansen TB, Kjems J, Damgaard CK (2013) Circular RNA and miR-7 in cancer. Cancer Res 73(18):5609–5612
116. Kallen AN, Zhou XB, Xu J, Qiao C, Ma J, Yan L, Lu L, Liu C, Yi JS, Zhang H, Min W, Bennett AM, Gregory RI, Ding Y, Huang Y (2013) The imprinted H19 lncRNA antagonizes let-7 microRNAs. Mol Cell 52(1):101–112
117. Tay Y, Kats L, Salmena L, Weiss D, Tan SM, Ala U, Karreth F, Poliseno L, Provero P, Di Cunto F, Lieberman J, Rigoutsos I, Pandolfi PP (2011) Coding-independent regulation of the tumor suppressor PTEN by competing endogenous mRNAs. Cell 147(2):344–357
118. Tay Y, Rinn J, Pandolfi PP (2014) The multilayered complexity of ceRNA crosstalk and competition. Nature 505(7483):344–352
119. Wang Y, Xu Z, Jiang J, Xu C, Kang J, Xiao L, Wu M, Xiong J, Guo X, Liu H (2013) Endogenous miRNA sponge lincRNA-RoR regulates Oct4, Nanog, and Sox2 in human embryonic stem cell self-renewal. Dev Cell 25(1):69–80
120. Zhang Z, Zhu Z, Watabe K, Zhang X, Bai C, Xu M, Wu F, Mo YY (2013) Negative regulation of lncRNA GAS5 by miR-21. Cell Death Differ 20(11):1558–1568
121. Bail S, Swerdel M, Liu H, Jiao X, Goff LA, Hart RP, Kiledjian M (2010) Differential regulation of microRNA stability. RNA (New York, NY) 16(5):1032–1039
122. Legnini I, Morlando M, Mangiavacchi A, Fatica A, Bozzoni I (2014) A feedforward regulatory loop between HuR and the long noncoding RNA linc-MD1 controls early phases of myogenesis. Mol Cell 53(3):506–514
123. Stevens SJ, van Ravenswaaij-Arts CM, Janssen JW, Klein Wassink-Ruiter JS, van Essen AJ, Dijkhuizen T, van Rheenen J, Heuts-Vijgen R, Stegmann AP, Smeets EE, Engelen JJ (2011) MYT1L is a candidate gene for intellectual disability in patients with 2p25.3 (2pter) deletions. Am J Med Genet A 155A(11):2739–2745
124. Rio M, Royer G, Gobin S, de Blois MC, Ozilou C, Bernheim A, Nizon M, Munnich A, Bonnefont JP, Romana S, Vekemans M, Turleau C, Malan V (2013) Monozygotic twins discordant for submicroscopic chromosomal anomalies in 2p25.3 region detected by array CGH. Clin Genet 84(1):31–36
125. Doco-Fenzy M, Leroy C, Schneider A, Petit F, Delrue MA, Andrieux J, Perrin-Sabourin L, Landais E, Aboura A, Puechberty J, Girard M, Tournaire M, Sanchez E, Rooryck C, Ameil A, Goossens M, Jonveaux P, Lefort G, Taine L, Cailley D, Gaillard D, Leheup B, Sarda P, Genevieve D (2014) Early-onset obesity and paternal 2pter deletion encompassing the ACP1, TMEM18, and MYT1L genes. Eur J Hum Genet 22(4):471–479

126. Bonaglia MC, Giorda R, Zanini S (2014) A new patient with a terminal de novo 2p25.3 deletion of 1.9 Mb associated with early-onset of obesity, intellectual disabilities and hyperkinetic disorder. Mol Cytogenet 7:53. doi:10.1186/1755-8166-7-53
127. Gaiano N, Nye JS, Fishell G (2000) Radial glial identity is promoted by Notch1 signaling in the murine forebrain. Neuron 26(2):395–404
128. Shimojo H, Ohtsuka T, Kageyama R (2008) Oscillations in notch signaling regulate maintenance of neural progenitors. Neuron 58(1):52–64
129. Yoon KJ, Koo BK, Im SK, Jeong HW, Ghim J, Kwon MC, Moon JS, Miyata T, Kong YY (2008) Mind bomb 1-expressing intermediate progenitors generate notch signaling to maintain radial glial cells. Neuron 58(4):519–531
130. Gordon FE, Nutt CL, Cheunsuchon P, Nakayama Y, Provencher KA, Rice KA, Zhou Y, Zhang X, Klibanski A (2010) Increased expression of angiogenic genes in the brains of mouse meg3-null embryos. Endocrinology 151(6):2443–2452
131. Sun S, Payer B, Namekawa S, An JY, Press W, Catalan-Dibene J, Sunwoo H, Lee JT (2015) Xist imprinting is promoted by the hemizygous (unpaired) state in the male germ line. Proc Natl Acad Sci U S A 112(47):14415–14422
132. Alfano G, Vitiello C, Caccioppoli C, Caramico T, Carola A, Szego MJ, McInnes RR, Auricchio A, Banfi S (2005) Natural antisense transcripts associated with genes involved in eye development. Hum Mol Genet 14(7):913–923
133. Young TL, Matsuda T, Cepko CL (2005) The noncoding RNA taurine upregulated gene 1 is required for differentiation of the murine retina. Curr Biol 15(6):501–512
134. Krol J, Krol I, Alvarez CP, Fiscella M, Hierlemann A, Roska B, Filipowicz W (2015) A network comprising short and long noncoding RNAs and RNA helicase controls mouse retina architecture. Nat Commun 6:7305
135. Rapicavoli NA, Poth EM, Blackshaw S (2010) The long noncoding RNA RNCR2 directs mouse retinal cell specification. BMC Dev Biol 10:49. doi:10.1186/1471-213X-10-49
136. Rapicavoli NA, Poth EM, Zhu H, Blackshaw S (2011) The long noncoding RNA Six3OS acts in trans to regulate retinal development by modulating Six3 activity. Neural Dev 6:32. doi:10.1186/1749-8104-6-32
137. Khalil AM, Guttman M, Huarte M, Garber M, Raj A, Rivea Morales D, Thomas K, Presser A, Bernstein BE, van Oudenaarden A, Regev A, Lander ES, Rinn JL (2009) Many human large intergenic noncoding RNAs associate with chromatin-modifying complexes and affect gene expression. Proc Natl Acad Sci U S A 106(28):11667–11672
138. Ishii N, Ozaki K, Sato H, Mizuno H, Saito S, Takahashi A, Miyamoto Y, Ikegawa S, Kamatani N, Hori M, Saito S, Nakamura Y, Tanaka T (2006) Identification of a novel non-coding RNA, MIAT, that confers risk of myocardial infarction. J Hum Genet 51(12):1087–1099
139. Sone M, Hayashi T, Tarui H, Agata K, Takeichi M, Nakagawa S (2007) The mRNA-like noncoding RNA Gomafu constitutes a novel nuclear domain in a subset of neurons. J Cell Sci 120(Pt 15):2498–2506
140. Barry G, Briggs JA, Vanichkina DP, Poth EM, Beveridge NJ, Ratnu VS, Nayler SP, Nones K, Hu J, Bredy TW, Nakagawa S, Rigo F, Taft RJ, Cairns MJ, Blackshaw S, Wolvetang EJ, Mattick JS (2014) The long non-coding RNA Gomafu is acutely regulated in response to neuronal activation and involved in schizophrenia-associated alternative splicing. Mol Psychiatry 19(4):486–494
141. Zhao J, Ohsumi TK, Kung JT, Ogawa Y, Grau DJ, Sarma K, Song JJ, Kingston RE, Borowsky M, Lee JT (2010) Genome-wide identification of polycomb-associated RNAs by RIP-seq. Mol Cell 40(6):939–953
142. Meola N, Pizzo M, Alfano G, Surace EM, Banfi S (2012) The long noncoding RNA Vax2os1 controls the cell cycle progression of photoreceptor progenitors in the mouse retina. RNA (New York, NY) 18(1):111–123
143. Ramos AD, Andersen RE, Liu SJ, Nowakowski TJ, Hong SJ, Gertz CC, Salinas RD, Zarabi H, Kriegstein AR, Lim DA (2015) The long noncoding RNA Pnky regulates neuronal differentiation of embryonic and postnatal neural stem cells. Cell Stem Cell 16(4):439–447

144. Kriegstein A, Alvarez-Buylla A (2009) The glial nature of embryonic and adult neural stem cells. Annu Rev Neurosci 32:149–184
145. Zheng S, Gray EE, Chawla G, Porse BT, O'Dell TJ, Black DL (2012) PSD-95 is post-transcriptionally repressed during early neural development by PTBP1 and PTBP2. Nat Neurosci 15(3):381–388, S1.
146. Yap K, Lim ZQ, Khandelia P, Friedman B, Makeyev EV (2012) Coordinated regulation of neuronal mRNA steady-state levels through developmentally controlled intron retention. Genes Dev 26(11):1209–1223
147. Keppetipola N, Sharma S, Li Q, Black DL (2012) Neuronal regulation of pre-mRNA splicing by polypyrimidine tract binding proteins, PTBP1 and PTBP2. Crit Rev Biochem Mol Biol 47(4):360–378
148. Shibasaki T, Tokunaga A, Sakamoto R, Sagara H, Noguchi S, Sasaoka T, Yoshida N (2013) PTB deficiency causes the loss of adherens junctions in the dorsal telencephalon and leads to lethal hydrocephalus. Cereb Cortex (New York, NY: 1991) 23(8):1824–1835
149. Boutz PL, Stoilov P, Li Q, Lin CH, Chawla G, Ostrow K, Shiue L, Ares M Jr, Black DL (2007) A post-transcriptional regulatory switch in polypyrimidine tract-binding proteins reprograms alternative splicing in developing neurons. Genes Dev 21(13):1636–1652
150. Licatalosi DD, Yano M, Fak JJ, Mele A, Grabinski SE, Zhang C, Darnell RB (2012) Ptbp2 represses adult-specific splicing to regulate the generation of neuronal precursors in the embryonic brain. Genes Dev 26(14):1626–1642
151. Bond AM, Vangompel MJ, Sametsky EA, Clark MF, Savage JC, Disterhoft JF, Kohtz JD (2009) Balanced gene regulation by an embryonic brain ncRNA is critical for adult hippocampal GABA circuitry. Nat Neurosci 12(8):1020–1027
152. Berghoff EG, Clark MF, Chen S, Cajigas I, Leib DE, Kohtz JD (2013) Evf2 (Dlx6as) lncRNA regulates ultraconserved enhancer methylation and the differential transcriptional control of adjacent genes. Development 140(21):4407–4416
153. Cajigas I, Leib DE, Cochrane J, Luo H, Swyter KR, Chen S, Clark BS, Thompson J, Yates JR 3rd, Kingston RE, Kohtz JD (2015) Evf2 lncRNA/BRG1/DLX1 interactions reveal RNA-dependent inhibition of chromatin remodeling. Development 142(15):2641–2652
154. Kraus P, Sivakamasundari V, Lim SL, Xing X, Lipovich L, Lufkin T (2013) Making sense of Dlx1 antisense RNA. Dev Biol 376(2):224–235
155. Cobos I, Calcagnotto ME, Vilaythong AJ, Thwin MT, Noebels JL, Baraban SC, Rubenstein JL (2005) Mice lacking Dlx1 show subtype-specific loss of interneurons, reduced inhibition and epilepsy. Nat Neurosci 8(8):1059–1068
156. Jones DL, Howard MA, Stanco A, Rubenstein JL, Baraban SC (2011) Deletion of Dlx1 results in reduced glutamatergic input to hippocampal interneurons. J Neurophysiol 105(5):1984–1991
157. Petryniak MA, Potter GB, Rowitch DH, Rubenstein JL (2007) Dlx1 and Dlx2 control neuronal versus oligodendroglial cell fate acquisition in the developing forebrain. Neuron 55(3):417–433
158. Niwa H, Yamamura K, Miyazaki J (1991) Efficient selection for high-expression transfectants with a novel eukaryotic vector. Gene 108(2):193–199
159. Stuhmer T, Anderson SA, Ekker M, Rubenstein JL (2002) Ectopic expression of the Dlx genes induces glutamic acid decarboxylase and Dlx expression. Development 129(1):245–252
160. Sauvageau M, Goff LA, Lodato S, Bonev B, Groff AF, Gerhardinger C, Sanchez-Gomez DB, Hacisuleyman E, Li E, Spence M, Liapis SC, Mallard W, Morse M, Swerdel MR, D'Ecclessis MF, Moore JC, Lai V, Gong G, Yancopoulos GD, Frendewey D, Kellis M, Hart RP, Valenzuela DM, Arlotta P, Rinn JL (2013) Multiple knockout mouse models reveal lincRNAs are required for life and brain development. elife 2:e01749
161. Grote P, Herrmann BG (2013) The long non-coding RNA Fendrr links epigenetic control mechanisms to gene regulatory networks in mammalian embryogenesis. RNA Biol 10(10):1579–1585

162. Grote P, Wittler L, Hendrix D, Koch F, Wahrisch S, Beisaw A, Macura K, Blass G, Kellis M, Werber M, Herrmann BG (2013) The tissue-specific lncRNA Fendrr is an essential regulator of heart and body wall development in the mouse. Dev Cell 24(2):206–214
163. McEvilly RJ, de Diaz MO, Schonemann MD, Hooshmand F, Rosenfeld MG (2002) Transcriptional regulation of cortical neuron migration by POU domain factors. Science (New York, NY) 295(5559):1528–1532
164. Sugitani Y, Nakai S, Minowa O, Nishi M, Jishage K, Kawano H, Mori K, Ogawa M, Noda T (2002) Brn-1 and Brn-2 share crucial roles in the production and positioning of mouse neocortical neurons. Genes Dev 16(14):1760–1765
165. Dominguez MH, Ayoub AE, Rakic P (2013) POU-III transcription factors (Brn1, Brn2, and Oct6) influence neurogenesis, molecular identity, and migratory destination of upper-layer cells of the cerebral cortex. Cereb Cortex (New York, NY: 1991) 23(11):2632–2643
166. Goff LA, Groff AF, Sauvageau M, Trayes-Gibson Z, Sanchez-Gomez DB, Morse M, Martin RD, Elcavage LE, Liapis SC, Gonzalez-Celeiro M, Plana O, Li E, Gerhardinger C, Tomassy GS, Arlotta P, Rinn JL (2015) Spatiotemporal expression and transcriptional perturbations by long noncoding RNAs in the mouse brain. Proc Natl Acad Sci U S A 112(22):6855–6862
167. Eissmann M, Gutschner T, Hammerle M, Gunther S, Caudron-Herger M, Gross M, Schirmacher P, Rippe K, Braun T, Zornig M, Diederichs S (2012) Loss of the abundant nuclear non-coding RNA MALAT1 is compatible with life and development. RNA Biol 9(8):1076–1087
168. Nakagawa S, Ip JY, Shioi G, Tripathi V, Zong X, Hirose T, Prasanth KV (2012) Malat1 is not an essential component of nuclear speckles in mice. RNA (New York, NY) 18(8):1487–1499
169. Zhang B, Arun G, Mao YS, Lazar Z, Hung G, Bhattacharjee G, Xiao X, Booth CJ, Wu J, Zhang C, Spector DL (2012) The lncRNA Malat1 is dispensable for mouse development but its transcription plays a cis-regulatory role in the adult. Cell Rep 2(1):111–123
170. Oliver PL, Chodroff RA, Gosal A, Edwards B, Cheung AF, Gomez-Rodriguez J, Elliot G, Garrett LJ, Lickiss T, Szele F, Green ED, Molnar Z, Ponting CP (2015) Disruption of Visc-2, a brain-expressed conserved long noncoding RNA, does not elicit an overt anatomical or behavioral phenotype. Cereb cortex (New York, NY: 1991) 25(10):3572–3585
171. Sunwoo H, Dinger ME, Wilusz JE, Amaral PP, Mattick JS, Spector DL (2009) MEN epsilon/beta nuclear-retained non-coding RNAs are up-regulated upon muscle differentiation and are essential components of paraspeckles. Genome Res 19(3):347–359
172. Sasaki YT, Ideue T, Sano M, Mituyama T, Hirose T (2009) MENepsilon/beta noncoding RNAs are essential for structural integrity of nuclear paraspeckles. Proc Natl Acad Sci U S A 106(8):2525–2530
173. Clemson CM, Hutchinson JN, Sara SA, Ensminger AW, Fox AH, Chess A, Lawrence JB (2009) An architectural role for a nuclear noncoding RNA: NEAT1 RNA is essential for the structure of paraspeckles. Mol Cell 33(6):717–726
174. Fox AH, Bond CS, Lamond AI (2005) P54nrb forms a heterodimer with PSP1 that localizes to paraspeckles in an RNA-dependent manner. Mol Biol Cell 16(11):5304–5315
175. Chen LL, Carmichael GG (2009) Altered nuclear retention of mRNAs containing inverted repeats in human embryonic stem cells: functional role of a nuclear noncoding RNA. Mol Cell 35(4):467–478
176. Nakagawa S, Naganuma T, Shioi G, Hirose T (2011) Paraspeckles are subpopulation-specific nuclear bodies that are not essential in mice. J Cell Biol 193(1):31–39
177. Nakagawa S, Shimada M, Yanaka K, Mito M, Arai T, Takahashi E, Fujita Y, Fujimori T, Standaert L, Marine JC, Hirose T (2014) The lncRNA Neat1 is required for corpus luteum formation and the establishment of pregnancy in a subpopulation of mice. Development 141(23):4618–4627
178. West JA, Mito M, Kurosaka S, Takumi T, Tanegashima C, Chujo T, Yanaka K, Kingston RE, Hirose T, Bond C, Fox A, Nakagawa S (2016) Structural, super-resolution microscopy analysis of paraspeckle nuclear body organization. J Cell Biol 214(7):817–830

179. Nishimoto Y, Nakagawa S, Hirose T, Okano HJ, Takao M, Shibata S, Suyama S, Kuwako K, Imai T, Murayama S, Suzuki N, Okano H (2013) The long non-coding RNA nuclear-enriched abundant transcript 1_2 induces paraspeckle formation in the motor neuron during the early phase of amyotrophic lateral sclerosis. Mol Brain 6:31. doi:10.1186/1756-6606-6-31
180. Ip JY, Sone M, Nashiki C, Pan Q, Kitaichi K, Yanaka K, Abe T, Takao K, Miyakawa T, Blencowe BJ, Nakagawa S (2016) Gomafu lncRNA knockout mice exhibit mild hyperactivity with enhanced responsiveness to the psychostimulant methamphetamine. Sci Rep 6:27204
181. Tsuiji H, Yoshimoto R, Hasegawa Y, Furuno M, Yoshida M, Nakagawa S (2011) Competition between a noncoding exon and introns: Gomafu contains tandem UACUAAC repeats and associates with splicing factor-1. Genes Cells 16(5):479–490
182. Ishizuka A, Hasegawa Y, Ishida K, Yanaka K, Nakagawa S (2014) Formation of nuclear bodies by the lncRNA Gomafu-associating proteins Celf3 and SF1. Genes Cells 19(9):704–721

Chapter 10
Technological Developments in lncRNA Biology

Sonali Jathar, Vikram Kumar, Juhi Srivastava, and Vidisha Tripathi

Abstract It is estimated that more than 90% of the mammalian genome is transcribed as non-coding RNAs. Recent evidences have established that these non-coding transcripts are not junk or just transcriptional noise, but they do serve important biological purpose. One of the rapidly expanding fields of this class of transcripts is the regulatory lncRNAs, which had been a major challenge in terms of their molecular functions and mechanisms of action. The emergence of high-throughput technologies and the development in various conventional approaches have led to the expansion of the lncRNA world. The combination of multidisciplinary approaches has proven to be essential to unravel the complexity of their regulatory networks and helped establish the importance of their existence. Here, we review the current methodologies available for discovering and investigating functions of long non-coding RNAs (lncRNAs) and focus on the powerful technological advancement available to specifically address their functional importance.

Keywords lncRNA • Chromatin • lncRNA interactions • Secondary structure • Functional characterization • Genome-wide characterization • SAGE • RNA-Seq • CLIP • CHART • ChIRP

10.1 Introduction

A large portion of the eukaryotic genome is transcribed into RNAs that do not code for proteins. These interesting molecules, formally known as non-coding RNAs (ncRNAs) that bypass the central dogma for flow of genetic information in cells, have been under constant scrutiny for their existence for several decades [1]. The notable exceptions during the initial discovery of this class of molecules have been the tRNAs and rRNAs owing to their abundance, which were further identified to have structural and regulatory roles in translational machinery [2]. These were

S. Jathar • V. Kumar • J. Srivastava • V. Tripathi (✉)
National Centre for Cell Science, Pune University Campus, Ganeshkhind, Pune 411007, India
e-mail: tvidisha@nccs.res.in

M.R.S. Rao (ed.), *Long Non Coding RNA Biology*, Advances in Experimental Medicine and Biology 1008, DOI 10.1007/978-981-10-5203-3_10

further accompanied by the discovery of a substantial number of other transcripts, e.g. miRNAs, siRNAs, etc. [3, 4]. Although the functions of these molecules were being deciphered gradually, a large part of the non-coding transcriptome was considered to be junk or transcriptional noise [2]. Beyond the abundant class, several large non-coding transcripts like RNaseP, H19, Xist and MALAT1 were identified much earlier, their prominence was not fully recognized and the possibility that these molecules could possess biological purpose was doubtful [5].

The emergence of high-throughput sequencing technologies had the greatest impact on the expanding world of non-coding RNAs. It is estimated that more than 90% of the mammalian genome is transcribed, out of which only approximately 2% contributes towards the protein-coding function [6, 7]. The remaining non-coding genome can be broadly classified into small (18–200 nts) and long (200 nts to >100 kb) non-coding RNAs (lncRNAs) [8]. Although the discovery and functional characterization of small ncRNAs have dominated the field of RNA biology over past several decades, long non-coding RNAs (lncRNAs) are the least explored emerging regulatory molecules.

Although the physical and functional classification of lncRNAs has been very challenging, it has been observed that they share many common characteristics with mRNAs. For instance, majority of the lncRNAs display epigenetic marks like increased H3K4me3 at promoters and RNA Pol II binding sites similar to mRNAs, indicative of active transcription [9]. Additionally, several lncRNAs are polyadenylated and have 5′caps similar to mRNAs [8]. One of the unique features of many lncRNAs is presence of H3K4me1 marks suggestive of enhancer elements [8]. A large number of lncRNAs are developmentally and temporally regulated or restricted to particular tissues or organs indicating their role in specific cellular processes, rendering their discovery more difficult [10–12]. Most importantly, lncRNAs generally show least sequence conservation across different species [13] unlike many protein-coding genes. Therefore, their functional categorization based on functional domains or sequence conservation has not been possible so far. Apart from carrying information in the sequence, RNAs can fold into intricate secondary or tertiary structures influencing their biological functions [14]. With the advent of computational structure prediction tools coupled with ultrahigh-throughput sequencing technologies, genome-scale RNA structural information appears feasible now [15–18]. This could provide great insights into the structural similarities of lncRNAs across species and could provide clues to evolutionary conservation of functional domains of lncRNAs.

Although the function of a large number of lncRNAs is still not known, there is clear evidence for their importance in physiology, embryology and development with numerous novel gene regulatory functions, including their role in contribution to high degree of complexity observed in multicellular organisms [3, 19]. Various studies have revealed active role of lncRNAs in controlling multiple regulatory layers including chromosome architecture, chromatin modulation and epigenetic modification, transcription, RNA maturation, splicing and translation [20–24]. In contrast to other types of regulatory ncRNAs, e.g. miRNAs, lncRNAs localize both in the cytoplasm and nucleus, which further suggests their important role in epigenetic modification and gene regulation [19]. Furthermore, their aberrant expression has been linked with a wide spectrum of disorders including cancer [25, 26]. Thus a comprehensive knowledge of their function would greatly facilitate our current understanding of various cell regulatory networks and disease mechanisms.

Based on the current evidences, lncRNAs can perform their function by physically interacting with DNA (chromatin), RNA (mRNAs, miRNAs, circular RNAs) and proteins, thereby regulating complex network of gene expression by acting as signals (for integrating spatiotemporal, developmental and stimulus-specific cellular information), decoys (the ability to sequester a range of RNA-dependent effectors and protein partners), guides (for proper localization of chromatin-modifying complexes and other nuclear proteins to specific genomic loci to exert effects) and scaffolds (for bringing two or more proteins into discrete complexes) [27]. In short, lncRNAs are emerging as new exciting players in gene regulatory network, and their deregulation may provide opportunity for their use as prognostic markers for many complex diseases.

Elucidating the different mechanisms of action of lncRNAs will not only provide the basic biological understanding of cellular function but also a critical nexus for revealing the basis of lncRNAs in disease aetiology and their use as targets in subsequent drug design. Most importantly, with the fact that mammalian transcriptome comprises several thousand lncRNAs with diverse signatures, the question that whether all of them have biological purpose still stands unanswered.

Extensive research and development of lncRNA identification and functional annotation tools have led to many successes and discoveries, the applicability of which has been still limited; however, over the past decade, with the emergence of ultrahigh-throughput technologies, development of newer strategies, accumulation of genome-wide experimental data and better optimization of existing algorithms have been able to bridge the knowledge gap. Here, we review the current methodologies available for discovering and investigating functions of lncRNAs. The attempt is to explain how to investigate an uncharacterized transcriptome systematically and to have a complete mechanistic understanding of the function of long non-coding RNAs.

10.2 Identification and Characterization of Long Non-coding RNAs (lncRNAs)

An explosion of technologies over the recent years has greatly made it possible to discover and functionally analyse the plethora of large non-coding transcripts. Owing to their variable relative abundance as well as cell and tissue type specificity in expression, identification and subsequent characterization of lncRNAs was an extremely difficult task. However, with new tools in hand, a substantial progress has been made in uncovering the diverse and dynamic lncRNA world (Fig. 10.1). In simple terms, according to the relative position of protein-coding genes, lncRNAs have been categorized into four groups: intergenic, intronic/intragenic, antisense and overlapping [26]. There have been several modes of categorization of lncRNAs based on their genomic location, expression, mechanism of action, etc. However, the initial step towards the functional and mechanistic exploration of lncRNAs is individual lncRNA identification, which completely depends on the appropriate experimental strategy.

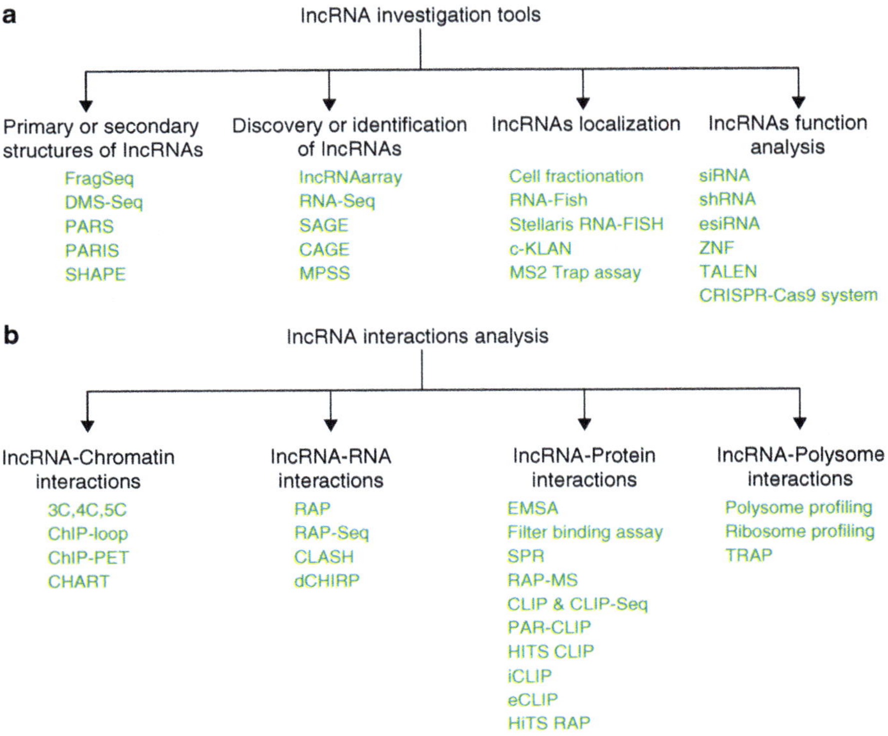

Fig. 10.1 Overview of different tools for lncRNA investigation

10.2.1 Transcript Sequence Assembly: Tiling Arrays and SAGE

The Human Genome Project and development of high-throughput sequencing technologies had the greatest impact in the expanding world of non-coding RNAs. Numerous transcripts without protein-coding potential have been identified in several species. During the initial period of lncRNA era, major discoveries of novel lncRNAs were facilitated by using cDNA cloning followed by Sanger sequencing (FANTOM project) or identifying chromatin signatures such as presence of K4-K36 trimethylation marks and utilizing tiled microarrays across the non-coding region of the genome in a tissue- and cell-specific manner (Fig. 10.2) [8, 28–30]. The FANTOM project at RIKEN provided the first large-scale catalog of approximately 16,000 novel transcripts in mouse (70% putative ncRNAs) by generating and sequencing cDNA libraries with almost 60,770 full-length cDNAs [28]. Additionally, this catalog also represented over 2000 sense-antisense pairs of transcripts [31]. Subsequently, the catalog grew with increasing number of novel transcripts in the list. The lncRNA-Uchl1 is among one of the first sense-antisense

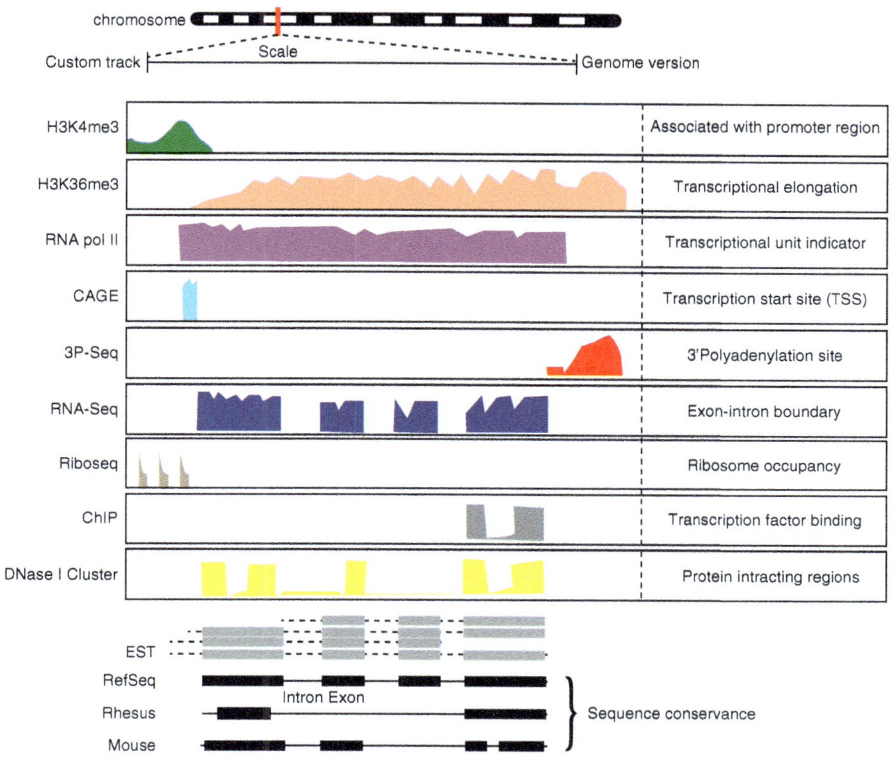

Fig. 10.2 Prediction of lncRNA based on NGS technology

transcript pairs discovered in this approach. It is expressed in an antisense orientation from the Uchl1 locus and overlaps with the 5′end of the Uchl1 mRNA. It regulates the translation efficiency of Uchl1 RNA by binding to it without altering its expression level [24].

One of the first methods developed for global transcriptome assembly and analysis was SAGE, which is based on the generation of short stretches of cDNA sequence tags containing restriction enzyme sites at the 3′end. The tags are concatenated followed by Sanger sequencing [29]. This methodology allowed for the identification of several unannotated transcripts and also revealed that lncRNAs are broadly distributed across all chromosomes in humans [12] and show high tissue specificity. However, this method was soon replaced by more sensitive sequencing technologies that profile a larger number of transcripts in greater depth.

One of the classical discoveries in the lncRNA field has been the identification of HOX antisense intergenic RNA (HOTAIR), which was possible through probing transcripts on tiling arrays. This methodology employs hybridization of cDNAs to microarray slides carrying tiled oligonucleotides designed to cover an entire chromosome or the entire genome [32]. HOTAIR has been shown to negatively regulate transcription across 40 kb of the HoxD locus in *cis* by interacting with polycomb repressive complex 2 (PRC2) [32]. Further, Cheng et al. utilized a higher resolution approach to provide in-depth analysis by using arrays of 25mer oligonucleotides spaced every 5 bp to create transcriptional maps of ten human chromosomes [30]. This approach provided a high-resolution mapping of the RNAs. Additionally, by separating the poly (A)+ and (A)− RNA fractions from the cellular compartments during initial sample preparation, they further demonstrated the existence of a large number of unannotated transcripts, many of which were poly (A)− and enriched in the nucleus [30].

A major limitation of these array-based techniques is their reliance on existing knowledge of genome sequences, restricting the discovery of novel transcripts. Additionally, they also require a large amount of starting material (RNA) to perform experiments, and expression dynamic detection becomes difficult after a saturation point. Most importantly, weak binding or cross hybridization of transcripts to probe renders significant noise in the data, affecting the study of repetitive sequences.

10.2.1.1 RNA Sequencing (RNA-Seq)

The idea of transcriptional units, transcriptional complexity and isoforms emerged and strengthened majorly due to the advent of RNA-Seq technology. RNA-Seq is a quantitative and exceptionally sensitive technique that revolutionized entire major discoveries in the lncRNA biology. It is based on the conversion of transcripts into a pool of cDNAs that will constitute the sequencing library. The library is prepared by RNA fragmentation, adapter ligation, cDNA synthesis, size selection and limited cycles of amplification. The libraries can be prepared from poly (A)+ enriched RNA by selectively depleting rRNA. However, this approach excludes non-polyadenylated or partially degraded RNAs, which may also result in exclusion of several novel transcripts from the sample under study. This drawback can be taken care of by using random priming methods or by selective exclusion of rRNA from the sample. Based upon the requirement, the sequencing can be ‘single end’ or ‘paired end’. Paired-end libraries allow sequencing from both the ends of the same molecule, providing additional information like splicing events and identifying the boundaries of transcription units especially for alternative start and polyadenylation sites. A high-throughput sequencing method provides approximately 20–40 million reads that is sufficient for transcript detection. However, majority of lncRNAs are low in abundance; therefore, the standard approaches provide a constraint for identification of novel low-abundance transcripts. To overcome this issue, RNA can be fractionated from different cellular compartments prior to sequencing, which increases the relative abundance of unique transcripts. In addition to transcriptome profiling, RNA-Seq has proven to be powerful approach for identification of alternatively spliced isoforms, novel splice junctions, SNPs and gene fusion events [33–36]. Using data from RNA-Seq, Guttman and colleagues, for the first time, developed a method to reconstruct the transcriptome of the mammalian cell without using the existing annotations and applied it to mESCs, neuronal precursor cells and lung fibroblasts. Using this approach, they identified a substantial variation in the protein-coding gene structure and also identified more than 1000 large intergenic ncRNAs (lincRNAs), majority of which were not reported earlier [37]. In a subsequent parallel study by Rinn's group, a reference catalog of more than 8000 lincRNAs was generated using an integrative approach of assembling data collected from approximately four billion RNA-Seq reads across 24 tissues and cell types [38]. This study also revealed the tissue-specific expression pattern of lncRNAs.

Several derivative approaches of sequencing-based genome-wide transcriptome analysis have been extensively used to identify thousands of lncRNAs from several species. For example, Rinn and Mattick's group utilized targeted RNA sequencing (CaptureSeq) by combining tiling arrays with RNA-Seq technology to refine transcript annotations and achieve enriched read coverage, accurate measurement of gene expression and quantitative expression data [39, 40]. During transcript assembly from large amount of short-read RNA-Seq data, it is likely that highly expressed transcripts will be represented more, and the transcripts with restricted or low expression might have a chance to be ignored. This problem was resolved by the use of targeted RNA capture sequencing [30, 34]. In a recent study, the members of Michigan Centre for Translational Pathology developed a database, MiTranscriptome, which is a catalog of human transcripts derived from computational analysis of high-throughput RNA-Seq data from over 6500 samples spanning diverse cancer and tissue types [41]. RNA-Seq libraries were curated from tumours, normal tissues and cell lines from 25 independent studies; ab initio assembly method was applied yielding approximately 91,000 expressed genes. More than 68% of these genes were lncRNAs, out of which almost 80% were unannotated. This catalog also revealed approximately 8000 lineage- or cancer-associated lncRNAs genes [41]. Since most lncRNAs share similarities in their characteristics with mRNAs, similar approaches have been adopted for their identification. Moreover, based on the experimental need, variations like ChIP/RNA-Seq (for transcription and epigenetic regulators), ChIA/RNA-Seq (Chromatin Interaction Analysis) and DNA/RNA-Seq (for genome assembly) have evolved and extensively used [42].

Another alternative approach based on RNA-Seq is Cap Analysis of Gene Expression (CAGE) that involves the simultaneous mapping and quantification of 5′capped RNA expression [43, 44]. This is an effective approach for identifying Pol II-driven transcription start sites and active promoter regions. However, it requires a relatively large quantity of starting material and excludes transcripts driven through Pol III or devoid of 5′ caps. To overcome some of these limitations, Carninci's group devised an upgraded version of the technology that could be used on highly refined samples obtained from tissue microdissections and subcellular fractions, referred to as nanoCAGE and CAGEscan [45]. These methods can be used to capture the 5′ends of RNAs from as little as 10 ng of total RNA. CAGEscan is a mate-pair adaptation of nanoCAGE that captures the transcript 5′ends linked to a downstream region and compensates for CAGE's inability to detect the 3′end of transcripts. Carninci's group in conjunction with the FANTOM consortium performed a thorough profiling of nucleus- and cytoplasmic-enriched RNA fractions from a representative set of human and mouse stem cell and differentiated cell lines using four complementary sequencing technologies (deep sequencing, CAGE, nanoCAGE and CAGEscan) and identified several thousand novel lncRNAs, which were principally enriched in nucleus. They also discovered that a large number of these novel transcripts originated from LTR elements [23]. Nevertheless, none of the high-throughput sequencing technologies are considered complete without further confirmation of transcript ends and structures through methods like quantitative RT-PCR, 5′ and 3′ RACE and cloning. Additionally, chromatin signatures play a very important role in further understanding of the accuracy of transcript assemblies [46].

10.2.1.2 Single-Cell RNA-Seq

Most transcriptomic studies are performed on a population level in bulk cell samples where gene expression data is a result of both the expression levels of genes in each cell type and the relative abundance of each cell in the population. Since, majority of the lncRNAs show restricted expression pattern based on particular cell type/stage, current sequencing technologies, though highly powerful, may not be able to capture the subtle but potentially biologically meaningful differences between cells. Moreover, in some instances, sample size is insufficient for such large-scale analysis, like stem cells, patient's samples, circulating tumour cells or flow-sorted cells for certain applications. In this scenario, single-cell transcriptome analysis provides tools for a new dimension and higher resolution for identifying and studying lncRNAs. Pioneering studies using single-cell RNA-Seq revealed additional layers of transcriptional differences between individual cells, like splicing patterns and allelic random expression, which suggests that it could provide a great tool towards reconstituting temporal transcription networks during developmental processes [37–39, 47]. For instance, using single-cell transcriptome analysis augmented by single-molecule RNA-FISH (smFISH), Kim et al. demonstrated that approximately 400 lncRNAs are differentially expressed during defined stages of reprogramming to pluripotency and showed how dynamic changes in the transcriptome reprogram cell stage [48]. Similarly, single-cell profiling of early stages of embryonic development in *C. elegans* revealed extensive transcriptional activity at stages that were previously considered inactive [49]. In mouse blastocysts, it was found that the heterogeneity in the transcriptomes of individual cells was a prerequisite for them to segregate into different lineages [50]. Transcriptome profiling of individual neocortex cells of developing brain revealed that several lncRNAs are abundantly expressed in individual cells and are cell-type specific. This discrete expression pattern of lncRNAs in individual cells reflects their important biological function in the development of particular neural cell types [51].

lncRNAs being remarkably cell type or tissue specific, single-cell transcriptomics can be the ultimate approach for a comprehensive understanding of the transcriptional landscape of individual cells. However, this approach has certain limitations and still requires significant development before it can be used for comprehensive understanding. Since it targets only poly (A)+ RNAs, the non-polyadenylated RNAs are currently underrepresented. Additionally, it is not possible to maintain strand specificity and detect isoform variants simultaneously. Most importantly, the sensitivity needs to be improved, as currently it is difficult to distinguish between noise and biological variability for low-abundance transcripts [52].

10.2.2 Cellular Localization of lncRNAs

Understanding the mechanism of action of lncRNAs often relies on their subcellular localization or distribution. Information about tissue specificity or cellular distribution could provide important circumstantial evidence for experimental planning, for example, in what type of cells do particular lncRNAs express and in which cellular compartment they act. For instance, lncRNAs associated with specific subnuclear domains or chromatin are more likely to have regulatory or structural roles, like many chromatin-associated lncRNAs which have been found to be *cis*-acting transcriptional regulators. MALAT1 is a nuclear retained lncRNA that localizes to nuclear speckles, subnuclear domains involved in pre-mRNA processing [53]. In speckles, MALAT1 regulates the activity and levels of SR splicing factors, thereby modulating alternative splicing of a subset of gene pre-mRNAs [53]. Additionally, MALAT1 levels are regulated during normal cell cycle progression, and it is required for proper G1/S and mitotic progression [54]. NEAT1 lncRNA that localizes to paraspeckles acts as a structural component of these nuclear domains, and it is essential for the nucleation of paraspeckles in cells [55]. The identification of MALAT1 as a speckle component was the first step towards exploring its function in regulating alternative splicing. Similarly, cytoplasmic lncRNAs are mainly found to be involved in post-transcriptional gene regulation, translation or protein localization. Most importantly, unlike protein-coding genes, the transcribed lncRNAs are the final product of their genes; their subcellular localization is directly linked to their physiological functions. Therefore, determining the cellular distribution of lncRNAs should be considered as the first step towards exploring the function in the cell.

10.2.2.1 Fluorescence In Situ Hybridization (FISH)

Fluorescence in situ hybridization (FISH) is a powerful method to examine the spatial distribution of lncRNAs at subcellular level, and it is extensively used for studying structural and dynamic properties of cells and subcellular entities. RNA-FISH technique is utilized to visualize transcripts localized in the nucleus or cytoplasm, therefore allowing to monitor gene expression and transcriptional activity at an individual cell level [56–61]. Basically, RNA-FISH involves the hybridization of fluorochrome- or enzyme-labelled nucleic acid probes to RNA target sequences in cells followed by microscopic visualization. Over the years, this technique has been utilized to answer various aspects of nuclear organization and gene expression at a single-cell level, facilitated through advancement in fluorescence microscopes, development of various labelling procedures and sensitive immunocytochemical detection systems. Using this method, it is possible to detect different RNA target sequences together with proteins or other cellular components simultaneously in the same cell; different components can be labelled using fluorochromes with different excitation and emission spectra. This combined detection of RNA and proteins has been widely used to study transcription, RNA processing and translation. RNA-FISH analysis of MALAT1 lncRNA demonstrated that it localizes to nuclear speckles and co-localizes to bonafide speckle components like SRSF1 [21]. It modulates SR-protein-mediated alternative splicing in cells. Additionally, it was also shown that depletion of SON protein, a core component of the speckles, leads to redistribution of MALAT1 in the nucleus suggesting that its localization to speckles is dependent on SON protein levels in the cell [21]. Similarly, RNA-FISH analysis for NEAT1 lncRNA confirmed its presence in paraspeckles and co-localization with paraspeckle proteins; however, depletion of NEAT1 RNA resulted in loss of paraspeckles from the nucleus as confirmed by immunolocalization of various speckle components [62]. Similarly, combining RNA/DNA-FISH in placental and embryo sections, it was shown that Kcnq1ot1 lncRNA establishes a nuclear domain to epigenetically inactivate genes in *cis*, whereas genes that are not regulated by this lncRNA are localized outside of the domain [63]. This was the first study to show that apart from Xist lncRNA, autosomal Kcnq1ot1 lncRNA can also establish nuclear domains that might create a repressive environment for epigenetic silencing of adjacent genes. Since most lncRNAs are not very abundant in the cell, RNA-FISH cannot provide deep mechanistic insights of lncRNA localization or function due to its limited resolution.

Another finer variation of this technique, termed as single-molecule FISH (smFISH), has become a more popular tool to study gene expression by direct visualization of individual RNAs [64–66]. This technique involves hybridization of a large number of (typically more than 30) fluorescently labelled short DNA oligonucleotides to different regions of target RNA, such that spatially separated RNA transcripts appear as individual diffraction-limited spots that are readily detectable using conventional wide-field fluorescence microscope.

10.3 Methodologies for Exploring Functions and Mechanisms of Action of lncRNAs

In recent years, newer tools for probing the complexity of lncRNA research have made a substantial progress in unravelling the potential regulatory functions of lncRNAs. A synergistic convergence of cutting-edge technologies from different areas of expertise has shown potential in fast-paced characterization of lncRNA function. New RNA interactome methods, cross-linking in vivo purification strategies, transcriptome-wide structure identification approaches and other RNA centric methodologies have opened up the new layers of complexity in lncRNA functions.

Once the annotation of an uncharacterized transcriptome is done, it is necessary to determine if the transcripts indeed possess biological functions. There is no universal experimental approach for characterization of the function of lncRNAs as it is a multistep process where one has to first identify a candidate that contributes to a physiological function in the cell and further confirm that the observed effect is mediated through an RNA-based mechanism. One of the first steps towards understanding this is to determine if a target RNA has any functional consequence in a particular context by gain- or loss-of-function analysis. Several approaches have been adopted to effectively deplete target lncRNAs and further understand the outcome of the depletion [53].

10.3.1 Loss-of-Function Strategies for lncRNAs

One of the ways to study lncRNA function is by suppressing its expression and observing the resulting phenotypes. Two main approaches are followed to experimentally achieve successful knockdown of a particular lncRNA: transient methods like RNA interference (RNAi)- and antisense oligonucleotide (ASO)-mediated depletion or in vivo loss-of-function strategies. However, certain considerations are important during the design of the experiment, for instance, the proximity of the lncRNA locus to protein-coding genes, its stability and copy number, its chromatin signatures and tissue expression profiles. Additionally, the cellular localization of the lncRNA could play an important role for the design of correct loss-of-function strategy. If the lncRNA is chromatin associated, it might be *cis*-acting, whereas if it's nucleoplasmic or cytoplasmic, it might be trans-acting RNA. For a trans-acting RNA, RNAi- or ASO-mediated depletion would work efficiently. However, for a *cis*-acting RNA careful consideration of the genes in proximity should be given. Antisense oligonucleotide (ASO)-mediated depletion of lncRNAs has been widely used to understand lncRNA function in cell-based assays. They act by forming a DNA/RNA hybrid with the nascent RNA and triggering an RNase H-dependent degradation of the RNA in the nucleus. However, the extent of off-target effects is highly likely [67]. Therefore, a successful rescue of the phenotype is extremely important to prove that the phenotype is caused by the alteration of the lncRNA transcript.

Various studies have used common strategies of using RNAi-mediated loss-of-function analyses to understand phenotypes mediated by functional lncRNAs (Table 10.1). The primary advantage of using these cell-based assays is the ability to directly test for the degradation of the target RNA itself. In one of the first large-scale screens, Guttman and colleagues used lentiviral shRNA-mediated analysis to understand the depletion of 147 lncRNAs in mESCs [68]. In this study, almost 15% of the lncRNAs displayed significant downregulation of pluripotency marker genes upon knockdown, indicating their involvement in maintaining the pluripotency state. Similarly, in an effort to understand the role of Xist RNA in X-chromosome inactivation (Xi), Locked nucleic acids (LNAs) were used to block the binding ofX-

Table 10.1 Role of lncRNAs in different biological processes

Biological process	lncRNAs	Mode of action	References
DNA damage	Linc p21	Gene repression by interaction with hnRNP-K	[133, 134]
	PANDA	Binds to NF-YA and prevents its binding to chromatin	[133, 135]
	DINO	Stabilizes p53 through RNA-protein interaction	[136]
	DDSR1	Sequesters the BRCA1/RAP80 complex	[137]
	NORAD	Sequesters PUMILIO protein	[138]
	lncRNA-JADE	Upregulates expression of jade1 resulting in elevated histone H4 acetylation	[133, 139]
	ANRIL	Suppresses the expression of INK4a, INK4b and ARF	[133, 140]
Pluripotency	lncRNA-ROR	Transcriptional regulator	[141]
	Gomafu	Binds to the SF1 and leads to the post-transcriptional regulation of splicing efficiency	[142]
	Mira	Epigenetic regulator of Hoxa6 and Hoxa7 gene	[142]
	HOTAIRM1	Regulates the genes from HOXA cluster	[142]
	Xist	Binds to PRC2 complex via RepA and in result inactivates X chromosome	[142, 143]
	Tsix	Epigenetic silencing of Xist expression	[142, 143]
	Xite	Interacts with Sox2 and Tsix and results in epigenetic reprogramming	[142]
	Jpx	Removes repressive RBP CTCF from Xist promoter and activates Xist	[142]
	NEAT1	Post-transcriptional regulator	[142]
	Rian	Binds to chromatin-modifying complex PRC2	[142]
	lincRNA-ES1–3	Regulates ESC differentiation	[144]
	lincRNA-SFMBT2	Regulates embryoid body differentiation	[145]
Cellular proliferation and senescence	ANRIL	Regulates the level of cdk inhibitor p15, p16	[146]
	MALAT1	Regulates the expression of transcription factor B-MYB	[54]
	$NcRNA_{CCND1}$	Regulates cyclin D1	[135, 146]
	SRA	Regulates cdki p21 and p27 and phosphorylation of CDK1	[147]
	HEIH	Suppresses the action of cdki p16, p21, p27 and p57	[147]
	HULC	Regulates the expression of cdki18	[148]
	MEG3	Reduces the level of MDM2	[149]
	Gadd7	CDK6 mRNA decay by preventing its interaction with TDP-43	[150]
	UCA1	Regulates cdk1 p57 level	[147, 151]
	7SL	Regulates p53 mRNA level	[147]
	H19	Regulates p57 level	[147]

(continued)

Table 10.1 (continued)

Biological process	lncRNAs	Mode of action	References
Inflammation associated	Lethe	Binds to RelA (subunit of NF-KB) and reduces the production of inflammatory protein	[152]
	THRIL	Interacts with hnRNP-L and binds with TNF-a promoter and regulates its expression	[153]
	Lnc-IL7R	Regulates the level of inflammatory mediators IL-6 and IL-8	[147]
	17A	Regulates GABA B alternative splicing and signalling	[147]
	lincRNA-EPS	Potent repressor of IRG	[147, 154]
Telomere stability	TERC	Promotes telomere extension	[155]
	TERRA	Suppresses telomere extension	[156]
Proteostasis	HULC	Promotes cell proliferation and invasion and inhibits apoptosis	[147]
	MEG3	Regulates p53 level by lowering MDM2 expression level	[149]
	7SL	Regulates p53 mRNA translation	[157]
	GAS5	Inhibits glucocorticoid receptor-mediated gene expression	[158]
	PANDA	Interferes transcriptional activity of NF-YA	[159]
	Gadd7	Destabilizes cdk6 mRNA by interfering the interaction of TDP-43 and cdk6 mRNA	[150]
	HOTAIR	Promotes ubiquitination and degradation of Ataxin1 and snurportin1	[160]
	lincRNA-p21	Inhibits the translation of CCNNB(encodes beta catenin) and JUNB mRNA (encodes JunB)	[147]
Epigenetic regulation	Xist	Binds to PRC2 complex via RepA in result inactivate X chromosome	[143]
	H19	Represses Igf2,Slc38a4 and Peg-1 by interacting with MBD1	[143]
	PTENpg1-AS	Represses expression of tumour suppressor gene phosphatase and PTEN	[161]
	ANRIL	Regulates CDKN2A/B(tumour suppressor)	[140]
	PAPAS	Increases trimethylation of H4K20	[162]
	TARID	Regulates the expression of tumour suppressor protein TCF21 by inducing promoter demethylation	[163]
	pRNA	Regulates transcription of rRNA through interacting with DNA at TTF1 target site	[99]
Haematopoiesis	Linc HSCs	Regulates HSC self-renewal	[164]
Muscle differentiation	Linc-MD1	Functions as ceRNA for miR133 and miR135 and regulates expression of MAML1and MEF2C	[165]

(continued)

Table 10.1 (continued)

Biological process	lncRNAs	Mode of action	References
Cardiac development	Bvht	Interacts with SUZ12 and competes for PRC2	[166, 167]
	Fendrr	Functions as chromatin signature modulator upon its interaction which binds to PRC2 and TrxG/MLL complexes	[11, 167]
	KCNQ1OT1	Silences UIGs and PIGs in *cis* by recruiting PCR2 complex members (Ezh2 and Suz12), G9a HMT and Dnmt	[135]
	Ak011347	Targets the gene Map3k7	[168]
Neuronal differentiation	Pnky	Interacts with PTPB1(splicing regulator)	[169]
EMT	lncRNA-HIT	Regulates TGF-β	[170]
	UCA1	Enhances wnt/β catenin signalling pathway	[171]
Apoptosis	GAS5	Riborepressor of glucocorticoid receptor	[172]
	SPRY4-IT1	Regulates MAPK signalling pathway	[173]

ist to its target regions. This study showed if the Xist RNA is displaced from the Xi, it leads to loss of X inactivation [69]. Additionally, in a large-scale analysis to examine the effect of lncRNA depletion using siRNAs, Shiekhattar's group revealed an unanticipated role for a class of lncRNAs in activation of critical regulators of development and differentiation. They showed that a subset of intergenic lncRNAs potentiated the expression of a protein-coding gene within 300 kb of the lncRNA, suggesting an enhancer-like function of lncRNAs. It was also proposed that the candidates that did not display any positive effect in the study might exert their action over longer distances, which was not assessed in the analysis [70]. This was the first study to report a new biological function in positive regulation of gene expression for a class of lncRNAs in human cells. The identification of scaffold function of HOTAIR lncRNA for distinct histone modification complexes was also demonstrated through RNAi-mediated depletion of HOTAIR from cells. HOTAIR RNA has distinct binding domains for PRC2 and LSD1 complexes, and its depletion abrogates the interaction between the two suggesting that HOTAIR is required to bridge this interaction and serves as scaffold [20].

Recent developments in technologies for altering RNA levels in the cell, like CRISPR interference (CRISPRi) and CRISPR activation (CRISPRa), have facilitated control over lncRNA transcription inhibition or activation efficiently [71–74]. This involves Cas9-mediated modulation of gene transcription levels through coupling of nuclease-deficient Cas9 (dCas9) with a transcriptional repressor (e.g. KRAB) or an activator (e.g. VP64/p65/Rta). This approach is highly promising as it does not require any manipulation on the RNA locus or the RNA itself; however, care must be taken to avoid the effect of the factors on local transcription. Different strategies have been employed for determining the physiological function of lncRNAs in vivo. One of the approaches has been to genetically manipulate the lncRNA locus by inserting transcriptional terminator sequences or deleting the regulatory elements. Similarly, various studies have employed deleting the complete lncRNA locus or mutating the putative functional domains or targeted inversions between the promoter and RNA. A large number of knockout models for lncRNAs have been tested for in vivo functional roles; however, it appears that in many cases lncRNAs appear dispensable. For instance, MALAT1 is a nuclear lncRNA that when depleted in cells shows deleterious phenotype in cells; however, the mouse knockout model for MALAT1 is viable and does not show any phenotype [75, 76]. Similarly, HOTAIR lncRNA that is transcribed from HOX C cluster guides the recruitment of chromatin modifiers to specific chromatin in HOX D cluster and inhibits transcription. However, a mouse in which the region of HOX C cluster spanning the entire HOTAIR was deleted showed normal viability and no defect in transcription or chromatin modifications [77]. To summarize, multiple approaches to achieve a significant depletion and carefully planned rescue experiments are extremely important to clearly demonstrate the physiological function of a particular lncRNA. For instance, using a premature termination signal in the Fendrr RNA sequence, it was shown that Fendrr lncRNA is required for viability in mice, and this phenotype could be rescued by a transgene expression [11]. In a parallel study, Sauvageau et al. (2013) performed a whole gene ablation strategy and identified similar phenotype of decreased viability [78].

Achieving a significant knockdown of lncRNAs is not always easy and straightforward; however, there are many factors that can lead to incorrect or uninterpretable results. The most crucial factor for achieving successful depletion of a lncRNA is directing well-designed reagents to the most effective target regions. It is challenging to identify the most potent target sites for oligos due to unavailability of predictive algorithms to determine target accessibility. Additionally, verification of off-target sites is extremely important during the selection of oligos. Some of the other considerations are secondary structure of the target lncRNAs, effective target length and splice regions.

10.3.2 lncRNA-Interactome Analyses

To gain insight into the relationship between the function of lncRNAs and the complement of cellular factors with which they interact, several in vitro and in vivo approaches have evolved over the recent years. With the advancement in various technologies, newer approaches exploiting the biophysical and biochemical characteristics of nucleic acids and proteins are being employed, which has made it possible to address 'how' of lncRNA function. Based on evidences, lncRNAs are suggested to act as signals, guides, decoys or scaffolds by physically interacting with chromatin, RNA or proteins, thereby facilitating various biological processes.

10.3.2.1 RNA-Protein Interaction

Long non-coding RNAs perform their regulatory functions in association with proteins, like chromatin-modifying complexes, transcription factors and RNP complexes. The paramount factor in understanding the function of a lncRNA is the identification of its interacting protein partners. Most of the RNAs have structured and unstructured region that are bound with specific set of proteins. Based on presence of consensus primary or secondary or tertiary structure motifs, several sets of RNA-binding proteins have evolved to form ribonucleoprotein (RNP) complexes. Since lncRNAs are extremely diverse in function and sequence conservation, obtaining the complete interaction map of lncRNA-bound proteins is critical to our understanding of their function. Most importantly, this could also help in categorization of lncRNAs based on their specificities to proteins linked with particular pathways. Various approaches have been employed to describe lncRNA-protein interactions for the characterization of their function in different cellular processes, which can be classified into two groups based on the target molecule (Fig. 10.3). In the first approach, the focus is on the particular lncRNA, and methods are devised to identify all the interacting proteins with the same. This can be achieved by using in vitro transcribed RNA to retrieve proteins from cell lysates, tagging the endogenous RNA with affinity aptamers or using immobilized oligos to capture RNA-protein complexes under native conditions [79–83]. The second approach is protein centric, where the protein of interest is immunoprecipitated using specific antibodies and further RNAs associated with it are analysed through high-throughput technologies. Ideally, the method that allows the capture of in vivo lncRNA-protein interactions with high yield and specificity should be followed, and strategies to achieve more comprehensive portraits of these interactions at a global scale should be devised.

One of the pioneering discoveries towards understanding protein partners of a ncRNA was the identification of spliceosomes associated with snoRNAs, which was performed in vitro by affinity pulldown of RNA as a bait and subsequent elution of associated proteins followed by mass spectrometry (MS) analysis [84]. Since then, numerous studies have often used cell lysates to incubate complete lncRNA sequence or an important region of the lncRNA as bait, on which RNPs can assemble. Further, the associated proteins are eluted and subjected to MS for identification [85, 86]. However, there are certain limitations to this approach that need to be taken care of before any conclusion towards function is drawn. Firstly, most RNA-protein interactions are not merely sequence-specific interactions; rather the secondary and tertiary conformations of RNAs impart modular domains, which dictate its interaction with proteins. Therefore, before using the lncRNA as bait, it must be conformed that it is accurately structured, which is often difficult to achieve under in vitro conditions. Secondly, usually the concentration of the RNA bait far exceeds the physiological levels of the RNA; therefore, the chance of post-lysis association of RNA with protein increases resulting in false-positive outcomes. It becomes essential to confirm any interaction by performing a reverse pulldown by incubating respective antibody with cellular lysate containing endogenous lncRNA.

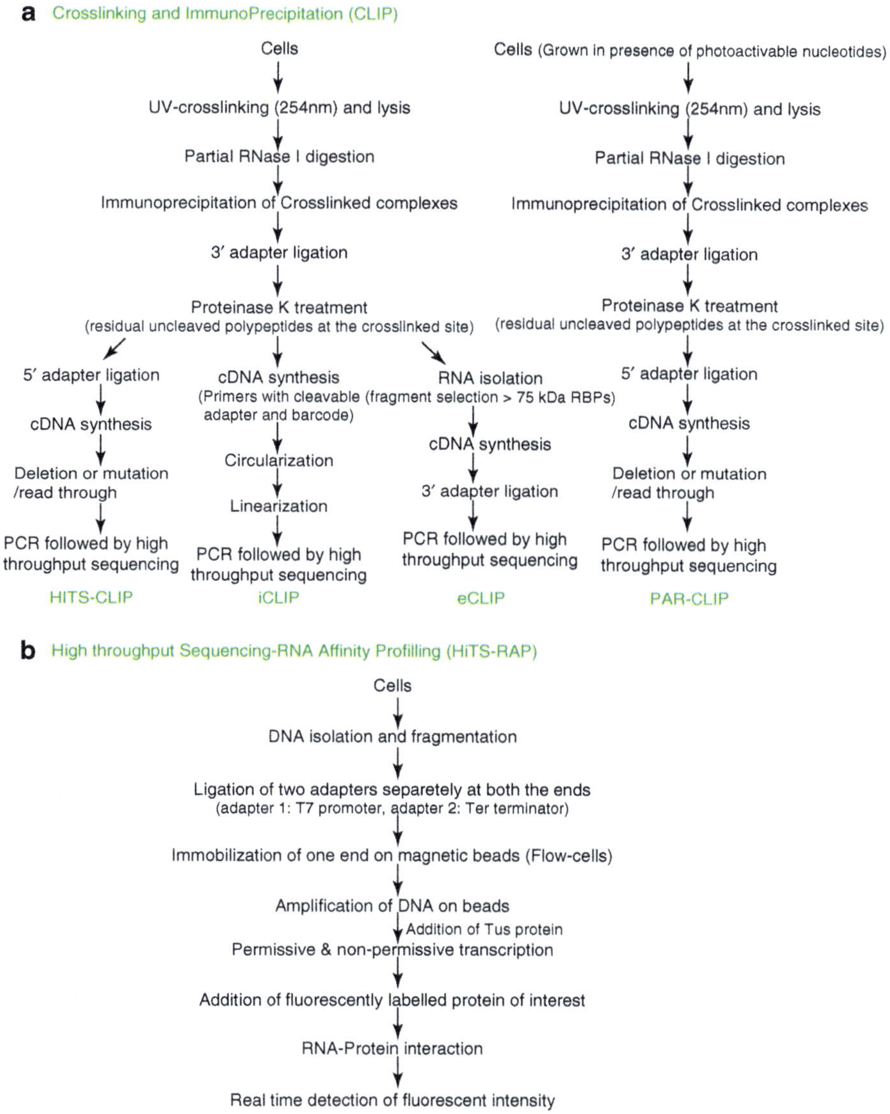

Fig. 10.3 Schematic representation of lncRNA-protein interaction techniques

RNA Immunoprecipitation (RIP)

This technique is exactly similar to any other protein immunoprecipitation reaction where an antibody specific to a particular protein is incubated with a nuclear/cellular lysate to pull down the protein of interest along with the other proteins that interact with it in a complex. However, the major difference is the reaction conditions are maintained to keep the cellular RNA intact. It is recommended to cross-link the endogenous RNA-protein interactions prior to lysate preparation, which helps to maintain the association of RNA-proteins intact during the stringent washing conditions. Post-lysis chance binding of potential artefacts is reduced with stringent washing conditions before elution. Pretreatment of the lysate with DNase I or RNase H is suggested, which helps in distinguishing between indirect and direct binding. Even though RIP has become one of the conventional methods for identifying protein partners of lncRNAs, it was first used to investigate how the Xist lncRNA mediates/initiates X-chromosome inactivation in females. Zhao and colleagues identified that the polycomb complex PRC2 is the direct target of RepA, which is a 1.6 kb lncRNA within Xist region. They further demonstrated that Ezh2, a component of PRC2, directly binds to the RNA and facilitates X inactivation as PRC2 deficiency compromises Xist upregulation [87].

To reveal the composition and dynamics of specific non-coding RNA-protein complexes in vivo, Chu and colleagues developed a comprehensive identification of RNA-binding proteins by mass spectrometry (ChIRP-MS) [88]. This involves extensive cross-linking of cells with formaldehyde followed by retrieval of target RNA with biotinylated tiling oligo hybridization. The RNA-protein complexes are further captured on streptavidin beads and liberated using gently biotin-elution step. The enriched proteins are finally subjected to MS. Using this method to pull down Xist RNA-protein complex, it was discovered that Xist interacts with 81 proteins, many of which are involved in chromatin modification, nuclear matrix and RNA remodelling pathways. Specific interactors like hnRNP-K participate in Xist-mediated gene silencing and histone modification, and Spen is required for gene silencing. Further characterization of interactions revealed that Xist lncRNA associates with different proteins in a modular and developmentally controlled manner to coordinate chromatin spreading and silencing [88].

Another classical method is RNA affinity in tandem (RAT), which is also a highly efficient RNA tag-based method for affinity purification of endogenously assembled RNP complexes. It involves genetically tagging RNAs with affinity sequences like naturally occurring RNA sequences that strongly bind to bacteriophage MS2 viral coat protein or addition of RNA aptamer tags in tandem with the lncRNA of interest. This construct is expressed in cells, and further endogenously assembled RNPs from crude cell extract are obtained through RNA tag-based affinity purification. Hogg and Collins developed this approach to purify 7SK lncRNA-associated protein complexes using RNA hairpins that bind PP7 coat protein for the first time [89]. Similarly, other derivatives of this approach include the use of MS2 repeats as tag in the lncRNA of interest, which is purified with the help of MS2 binding coat protein [90, 91]. The advantage of this technique is that the tagged RNAs are transcribed in vivo; therefore, the association of different proteins with these transcripts occurs in a more favourable environment biochemically. The only disadvantage with this approach is that tagging RNAs with exogenous sequence might interfere with its endogenous structure and function.

Cross-Linking Immunoprecipitation (CLIP)

CLIP, HITS-CLIP or CLIP-Seq is a method to identify RNA-protein interaction by cross-linking cells using UV light. In this process, using UV radiation, RNA-protein complexes are covalently cross-linked in vivo at the site where proteins are in direct contact with RNA. After cross-linking and immunoprecipitation using antibody against known protein, the immunoprecipitate is subjected to RNase treatment followed by proteinase K digestion and purification. Finally, the purified RNA fragments are adapter ligated and sequenced following the same procedures as RNA-Seq. Further, the reads obtained are mapped to the reference genome, and the binding sites of the respective protein on the total transcriptome are analysed. Therefore, this technique identifies the RNA footprint of protein-binding sites on the basis of the prec*ise* site of RNA-protein cross-linking. Guil et al. used this method to detect and identify direct PRC2-RNA interactions in human cancer cells. They observed that the PRC2 core component Ezh2 binds to a large number of intronic RNA sequences and regulates the transcriptional output of its genomic counterpart. They further demonstrated that overexpression of Ezh2-bound intronic RNA for SMYD3, which is a H3K4 methyltransferase gene, leads to a concomitant increase in Ezh2 occupancy throughout the corresponding genomic fragment resulting in reduced levels of the endogenous transcript and protein [92]. One of the most interesting recent findings is the exon-intron circRNAs or ElciRNAs, which were discovered through RNA Pol II CLIP followed by RNA-Seq in human cells. This class of circRNAs with exons 'circularized' and introns retained between exons is associated with RNA pol II in the nucleus. They interact with U1 snRNP and promote transcription of their parental genes [93]. One of the limitations of HITS-CLIP is that it does not provide a quantitative or structural understanding of RNA-protein interfaces owing to its low resolution and inability to provide full-length sequence.

Photoactivable ribonucleoside-enhanced cross-linking and immunoprecipitation (PAR-CLIP) is a modified version of CLIP with improved cross-linking efficiency and higher resolution mediated through photoreactive ribonucleoside analogs, 4-thiouridine (4SU) and 6-thioguanosine (6SG). The photoreactive analogs are applied to living cells that get incorporated in the newly synthesized RNAs and in response to cross-linking with UV lights, undergo a structural change and develop specific sequence mutations like T to C in 4SU and G to A in 6SG [94]. Using this approach, specific sites for RNA-binding proteins can be detected with greater resolution. The presence of the modified ribonucleotides provides an internal control for the binding events. Using this approach, it was found that the RNA-binding protein AUF1 associated with highly U- and GU-rich regions in 3′ UTRs of mRNAs [95]. Combining PAR-CLIP with several high-throughput analyses, it was further revealed that AUF1 influences the steady-state levels of mRNAs and several lncRNAs. Similarly, in order to understand the regulation of PRC2-mediated deposition of repressive H3K27me3 marks on specific genes before and during differentiation in ESCs, Kaneko et al. reported that the catalytic subunit of PRC2, EZH2, directly binds the 5′ region of nascent RNAs transcribed from a subset of gene promoters and this binding event correlates with decreased H3K27me3 [96].

CLIP can be considered to be a powerful tool for studying protein-RNA interactions on a genome-wide scale when combined with high-throughput sequencing technologies. However, despite the high specificity, CLIP experiments often generate cDNA libraries of limited sequence complexity due to the restricted amount of co-purified RNA. It has also been seen that several cDNAs truncate prematurely at the cross-linked nucleotide, which are lost in further experimental steps.

A recently developed, finer version of this method, individual-nucleotide resolution CLIP (iCLIP), captures the truncated cDNAs by introducing an intramolecular DNA circularization step. The sequencing of truncated cDNAs provides information about the cross-link sites at nucleotide resolution. In this method, immunoprecipitation is directly followed by proteinase K treatment leaving the only amino acids that are cross-linked to the binding site. Unlike CLIP, RNAs are not subjected to 5′ RNA adaptor ligation but instead undergo RT directly. During RT, amino acids bound to the RNAs truncate the cDNA synthesis at their binding sites. RNAs that are not bound with proteins get converted into read-through cDNAs. Amplification and sequencing of both types of cDNAs and their further comparison provide insights in determining protein-binding sites at single-nucleotide resolution [97]. The precision and sensitivity of this method to provide nucleotide resolution information were first demonstrated by determining the positions within uridine tracts that cross-link to hnRNP-C and the positions downstream of 5′ splice sites that cross-link to cytotoxic granule-associated RNA-binding proteins (TIA1 and TIAL1) [97, 98]. Although this method is highly sensitive, it is very difficult to identify a problem if a particular experiment fails due to a diverse range of enzymatic reactions and purification steps involved.

10.3.2.2 RNA-Chromatin Interactions

It is established that lncRNAs can regulate gene expression through interactions with various chromatin-modifying complexes in order to alter chromatin at promoters of specific genes, thereby affecting their transcriptional output. In many cases, lncRNAs act like scaffolds to bring various complexes together at specific chromatin sites [83]. However, how lncRNAs target specific chromatin regions or globally where lncRNAs bind at chromatin, what are the protein complexes that mediate this interaction and how their structure imparts their functionality in chromatin interaction, is still not completely understood. Specific lncRNAs have been observed to be physically located to chromatin regions at low resolution through hybridization techniques like DNA-RNA FISH. However, this method cannot be used for all lncRNAs due to their variability in the expression levels. Moreover, this approach does not provide mechanistic insights into the lncRNA-chromatin interaction. Several technologies for understanding the association of lncRNAs with chromatin have emerged over recent years, which have been instrumental in uncovering the mechanism of action of lncRNAs (Figs. 10.4 and 10.5). However, these methods

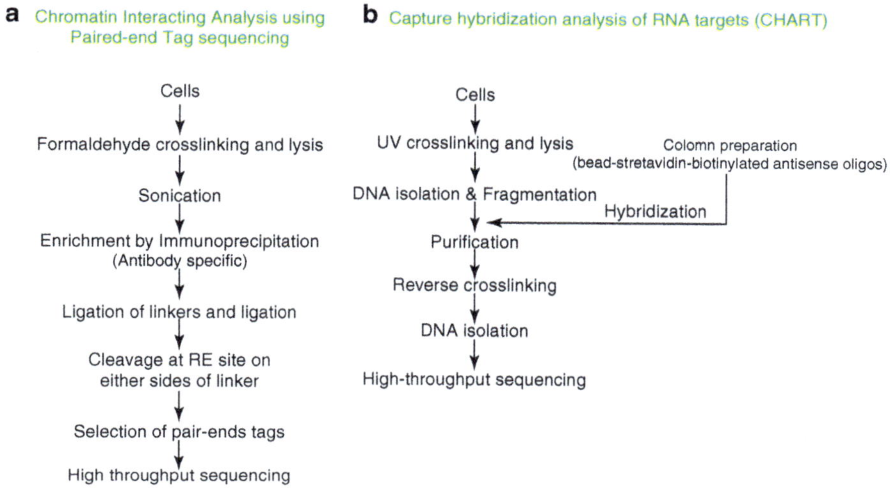

Fig. 10.4 Schematic representation of methods to identify lncRNA-chromatin interaction

rely on mapping interactions of a particular lncRNA and chromatin, and it is still not possible to map such interactions in a genome-wide context [99–101].

Chromatin Isolation by RNA Purification (ChIRP)

lncRNAs harbour less sequence conservancy, and majority of their functional interactions with their partners is imparted by their conformational domains. Therefore, identifying the functional domains of lncRNAs might provide a common theme of action of a subset of lncRNAs or domain-based functional classification of lncRNAs. The methodologies discussed above focus on the lncRNA-protein interactions where the goal is to determine lncRNAs bound to a protein of interest. However, in cases when the protein complexes associated with a particular lncRNA are not known, ChIRP has been employed that can identify associations between an RNA of interest and chromatin [102]. This method uses RNA domain-specific, non-overlapping, biotinylated antisense oligonucleotides to identify RNA-RNA interacting domains, protein-binding domains and chromatin-associated domains for a lncRNA. In this process, macromolecular interactions are cross-linked with formaldehyde, nuclei isolated and lysed followed by sonication to generate smaller fragments. The sonicated lysate is passed through streptavidin bead-bound biotinylated antisense oligonucleotides, which specifically recognize target lncRNA under stringent hybridization condition to avoid non-specific interaction. To ensure high specificity and sensitivity for the target lncRNA, the tiled oligonucleotides are designed to cover the entire length of the sequence of lncRNA. The lncRNA-DNA-protein complexes are then washed and purified and further subjected to sequencing

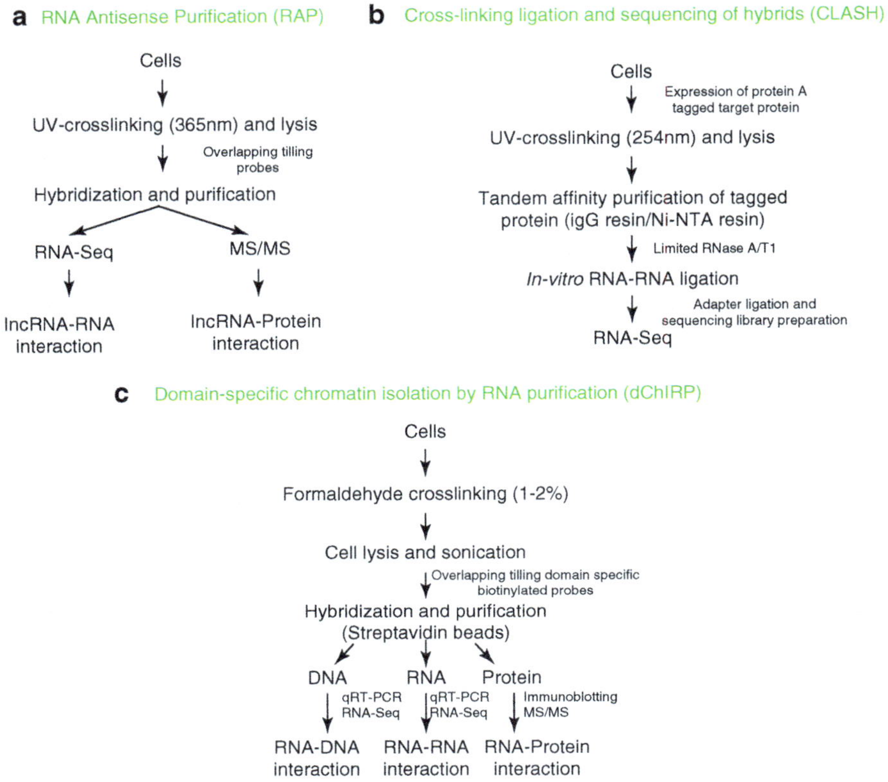

Fig. 10.5 Schematic representation of methods to identify lncRNA-RNA interaction

platforms to identify genomic regions associated with lncRNAs. Proteins associated with the target lncRNAs can be analysed using immunoblotting or mass spectrometry [88, 102]. Quinn et al. employed this technique to study the interaction of a *Drosophila melanogaster*-specific lncRNA, roX1, with MSL proteins, chromatin and CLAMP [103]. roX1 plays an essential role in X-chromosome dosage compensation. dChIRP revealed a 'three-fingered hand' ribonucleoprotein topology, wherein each RNA finger binds chromatin and MSL protein complex and can individually rescue male lethality in roX-null flies, thus defining a minimal RNA domain for chromosome-wide dosage compensation [103]. When compared with other techniques, dChIRP was able to enhance the RNA genomic localization signal by >20-fold indicating the high precision and sensitivity of this technique. This technique was also used to identify HOTAIR-associated DNA regions by Chu et al., where they demonstrated that HOTAIR commonly nucleates at GA-rich DNA regions [102]. Additionally, Xist lncRNA-associated 81 new proteins were also identified using this technique [88].

RNA Antisense Purification (RAP)

This technique is almost similar to ChIRP and can be used to identify lncRNA-associated genomic regions; however, the antisense oligonucleotide probes used are approximately 120 nt long that ensure high-affinity binding to the target lncRNA and a reduced signal-to-noise ratio [104]. In this process, a library of overlapping tilling probes (DNA or RNA oligos) is designed throughout the RNA and is biotinylated. Macromolecular interactions are captured using in vivo cross-linking methods. Cells are lysed and cellular lysate are mixed with antisense probes. Hybridization is performed under the stringent conditions. Probes are designed to have maximum interactions and reduce non-specific binding. Further, entire mixture is passed through streptavidin affinity column to purify interacting complexes. Purified complex might have all possible interacting partners (RNA-RNA, RNA-DNA, RNA-protein). RAP can be further conjugated with high-throughput sequencing technology as well with tandem mass spectrometry (MS/MS) to find novel interacting RNA or protein partners. Based on its conjugation with new technologies, RAP assays have several other variant names like RAP conjugated with RNA-Seq, and mass spectrometry is known as *RAP-Seq* and *RAP-MS*, respectively [105]. With the help of this technique, McHugh et al. were able to identify ten proteins that specifically associate with Xist, three of these proteins—SHARP, SAF-A and LBR—are required for Xist-mediated transcriptional silencing [106]. The pre*cise* binding of Xist RNA to X chromosome during its inactivation was identified using this technique in mouse ES cells [104]. Further, by using a time course RAP analysis combined with three-dimensional chromatin conformation capture data, it was shown that X-chromosome inactivation is initiated at distal regions of the X chromosomes and gradually spreads from each contact point [107, 108].

Capture Hybridization Analysis of RNA Targets (CHART)

CHART is a method to experimentally identify chromatin-associated RNA similar to ChIRP or RAP [109]. This method uses short affinity-tagged oligonucleotides (C-oligos) around the region of potential open binding sites instead of covering the entire length of the target lncRNA. These are detected beforehand by hybridization of oligos and RNase H mapping. So, basically the lncRNA-protein- or lncRNA-chromatin-associated regions are inaccessible to the C-oligos, and they will only bind to open regions and will be further subjected to RNase H digestion. In this process, first RNA-binding sites on chromatin are identified either from computation prediction or from high-throughput RNA-DNA pull-down experiments. RNA-chromatin interactions are UV cross-linked in vivo and cells are lysed. Cross-linked chromatin-associated RNA complexes are sonicated and fragmented into smaller fragments and passed through streptavidin bead-bound biotinylated antisense oligonucleotides of about 22–28 nt length (C-oligo-linker-biotin). Purified complex is treated with RNase H. RNase H unwinds RNA-DNA hybrid and separates them.

RNA, DNA and associated proteins are isolated and can be analysed separately using qRT-PCR, reverse cross-linking and Western blotting, respectively [109, 110]. This technique was used to identify genome-wide localization of a well-known lncRNA-roX2, which is involved in dosage compensation in *D. melanogaster*. The CHART analysis demonstrated the binding of roX2 to specific genomic sites that overlap with the binding sites of protein from the male-specific lethal complex that affects dosage compensation [109]. CHART-Seq was employed to map the Xist-binding sites on the X chromosome throughout the development. With the help of this technique, it was observed that during X-chromosome inactivation in the female cells, Xist functions in a two-step mechanism. First, it targets gene-rich islands and then spreads to the intervening gene-poor regions [111]. In another study, CHART pointed out the extensive genomic binding of NEAT1 and MALAT1 around hundreds of genomic sites. Also, both the lncRNAs showed co-localization at most of the loci, but exhibited dissimilar binding patterns, implying independent but balancing roles [112].

10.3.2.3 Interaction: RNA-RNA

RNA-RNA interactions could be direct or indirect via several accessory-bridging proteins. The functional role played by directly interacting RNA is different from the indirectly interacting RNAs. The direct RNA-RNA interactions are mediated through base pairing, like miRNA-mRNA interactions or mRNA-lncRNA sense-antisense interactions. The indirect interactions are mediated through protein intermediates. Several lncRNAs have been found to be associated with RNA processing factors [21, 113, 114], which suggests that these lncRNAs might target other RNAs in order to regulate them post-transcriptionally.

RAP-RNA

Several modifications of cross-linking and affinity capture techniques have been utilized to detect RNA-RNA interactions. By using different cross-linking methods, direct and indirect interactions between transcripts can be detected. For example, by utilizing the cross-linker 4′aminomethyltrioxalen (AMT), which generates specific uridine bases cross-links, RNAs that bind directly to each other without a protein intermediate could be captured [105]. On the other hand, using formaldehyde, RNA-protein interactions can be cross-linked allowing the capture of additional RNA species that bind to the target RNA in an indirect fashion. Similarly, combining formaldehyde along with disuccinimidylglutarate leads to stronger cross-links between proteins and allows the capture of RNAs bound to multiple protein intermediates in larger complexes. Cross-linking is followed by sonication and protein digestion and, finally, captured with biotinylated antisense oligonucleotides and sequencing. Engreitz et al. developed this method to systematically map RNA-RNA interactions using two lncRNAs U1 snRNA and MALAT1 as targets. Using different cross-linking strategies, they demonstrated that U1 RNA directly binds to 5′ splice sites and 5′ splice site motifs throughout introns and confirmed that MALAT1 interacts with pre-mRNAs indirectly through protein intermediates. These interactions with nascent pre-mRNAs cause U1 and MALAT1 to localize proximally to chromatin at active genes [105].

Cross-Linking, Ligation and Sequencing of Hybrids (CLASH)

CLASH is an in vivo technique to identify RNA-RNA interactions globally. In this experiment, cellular RNA-protein interactome is cross-linked in vivo using UV irradiation (254 nm), and interacting RNA-RNA complexes are pulled under denaturing condition by an affinity-tagged protein (which binds to either of them) using affinity chromatography. Limited RNase A/T1 is used for trimming RNA-RNA hybrids. Trimmed ends are prepared for ligation using T4 PNK (5′ phosphorylation). Hybrid RNAs are ligated to form chimeric guide (RNA1)-target (RNA2) molecules. Samples are treated with proteinase K and RNA is isolated. Further, RNA samples are prepared for RNA-Seq experiment. Reads obtained from RNA-Seq are mapped back to genome, and individual chimeric RNA is separated into individual RNA. Individual RNA obtained from a chimeric reads are considered to be interacting RNAs. The main advantages of CLASH are it is an in vivo method therefore and kinetics of RNA-RNA interaction can be analysed by limiting UV cross-linking timing. RNA-RNA hybrids are pulled under the denaturing condition that denatures many other proteins, thereby increasing signal-to-noise ratio [115]. Identification of novel snoRNA-rRNA interactions in yeast was possible due to this technique [116]. Helwak et al. applied this technique to study the human miRNA interactome associated with AGO1. The CLASH analysis, in an unbiased manner, helped in uncovering the occurrence of seed and non-seed miRNA-target interaction [117]. This technique has not still been used to map transcriptomes for lncRNA interactions due to the low ligation rate of RNAs, which appears to be the limiting step.

10.3.3 Identification of lncRNAs Associated with Translation Machinery

The fact that lncRNAs share highly similar features with mRNAs has always created a possibility to investigate whether lncRNAs are truly non-coding in nature and that no viable protein product could result from lncRNAs. Moreover, the observation of lncRNAs being associated with ribosome subunits further created a doubt as to if lncRNAs are truly non-coding in nature, then why do they associate with the translation machinery?

Ribosome profiling: Ribosome profiling is a method that enables global translational analysis and allows for the visualization of direct binding of transcripts to

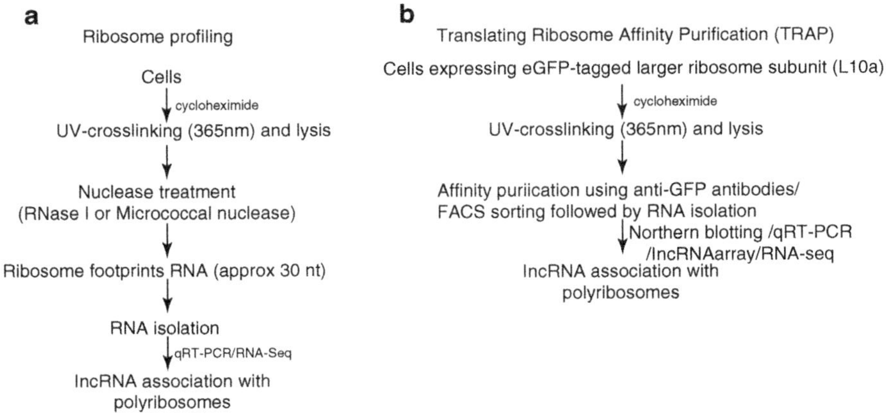

Fig. 10.6 Schematic representation of ribosome profiling and TRAP assay

ribosomes by exactly demarking the ORF boundary within a transcript using ribosome footprints (Fig. 10.6) [118]. Ribosome footprints are the regions (around 30 nt) of a transcript that are protected by translating ribosomes upon nuclease treatment. In this process, translating RNA-ribosome interactions are stabilized with cycloheximide treatment followed by digestion of free RNAs that are not engaged by ribosomes. The ribosome-protected RNA fragments, as ribosome footprints are further sedimented through a sucrose gradient based on size and density. Finally, the RNA fragments are sequenced. Ribosome profiling shows pre*cis*ely where ribosomes are occupied on the mRNA and which open reading frames are translated. It gives an accurate average number of ribosomes occupied on single RNA (ribosome occupancy) [118, 119]. The drawback of this methodology is that the ribosome footprints are short fragments, which results in mapping the accuracy extremely difficult. Additionally, due to the size limitation of the footprints, it is difficult to exclude RNA fragments protected by proteins other than ribosomal proteins. In a few studies, several lncRNAs have also been found associated with ribosomes and observed to be encoding short peptides [120, 121]. However, it was later shown that the association of lncRNAs with ribosomes is comparable to that of 5′ UTRs of protein-coding genes [122] suggesting that mere occupancy of ribosomes to any RNA does not imply its translation. For example, the antisense RNA of Uchl1 interacts with its sense counterpart to enhance its translational rate by directing it to the polysomal fraction. Therefore, the association of several lncRNAs with polysomes does not necessarily implicate their ongoing translation; it could also suggest that they might regulate translation. However, the exact mechanism and the relevance of this association is still not understood. One of the studies supporting this argument was done by FANTOM3, where it was observed that many lncRNAs produced short ORFs; however, the peptides produced through these ORFs were without biologically active structures [3]. This suggests that even if lncRNAs may encode some peptides, the likelihood of them having biological functions is debatable. In contrast to this, there have been a few studies highlighting the existence of functional peptides encoded by RNAs otherwise thought to be non-coding. For example, a muscle-specific lncRNA was found to encode a highly conserved small peptide myoregulin (MLN), which act as a dominant negative regulator of the calcium pump SERCA in the sarcoplasmic reticulum [123].

Translating Ribosome Affinity Purification (TRAP): TRAP is a method for identifying translation rate of transcripts in cell-type-specific manner from a complex tissue. In this process, larger subunit of ribosomes L10a is tagged by enhanced green fluorescent protein (EGFP) expressed using cell-type-specific regulatory elements. Translating transcripts are halted using Cycloheximide treatment and cells are lysed (Fig. 10.6). EGFP-ribosome associated transcripts are pulled down using anti-GFP antibodies. Alternatively, cells can be sorted manually or using FACS, then RNA is isolated. Further, expression of transcript can be measured using Northern blotting, qRT-PCR, microarray or RNA-Seq [124]. An extensive and thorough analysis of the proteome and transcriptome of embryonic brain and kidneys revealed that 85 previously thought ncRNAs from the brain and 60 from the kidney were indeed translated suggesting the importance of the approach towards assigning a non-coding function to a potentially coding RNA [125].

10.4 Secondary Structure of lncRNA

To understand the function of lncRNAs, it is important to identify their secondary and tertiary structures because the specific interactions of lncRNAs with their protein partners or chromatin are not only dictated by their sequences but also by their conformations imparted by their structures [126]. Computational prediction to determine the secondary structure of lncRNAs is never successful because of the limitation of these approaches with the length of the transcripts. For example, a lncRNA of around 2.0 kb can have up to 10,000 possible secondary structures [127], suggesting that these calculations are practically difficult not only on a global scale but even at the level of an individual RNA. Recently, several genome-wide approaches, using either chemical probing to acylate flexible RNA bases or specific enzymes to cleave structured and unstructured regions of RNAs, have been introduced for identification of RNA secondary structures [14].

Fragment Sequencing (Frag-Seq) Technology: At a particular time, only a handful of RNA structures can be deduced using classical RNA structure determination. Frag-Seq technology uses combination of high-throughput sequencing technology and computation algorithms for global mapping of ssRNA regions in thousands of RNAs. Due to its high-throughput nature, it has extensively been used for determining lncRNAs structures and conformational changes. In this process, cells are treated with stimuli, and RNAs are isolated followed by limited treatment with P1 endonuclease, which preferentially digests single-stranded regions of RNA and DNA. RNA fragments are resolved and size fractionated on Bioanalyzer. Endonuclease removes 5′ phosphate and 3′ OH; therefore, it is being added chemically using polynucleotide kinase (PNK) and ATP. Size-selected RNAs (20–200 nt) are ligated with 5′ and 3′ universal sequencing adapters and reverse transcribed to form cDNA. cDNA are PCR amplified using universal primer-specific barcodes and sequenced using high-throughput sequencing technology. Sequencing reads are mapped to the corresponding genome sequence and compared between two conditions. Reads aligned to genomic coordinates corresponding to RNA of interest are calculated and cutting scores are assigned. A cutting score is a numerical value that depicts preferential P1 endonuclease activity. Higher cutting score corresponds to high probability of single-stranded region. P1 endonuclease is thermostable thereby; it is being used for determination of temperature-dependent conformation change in the RNA structure. Using this technique, the presence of single-stranded regions of previously studied ncRNAs has been identified [15]. Additionally, the structure of U15b C/D box snoRNA was also determined using Frag-Seq technology. Although Frag-Seq is a powerful tool for a genome-wide prediction, the major limitation is that it does not generate uniform size transcript that creates problems during calculation of transcript abundance and splice variant characterization [15].

Dimethyl Sulphate (DMS)-Seq: In vitro RNA structure determination does not account for the presence of various RNA-binding proteins (RBPs). Therefore, in vivo methods of RNA structure determination are required. Dimethyl sulphate (DMS)-Seq is an in vivo high-throughput method for determining conformation changes in the RNA structure. DMS enters into the cells rapidly and binds only with unpaired bases (ssRNA). Binding of DMS to adenine (A) or cytosine (C) blocks reverse transcriptase resulting in premature cDNA formation. Initially cells are grown at various conditions to determine conformation of the RNA. Cells are treated with DMS, RNAs are isolated, size is fractionated before adapter ligation (60–70 nt), and sequencing adapter is ligated at both ends and reverse transcribed. DMS bound to adenine and cytosine prematurely aborts reverse transcription; therefore, cDNA is again size selected. These fragments are ligated to form a circularized RNA, amplified and sequenced using RNA-Seq. Reads mapped to genome containing modified adenine and cytosine are calculated and called as DMS signals. The advantage of this method is that it is robust and reproducible and has less signal-to-noise ratio. This technique was employed to review yeast and mammalian mRNAs, which were in accord with the well-studied mRNA structures. In addition, the technique suggested that as compared to the in vitro conditions, the rapidly dividing cells harbour limited structured mRNA under in vivo conditions [16].

Parallel Analysis of RNA Structure (PARS): PARS is another global approach for RNA structure determination at high resolution like Frag-Seq; however, it differs with Frag-Seq as to how the transcript folding is mapped [128]. In this process, RNA is isolated, and samples are enriched for mRNA or non-coding RNA and digested with RNase VI and S1 nuclease. RNase VI and S1 nuclease digest RNA at single-stranded and double-stranded region, respectively. Further these larger fragments are randomly fragmented into smaller fragments, adapters are ligated and converted into cDNA, and PCR is amplified and sequenced using parallel high-throughput sequencing technologies. The reads are mapped back to their respective genome and PARS score are calculated. PARS score is the frequency of digestion at each nucleotide by RNase VI and S1 nuclease separately. These two PARS scores are compared throughout the genome and plotted to demark various secondary structure profiles of the transcriptome. High PARS score is inferred as highly structured region and low PARS score is unstructured region. In an open reading frame (ORF), start and stop codon tends to have lower

PARS score as compared to other regions. 5′ and 3′ UTRs also tend to have lower PARS score which suggests that these regions are less structured or unstructured, further providing clues that these regions are more accessible for RNA-Protein, RNA-RNA and RNA-DNA interactions. In this way RNA secondary structure demarcation helps in determining coding nature of the transcripts and also separates non-coding RNAs. The major disadvantage associated with this technique is that low abundant transcripts are barely detectable. PARS is an in vitro method; therefore, the role of RNA-binding proteins (RBPs) and various cellular stimuli that changes RNA secondary structure has not been accounted. It is evident from the various PARS experimental analyses that it fails to correctly determine secondary structure profile of most of physiologically regulatory RNAs like 18sRNA and tRNAs [17].

Selective 2′-Hydroxyl Acylation Analysed by Primer Extension (SHAPE): RNAs have an intrinsic propensity to form unique structures, and these structures play a key role in RNA regulatory mechanisms. SHAPE is an in vitro chemical probing method for the determination of RNA secondary structure, which employs an electrophilic agent (NMIA) to acylate the nucleophilic 2′-OH group of the RNA nucleotides yielding a 2′-O-adduct without being influenced by the base identity. In principle, the nucleophilicity of the 2′-OH group is affected by the adjoining 3′-phosphodiester group. As compared to the conformationally constrained nucleotides, the flexible nucleotides show a higher propensity to form a stable 2′-O-adduct. After the completion of the acylation reaction, the treated RNA is reverse transcribed by priming it with an end-labelled cDNA. The modified sites in the treated RNA stop the primer extension, which is then determined by high-resolution gel electrophoresis, thus facilitating the determination of the local RNA backbone flexibility [129]. This technique was utilized to determine the secondary structure of one of the most studied lncRNA-HOTAIR, suggesting that HOTAIR forms four independent structural domains with two of them to be the predicted protein-binding domains [130]. icSHAPE is a modified version of SHAPE (selective 2′-hydroxyl acylation analysed by primer extension) which catches the secondary structures of RNAs at a transcriptome-wide level in a living cell for all four bases. In icSHAPE, the previously used SHAPE reagent—2-methylnicotinic acid imidazolide (NAI)—is modified to NAI-N3 by the addition of an azide group, rendering it suitable for probing RNA structure in vivo. The azide group is important for 'click' of biotin moiety to SHAPE reagent, facilitating the purification of the NAI-N3-modified RNA via streptavidin beads. The end result is the increased signal-to-noise ratio in sequence analysis of the modified RNAs. Methodologically, the cells are given a treatment of the icSHAPE reagent—NAI-N3—which leads to in vivo acetylation of flexible RNA nucleobases throughout the transcriptome. The modified (and mock-modified) RNA is isolated, and further 'click' reactions add DIBO-biotin to the modified RNAs. The purified RNA is then fragmented and this is followed by the synthesis of cDNA. The reverse transcriptase reaction 'reads out' the NAI-N3 modification; subsequently only NAI-N3-modified molecules are selectively isolated and sequenced. With the help of computational analysis, flexibility scores are calculated for each base throughout the RNA population [18].

10.5 Bioinformatics Tools for Studying lncRNA Function

The advancement in various transcriptomic approaches and high-throughput sequencing technologies has facilitated the fast accumulation of a large amount of lncRNA dataset. Novel computing tools and information resources are required to enable the generation of new hypotheses about lncRNA functions and their association with different disease phenotypes, which would greatly facilitate the unmet knowledge discovery needs of the research community. For the identification and characterization of lncRNAs in genomic studies, multiple high-quality resources of annotations are required which can also computationally predict relevant associations. For example, the Cancer Genome Atlas (TCGA) data analysis identified potentially clinically relevant non-coding transcripts. This comprehensive dataset explains the association of specific lncRNA expression with patient survival, copy number alteration or histological subgrouping in various cancers [131]. In continuation, several databases have emerged that compile and integrate different types of lncRNA-related information [132]. Since the evaluation, usages, applications and key features of all the relevant resources and bioinformatics tools are beyond the scope of this review, we describe a few relevant resources in Table 10.2.

Table 10.2 Resources/databases for lncRNA studies

S. no.	Database	Website/link	Application	References
01	LNCipedia	http://www.lncipedia.org/	It provides information of structure, protein-coding potential, microRNA-binding site and secondary structure of lncRNAs	[174, 175]
02	lncRNAdb	http://lncrnadb.com/	This database contains information of RNA related to nucleotide sequence, genomic context, gene expression data derived from the Illumina Body Atlas, structural information, subcellular localization, conservation, function with referenced literature	[176, 177]
03	starBase	http://starbase.sysu.edu.cn/	This database contains information related to Pan-Cancer and Interaction Networks of lncRNAs, miRNAs, competing endogenous RNAs(ceRNAs), RNA-binding proteins (RBPs) and mRNAs from large-scale CLIP-Seq (HITS-CLIP, PAR-CLIP, iCLIP, CLASH) data and tumour samples. It also provides information related to protein-lncRNA, miRNA-lncRNA, miRNA-mRNA, miRNA-sncRNA, protein-sncRNA, protein-mRNA, protein-pseudogene, miRNA-circRNA, miRNA-pseudogene interactions and ceRNA networks from 108 CLIP-Seq (HITS-CLIP, PAR-CLIP, iCLIP, CLASH) datasets	[178, 179]

(continued)

Table 10.2 (continued)

S. no.	Database	Website/link	Application	References
04	miRCode	http://www.mircode.org/	This database is useful in microRNA target prediction based on GENCODE annotation	[180]
05	TargetScan	http://www.targetscan.org/	This database predicts the microRNA target sites by presence of 8mer, 7mer and 6mer that match the seed region of each miRNA	[181]
06	Ensembl	http://www.ensembl.org/index.html	This browser includes the BLAST/BLAT BioMart and VEP tools. This database is helpful in comparative genomics, evolution, sequence variation and transcriptional regulation	[182–184]
07	CLIPdb	http://lulab.life.tsinghua.edu.cn/clipdb/	It contains large data by integrating data from CLIP-Seq which helps to characterize the regulatory network between RNA-binding proteins and various RNA transcripts	[185, 186]
08	PARma	https://drupal.bio.ifi.lmu.de/PARma/	This is complete data analysis software for AGO-PAR-CLIP experiments to identify the target site of miRNA as well as microRNA binding to these sites	[187]
09	lncRNAtor	http://lncrnator.ewha.ac.kr/index.htm	It contains data collected from TCGA, GEO, ENCODE and modENCODE. It contains information related to expression profile, interacting (binding) protein, integrated sequence curation, evolutionary scores and coding potential	[188]
10	NONCODE	http://www.noncode.org/	This contains large data which provides details on annotation of lncRNA	[189, 190]
12	lnCeDB	http://gyanxet-beta.com/lncedb/index.php	This database provides information of lncRNA which can act as ceRNAs	[191]

10.6 Discussion/Perspective

The biggest contribution of the post-genomic deep-sequencing technologies has been the revelation that the majority of the cellular transcriptome contributes to diverse types of ncRNAs. Among this group, lncRNAs emerged as subjects undergoing intense study due to the possibility of them being the key regulators employing multiple mechanisms to regulate mammalian homeostasis and development. Several thousand lncRNAs have been discovered till date and the list is still getting updated. This is because of the rapid development of the biochemical toolkit for the lncRNA research and the technological breakthroughs. However, the functional characterization of lncRNAs has remained challenging because of the diversity in their occurrence and association with cellular function and partly because of the shortage of genome-scale experimental techniques. The continuous increase in the discovery of newer lncRNAs also presents a challenge in terms of definition and annotation. This also requires more and more comprehensive transcriptome analyses and transcript assemblies. One of the ways to categorize functional lncRNAs from the list of thousands discovered is to divide them into different classes based on their molecular signatures and characteristics shared with already characterized lncRNAs. However, there is a long road ahead that will require a more comprehensive understanding of the molecular mechanisms of action and in vivo functions of most of them. Having said that, technological breakthroughs are occurring simultaneously, for example, newer imaging methods with unparalleled resolution and sensitivity, genome-scale high-throughput biochemical methods, gene knockout studies and, most importantly, update on newer and refined software for bioinformatics analyses. Finally, the convergence of cutting-edge methods from interdisciplinary areas to resolve the true in vivo functions of these enigmatic transcripts is creating tremendous synergy; however, the strengths and weaknesses of available technologies need to be acknowledged.

References

1. Waldron C, Lacroute F (1975) Effect of growth rate on the amounts of ribosomal and transfer ribonucleic acids in yeast. J Bacteriol 122:855–865
2. Comings DE (1972) The structure and function of chromatin. Adv Hum Genet 3:237–431
3. Fatica A, Bozzoni I (2014) Long non-coding RNAs: new players in cell differentiation and development. Nat Rev Genet 15:7–21
4. Cech TR, Steitz JA (2016) The non-coding RNA revolution-trashing old rules to forge new ones. Cell 157:77–94. doi:10.1016/j.cell.2014.03.008
5. Quinn JJ, Chang HY (2016) Unique features of long non-coding RNA biogenesis and function. Nat Rev Genet 17:47–62. doi:10.1038/nrg.2015.10
6. Carninci P, Kasukawa T, Katayama S et al (2005) The transcriptional landscape of the mammalian genome. Science 309:1559–1563. doi:10.1126/science.1112014
7. Djebali S, Davis CA, Merkel A et al (2012) Landscape of transcription in human cells. Nature 489:101–108. doi:10.1038/nature11233
8. Guttman M, Amit I, Garber M et al (2009) Chromatin signature reveals over a thousand highly conserved large non-coding RNAs in mammals. Nature 458:223–227. doi:10.1038/nature07672
9. De Santa F, Barozzi I, Mietton F et al (2010) A large fraction of extragenic RNA pol II transcription sites overlap enhancers. PLoS Biol 8:e1000384. doi:10.1371/journal.pbio.1000384
10. Tsoi LC, Iyer MK, Stuart PE et al (2015) Analysis of long non-coding RNAs highlights tissue-specific expression patterns and epigenetic profiles in normal and psoriatic skin. Genome Biol 16:24. doi:10.1186/s13059-014-0570-4
11. Grote P, Wittler L, Hendrix D et al (2013) The tissue-specific lncRNA Fendrr is an essential regulator of heart and body wall development in the mouse. Dev Cell 24:206–214. doi:10.1016/j.devcel.2012.12.012
12. Gibb EA, Vucic EA, Enfield KSS et al (2011) Human cancer long non-coding RNA transcriptomes. PLoS One 6:1–10. doi:10.1371/journal.pone.0025915
13. Derrien T, Johnson R, Bussotti G et al (2012) The GENCODE v7 catalog of human long non-coding RNAs: analysis of their gene structure, evolution, and expression. Genome Res 22:1775–1789. doi:10.1101/gr.132159.111

14. Wan Y, Kertesz M, Spitale RC et al (2011) Understanding the transcriptome through RNA structure. Nat Rev Genet 12:641–655. doi:10.1038/nrg3049
15. Underwood JG, Uzilov A V, Katzman S et al (2010) FragSeq: transcriptome-wide RNA structure probing using high-throughput sequencing. Nat Methods 7(12):995–1001. doi:10.1038/NMETH.1529
16. Rouskin S, Zubradt M, Washietl S et al (2014) Genome-wide probing of RNA structure reveals active unfolding of mRNA structures in vivo. Nature 505:701–705. doi:10.1038/nature12894
17. Mauger DM, Weeks KM (2010) Toward global RNA structure analysis. Nat Publ Gr 28:1178–1179. doi:10.1038/nbt1110-1178
18. Flynn RA, Zhang QC, Spitale RC et al (2016) Transcriptome-wide interrogation of RNA secondary structure in living cells with icSHAPE. Nat Protoc 11:273–290. doi:10.1038/nprot.2016.011
19. Li L, Chang HY (2014) Physiological roles of long non-coding RNAs: insight from knockout mice. Trends Cell Biol 24:594–602. doi:10.1016/j.tcb.2014.06.003
20. Tsai M-C, Manor O, Wan Y et al (2010) Long non-coding RNA as modular scaffold of histone modification complexes. Science 329:689–693. doi:10.1126/science.1192002
21. Tripathi V, Ellis JD, Shen Z et al (2010) The nuclear-retained non-coding RNA MALAT1 regulates alternative splicing by modulating SR splicing factor phosphorylation. Mol Cell 39:925–938. doi:10.1016/j.molcel.2010.08.011
22. Kapranov P, St Laurent G, Raz T et al (2010) The majority of total nuclear-encoded non-ribosomal RNA in a human cell is "dark matter" un-annotated RNA. BMC Biol 8:149. doi:10.1186/1741-7007-8-149
23. Fort A, Hashimoto K, Yamada D et al (2014) Deep transcriptome profiling of mammalian stem cells supports a regulatory role for retrotransposons in pluripotency maintenance. Nat Genet 46:558–566. doi:10.1038/ng.2965
24. Carrieri C, Cimatti L, Biagioli M et al (2012) Long non-coding antisense RNA controls Uchl1 translation through an embedded SINEB2 repeat. Nature 491:454–457. doi:10.1038/nature11508
25. Ling H, Vincent K, Pichler M et al (2015) Junk DNA and the long non-coding RNA twist in cancer genetics. Oncogene 34:5003–5011. doi:10.1038/onc.2014.456
26. Kung JTY, Colognori D, Lee JT (2013) Long non-coding RNAs: past, present, and future. Genetics 193:651–669. doi:10.1534/genetics.112.146704
27. Rinn JL, Chang HY (2012) Genome regulation by long non-coding RNAs. Annu Rev Biochem 81:145–166. doi:10.1146/annurev-biochem-051410-092902
28. Okazaki Y, Furuno M, Kasukawa T et al (2002) Analysis of the mouse transcriptome based on functional annotation of 60,770 full-length cDNAs. Nature 420:563–573. doi:10.1038/nature01266
29. Velculescu VE, Zhang L, Vogelstein B, Kinzler KW (1995) Serial analysis of gene expression. Science 270:484–487
30. Cheng J, Kapranov P, Drenkow J et al (2005) Transcriptional maps of 10 human chromosomes at 5-nucleotide resolution. Science 308:1149–1154. doi:10.1126/science.1108625
31. Katayama S, Tomaru Y, Kasukawa T et al (2005) Antisense transcription in the mammalian transcriptome. Science 309:1564–1566. doi:10.1126/science.1112009
32. Rinn JL, Kertesz M, Wang JK et al (2007) Functional demarcation of active and silent chromatin domains in human HOX loci by non-coding RNAs. Cell 129:1311–1323. doi:10.1016/j.cell.2007.05.022
33. Trapnell C, Pachter L, Salzberg SL (2009) TopHat: discovering splice junctions with RNA-Seq. Bioinformatics 25:1105–1111. doi:10.1093/bioinformatics/btp120
34. Sultan M, Schulz MH, Richard H et al (2008) A global view of gene activity and alternative splicing by deep sequencing of the human transcriptome. Science 321:956–960. doi:10.1126/science.1160342
35. Quinn EM, Cormican P, Kenny EM et al (2013) Development of strategies for SNP detection in RNA-seq data: application to lymphoblastoid cell lines and evaluation using 1000 genomes data. PLoS One 8:e58815. doi:10.1371/journal.pone.0058815

36. Edgren H, Murumagi A, Kangaspeska S et al (2011) Identification of fusion genes in breast cancer by paired-end RNA-sequencing. Genome Biol 12:R6. doi:10.1186/gb-2011-12-1-r6
37. Guttman M, Garber M, Levin JZ et al (2010) Ab initio reconstruction of cell type-specific transcriptomes in mouse reveals the conserved multi-exonic structure of lincRNAs. Nat Biotechnol 28:503–510. doi:10.1038/nbt.1633
38. Cabili MN, Trapnell C, Goff L et al (2011) Integrative annotation of human large intergenic non-coding RNAs reveals global properties and specific subclasses. Genes Dev 25:1915–1927. doi:10.1101/gad.17446611
39. Mercer TR, Gerhardt DJ, Dinger ME et al (2011) Targeted RNA sequencing reveals the deep complexity of the human transcriptome. Nat Biotechnol 30:99–104. doi:10.1038/nbt.2024
40. Bussotti G, Leonardi T, Clark MB et al (2016) Improved definition of the mouse transcriptome via targeted RNA sequencing. Genome Res 26:705–716. doi:10.1101/gr.199760.115
41. Iyer MK, Niknafs YS, Malik R et al (2015) The landscape of long non-coding RNAs in the human transcriptome. Nat Genet 47:199–208. doi:10.1038/ng.3192
42. Zhang Y, Wong C-H, Birnbaum RY et al (2013) Chromatin connectivity maps reveal dynamic promoter-enhancer long-range associations. Nature 504:306–310. doi:10.1038/nature12716
43. Shiraki T, Kondo S, Katayama S et al (2003) Cap analysis gene expression for high-throughput analysis of transcriptional starting point and identification of promoter usage. Proc Natl Acad Sci U S A 100:15776–15781. doi:10.1073/pnas.2136655100
44. Lasda E, Parker R (2014) Circular RNAs: diversity of form and function. RNA 20:1829–1842. doi:10.1261/rna.047126.114
45. Plessy C, Bertin N, Takahashi H et al (2010) Linking promoters to functional transcripts in small samples with nanoCAGE and CAGEscan. Nat Methods 7:528–534. doi:10.1038/nmeth.1470
46. Guttman M, Rinn JL (2012) Modular regulatory principles of large non-coding RNAs. Nature 482:339–346. doi:10.1038/nature10887
47. Yang L, Duff MO, Graveley BR et al (2011) Genomewide characterization of non-polyadenylated RNAs. Genome Biol 12:R16. doi:10.1186/gb-2011-12-2-r16
48. Kim JK, Kolodziejczyk AA, Ilicic T et al (2015) Characterizing noise structure in single-cell RNA-seq distinguishes genuine from technical stochastic allelic expression. Nat Commun 6:8687. doi:10.1038/ncomms9687
49. Finn RD, Mistry J, Tate J et al (2010) The Pfam protein families database. Nucleic Acids Res 38:D211–D222. doi:10.1093/nar/gkp985
50. Ohnishi Y, Huber W, Tsumura A et al (2014) Cell-to-cell expression variability followed by signal reinforcement progressively segregates early mouse lineages. Nat Cell Biol 16:27–37. doi:10.1038/ncb2881
51. Liu SJ, Nowakowski TJ, Pollen AA et al (2016) Single-cell analysis of long non-coding RNAs in the developing human neocortex. Genome Biol 17:67. doi:10.1186/s13059-016-0932-1
52. Saliba A-E, Westermann AJ, Gorski SA, Vogel J (2014) Single-cell RNA-seq: advances and future challenges. Nucleic Acids Res 42:8845–8860. doi:10.1093/nar/gku555
53. Mattick JS (2009) The genetic signatures of non-coding RNAs. PLoS Genet 5:e1000459. doi:10.1371/journal.pgen.1000459
54. Tripathi V, Shen Z, Chakraborty A et al (2013) Long non-coding RNA MALAT1 controls cell cycle progression by regulating the expression of oncogenic transcription factor B-MYB. PLoS Genet 9(3):e1003368. doi:10.1371/journal.pgen.1003368
55. Chen L-L, Carmichael GG (2009) Altered nuclear retention of mRNAs containing inverted repeats in human embryonic stem cells: functional role of a nuclear non-coding RNA. Mol Cell 35:467–478. doi:10.1016/j.molcel.2009.06.027
56. Hutchinson JN, Ensminger AW, Clemson CM et al (2007) A screen for nuclear transcripts identifies two linked non-coding RNAs associated with SC35 splicing domains. BMC Genomics 8:39. doi:10.1186/1471-2164-8-39
57. Femino AM, Fay FS, Fogarty K, Singer RH (1998) Visualization of single RNA transcripts *in-situ*. Science 280:585–590
58. Fan Y, Braut SA, Lin Q et al (2001) Determination of transgenic loci by expression FISH. Genomics 71:66–69. doi:10.1006/geno.2000.6403

59. Dirks RW, Raap AK (1995) Cell-cycle-dependent gene expression studied by two-colour fluorescent detection of a mRNA and histone mRNA. Histochem Cell Biol 104:391–395
60. Clemson CM, McNeil JA, Willard HF, Lawrence JB (1996) XIST RNA paints the inactive X chromosome at interphase: evidence for a novel RNA involved in nuclear/chromosome structure. J Cell Biol 132:259–275
61. Bridger JM, Kalla C, Wodrich H et al (2005) Nuclear RNAs confined to a reticular compartment between chromosome territories. Exp Cell Res 302:180–193. doi:10.1016/j.yexcr.2004.07.038
62. Clemson CM, Hutchinson JN, Sara SA et al (2009) An architectural role for a nuclear noncoding RNA: NEAT1 RNA is essential for the structure of paraspeckles. Mol Cell 33:717–726. doi:10.1016/j.molcel.2009.01.026
63. Redrup L, Branco MR, Perdeaux ER et al (2009) The long non-coding RNA Kcnq1ot1 organises a lineage-specific nuclear domain for epigenetic gene silencing. Development 136:525–530. doi:10.1242/dev.031328
64. Raj A, van den Bogaard P, Rifkin SA et al (2008) Imaging individual mRNA molecules using multiple singly labeled probes. Nat Methods 5:877–879. doi:10.1038/nmeth.1253
65. Raj A, Tyagi S (2010) Detection of individual endogenous RNA transcripts *in-situ* using multiple singly labeled probes. Methods Enzymol 472:365–386. doi:10.1016/S0076-6879(10)72004-8
66. Raj A, Rifkin SA, Andersen E, van Oudenaarden A (2010) Variability in gene expression underlies incomplete penetrance. Nature 463:913–918. doi:10.1038/nature08781
67. Sahu NK, Shilakari G, Nayak A, Kohli D V (2007) Antisense technology: a selective tool for gene expression regulation and gene targeting. Curr Pharm Biotechnol 8:291–304
68. Guttman M, Donaghey J, Carey BW et al (2011) lincRNAs act in the circuitry controlling pluripotency and differentiation. Nature 477:295–300. doi:10.1038/nature10398
69. Sarma K, Levasseur P, Aristarkhov A, Lee JT (2010) Locked nucleic acids (LNAs) reveal sequence requirements and kinetics of Xist RNA localization to the X chromosome. Proc Natl Acad Sci U S A 107:22196–22201. doi:10.1073/pnas.1009785107
70. Orom UA, Derrien T, Guigo R, Shiekhattar R (2010) Long non-coding RNAs as enhancers of gene expression. Cold Spring Harb Symp Quant Biol 75:325–331. doi:10.1101/sqb.2010.75.058
71. Zalatan JG, Lee ME, Almeida R et al (2015) Engineering complex synthetic transcriptional programs with CRISPR RNA scaffolds. Cell 160:339–350. doi:10.1016/j.cell.2014.11.052
72. Qi LS, Larson MH, Gilbert LA et al (2013) Repurposing CRISPR as an RNA-guided platform for sequence-specific control of gene expression. Cell 152:1173–1183. doi:10.1016/j.cell.2013.02.022
73. Gilbert LA, Larson MH, Morsut L et al (2013) CRISPR-mediated modular RNA-guided regulation of transcription in eukaryotes. Cell 154:442–451. doi:10.1016/j.cell.2013.06.044
74. Gilbert LA, Horlbeck MA, Adamson B et al (2014) Genome-scale CRISPR-mediated control of gene repression and activation. Cell 159:647–661. doi:10.1016/j.cell.2014.09.029
75. Nakagawa S, Ip JY, Shioi G et al (2012) Malat1 is not an essential component of nuclear speckles in mice. RNA 18:1487–1499. doi:10.1261/rna.033217.112
76. Eissmann M, Gutschner T, Hammerle M et al (2012) Loss of the abundant nuclear non-coding RNA MALAT1 is compatible with life and development. RNA Biol 9:1076–1087. doi:10.4161/rna.21089
77. Schorderet P, Duboule D (2011) Structural and functional differences in the long non-coding RNA hotair in mouse and human. PLoS Genet 7:e1002071. doi:10.1371/journal.pgen.1002071
78. Sauvageau M, Goff LA, Lodato S et al (2013) Multiple knockout mouse models reveal lincRNAs are required for life and brain development. Elife 2:e01749. doi:10.7554/eLife.01749
79. McFadden EJ, Hargrove AE (2016) Biochemical methods to investigate lncRNA and the influence of lncRNA:protein complexes on chromatin. Biochemistry 55:1615–1630. doi:10.1021/acs.biochem.5b01141
80. Lai F, Blumenthal E, Shiekhattar R (2016) Detection and analysis of long non-coding RNAs. Methods Enzymol 573:421–444. doi:10.1016/bs.mie.2016.03.010

81. Kashi K, Henderson L, Bonetti A, Carninci P (2016) Discovery and functional analysis of lncRNAs: methodologies to investigate an uncharacterized transcriptome. Biochim Biophys Acta 1859:3–15. doi:10.1016/j.bbagrm.2015.10.010
82. Ferre F, Colantoni A, Helmer-Citterich M (2016) Revealing protein-lncRNA interaction. Brief Bioinform 17:106–116. doi:10.1093/bib/bbv031
83. Chu C, Spitale RC, Chang HY (2015) Technologies to probe functions and mechanisms of long non-coding RNAs. Nat Struct Mol Biol 22:29–35. doi:10.1038/nsmb.2921
84. Eddy SR (2001) Non-coding RNA genes and the modern RNA world. Nat Rev Genet 2:919–929. doi:10.1038/35103511
85. Marin-Bejar O, Huarte M (2015) RNA pulldown protocol for in vitro detection and identification of RNA-associated proteins. Methods Mol Biol 1206:87–95. doi:10.1007/978-1-4939-1369-5_8
86. Bai Q, Bai Z, Sun L (2016) Detection of RNA-binding proteins by in vitro RNA pull-down in adipocyte culture. J Vis Exp. doi:10.3791/54207
87. Zhao J, Sun BK, Erwin JA et al (2008) Polycomb proteins targeted by a short repeat RNA to the mouse X chromosome. Science 322:750–756. doi:10.1126/science.1163045
88. Chu C, Zhang QC, da Rocha ST et al (2015) Systematic discovery of Xist RNA binding proteins. Cell 161:404–416. doi:10.1016/j.cell.2015.03.025
89. Hogg JR, Collins K (2007) RNA-based affinity purification reveals 7SK RNPs with distinct composition and regulation. RNA 13:868–880. doi:10.1261/rna.565207
90. Yoon J-H, Srikantan S, Gorospe M (2012) MS2-TRAP (MS2-tagged RNA affinity purification): tagging RNA to identify associated miRNAs. Methods 58:81–87. doi:10.1016/j.ymeth.2012.07.004
91. Yoon J-H, Gorospe M (2016) Identification of mRNA-interacting factors by MS2-TRAP (MS2-tagged RNA affinity purification). Methods Mol Biol 1421:15–22. doi:10.1007/978-1-4939-3591-8_2
92. Guil S, Soler M, Portela A et al (2012) Intronic RNAs mediate EZH2 regulation of epigenetic targets. Nat Struct Mol Biol 19:664–670. doi:10.1038/nsmb.2315
93. Xi L, Cech TR (2015) Protein-RNA interaction restricts telomerase from running through the stop sign. Nat Struct Mol Biol 22:835–836. doi:10.1038/nsmb.3118
94. Spitzer J, Hafner M, Landthaler M et al (2014) PAR-CLIP (photoactivatable ribonucleoside-enhanced crosslinking and immunoprecipitation): a step-by-step protocol to the transcriptome-wide identification of binding sites of RNA-binding proteins. Methods Enzymol 539:113–161. doi:10.1016/B978-0-12-420120-0.00008-6
95. Yoon J-H, De S, Srikantan S et al (2014) PAR-CLIP analysis uncovers AUF1 impact on target RNA fate and genome integrity. Nat Commun 5:5248. doi:10.1038/ncomms6248
96. Kaneko S, Son J, Shen SS et al (2013) PRC2 binds active promoters and contacts nascent RNAs in embryonic stem cells. Nat Struct Mol Biol 20:1258–1264. doi:10.1038/nsmb.2700
97. Konig J, Zarnack K, Rot G et al (2011) iCLIP—transcriptome-wide mapping of protein-RNA interactions with individual nucleotide resolution. J Vis Exp. doi:10.3791/2638
98. Wang Z, Kayikci M, Briese M et al (2010) iCLIP predicts the dual splicing effects of TIA-RNA interactions. PLoS Biol 8:e1000530. doi:10.1371/journal.pbio.1000530
99. Schmitz K-M, Mayer C, Postepska A, Grummt I (2010) Interaction of non-coding RNA with the rDNA promoter mediates recruitment of DNMT3b and silencing of rRNA genes. Genes Dev 24:2264–2269. doi:10.1101/gad.590910
100. Martianov I, Ramadass A, Serra Barros A et al (2007) Repression of the human dihydrofolate reductase gene by a non-coding interfering transcript. Nature 445:666–670. doi:10.1038/nature05519
101. Jeon Y, Lee JT (2011) YY1 tethers Xist RNA to the inactive X nucleation center. Cell 146:119–133. doi:10.1016/j.cell.2011.06.026
102. Chu C, Qu K, Zhong FL et al (2011) Genomic maps of long non-coding RNA occupancy reveal principles of RNA-chromatin interactions. Mol Cell 44:667–678. doi:10.1016/j.molcel.2011.08.027
103. Quinn JJ, Ilik IA, Qu K et al (2014) Revealing long non-coding RNA architecture and functions using domain-specific chromatin isolation by RNA purification. Nat Biotechnol 32:933–940. doi:10.1038/nbt.2943

104. Engreitz JM, Pandya-Jones A, McDonel P et al (2013) The Xist lncRNA exploits three-dimensional genome architecture to spread across the X chromosome. Science 341:1237973. doi:10.1126/science.1237973
105. Engreitz JM, Sirokman K, Mcdonel P et al (2014) Resource RNA-RNA interactions enable specific targeting of non-coding RNAs to nascent pre-mRNAs and chromatin sites. Cell 159:188–199. doi:10.1016/j.cell.2014.08.018
106. Mchugh CA, Chen C, Chow A et al (2015) The Xist lncRNA interacts directly with SHARP to silence transcription through HDAC3. Nature 521:232–236. doi:10.1038/nature14443
107. Lieberman-Aiden E, van Berkum NL, Williams L et al (2009) Comprehensive mapping of long-range interactions reveals folding principles of the human genome. Science 326:289–293. doi:10.1126/science.1181369
108. Dixon JR, Selvaraj S, Yue F et al (2012) Topological domains in mammalian genomes identified by analysis of chromatin interactions. Nature 485:376–380. doi:10.1038/nature11082
109. Simon MD, Wang CI, Kharchenko PV et al (2011) The genomic binding sites of a non-coding RNA. Proc Natl Acad Sci U S A 108(51):20497–20502. doi:10.1073/pnas.1113536108
110. Davis CP, West JA (2015) Purification of specific chromatin regions using oligonucleotides: capture hybridization analysis of RNA targets (CHART). Methods Mol Biol 1262:167–182. doi:10.1007/978-1-4939-2253-6_10
111. Simon MD, Pinter SF, Fang R et al (2013) Spreading during X-chromosome inactivation. Nature 504:465–469. doi:10.1038/nature12719
112. West JA, Davis CP, Sunwoo H et al (2014) Resource the long non-coding RNAs NEAT1 and MALAT1 bind active chromatin sites. Mol Cell 55:791–802. doi:10.1016/j.molcel.2014.07.012
113. Tollervey JR, Curk T, Rogelj B et al (2011) Characterizing the RNA targets and position-dependent splicing regulation by TDP-43. Nat Neurosci 14:452–458. doi:10.1038/nn.2778
114. Wang G, Chen H-W, Oktay Y et al (2010) PNPASE regulates RNA import into mitochondria. Cell 142:456–467. doi:10.1016/j.cell.2010.06.035
115. Helwak A, Tollervey D (2014) Mapping the miRNA interactome by cross-linking ligation and sequencing of hybrids (CLASH). 9:711–728. doi: 10.1038/nprot.2014.043
116. Kudla G, Granneman S, Hahn D et al (2011) Cross-linking, ligation, and sequencing of hybrids reveals RNA-RNA interactions in yeast. Proc Natl Acad Sci U S A 108:10010–10015. doi:10.1073/pnas.1017386108
117. Helwak A, Kudla G, Dudnakova T, Tollervey D (2013) Mapping the human miRNA interactome by CLASH reveals frequent noncanonical binding. Cell 153:654–665. doi:10.1016/j.cell.2013.03.043
118. Ingolia NT, Lareau LF, Weissman JS (2011) Ribosome profiling of mouse embryonic stem cells reveals the complexity and dynamics of mammalian proteomes. Cell 147:789–802. doi:10.1016/j.cell.2011.10.002
119. Ingolia NT, Ghaemmaghami S, Newman JRS, Weissman JS (2009) Genome-wide analysis in vivo of translation with nucleotide resolution using ribosome profiling. Science 324:218–223. doi:10.1126/science.1168978
120. Ruiz-Orera J, Messeguer X, Subirana JA, Alba MM (2014) Long non-coding RNAs as a source of new peptides. Elife 3:e03523. doi:10.7554/eLife.03523
121. Bazzini AA, Johnstone TG, Christiano R et al (2014) Identification of small ORFs in vertebrates using ribosome footprinting and evolutionary conservation. EMBO J 33:981–993. doi:10.1002/embj.201488411
122. Guttman M, Russell P, Ingolia NT et al (2013) Ribosome profiling provides evidence that large non-coding RNAs do not encode proteins. Cell 154:240–251. doi:10.1016/j.cell.2013.06.009
123. Anderson DM, Anderson KM, Chang C-L et al (2015) A micropeptide encoded by a putative long non-coding RNA regulates muscle performance. Cell 160:595–606. doi:10.1016/j.cell.2015.01.009
124. Heiman M, Kulicke R, Fenster RJ et al (2014) Cell type—specific mRNA purification by translating ribosome affinity purification (TRAP). Nat Protoc 9:1282–1291. doi:10.1038/nprot.2014.085

125. Hupe M, Li MX, Gertow Gillner K et al (2014) Evaluation of TRAP-sequencing technology with a versatile conditional mouse model. Nucleic Acids Res 42:e14. doi:10.1093/nar/gkt995
126. Maenner S, Blaud M, Fouillen L et al (2010) 2-D structure of the A region of Xist RNA and its implication for PRC2 association. PLoS Biol 8:e1000276. doi:10.1371/journal.pbio.1000276
127. Novikova IV, Dharap A, Hennelly SP, Sanbonmatsu KY (2013) 3S: shotgun secondary structure determination of long non-coding RNAs. Methods 63:170–177. doi:10.1016/j.ymeth.2013.07.030
128. Kertesz M, Wan Y, Mazor E et al (2010) Genome-wide measurement of RNA secondary structure in yeast. Nature 467:103–107. doi:10.1038/nature09322
129. Wilkinson KA, Merino EJ, Weeks KM (2006) Selective 2'-hydroxyl acylation analyzed by primer extension (SHAPE): quantitative RNA structure analysis at single nucleotide resolution. Nat Prot 1:1610–1616. doi:10.1038/nprot.2006.249
130. Somarowthu S, Legiewicz M, Liu F et al (2015) HOTAIR forms an intricate and modular secondary structure. Mol Cell 58:353–361
131. Du Z, Fei T, Verhaak RGW et al (2013) Integrative genomic analyses reveal clinically relevant long non-coding RNAs in human cancer. Nat Struct Mol Biol 20:908–913. doi:10.1038/nsmb.2591
132. Fritah S, Niclou SP, Azuaje F (2014) Databases for lncRNAs: a comparative evaluation of emerging tools. RNA 20:1655–1665. doi:10.1261/rna.044040.113
133. Liu Y, Lu X (2012) Non-coding RNAs in DNA damage response. Am J Cancer Res 2:658–675
134. Yoon J, Abdelmohsen K, Srikantan S et al (2012) LincRNA-p21 suppresses target mRNA translation. Mol Cell 47:648–655. doi:10.1016/j.molcel.2012.06.027
135. Hung T, Wang Y, Lin MF et al (2011) Extensive and coordinated transcription of noncoding RNAs within cell-cycle promoters. Nat Genetics 43(7):621–9. doi:10.1038/ng.848
136. Schmitt AM, Garcia JT, Hung T et al (2016) An inducible long non-coding RNA amplifies DNA damage signaling. Nat Genet 48(11):1370–1376. doi:10.1038/ng.3673
137. Lukas J, Altmeyer M (2015) A lncRNA to repair DNA. EMBO Rep 16:1413–1414
138. Lee S, Kopp F, Chang T et al (2016) Non-coding RNA NORAD regulates genomic stability by sequestering PUMILIO proteins non-coding RNA NORAD regulates genomic stability by sequestering PUMILIO proteins. Cell 164:69–80. doi:10.1016/j.cell.2015.12.017
139. Wan G, Hu X, Liu Y et al (2013) A novel non-coding RNA lncRNA-JADE connects DNA damage signalling to histone H4 acetylation. EMBO J 32:2833–2847. doi:10.1038/emboj.2013.221
140. Wan G, Mathur R, Hu X et al (2013) Long non-coding RNA ANRIL (CDKN2B-AS) is induced by the ATM-E2F1 signaling pathway. Cell Signal 25:1086–1095. doi:10.1016/j.cellsig.2013.02.006
141. Loewer S, Cabili MN, Guttman M et al (2010) Large intergenic non-coding RNA-RoR modulates reprogramming of human induced pluripotent stem cells. Nat Genet 42(12):1113–1117. doi:10.1038/ng.710
142. Ghosal S, Das S, Chakrabarti J (2013) Long non-coding RNAs: new players in the molecular mechanism for maintenance and differentiation of pluripotent stem cells. Stem Cells Dev 22:2240–2253. doi:10.1089/scd.2013.0014
143. Hung T, Chang HY (2010) Long noncoding RNA in genome regulation: prospects and mechanisms. RNA Biol 7:582–585
144. Eades G, Zhang Y, Li Q et al (2014) Long non-coding RNAs in stem cells and cancer. World J Clin Oncol 5:134–142. doi:10.5306/wjco.v5.i2.134
145. Loewer S, Cabili MN, Guttman M et al (2010) Large intergenic non-coding RNA-RoR modulates reprogramming of human induced pluripotent stem cells. Nature Genetics 42(12):1113–7. doi:10.1038/ng.710
146. Wang KC, Chang HY (2011) Molecular Mechanisms of Long Noncoding RNAs. Mol Cell 43(6):904–14. doi:10.1016/j.molcel.2011.08.018

147. Grammatikakis I, Panda AC, Abdelmohsen K, Gorospe M (2014) Non-coding RNAs (lncRNAs) and the molecular hallmarks of aging. Aging (Albany NY) 6:992–1009
148. Zhao YAN, Guo Q, Chen J et al (2014) Role of long non-coding RNA HULC in cell proliferation, apoptosis and tumor metastasis of gastric cancer: a clinical and in vitro investigation. Oncol Rep 31:358–364. doi:10.3892/or.2013.2850
149. Ying L, Huang Y, Chen H et al (2013) Downregulated MEG3 activates autophagy and increases cell proliferation in bladder cancer. Mol Biosyst 9:407–411. doi:10.1039/c2mb25386k
150. Liu X, Li D, Zhang W et al (2012) Long non-coding RNA gadd7 interacts with TDP-43 and regulates Cdk6 mRNA decay. EMBO J 31:4415–4427. doi:10.1038/emboj.2012.292
151. Huang J, Zhou N, Watabe K et al (2014) Long non-coding RNA UCA1 promotes breast tumor growth by suppression of p27 (Kip1). Cell Death Dis 27:1–10. doi:10.1038/cddis.2013.541
152. Rapicavoli NA, Qu K, Zhang J, Mikhail M (2013) A mammalian pseudogene lncRNA at the interface of inflammation and anti-inflammatory therapeutics. eLife 50:1–16. doi:10.7554/eLife.00762
153. Li Z, Chao T, Chang K et al (2013) The long non-coding RNA THRIL regulates TNF α expression through its interaction with hnRNPL. Proc Natl Acad Sci U S A 111:1002–1007. doi:10.1073/pnas.1313768111
154. Atianand MK, Hu W, Satpathy AT et al (2016) A long non-coding RNA lincRNA-EPS acts as a transcriptional brake to restrain inflammation article a long non-coding RNA lincRNA-EPS acts as a transcriptional brake to restrain inflammation. Cell 165:1672–1685. doi:10.1016/j.cell.2016.05.075
155. Samper E, Flores JM, Blasco MA (2001) Restoration of telomerase activity rescues chromosomal instability and premature aging in Terc$^{-/-}$ mice with short telomeres. EMBO Rep 2:800–807
156. Porro A, Feuerhahn S, Lingner J (2014) TERRA-reinforced association of LSD1 with MRE11 promotes processing of uncapped telomeres. Cell Rep 6:765–776. doi:10.1016/j.celrep.2014.01.022
157. Abdelmohsen K, Panda AC, Kang M et al (2014) 7SL RNA represses p53 translation by competing with HuR. Nucleic Acids Res 42:10099–10111. doi:10.1093/nar/gku686
158. Mourtada-Maarabouni M, Pickard MR, Hedge VL, Farzaneh F, Williams GT (2009) GAS5, a non-protein-coding RNA, controls apoptosis and is downregulated in breast cancer. Oncogene 28:195–208. doi:10.1038/onc.2008.373
159. Di Agostino S, Strano S, Emiliozzi V et al (2006) Gain of function of mutant p53: the mutant p53/NF-Y protein complex reveals an aberrant transcriptional mechanism of cell cycle regulation. Cancer Cell 10:191–202. doi:10.1016/j.ccr.2006.08.013
160. Yoon J, Abdelmohsen K, Kim J et al (2013) Scaffold function of long non-coding RNA HOTAIR in protein ubiquitination. Nat Commun 4:1–14. doi:10.1038/ncomms3939
161. Johnsson P, Ackley A, Vidarsdottir L et al (2013) A pseudogene long-non-coding-RNA network regulates PTEN transcription and translation in human cells. Nat Struct Mol Biol 20:440–446. doi:10.1038/nsmb.2516
162. Bierhoff H, Dammert MA, Brocks D et al (2014) Short article. Quiescence-induced LncRNAs trigger H4K20 trimethylation and transcriptional silencing. Mol Cell 54:675–682. doi:10.1016/j.molcel.2014.03.032
163. Oakes C, Weichenhan D, Arab K et al (2013) Long non-coding RNA TARID directs demethylation and activation of the tumor suppressor TCF21 via GADD45A. Mol Cell 55:604–614. doi:10.1016/j.molcel.2014.06.031
164. Luo M, Jeong M, Li W et al (2015) Long non-coding RNAs control hematopoietic stem cell function. Cell Stem Cell 16:426–438. doi:10.1016/j.stem.2015.02.002
165. Cesana M, Cacchiarelli D, Legnini I et al (2011) A long non-coding RNA controls muscle differentiation by functioning as a competing endogenous RNA. Cell 147:358–369. doi:10.1016/j.cell.2011.09.028

166. Klattenhoff CA, Scheuermann JC, Surface LE et al (2013) Braveheart, a long non-coding RNA required for cardiovascular lineage commitment. Cell 152:570–583. doi:10.1016/j.cell.2013.01.003
167. Grote P, Herrmann BG (2015) Long non-coding RNAs in organogenesis: making the difference. Trends Genet 31:329–335. doi:10.1016/j.tig.2015.02.002
168. Zhu JG, Shen YH, Liu HL et al (2014) Long non-coding RNAs expression profile of the developing mouse heart. J Cell Biochem 115:910–918. doi:10.1002/jcb.24733
169. Ramos AD, Andersen RE, Kriegstein AR et al (2015) The long non-coding RNA Pnky regulates neuronal stem cells short article the long non-coding RNA Pnky regulates neuronal differentiation of embryonic and postnatal neural stem cells. Stem Cell 16:439–447. doi:10.1016/j.stem.2015.02.007
170. Richards EJ, Zhang G, Li Z et al (2015) Long non-coding RNAs (LncRNA) regulated by transforming growth factor (TGF). J Biol Chem 290:6857–6867. doi:10.1074/jbc.M114.610915
171. Xiao C, Wu C, Hu H (2016) LncRNA UCA1 promotes epithelial-mesenchymal transition (EMT) of breast cancer cells via enhancing Wnt/beta-catenin signaling pathway. Eur Rev Med Pharmacol Sci 7:2819–2824
172. Pickard MR, Williams GT (2013) Long non-coding RNA GAS5 regulates apoptosis in prostate cancer cell lines. Biochim Biophys Acta 1832:1613–1623. doi:10.1016/j.bbadis.2013.05.005
173. Khaitan D, Dinger ME, Mazar J et al (2011) The melanoma-upregulated long non-coding RNA SPRY4-IT1 modulates apoptosis and invasion. Cancer Res 15:3852–3862. doi:10.1158/0008-5472.CAN-10-4460
174. Volders PJ, Helsens K, Wang X et al (2013) LNCipedia: a database for annotated human IncRNA transcript sequences and structures. Nucleic Acids Res 41:1–6. doi:10.1093/nar/gks915
175. Volders PJ, Verheggen K, Menschaert G et al (2015) An update on LNCipedia: a database for annotated human lncRNA sequences. Nucleic Acids Res 43:D174–D180. doi:10.1093/nar/gku1060
176. Amaral PP, Clark MB, Gascoigne DK et al (2011) LncRNAdb: a reference database for long non-coding RNAs. Nucleic Acids Res 39:146–151. doi:10.1093/nar/gkq1138
177. Quek XC, Thomson DW, Maag JLV et al (2015) lncRNAdb v2.0: expanding the reference database for functional long non-coding RNAs. Nucleic Acids Res 43:D168–D173. doi:10.1093/nar/gku988
178. Li JH, Liu S, Zhou H et al (2014) StarBase v2.0: decoding miRNA-ceRNA, miRNA-ncRNA and protein-RNA interaction networks from large-scale CLIP-Seq data. Nucleic Acids Res 42:92–97. doi:10.1093/nar/gkt1248
179. Yang JH, Li JH, Shao P et al (2011) StarBase: a database for exploring microRNA-mRNA interaction maps from Argonaute CLIP-Seq and Degradome-Seq data. Nucleic Acids Res 39:202–209. doi:10.1093/nar/gkq1056
180. Jeggari A, Marks DS, Larsson E (2012) miRcode: a map of putative microRNA target sites in the long non-coding transcriptome. Bioinformatics 28:2062–2063. doi:10.1093/bioinformatics/bts344
181. Agarwal V, Bell GW, Nam JW, Bartel DP (2015) Predicting effective microRNA target sites in mammalian mRNAs. Elife 4:1–38. doi:10.7554/eLife.05005
182. McLaren W, Gil L, Hunt SE et al (2016) The Ensembl variant effect predictor. Genome Biol 17:42374. doi:10.1101/042374
183. Zerbino DR, Johnson N, Juetteman T et al (2016) Ensembl regulation resources. Database 2016:1–13. doi:10.1093/database/bav119
184. Aken BL, Ayling S, Barrell D, et al (2016) The Ensembl gene annotation system. Database. 1–19. doi: 10.1093/database/baw093
185. Yang Y-CT, Di C, Hu B et al (2015) CLIPdb: a CLIP-seq database for protein-RNA interactions. BMC Genomics 16:51. doi:10.1186/s12864-015-1273-2
186. Hu B, Yang Y-CT, Huang Y et al (2016) POSTAR: a platform for exploring post-transcriptional regulation coordinated by RNA-binding proteinsgkw888. Nucleic Acids Res 45:D104–D114. doi:10.1093/nar/gkw888

187. Erhard F, Dölken L, Jaskiewicz L, Zimmer R (2013) PARma: identification of microRNA target sites in AGO-PAR-CLIP data. Genome Biol 14:R79. doi:10.1186/gb-2013-14-7-r79
188. Park C, Yu N, Choi I et al (2014) LncRNAtor: a comprehensive resource for functional investigation of long non-coding RNAs. Bioinformatics 30:2480–2485. doi:10.1093/bioinformatics/btu325
189. He S, Liu C, Skogerbø G et al (2008) NONCODE v2.0: decoding the non-coding. Nucleic Acids Res 36:2007–2009. doi:10.1093/nar/gkm1011
190. Liu C, Bai B, Skogerbø G et al (2005) NONCODE: an integrated knowledge database of non-coding RNAs. Nucleic Acids Res 33:112–115. doi:10.1093/nar/gki041
191. Das S, Ghosal S, Sen R, Chakrabarti J (2014) lnCeDB: database of human long non-coding RNA acting as competing endogenous RNA. PLoS One 9:e98965. doi:10.1371/journal.pone.0098965

The manufacturer's authorised representative in the EU is Springer Nature Customer Service Centre GmbH, Europaplatz 3, 69115 Heidelberg, Germany. If you have any concerns regarding our products, please contact ProductSafety@springernature.com

Printed and bound by CPI Group (UK) Ltd, Croydon, CR0 4YY
30/09/2026
02225587-0005